THE STEPHEN BECHTEL FUND

IMPRINT IN ECOLOGY AND THE ENVIRONMENT

The Stephen Bechtel Fund has

established this imprint to promote

understanding and conservation of

our natural environment.

The publisher gratefully acknowledges the generous
contribution to this book provided by the Stephen Bechtel Fund.

ECOLOGY *of* NORTH AMERICAN FRESHWATER FISHES

ECOLOGY *of* NORTH AMERICAN FRESHWATER FISHES

Stephen T. Ross

UNIVERSITY OF CALIFORNIA PRESS
Berkeley   Los Angeles   London

I dedicate this book to Yvonne Y. Ross, my wife, best friend, willing editor, and enthusiastic field partner for more than 43 years. This book would not have been possible without her constant encouragement and support.

University of California Press, one of the most distinguished university presses in the United States, enriches lives around the world by advancing scholarship in the humanities, social sciences, and natural sciences. Its activities are supported by the UC Press Foundation and by philanthropic contributions from individuals and institutions. For more information, visit www.ucpress.edu.

University of California Press
Berkeley and Los Angeles, California

University of California Press, Ltd.
London, England

Library of Congress Cataloging-in-Publication Data

Ross, Stephen T.
    Ecology of North American freshwater fishes / Stephen T. Ross.
       pages   cm
    Includes bibliographical references and index.
    ISBN 978-0-520-24945-5 (cloth : alkaline paper)
       1. Freshwater fishes—Ecology—North America.   I. Title.
QL625.R67   2013
597.176—dc23   2012043368

19 18 17 16 15 14 13
10 9 8 7 6 5 4 3 2 1

The paper used in this publication meets the minimum requirements of ANSI/NISO Z39.48-1992 (R 2002) (*Permanence of Paper*).

Cover design: Glynnis Koike.
Cover images: Tombigbee Darter (*Etheostoma lachneri*), top; photo by author. Banded Sculpin (*Cottus carolinae*), bottom; photo by Tom Kennedy.

CONTENTS

PREFACE

The North American freshwater fish fauna, the most diverse temperate fish fauna in the world, is also one of the best-studied fish faunas. The taxonomy, systematics, and biology of the fish fauna are treated in a number of excellent books, largely organized by country, province, or state, and by a forthcoming three-volume series treating the entire North American fish fauna on a family-by-family basis (Warren and Brooks, in press). Given the amount of information available, the time is right for an ecological text that also focuses on the North American ichthyofauna. There have been excellent books dealing with ecology of fishes, but none have had their primary emphasis on the North American freshwater ichthyofauna (the fishes occurring in Canada, the continental United States, and the temperate regions of Mexico).

Foremost in writing this book, I have tried to convey the amazingly interesting and exciting things that we know about fishes. In doing this, however, it soon becomes apparent that most of what is known about a particular aspect of fish biology is often based on only a small fraction of the total number of fish species, or only on a small part of a species' range. Part 1 of the text provides a broad picture, both spatially and temporally, of the derivation of the North American freshwater fish fauna.

This includes examples of how global as well as North American geological and climatological processes have shaped the fauna. Part 2 focuses on how local populations and assemblages are formed and how they persist, or not, through time. Part 3 deals with the relationship of body-form and life-history patterns as they are related to ecological functions. The wide range and complexity of interactions among individuals and species through communication, competition, predation, mutualism, and facilitation are the topics of Part 4. Part 5 focuses on several primary conservation issues concerning fish populations. It also integrates much of the information presented in the first four parts. The organization of the book is intended to lead the reader from a broadscale appreciation of why specific species and assemblages occur in particular places to a finer-scale look at how individuals and species interact with each other and with their environments, how such interactions have been altered by anthropogenic impacts, and the relative success of efforts to restore damaged ecosystems.

Even though the emphasis is on studies of North American fishes, I have included pertinent examples from other continents. Because the methods or concepts that form the basis for understanding fish ecology and conservation come from diverse areas (systematics, statistics, population ecology, molecular genetics, stable isotope analysis, environmental ethics, etc.), the chapters also include text boxes that help to provide additional background information.

The intended audience for this book includes upper-level undergraduate students, graduate students, professional fish biologists, and anyone else curious about or with a passion for fishes. Students using this book would benefit from having taken a general biology sequence. A general ecology course, although not absolutely necessary, would also provide helpful background. The book should be appropriate for use in one- or two-semester undergraduate and/or graduate courses in fish ecology.

ACKNOWLEDGMENTS

My initial involvement with this book happened at the suggestion of William J. Matthews. I thank Bill for encouraging me to take on this task, for his confidence in my ability to complete it, for his many helpful suggestions along the way, and for providing the foundation for this work, and many others, through his monumental book *Patterns in Freshwater Fish Ecology*. I especially appreciate Chuck Crumly at the University of California Press for inviting me to write a textbook on the ecology of North American freshwater fishes. I thank Lynn Meinhardt, also of the University of California Press, for her helpful comments and encouragement throughout.

During my work on this book I have benefited greatly from the support of my colleagues at the University of New Mexico, Museum of Southwestern Biology and the Department of Biology, particularly Tom Turner, Manuel Molles, and Alexandra (Lex) Snyder. I thank Tom for his helpful comments and suggestions during the course of this project, and for the opportunity to present some of the chapter contents for discussions with the Turner lab group. Manuel, a textbook author himself, has been more than patient in guiding me through the intricacies of textbook writing and in vetting my use of ecological terms and concepts. I appreciate Lex, who has always been supportive and has made

resources of the Division of Fishes readily available to me. I am grateful to Steve Platania, Rob Dudley, Howard Brandenburg, and Mike Farrington of the Museum of Southwestern Biology (MSB) and American Southwest Ichthyological Researchers (ASIR) for providing help with literature and for giving me access to unpublished data. Howard Brandenburg also provided slides of the San Juan River, and I thank both Howard and David Propst, formerly of New Mexico Game and Fish, for giving me permission to use Howard's outstanding drawings of New Mexico fishes. I also thank Tom Kennedy for generously allowing me to use his photographs of southwestern fishes and Trevor Krabbenhoft for providing larval fish pictures. The faculty and staff of the Interlibrary Loan Department of the University of New Mexico have provided exceptional help throughout the duration of this book. Indeed, without their assistance this book would not have been possible.

I am grateful to colleagues around the country who have given such tremendous support, sharing information on their research and findings. I thank John Baker always for his enthusiastic help, particularly on stickleback biology, and Ron Bliesner, of Keller-Bliesner Engineering, for allowing me to use his GIS analysis of the San Juan River. Other colleagues have also been generous in allowing me to use their photographs of fishes and habitats, although page limitations precluded using many of the images. For this I thank John Baker and Anna Mazzarella (Clark University), Peter Bisson (Pacific Northwest Research Station, USDA Forest Service), Mollie Cashner (Southeastern Louisiana University), Katie May Laumann (Virginia Institute of Marine Science), Gary Meffe (University of Florida), James Morel (Navajo Nation Department of Fish and Wildlife), and W. Todd Slack and Steven G. George (Waterways Experiment Station, USACE).

I thank Tim Modde (USFWS) for including me on research expeditions on the Yampa and Green rivers of the Colorado system years ago, and for providing input on the Razorback Sucker and Colorado Pikeminnow. I am also grateful to Marlis and Mike Douglas for including me on a research expedition through the Grand Canyon of the Colorado River.

I am forever indebted to colleagues who devoted the time and effort to review entire chapters: Chapter 7, Trevor Krabbenhoft and Evan Carson, University of New Mexico; Chapter 8, Trevor Krabbenhoft and Tracy Diver, University of New Mexico; Chapter 9, Eric Charnov, University of New Mexico, and David Heins, Tulane University; Chapter 10, Trevor Krabbenhoft; Chapter 11, Tom Turner, University of New Mexico; Chapter 12, Chet Rakocinski, University of Southern Mississippi; Chapter 13, Mike Farrington, University of New Mexico and ASIR; Chapter 14, Megan Osborne, University of New Mexico; Chapter 15, Corey Krabbenhoft and Nathan Franssen, University of New Mexico.

As in all my previous endeavors throughout my career in biology, work on this book has been greatly aided, indeed made possible, by Yvonne Ross, my wife, editor, and best friend for more than 40 years, who read and commented on the entire text. Our son, Derek G. Ross (Auburn University), has helped tremendously by leading me into twenty-first-century technology in my writing, bibliographic research, and in organizing information. He also lent his expertise in technical communication, scientific rhetoric, and ethics by contributing one of the text boxes. Finally, travels across North America for habitat pictures as well as day-to-day writing have all been helped by the constant supervision and diversion of Bergen, our indefatigable canine companion.

PART ONE

Faunal Origins, Evolution, and Diversity

Fishes represent the most diverse and species-rich group of craniate organisms. In this part, the focus is on introducing the freshwater fish faunas of the world and on the unique nature of the North American freshwater fish fauna (Chapter 1). Chapters 2 and 3 deal with the very basic issues of where the North American freshwater fish fauna came from, how long various fish lineages have occurred in North America, and how the fauna has been shaped by large scale geologic and climatic events.

A useful way of thinking about why certain fish species occur in particular regions, water bodies, and local habitats is to envision an ancestral fauna being shaped by a series of filters. Some filters can be stronger than others and, of course, some taxa might be better at surmounting the challenges imposed by the filters. The broadscale filters are the concern of Part 1 (see the following figure), the end result being an understanding of the origins and ages of North American regional fish faunas. Subsequent parts will move through smaller scale, more local filters, leading to an understanding of factors shaping the occurrence of fishes in local habitats such as a particular pond, lakeshore, riffle, or run.

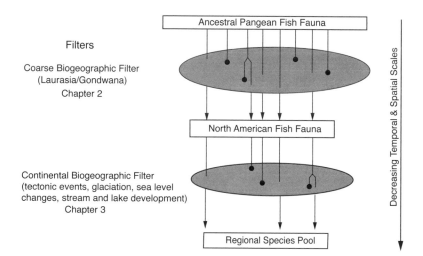

Filters

Coarse Biogeographic Filter
(Laurasia/Gondwana)
Chapter 2

Continental Biogeographic Filter
(tectonic events, glaciation, sea level
changes, stream and lake development)
Chapter 3

Ancestral Pangean Fish Fauna

North American Fish Fauna

Regional Species Pool

Decreasing Temporal & Spatial Scales

A conceptual model of the formation of fish assemblages through progressive loss
(i.e., filtration) and through the progressive addition (i.e., speciation) of lineages. Adapted
in part from Smith and Powell (1971); Poff (1997); Matthews (1998); Tonn et al. (1990);
Rahel (2002); Ross and Matthews (in press).

ONE

Introduction

U NDERSTANDING FISH ECOLOGY IS both exciting and challenging. Fishes have the greatest diversity among the craniate organisms, composing more than half of all extant species, with an estimated 27,977 species of fishes worldwide (Nelson 2006), and with the descriptions of "new" (newly documented) species continuing at approximately 200 species per year (Eschmeyer 1998). As further emphasis of fish diversity, unlike other recognized groups within the phylum Chordata (amphibians, reptiles, birds, and mammals), organisms commonly referred to as fishes comprise five living classes (hagfishes, Myxini; lampreys, Petromyzontidae; sharks and rays, Chondrichthyes; ray-finned fishes, Actinopterygii;

and lobe-finned fishes, Sarcopterygii). Three of these groups, petromyzontids, chondrichthyans, and actinopterygians, are represented in North American fresh water. The Myxini are restricted to the marine environment, whereas the fishes included within the class Sarcopterygii occur in the marine environment (Coelacanthimorpha, coelacanths) or fresh waters of Australia, Africa, and South America (Dipnoi, lungfishes). Tetrapods, including, of course, readers of this book, are also included in the Sarcopterygii (Nelson 2006).

Despite the abundance of organisms to study, the aquatic environment provides challenges to work in, making ecological studies difficult or even dangerous. Especially in many North American streams, visual observations are often not possible because of high flows or turbidity, necessitating indirect approaches, such as seining, electrofishing, or using electronic tags, to infer information on fish habitat use, behavior, and interactions.

This is nonetheless an exciting, as well as challenging, time to write about North American freshwater fish ecology. Modern computers and the rapidly increasing quality of electronic libraries and data searches have made the literature much more available, and the rate of publications has greatly accelerated. In fact, many of the cited references were published after I started work on this book in 2005.

DISTRIBUTION OF FRESH WATER •
Global Patterns

Surprisingly, the distribution of fish species among major habitats is not at all predicted by the habitat area or volume. Even though liquid fresh water makes up only 0.0142% of the water on our planet (Figure 1.1A), there are an estimated 11,952 freshwater species worldwide, which make up 43% of all known fishes (Cohen 1970; Shiklomanov 1993; Nelson 2006). In addition, the proportionately small amount of fresh water (including inland saline lakes) is not distributed equally among habitats but occurs primarily in lakes, something that will be considered in more detail later.

Of course, fishes and other aquatic organisms are not the only organisms dependent on the 0.01% of liquid fresh water—the same limited resource supports most of the world's human population, including human industry, farming, and ranching. Consequently, it is not surprising that freshwater biodiversity is at risk worldwide—indeed, freshwater ecosystems are among the most endangered ecosystems on our planet (Stiassny 1996; Dudgeon et al. 2006).

Diversity of freshwater fishes varies greatly among major zoogeographic regions of the world (Figure 1.2). Considering the six major zoogeographic realms (reviewed in Berra 2001 and Cox 2001), the greatest freshwater fish diversity occurs in the Neotropical realm (Central and South America and tropical Mexico), with 5,000–8,000 species, followed by the Oriental realm (India and southeast Asia), with approximately 3,000 species; the Ethiopian realm (Africa and southern Arabia), with an estimated 2,850 species; the Nearctic realm (North America except tropical Mexico), with at least 1,116 species; the Palearctic (Europe and Asia north of the Himalayas), with 552 species; and the Australian realm, with 500 species, including those marine fishes that enter fresh water (Burr and Mayden 1992; Matthews 1998; Lundberg et al. 2000; Berra 2001; Moyle and Cech 2004; G. R. Smith et al. 2010).

NORTH AMERICAN FRESHWATER FISHES

Although overshadowed by tropical regions, the Nearctic (North American) realm is by far the most speciose of the temperate zoogeographic regions. There are different views on what constitutes the biogeographic extent of North America. I follow Burr and Mayden (1992) in considering the southern boundary of North America to be 18° N latitude on the Atlantic coast (the Río Papaloapan drainage) and 16° N on the Pacific coast (the Río Verde/Atoyac drainage), although southward penetration of Nearctic fish groups varies among families with two gar species, *Atractosteus spatula* and *A. tropicus*, reaching the Rio San Juan drainage of Costa Rica (Minckley et al. 2005). Along the eastern coast of Mexico, the boundary between Nearctic and Neotropical fishes has been further examined. Based on the family Cichlidae, the formation of the Mexican Neovolcanic Plateau, some five million years ago, has served as a barrier separating the Neotropical and Nearctic fish faunas (Hulsey et al. 2004).

Patterns of North American Diversity

North American fish species are included in some 201 genera, 50 families, and 24 orders (Figure 1.3) (Burr and Mayden 1992). However, only about half of the families could be considered as major components based on their number of species and/or their breadth of distribution. Families with the greatest number of species include minnows (Cyprinidae, 297 species), perches (Percidae, 186 species), suckers (Catostomidae, 71 species), livebearers (Poeciliidae,

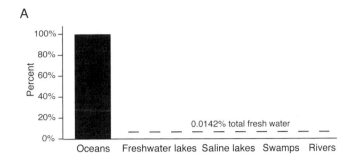

A

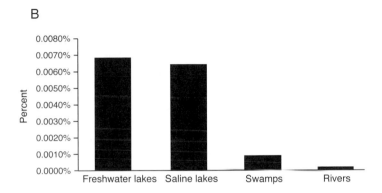

B

FIGURE 1.1 A. The distribution of liquid water in the biosphere. Percents of freshwater habitats are too small to show above the x-axis. B. The distribution of liquid freshwater in inland systems. Percentage data in both panels are based on total liquid water in the biosphere; data are from Shiklomanov (1993).

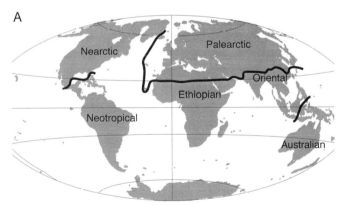

A

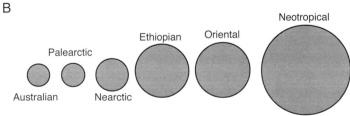

B

FIGURE 1.2 A. Zoogeographic realms of the world based on Berra (2001) and Cox (2001). B. The relative sizes of the zoogeographic realms based on the diversity of fishes.

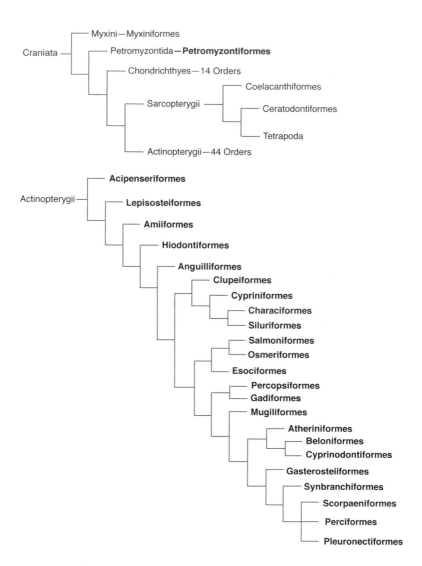

FIGURE 1.3. Relationships of orders that have species of North American freshwater fishes (bold). Based primarily on Nelson (2006).

69 species), and North American catfishes (Ictaluridae, 46 species) (Table 1.1). These 5 families alone make up 62% of the fauna, and along with an additional 10 families (Goodeidae through Clupeidae), they compose 90% of the North American freshwater ichthyofauna.

Patterns of freshwater fish diversity across North America vary widely, in part as a response to variation in landforms, watershed boundaries, historical and recent climates, and geological activity (Figure 1.4). Diversity is greatest in eastern North America, especially in the southeastern United States, which contains more than half (662 species) of the fauna (Warren et al. 2000;

G. R. Smith et al. 2010). In fact, the rich southeastern region has been referred to as a "piscine rainforest" and harbors some 662 native freshwater and diadromous fish species (Warren and Burr 1994; Warren et al. 2000). States with the greatest native fish diversity, in order of number of species, are Tennessee (297), Alabama (295), Kentucky (220), Georgia (219), and Mississippi (212) (Etnier and Starnes 1993; Warren and Burr 1994; Ross 2001 and additional unpublished material; Boschung and Mayden 2004).

Western fish diversity is about one-third that of overall eastern diversity, but endemism tends to be greater (Moyle and Herbold 1987; Burr and Mayden 1992). Considering just the United States and southern Canada, McAllister et al. (1986) determined that geographic grids of one degree latitude and longitude contained on average 10 or fewer species in western areas, with maximum values of 19 in Oregon, 14 in California, and 11 along the Colorado River. In contrast, the same-sized grids in the southeastern United States supported up to 73 species. The more recent treatment of North American diversity patterns that include Mexico (Figure 1.5) further illustrates this pattern (G. R. Smith, in Lundberg et al. 2000; G. R. Smith et al. 2010) and shows that a band of high-diversity grids, located in lower elevation areas, continues south into eastern Mexico to the Yucatan Peninsula. Because of the strong regional differences in species richness, streams of approximately similar sizes located across North America harbor drastically different numbers of native fishes. For example, four streams spanning the temperate region of North America (Figure 1.6) differ greatly in species richness, especially in relation to stream discharge. The four streams were chosen for their geographic locations and also because I have worked in or visited all of them. The western streams tend to have higher average annual discharge, primarily because of winter snowmelt, but far fewer species than the two eastern streams.

Lentic versus Lotic Systems

Somewhat akin to the near-equal split between freshwater and marine fishes, despite the overwhelming preponderance of marine habitats, the North American freshwater fish fauna is primarily a fauna of lotic (flowing water) rather than lentic (standing water) systems despite the much greater volume of lentic habitats (Figure 1.1). On a worldwide basis, inland lentic habitats make up 98.88% of liquid water available for aquatic organisms, compared to 1.12% for lotic systems (data from Shiklomanov 1993). Species richness in lakes is indeed higher in some parts of the world, and certain genera or families of fishes have radiated extensively to form

TABLE 1.1

Families Composing 95% of North American Freshwater Fish Species Ranked by the Number of Native Species

Family	Number of species	Cumulative percent
Cyprinidae	297	28
Percidae	186	45
Catostomidae	71	51
Poeciliidae	69	58
Ictaluridae	46	62
Goodeidae	45	66
Atherinopsidae	39	70
Salmonidae	38	74
Cyprinodontidae	35	77
Fundulidae	34	80
Centrarchidae	31	83
Cottidae	30	86
Petromyzontidae	21	88
Cichlidae	16	89
Clupeidae	10	90
Eleotridae	10	91
Acipenseridae	8	92
Osmeridae	6	92
Elassomatidae	6	93
Gobiidae	6	93
Amblyopsidae	6	94
Pimelodidae	6	94
Gasterosteidae	5	95

SOURCE: Compiled primarily from Mayden (1992), Nelson et al. (2004), and Miller and Norris (2005).

"species flocks"—groups of closely related species found in a restricted geographical area. Examples of species flocks include the cichlid species of African Lakes and the sculpin (Cottidae) fauna of Lake Baikal, Russia (Fryer and Iles 1972; Echelle and Kornfield 1984; G. R. Smith and Todd 1984). However, in contemporary North America, the number of species unique to lakes is much lower, primarily because of the young age of large North American lakes (G. R. Smith 1981). The silversides

FIGURE 1.4. North American topography illustrating the tectonically active regions of the west; the erosional areas to the east of the western mountains; and the lower, depositional areas primarily in eastern North America. Gray gradients indicate topographic elevation, from white at the highest elevations grading through darker grays for progressively lower elevations. The image of North America was generated with data from the Shuttle Radar Topography Mission (SRTM). Courtesy of NASA/JPL-Caltech. See also corresponding figure in color insert.

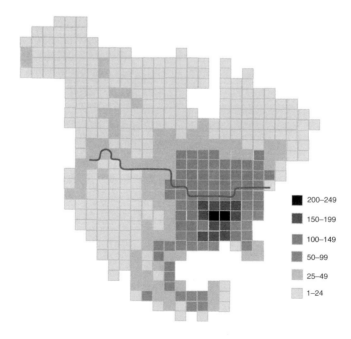

FIGURE 1.5. The number of fish species in equal-sized grids across North America, and the limit of Pleistocene glaciations (solid black line). Reproduced with permission from G. R. Smith. See also corresponding figure in color insert.

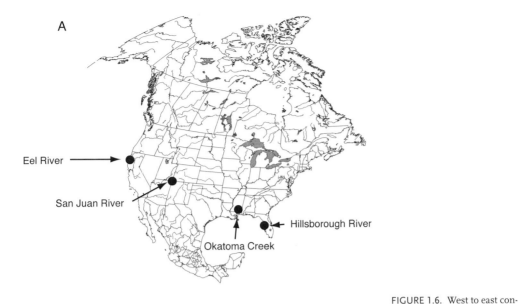

A

Eel River

San Juan River

Hillsborough River

Okatoma Creek

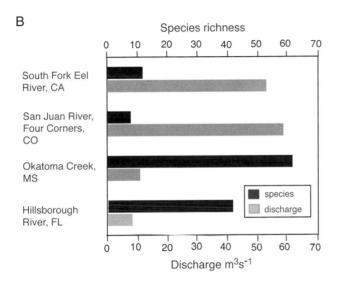

B

Species richness

South Fork Eel
River, CA

San Juan River,
Four Corners,
CO

Okatoma Creek,
MS

Hillsborough
River, FL

■ species
▨ discharge

Discharge m³s⁻¹

FIGURE 1.6. West to east contrasts in stream discharge and fish species diversity.
A. Locations of four streams along an east to west pattern in North America.
B. Average annual discharge (m³/s) and fish species richness. Average annual discharge in m³/s from USGS (2007) as determined from the USGS gauge. Species richness is for the entire river. Sources: Hillsborough River (fish diversity: unpublished data, B. S. Barnett 1972); Okatoma Creek (fish diversity: Ross et al. 1992a); San Juan River (fish diversity: Propst and Gido 2004); South Fork Eel River (fish diversity: Brown and Moyle 1997).

(genus *Menidia*) from Mexico's largest natural lake, Lake Chapala on the Mexican Plateau, provide an example of a small North American species flock with 12 species either restricted to the lake or also occurring in the surrounding streams (Barbour 1973; Miller 2005).

The Laurentian Great Lakes also support (or supported, because some taxa are now extinct or restricted to only a fraction of their former range) an incipient species flock of perhaps 8–9 species/morphotypes of ciscoes (family Salmonidae, trouts and salmons, subfamily Coregoninae) (G. R. Smith and Todd 1984; Underhill 1986; Turgeon et al. 1999; Cudmore-Vokey

and Crossman 2000; Etnier and Skelton 2003). Overall, the known native fish fauna of the Laurentian Great Lakes, excluding the St. Lawrence River and tributaries and including extirpated or extinct taxa, has comprised 126 species, but only 5% are endemic (primarily ciscoes, *Coregonus* spp.) (Cudmore-Vokey and Crossman 2000). In contrast, 17.9% of the rich lotic Tennessee River fish fauna (approximately 229 species and subspecies) is endemic (Etnier and Starnes 1993; Warren et al. 2000). The Great Lakes fauna is primarily derived from the upper Mississippi River drainage, streams of the Atlantic coastal plain, and the Beringian Refugium of the Yukon

Valley, following retreat of the Pleistocene glaciers (Underhill 1986; see also Chapter 3).

This is not to say that North America has never supported large lacustrine species flocks, but only that such faunas are presently uncommon. For instance, the fossil sculpin fauna (*Myoxocephalus* and *Kerocottus*) of the Pliocene Glenn's Ferry Formation in southwest Idaho provides one of the better examples of a North American lacustrine species flock (G. R. Smith 1981), as does the rich fossil semionotid gar fauna from the Mesozoic Newark lakes of eastern North America (McCune et al. 1984).

Because of the influence on lentic assemblages by species from lotic environments, Kitchell et al. (1977) proposed the term "River Analogy" to explain the distribution of large percid fishes. They argued that most North American and European lakes were of recent origin (i.e., Pleistocene or later), that lake-inhabiting fishes had a riverine ancestry, and that pool habitats in low-gradient rivers (sloughs, oxbows) were analogous to littoral lake habitats. Ross and Matthews (in press) suggested, as did Kitchell et al. (1977), that the river analogy applies to many other groups of lake-inhabiting fishes in North America. Furthermore, fishes of large rivers may use habitats in new lakes (or impoundments) similarly to their use of unimpounded, large-volume habitats, for example, Blue Catfish (*Ictalurus furcatus*) versus Channel Catfish (*I. punctatus*) in Lake Texoma (Edds et al. 2002). Thus, although habitats occupied by North American fishes can, at first glance, be separated into lentic and lotic categories, with a few exceptions, this may not be the most meaningful ecological axis along which to consider fish assemblages. In many cases, a more meaningful axis would be upland versus lowland fishes or fishes occupying habitats with lower to higher water retention times (Part 5; Ross and Matthews, in press).

SUMMARY

Fishes are the most diverse group of craniate organisms with nearly 28,000 species distributed among five classes. Freshwater fishes make up nearly half of all fish species (some 43%), yet liquid fresh water accounts for only 0.01% of all water on our planet. The greatest diversity of freshwater fishes is found in the New and Old World tropics, but North America harbors the richest temperate freshwater fish fauna with some 1,061 species. Within North America, more than half of all species occur in the east. The North American fish fauna is primarily a fauna of flowing water, with relatively few contemporary species unique to lakes.

SUPPLEMENTAL READING

Dudgeon, D., A. H. Arthington, M. O. Gessner, Z.-I. Kawabata, D. J. Knowler, C. Lévêque, R. J. Naiman, A.-H. Prieur-Richard, D. Soto, M. L. J. Stiassny, and C. A. Sullivan. 2006. Freshwater biodiversity: Importance, threats, status and conservation challenges. *Biological Reviews* 81:163–82. An overview of challenges facing aquatic organisms and those working to protect those organisms and their ecosystems.

Smith, G. R., C. Badgley, T. P. Eiting, and P. S. Larson. 2010. Species diversity gradients in relation to geological history in North American freshwater fishes. *Evolutionary Ecology Research* 12:693–726. A recent and thorough treatment of the interplay between geological processes and fish distributions and diversity.

Stiassny, M. L. J. 1996. An overview of freshwater biodiversity: With some lessons from African Fishes. *Fisheries* 21(9):7–13. Documents the interplay between demands for water and the protection of aquatic biodiversity.

Warren, M. L., Jr., B. M. Burr, S. J. Walsh, H. L. Bart, Jr., R. C. Cashner, D. A. Etnier, B. J. Freeman, B. R. Kuhajda, R. L. Mayden, H. W. Robison, S. T. Ross, and W. C. Starnes. 2000. Diversity, distribution, and conservation status of the native freshwater fishes of the southern United States. *Fisheries* 25:7–31. The southeastern United States contains the richest North American fish fauna and also contains the greatest number of species at risk.

WEB SOURCES

Digital image of North America. http://photojournal.jpl.nasa.gov/catalog/PIA03377.

TWO

Origin and Derivation of the North American Freshwater Fish Fauna

ASSEMBLING A FAUNA: FISH EVOLUTION AND PLATE TECTONICS

STUDIES OF FISH DISTRIBUTION and ecology are often initiated by making a series of collections in various aquatic habitats. In doing so, there is a tendency to consider fishes taken in each particular mesohabitat, such as a pond, lake shore, or stream riffle, to be part of a natural assemblage developed as a unit over evolutionary and ecological time through the interaction of local processes. However, even discounting the recent major role of humans in introducing fishes outside of their natural ranges, this assumption that species in the assemblage share a long history of coexistence may not be valid. As pointed out by Brooks and McLennan (1991), Matthews (1998), and others, in addition to local determinism in shaping assemblage composition, contemporary species assemblages may also be due to the association of the species' ancestors in that particular geographic region, or the species may have evolved within different assemblages and one or more entered the modern assemblage through dispersal.

For instance, among the resident species in the diverse (43 species) fish assemblage of the Piney Creek watershed of Arkansas, some species have affinities with faunas to the northeast, others to the east, and yet others to the southeast (Matthews 1998). This suggests that a more realistic view is that assemblages are composed of a mixture of species that have different origins and ages, and have been interacting for widely different periods of time. Of course, this situation is now made even more extreme by the rapid and widespread introduction of nonnative species (Courtenay et al. 1986; Fuller

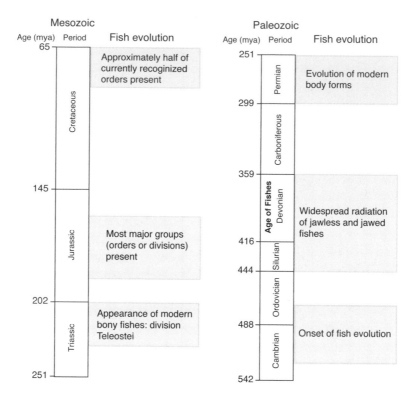

FIGURE 2.1. Paleozoic and Mesozoic landmarks in fish evolution. Based on Moy-Thomas and Miles (1971), Nelson (2006), and Helfman et al. (2009).

et al. 1999) and the resulting homogenization of faunas (Rahel 2000, 2002; Olden and Poff 2003). This chapter deals with the initial large-scale filters affecting the origin of the North American fish fauna (see figure in Part I).

Fish evolution began in the early Paleozoic Period in the late Cambrian or early Ordovician, approximately 500–470 million years ago (mya) (dating of geologic time periods follows Walker and Geissman 2009). During the Silurian and Devonian (444 to 359 mya), there was widespread radiation of both jawless and jawed fish lineages (Figure 2.1). Because of this, the Devonian is often referred to as the Age of Fishes. By the close of the Paleozoic, approximately 250 mya, body forms had evolved that differed very little from those living today (Moy-Thomas and Miles 1971). Modern bony fishes (fishes in the division Teleostei) appeared in the lower Mesozoic (middle or late Triassic, 245–202 mya), and representative forms of most major groups (i.e., orders or divisions) of modern fishes were

present at least by the middle Mesozoic (Jurassic, 202–145 mya [Figure 2.1]) (Nelson 2006; Helfman et al. 2009). In fact, almost half of the currently recognized orders of teleostean fishes have fossil records that reach to the Cretaceous Period of the Mesozoic, some 145 to 65 mya (Figure 2.1) (Helfman et al. 2009). Consequently, given the great age of many of the major fish lineages, any understanding of fish biogeography requires looking at a broad slice of the earth's dynamic geologic history.

A Dynamic Earth

During the Carboniferous Period of the late Paleozoic, a time of active radiation in fishes, precursors to the present-day continents collided to form the single large, but highly dynamic and ephemeral, landmass of Pangea (Figure 2.2; Box 2.1). Although subsequently reworked by geological processes such as uplift and erosion, Pangean geography included familiar elements, including an ancestral Mississippi River in central Pangea that flowed through a gap bounded by the Southern Appalachian Mountains and the Ouachita Mountains (Redfern 2001). The process of mountain building (orogeny) that formed the Ouachita and Alleghenian-Appalachian ranges was driven by the collision of the supercontinents Laurasia and Gondwana along the region now recognized as the Mississippi Embayment, as one of the final stages in the assembly of Pangea (Redfern 2001). During the maximum extent of Pangea, what is now the eastern continental margin of North America was adjacent to northwestern Africa and the northeastern region of North America abutted Western Europe (Figure 2.2). Pangea gradually began to break up during the middle Jurassic (176–161 mya) with the intrusion of the central Atlantic Ocean between the northern (Laurasia) and southern (Gondwana) landmasses (Torsvik and Van der Voo 2002). Continued rifting, especially through the late Jurassic to middle Tertiary, resulted in the present-day arrangement of continents (Cracraft 1974; Briggs 1987; Hocutt 1987). Because forms ancestral to most modern fish lineages (but generally not modern orders or families) were present prior to the breakup of Pangea, the subsequent movements of tectonic plates and their associated faunas were primary factors in the formation of fish assemblage composition (Figure 2.1), although in some cases, phylogenies are too poorly understood or fossil material is lacking to clearly establish an area of origin.

Ages of North American Fish Families

Understanding how long particular fish groups have inhabited North America is, at best, a difficult endeavor. The direct evidence is based on the fossil record, which is usually incomplete and requires access to rock layers of various ages (i.e., the rock formations of particular ages must be exposed by weathering or excavation such as during road construction). In addition, even if there is a rock outcrop of a particular age, it might only represent a particular kind of habitat

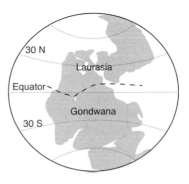

FIGURE 2.2. The configuration of the major continental landmasses at the close of the Paleozoic (250 mya) as Pangea approached its maximum extent. Based on Torsvik and Van der Voo (2002) and Torsvik and Cocks (2004).

from one particular region and thus show great ecological bias (Patterson 1981). This would be akin to trying to understand the entire modern North American fish fauna by sampling only a few low-gradient rivers where they enter into the sea or several isolated lakes. In addition, the ages represented by fossils are likely minimum ages because, from the standpoint of vicariance biogeography, groups are older than their oldest fossil (Box 2.2; Parenti 1981). However, in spite of its shortcomings, the fossil record is often the best evidence that is available.

Ages can also be determined indirectly from a well-developed phylogeny if there are fossils or geologic events available for calibration of molecular divergence times (Box 2.3; Lieberman 2003). For example, based on a calibrated molecular clock analysis, species' divergence times within logperches (a group of darters in the genus *Percina*) ranged from 4.20 to 0.42 mya, with most speciation events taking place in the Pleistocene (Near and Benard 2004). The divergence times were based on the assumption that there is a constant rate of gene substitutions over time, and that by comparing the degree of genetic divergence among darter lineages, it is possible to convert the degree of divergence to a time estimate—the "molecular clock." Because it is known that rates of gene substitution vary among taxonomic groups (Britten 1986; Avise 2004)

The idea that continents change position over time, or drift, is understood today as a scientific fact and is as important to geology and biogeography as evolution is to an understanding of biology and biogeography. However, it was not that long ago that continental drift was quite a controversial issue, in spite of the fact that a quick glance at a world map suggested that continental shapes could be fitted together like a crude jigsaw puzzle—a point recognized as early as 1620 by Francis Bacon (Cattermole 2000). The idea that continents move was proposed early in the twentieth century (1910 and 1912) by Frederick B. Taylor, H. D. Baker, and A. L. Wegener, with the German meteorologist Wegener generally recognized as the father of the modern theory of continental drift (Briggs 1995; Cattermole 2000). However, the reigning view at that time was that the earth's crust was solid and that crustal movement was therefore not possible, and that former land connections (so-called land bridges) were largely responsible for the movement of many organisms between continents. In fact, my first course in biogeography used the classic text by Darlington (1957) in which the issue of continental drift, although mentioned briefly, was largely discounted. By the late 1950s and 1960s the issue of continental drift was gaining increasing

attention, and by 1970 the theory was generally accepted as fact (Cattermole 2000).

The reasons for the fairly rapid turnaround had to do with advancement in the tools of physical geography, paleontology, and biogeography. The understanding that molten rocks can capture the orientation of the magnetic field at the time that they cool provided important insight into past continental orientations. If continents were fixed in their positions, then past magnetic orientations should be coincident with present-day orientations. However, this is obviously not the case. Indeed, magnetic anomalies (the nonalignment of the magnetic field in rocks with the present-day magnetic field) are an important data source in the reconstruction of the earth's surface (Torsvik et al. 2001). Today the earth's surface is understood to comprise a series of semirigid plates that are moving relative to one another, driven by the powerful convection currents of the underlying mantle. Because the continents are riding on the plates, the term continental drift has been replaced by plate tectonics (Cattermole 2000; Cox and Moore 2005). Additional evidence for plate tectonics comes from modern studies of the alignment of continents and the paleodistribution of organisms as shown by fossils.

and, consequently, that molecular clocks need to be calibrated using related taxa, Near and Benard (2004) used rates of gene substitution from the family Centrarchidae (in the same order, Perciformes, as the darters) and applied this to their analysis of logperches to generate their divergence time estimates.

Age information from fossils or calibrated molecular phylogenies is available for 27 families of North American freshwater fishes (Figure 2.3). One family, the lampreys (Petromyzontidae), likely dates to the Paleozoic, and 5 groups (bowfins, Amiidae; pikes, Esocidae; sturgeons, Acipenseridae; paddlefishes, Polyodontidae; and gars, Lepisosteidae) have been present since the Cretaceous Period of the late Mesozoic. The remaining 21 families all date within the Cenozoic. Although 6 of the 27 families (22%) were represented prior to the

Cenozoic, considering the current number of species per family, the ancestors of only 1.8% of the North American fish fauna occurred earlier than the Cenozoic. Within the Tertiary Period of the Cenozoic, 11 families are represented in the Paleogene Period (Paleocene to Oligocene epochs), and the remaining 10 families are represented in the Neogene Period (Miocene to Pliocene epochs).

Paleogene families include the ictalurids, percopsids, clupeids, salmonids, moronids, hiodontids, catostomids, centrarchids, aphredoderids, umbrids, and cyprinids. Neogene families are represented by the goodeids, poeciliids, percids, cichlids, fundulids, cyprinodontids, cottids, gasterosteids, atherinopsids, and sciaenids. Although the second most speciose family of North American freshwater fishes, the Percidae, is known from fossils only from the

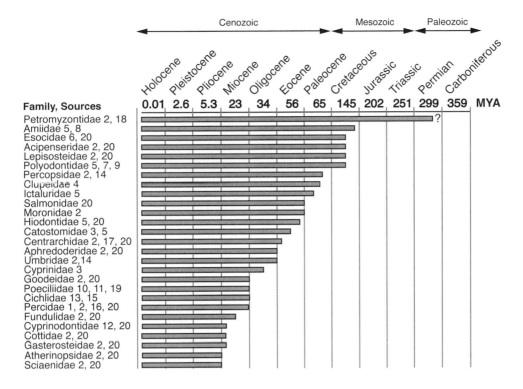

FIGURE 2.3. The earliest representation of major fish families in North America based on the first occurrence of fossils or from calibrated molecular phylogenics. Because the earliest fossils represent a minimal age of origin, families could be much older. Within the Cenozoic, geologic ages refer to epochs; within the Mesozoic and Paleozoic, ages refer to periods. Numbers at the top of each column are the beginning age (mya) of each geologic age or period. Numbers after families indicate sources; gaps in fossil record are not shown.

SOURCES: 1. Carlson et al. (2009), 2. Cavender (1986), 3. Cavender (1991), 4. Grande (1982), 5. Grande (1984), 6. Grande (1999), 7. Grande and Bemis (1991), 8. Grande and Bemis (1996), 9. Grande et al. (2002), 10. Mateos et al. (2002), 11. Meyer and Lydeard (1993), 12. Miller (1981), 13. Murray (2001a), 14. Murray and Wilson (1996), 15. Myers (1966), 16. Near and Keck (2005), 17. Near et al. (2005), 18. Nelson (2006), 19. Webb et al. (2004), 20. Wilson and Williams (1992).

Pleistocene, calibrated molecular phylogenies suggest a much earlier occurrence. The separation of darters from nondarter percids dates to 19.8 mya (Carlson et al. 2009) and within the darter genus *Nothonotus*, the age of the most recent common ancestor dates to 18.5 mya (Near and Keck 2005). Consequently, percids likely occurred in North America at least by the early Miocene (approximately 23 mya). Seventy-eight percent of the 27 major families were present in North America by the early Miocene (23–16 mya) and were thus affected by numerous geologic and climatic events of the late Tertiary.

In western North America, a freshwater fauna dominated by teleosts first appeared by the late Paleocene, followed by the expansion of an essentially modern fauna by the Oligocene and Miocene (Minckley et al. 1986). The western fauna during the Eocene (56–34 mya) and Oligocene (34–23 mya) shared forms with an eastern fauna, including paddlefishes, gars, sturgeons, bowfins, salmon and trout, mooneyes, suckers, catfishes, troutperch, and pickerel (Grande 1984; Minckley et al. 1986; Grande and Lundberg 1988; Grande 1999). The Oligocene fauna included elements from the earlier fauna, such as mooneyes, salmon and trout, and pickerel, as well as from more recent groups, such as minnows, atherinopsids, pupfishes, topminnows, sticklebacks, bass and sunfishes, surfperches, and sculpin (Minckley et al. 1986). Of the nonteleosts, sturgeon are

BOX 2.2 • Biogeographic Theory

From the standpoint of interpreting broadscale distribution patterns of organisms, there are two major paradigms of biogeography: dispersal explanations and vicariance explanations.

DISPERSAL EXPLANATIONS

There is a natural tendency of organisms to disperse within areas of suitable habitat, and certainly many organisms, including fishes, have well-defined long-distance dispersal patterns, achieved either through adult movement or through dispersal of larval stages. In the dispersal model, organisms are assumed to have migrated across preexisting barriers and there are many examples where this has occurred. However, when we consider the present-day distribution of related taxa over widely separated areas with no intervening populations (e.g., a disjunct distribution), then the dispersal explanation becomes more difficult.

The dispersal model was espoused by numerous early biogeographers, such as Darwin's contemporary, Alfred Russell Wallace (1876), and later William D. Matthew (1915) and Phillip J. Darlington (1957). Among ichthyologists, Briggs (1974, 1995, and included papers) has been a strong proponent for the importance of dispersal as a primary mechanism.

VICARIANCE EXPLANATIONS

The basic tenet is that organisms are passively transported by movement of tectonic plates or by other geological means. If this occurs, then several or more taxa should share common distribution patterns, where the distribution of each taxon is referred to as a track. As such, a major starting point in vicariance biogeography is the search for common patterns of distribution (i.e., generalized tracks) among different taxa. If a common pattern of distribution exists for two or more monophyletic taxa, then this suggests that the generalized

track may be due to geologic events (Croizat et al. 1974; Wiley 1981; Grande 1990). The emphasis in vicariance biogeography is on patterns generated by many, and not necessarily closely related, taxa. (In contrast, while not ignoring the generality of patterns, dispersalists have, at least historically, focused more on individual taxa.) The range of a species can be disrupted by the formation of a barrier (a vicariant event) so that a formerly contiguous population is split into separate populations (termed vicariance).

Pioneering studies by Croizat (1958; Croizat et al. 1974) helped form the basis for vicariance biogeography. For instance, Croizat's panbiogeography (1958) ultimately worked as "a major catalyst for change during the 1960s resurgence of interest in biogeographical thought" (Keast 1991). Croizat amassed distributional patterns of species (e.g., tracks) and stressed the importance of concordant patterns (e.g., generalized tracks), even though in his 1958 book he still discounted the role of continental drift. Strong ichthyological proponents of vicariance biogeography have included Gareth Nelson and Norman Platnick (e.g., Nelson and Platnick 1981), Edward Wiley (e.g., 1981), and the late Donn Rosen (e.g., Rosen 1978).

SYNOPSIS OF PARADIGMS IN BIOGEOGRAPHY

In a dispersal model, the barrier is older than at least one of the isolated populations and the age of the barrier is older than the disjunction in range. In the vicariance model, the populations predate the age of the barrier and ages of the barrier and the disjunction in the range are the same. While there has been considerable argument among biogeographers about the relative merits of each major explanation, a synthesis of views is, undoubtedly, required to understand the distribution of fishes (Wiley 1981; Briggs 1995; Moyle and Cech 2004).

---

represented by extant western forms, but gars, paddlefishes, and bowfin are now absent from the western fauna.

Although ecologists, until recently, have tended to focus more on current faunas and less on historical aspects, knowledge of the

varying ages of occupation of fish groups in North America is of paramount importance to our understanding of fish assemblages and the extent and duration of coevolutionary processes. In addition, recent studies have stressed the importance of incorporating information on

The study of evolutionary relationships of species or higher taxa is the field of systematic biology. The concepts and methodologies underlying how evolutionary relationships are studied, and how best to portray the resultant phylogenies, have been areas of considerable debate and advancement over the last several decades. (For an overview, see Mayden and Wiley [1992] and Mayden and Wood [1995].) Concurrent with the development of principles and procedures of systematic biology, molecular data, first as protein analyses and then broadened to include mitochondrial and genomic DNA information, have complemented morphological, ecological, and behavioral data that are used in developing phylogenies. Phylogenetic (i.e., evolutionary) relationships are shown hierarchically in branching diagrams referred to as dendrograms or cladograms, depending on the methodology used to construct them. For the most part, especially among ichthyologists, phylogenies follow methods proposed by the German biologist Willi Hennig (English translation, 1966)—the method of phylogenetic systematics. This approach has been further developed by Niles Eldredge and Joel Cracraft (1980), Gareth Nelson and Norman Platnick (1981), Edward Wiley (1981), and many others.

Basic principles of the cladistic method are that (1) relationships among groups are based primarily on branching points in evolution and not on degrees of divergence; (2) recognized groups should be derived from a single ancestral group (the principle of monophyly); and (3) taxa should be recognized on the basis of possessing shared, derived characters (synapomorphies) (Mayr and Ashlock 1991; Brooks and McLennan 1991). Sister groups are derived from the same common ancestor. A major goal of cladistic analysis is the recognition of synapomorphies, the derived (homologous) characters. Convergent evolution can result in organisms possessing structurally analogous but not evolutionarily related characters (termed homoplasies). Cladograms should be considered as hypotheses of evolutionary relationships. Support for cladistic hypotheses increases with the number of synapomorphies used in a study (i.e., studies based on few characters can be misleading), and by the number of studies, based on different characters, that come to the same or similar conclusions.

evolutionary relationships of component species (i.e., phylogeny) into studies of community ecology (Webb et al. 2002).

As is evident from Figure 2.3 and the previous paragraphs, fish groups vary widely in their ages of occupation in North America. Consequently, the forces shaping the evolution of morphology, physiology, and behavior of species making up present-day assemblages are likely not to be found by only looking within the contemporary assemblage. Instead, selective pressures leading to various traits may date to earlier time periods and may not even include the present-day assemblage. As an example, assemblages that have included pike have changed over time since the Paleocene (65–56 mya), when pike would have been part of fish assemblages including osteoglossomorphs, percopsiforms, amiids, gonorynchids, lepisosteids, asineopids (now extinct), osmerids, clupeids, cyprinoids (possibly catostomids), and ictalurids (Wilson and Williams 1992). Although feeding and morphological specializations show little change in pike, the community relationships have since changed a great deal, consequently, and "major adaptations of pike evolved before modern predator-prey systems existed" (Wilson and Williams 1992).

Origins of North American Fish Families

As with differences in ages, North American fish families exhibit a variety of origins, including archaic groups and some more recent groups whose origins can be traced to Pangean and Laurasian faunas (Figure 2.4). Of the 50 North American fish families listed by Burr and Mayden (1992), half show a marine origin. In some groups the radiation from the marine environment into fresh water occurred early, as with the bowfin subfamily Amiinae (family Amiidae) that has occupied freshwater habitats in the Northern

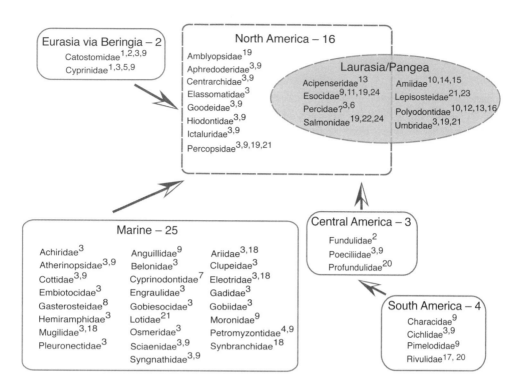

FIGURE 2.4. General origins of North American fish families. Phylogenetic and/or fossil information generally does not allow enough resolution to determine modern continental origins listed as arising in Laurasia/Pangea.

SOURCES: 1. Berra (2001), 2. Briggs (1986), 3. Burr and Mayden (1992), 4. Cavender (1986), 5. Cavender (1991), 6. Collette and Banarescu (1977), 7. Echelle and Echelle (1992), 8. S. A. Foster et al. (2003), 9. Gilbert (1976), 10. Grande (1984), 11. Grande (1999), 12. Grande and Bemis (1991), 13. Grande and Bemis (1996), 14. Grande and Bemis (1998), 15. Grande and Bemis (1999), 16. Grande et al. (2002), 17. Hrbek and Larson (1999), 18. Miller and Smith (1986), 19. Moyle and Cech (2004), 20. Parenti (1981), 21. Patterson (1981), 22. G. R. Smith and Stearley (1989), 23. Wiley (1976), 24. Wilson and Williams (1992).

Hemisphere since the late Cretaceous, some 90 mya (Grande and Bemis 1999). The second-largest group has a North American origin (including those originating in old landmasses of Pangea or Laurasia), followed by groups originating in Central and South America and Eurasia (Figure 2.4).

The 15 fish families containing 90% of the modern species have their origins in Eurasia (minnows and suckers); Central America (live-bearers and topminnows); North America, including Pangean/Laurasian elements (cat-fishes, trouts and salmons, goodeids, sunfishes, and perches); the marine environment (New World silversides, pupfishes, sculpins, lam-preys, herrings); and South America (cichlids). The histories of these groups are treated in more detail in the following section.

Numerically Dominant Families • Plate Tectonics, Ages, and Origins

This section examines in more detail how potential fish assemblages have changed over time by the addition of new taxa and relates the arrival of these taxa to the positions of major landmasses at various times in the past. It is primarily limited to families composing 90% of the North American fish fauna, although sturgeons and pickerels are included because of conservation interest and past or present economic importance. Families are listed by decreasing age of occurrence (largely as determined from fossils) in North America. For the goal of relating fish distribution to continental positions and connections over geological time, I have followed Cracraft

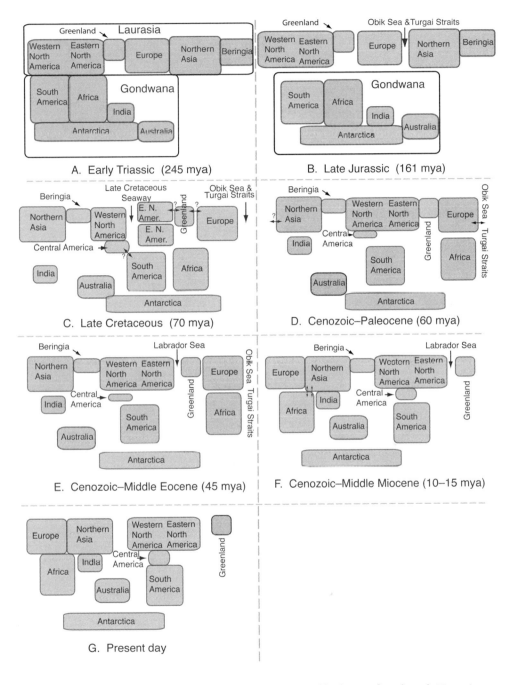

FIGURE 2.5. Schematic diagrams of the relative positions of oceans and landmasses from the early Mesozoic to the present. Arrows indicate possible connections between landmasses. Sources are given in the accompanying text. Landmasses and oceans are not drawn to scale.

(1974) and Matthews (1998) in portraying landmasses as a series of blocks (Figure 2.5). Although temperatures in Antarctica were generally too cold for the survival of freshwater fishes (Matthews 1998), I have included it in the figure to help with orientation. Simple diagrams might seem to suggest otherwise, but movements of landmasses were not necessarily unidirectional, so that connections between different elements may have been made and broken numerous times. A case in point is the union of eastern North America and Europe during the Jurassic and early Cretaceous when these landmasses were joined and then separated several times (A. G. Smith et al. 1994).

Continental Positions

MESOZOIC, EARLY TRIASSIC (245 MYA) The Pangean supercontinent reached its maximum extent in the early Triassic, then began to break apart by the late Triassic (Figure 2.5A). At its maximum, the supercontinents of Laurasia and Gondwana were joined, providing dispersal routes for numerous mobile terrestrial and freshwater organisms from pole to pole and east to west (A. G. Smith et al. 1994; Scotese 2002). During this time, tetrapod vertebrates were essentially cosmopolitan with no evidence of latitudinal variation (Briggs 1995).

MESOZOIC, LATE JURASSIC (161 MYA) The northern and southern supercontinents of Laurasia and Gondwana separated from each other as the young central Atlantic Ocean began to increase in size (Figure 2.5B). The southern landmasses were still grouped within Gondwana, but the components of Laurasia had begun to drift apart. Western and eastern North America were joined along the southern margin, although A. G. Smith et al. (1994) portrayed the northern elements as separated by a large, north-to-south-oriented inland sea (not shown by Scotese 2002). Greenland was part of eastern North America, whereas Europe had "recently" (e.g., approximately 20 million years earlier) separated from eastern North America and was also separated from northern Asia by the Obik Sea to the north, and by the Turgai Straits to the south (A. G. Smith et al. 1994; Zwick 2000; Scotese 2002).

MESOZOIC, LATE CRETACEOUS (70 MYA) High sea levels, resulting in shallow epicontinental seas, separated western and eastern North America as well as northern and southern sections of eastern North America (A. G. Smith et al. 1994) (Figure 2.5C). A large landmass comprising northern Asia, Beringia, and western North America dominated the Northern Hemisphere. Greenland was apparently close to, but perhaps separated from, eastern North America. There is disagreement about whether eastern North America, Greenland, and Europe were connected. Briggs (1986, 1995) and Rage and Rocek (2003) supported a connection, whereas other recent authors (e.g., A. G. Smith et al. 1994; Scotese 2002) show separations between these landmasses (as I have done in Figure 2.5C). Europe was separated from Asia by the Obik Sea to the north and the Turgai Straits to the south (the combined water body is the Uralian Sea [Rage and Rocek 2003]). There is also disagreement about the connection of North and South America during this period. Briggs (1995) and Rage and Rocek (2003) indicate a connection via Central America, whereas A. G. Smith et al. (1994) and Scotese (2002) do not.

CENOZOIC, EARLY TERTIARY

PALEOCENE (60 MYA) The Northern Hemisphere was largely ringed by a single landmass comprising Asia, Beringia, North America, Greenland, and Europe (Figure 2.5D). Asia was linked to western North America via Siberia and Beringia, the large inland sea separating eastern and western North America had partially receded, and eastern North America was connected to Europe via Greenland. Europe continued to be totally (Rage and Rocek 2003) or partially (A. G. Smith et al. 1994) separated from northern Asia, and North and South America were not connected.

MIDDLE EOCENE (45 MYA) The Northern Hemisphere was dominated by a landmass comprising northern Asia, Beringia, and North America (Figure 2.5E). There were possible northern connections between Greenland and North America, and Greenland and northern Europe (A. G. Smith et al. 1994; Briggs 1995), although more recent data seem to cast doubt on this since Torsvik et al. (2001) showed Greenland separated from North America by the Labrador Sea in the early Tertiary (54 mya) and also separated from northern Europe; this is reflected in Figure 2.5E. The Turgai Straits had reopened from the south and, along with the Obik Sea to the north, separated Europe from northern Asia. South America was separate from North America, with Central America existing as island archipelagos (A. G. Smith et al. 1994; Briggs 1995; Scotese 2002). Although less critical for understanding the North American fauna, the position of India during the Eocene is uncertain—Briggs (1989) argued, based on faunal evidence, that India must have already contacted the Asian continent by the early Eocene, whereas other studies based on geophysical evidence indicate that it did not make contact until the Miocene (e.g., A. G. Smith et al. 1994; Scotese 2002).

CENOZOIC, LATE TERTIARY, MIDDLE MIOCENE (10–15 MYA) The large, Northern Hemisphere landmass of northern Asia, Beringia, and North America not only continued to persist but had expanded with the closure of the Uralian Sea (Turgai Straits and Obik Sea) and the union of Europe with northern Asia (Figure 2.5F). South America was still separate, but Central America was joined with North America. By the early Miocene, Africa had contacted Asia along the Arabian Peninsula, allowing potential interchange of freshwater fishes and resulting in the formation of the Mediterranean Sea (A. G. Smith et al. 1994; Briggs 1995; Scotese 2002; Rage and Rocek 2003). The present configuration of landmasses differs from the middle Miocene by the submergence of Beringia, the

expansion of the Atlantic and Pacific oceans, and the union of South and Central America (Figure 2.5G).

Ages and Origins of Major Fish Families

PETROMYZONTIDAE (21 SPECIES) Lampreys represent one of the two surviving groups of jawless fishes, the other being the strictly marine hagfishes (Myxini). All lampreys have a prolonged larval stage (termed ammocoetes) during which they burrow into soft sediments of streams and feed on small organisms at the sediment-water interface. Some lampreys have a parasitic adult stage where they feed on body fluids of other fishes, whereas others do not feed after their metamorphosis to adults (Hardisty and Potter 1971). Petromyzontid lampreys likely represent the oldest living group of North American freshwater fishes, although there are no North American fossils that can be conclusively placed within the family (Cavender 1986). Lamprey fossils, described as *Mayomyzon pieckoensis*, were from Pennsylvanian marine shale deposits in Illinois, dating to approximately 310 mya (Bardack and Zangerl 1968), and another species, *Hardistiella montanensis*, was described from lower Carboniferous (ca. 350 mya) marine formations from what is now Montana (Janvier and Lund 1983). Although both were initially placed within the Petromyzontidae, *Mayomyzon* is now recognized as being the sole described species in the extinct family Mayomyzontidae, whereas the family relationship of *Hardistiella* is uncertain (Nelson 2006). A third fossil species, *Pipiscius*, also dates from Paleozoic formations in North America (Janvier 1997a). Of these extinct lampreys, *Mayomyzon* is most similar in body form to modern petromyzontids and certainly demonstrates that the lamprey body plan, and most likely mode of life, has been around since the Paleozoic. Recent work on lamprey phylogeny supports the notion that the fossil family Mayomyzontidae is as old or older than any extant families and that the North American Petromyzontidae are monophyletic (Gill et al. 2003). Freshwater lampreys are likely derived from marine ancestors (Gilbert 1976) and the

Petromyzontidae probably originated in Pangean North America (Figure 2.2) in the Paleozoic (Cavender 1986).

ESOCIDAE (4 SPECIES) The pike and pickerel are one of the five North American families dating from the Mesozoic, a group that includes two subclasses: Chondrostei (sturgeon, paddlefish) and Neopterygii (bowfins, gars, and pickerels) (Nelson 2006). Fossil and living pikes and pickerels are found only in the Northern Hemisphere, including North America, Europe, and Asia (Grande 1999; Berra 2001). The family includes major recreational species such as Muskellunge (*Esox masquinongy*) and Northern Pike (*E. lucius*). The earliest known North American fossils are from late Cretaceous deposits of the Green River Formation (Grande 1999), when western North America was separated from eastern North America by the Late Cretaceous Seaway but linked with eastern Asia via Beringia (Figure 2.5C). The oldest species, *Esox tiemani*, was described from Paleocene lake and river deposits in Alberta and Saskatchewan (Wilson and Williams 1992). The discovery of this fossil species demonstrated the highly conserved body plan of esocids. The family perhaps originated in Laurasian North America, although data remain inconclusive about possible origins in northern Europe or Asia (Patterson 1981; Grande 1999; Wilson and Williams 1992).

ACIPENSERIDAE (8 SPECIES) Sturgeons are also an ancient group with fossils from Asia, Europe, and North America. The family is, and has been, essentially limited to temperate regions (Bemis and Kynard 1997). The order Acipenseriformes, containing sturgeons and paddlefishes, originated during the Triassic in Western Europe at a time when Laurasian elements were beginning to separate (Figure 2.5A, B) and sturgeons likely had their earliest diversification in Central Asia (Bemis and Kynard 1997). North America fossils date from the Paleogene and Upper Cretaceous deposits in areas that are now Montana and Alberta (Cavender 1986; Bemis and Kynard 1997). There

is also a Miocene fossil from Virginia (Bemis and Kynard 1997). Sturgeon have a Laurasian/Pangean origin (Grande and Bemis 1996), although whether ancestral forms were present in Laurasian North America, or whether Sturgeon arrived via Beringia or from Western Europe is uncertain. Sturgeon body form is highly conserved, having changed little since the Mesozoic (G. R. Smith 1981; Sulak and Randall 2002). The family includes both freshwater and anadromous species, has a Holarctic distribution, and includes some of the largest and longest-lived species of freshwater fishes (Berra 2001). In fact, the Beluga Sturgeon (*Huso huso*) of the Black and Caspian seas reaches 9 m and 1300 kg and may live for nearly 100 years (Berra 2001). North American sturgeons in the genus *Acipenser* also reach large sizes (> 2m TL) and may live 100 years or more (Sulak and Randall 2002). Anadromous North American forms (all in the genus *Acipenser*) are derived from freshwater forms and were one of the first groups of fishes to solve the physiological challenges of moving from fresh to salt water (Bemis and Kynard 1997; Sulak and Randall 2002).

CLUPEIDAE (10 SPECIES) Although this family is primarily marine, there are a number of freshwater species worldwide, all considered to be derived from marine forms. In North America, clupeids first appeared in the fossil record in the middle Paleocene (ca. 60 mya) in what is now Montana and apparently occupied western North America until the middle Eocene (Grande 1982). After the middle Eocene, freshwater clupeid fossils did not show up again in the fossil record for approximately 40 million years until the Pliocene/Pleistocene (Cavender 1986). The Plio/Pleistocene fossil was identifiable as a modern species—the Threadfin Shad (*Dorosoma petenense*) (Miller 1982).

ICTALURIDAE (46 SPECIES) Ictalurid catfishes date from the Paleogene and are restricted to North America. The earliest North American fossils referable to the Ictaluridae date from the late Paleocene (60 mya) (Cavender 1986;

Lundberg 1992). The most complete specimens are of the extinct genus *Astephus* and are from Eocene lake deposits in the Green River Formation, a large system of lakes located in intermontane basins, in what is now Utah, Colorado, and Wyoming, that were formed by the uplift of the Rocky Mountains in the Tertiary (Grande 1984; Grande and Lundberg 1988). From the Paleocene through the late Eocene, the Green River system comprised one of the world's largest and longest-lived Great Lakes systems (Grande 2001). Ictalurids likely originated in North America (Gilbert 1976; Burr and Mayden 1992; Lundberg 1992) and, although there were connections with Asia and Europe in the Cretaceous and Paleocene (Figs. 2.5C, D), this group has never been found outside of North America. Of the modern genera, *Ictalurus*, *Ameiurus*, and *Trogloglanis* occurred in the early Oligocene, and *Pylodictis* fossils are known from the middle Miocene (Lundberg 1992). The genus *Noturus* (madtom catfishes) is younger, with fossil material dating only from the early Pleistocene (1–2 mya) (Cavender 1986).

SALMONIDAE (38 SPECIES) The oldest fossil salmonids are of the extinct genus *Eosalmo*, found in Eocene deposits located in what is now British Columbia and northern Washington (Wilson and Williams 1992). *Eosalmo* apparently occupied Pacific drainages and, based on phylogenetic analysis, was basal to all other members of the Salmonidae (i.e., considered a stem group) (Wilson 1992). Although Gilbert (1976), among many others, considered the Salmonidae to likely have a marine origin, later work on salmonid phylogeny and ecology points to a freshwater origin for the group. All primitive salmonids are restricted to freshwater habitats, whereas derived groups all contain anadromous species (i.e., those spawning in fresh water and then moving to the sea to feed) (Stearley 1992). The fossil species *Eosalmo* provides additional support for the freshwater origin hypothesis because of the discovery of a large size series of specimens, ranging from young to adult fish, in the same lake deposit (Stearley 1992; Wilson

and Williams 1992). The anadromous life cycle shown by some modern members of the family, such as Steelhead (*Oncorhynchus mykiss*), Pacific salmon species, and Atlantic Salmon (*Salmo salar*), is thus considered a secondary adaptation. Anadromy may have been triggered by increased seasonality caused by the cooling of the climate in the middle Cenozoic, such that the marine habitat offered greater productivity and more constant temperatures that would have favored increased growth in the marine compared to the freshwater habitats (Gross et al. 1988; Stearley 1992).

Based on the worldwide distribution of the family throughout northern Asia and also Europe (Berra 2001), and the presence of the stem-group fossils in western North America, the family likely evolved in the region of Laurasia that included western North America and perhaps northern Asia (Figure 2.5E). The modern species of Pacific trout (genus *Oncorhynchus*) likely originated in the Miocene and have had at least six million years of history (Stearley and Smith 1993). The Atlantic basin salmonids (genus *Salmo*) are primarily a European lineage, and the separation of the eastern *Salmo* and the western *Oncorhynchus* lineages likely occurred via a vicariant event across the northern coast of Asia in the Miocene (Stearley 1992).

CATOSTOMIDAE (71 SPECIES) Suckers represent the median in terms of age in North America (Figure 2.3). The oldest North American fossils, of the extinct genus *Amyzon*, date from the middle Eocene (ca. 49 mya). *Amyzon* is represented in various western fossil deposits, including the Green River Formation of Wyoming (Grande et al. 1982; Grande 1984; Cavender 1991). The habitat was likely swamp-like and included crocodiles and alligators (Grande et al. 1982). Like the modern subfamily Ictiobinae with which it is closely related, *Amyzon* was a fairly large-bodied fish. More derived species of suckers have tended toward smaller body sizes (G. R. Smith 1992). Catostomids occupied much of western North America and eastern Asia by the late Eocene, a time when the Asian and North America landmasses were connected via Beringia (Figure 2.5E) (Cavender 1986, 1991). Of the modern genera of suckers, *Ictiobus* occurred by the middle Miocene and *Chasmistes* by the late Miocene (Cavender 1986). Suckers are thought to have originated in Eurasia and then reached North America via the Pacific connection of Beringia (Figs. 2.4 and 2.5E) (Gilbert 1976; Briggs 1986; Burr and Mayden 1992; Berra 2001).

CENTRARCHIDAE (31 SPECIES) The centrarchids are endemic to North America and likely evolved there (Gilbert 1976; Burr and Mayden 1992). The earliest fossils of this primarily eastern North American family date from the Eocene epoch of northwestern Montana (ca. 45 mya) in drainages that flowed eastward from the continental divide (Cavender 1986). During this time, North America had separated from Europe but was still connected to Asia via Beringia (Figure 2.5E). Based on a fossil-calibrated molecular phylogeny, Near et al. (2005) estimated the age of the most recent common ancestor to the Centrarchidae to be 33.6 million years, providing another line of evidence supporting the Eocene age estimate for the group. Because the earliest fossils have not been linked to species, they could not be used in calibrating the molecular phylogeny. By the Miocene, modern genera including *Lepomis*, *Micropterus*, and *Pomoxis* were well represented and, especially by the early Pleistocene, centrarchids had become a dominant element in the North American freshwater fish fauna (G. R. Smith 1981). Ages of modern species, based on molecular phylogenies, are 8–11 million years for *Micropterus*, at least 11 million years for *Pomoxis*, and 14 million years for *Lepomis* (Near et al. 2003, 2005). Centrarchids also were widespread by the middle Miocene, based on fossils found west of the continental divide and including fossils of the extant western genus *Archoplites* (Cavender 1986).

CYPRINIDAE (297 SPECIES) The largest family of North American fishes has a Eurasian origin and likely reached North America via Beringia. The earliest fossil evidence in North America is from several Oligocene deposits in the northwestern United States in a region that in the middle Tertiary would have been near the western continental margin (Cavender 1986, 1991). By the late Miocene and Pliocene, cyprinids had taken their place as a major component of the North American fish fauna (Cavender 1986, 1991). In the New World, cyprinids are restricted to North America with no records, past or present, from South America. They are well represented both in lineage and species diversity throughout Europe, Asia, and sub-Saharan Africa (Howes 1991; Berra 2001). Although there are a variety of hypotheses of relationships within the Cyprinidae, there appear to be two main lineages (treated as subfamilies), the Leuciscinae and the Cyprininae (Cavender and Coburn 1992). North American minnows are all within the subfamily Leuciscinae, which is also well represented in Eurasia. Two phyletic groups are recognized within the subfamily Leuciscinae, the Phoxinini and the Leuciscini. The majority of North American minnows are phoxinins, with only the monotypic genus *Notemigonus* placed in the Leuciscini (Cavender 1991).

Cyprinids likely reached North America via Beringia during periods of lowered sea level that occurred coincident with a period of climatic cooling during the late Eocene to early Oligocene—a cooling event perhaps caused by changes in ocean currents related to the separation of Australia from Antarctica and the opening of a seaway between Greenland and Norway, allowing an exchange between North Atlantic and Arctic waters (Figs. 2.5D, E) (Cavender 1991). By the time that cyprinids reached North America, the Atlantic Ocean had filled the gap between the North American and European plates, precluding movement from eastern North America and western Europe (Figure 2.5E). As a group, the Cyprinidae likely originated in the Oriental region where all major cyprinid groups are represented (Cavender 1991).

From an ecological standpoint, the speciose cyprinids are relatively recent arrivals to North America and were thus a new element incorporated into fish assemblages already composed of older groups such as salmonids, esocids, ictalurids, and others (cf. Figure 2.3). The rapid radiation of minnows was perhaps related to the rise of many insect families such as the dipterans (Cavender 1991).

GOODEIDAE (45 SPECIES) The distribution of the goodeids includes the western Great Basin of the United States (subfamily Empetrichthyinae) and the Mexican plateau (subfamily Goodeinae) (Berra 2001; Webb et al. 2004). The oldest fossils are Miocene—the subfamily Goodeinae is represented by the extinct genus *Tapatia* from deposits in the state of Jalisco, Mexico (Cavender 1986), and the subfamily Empetrichthyinae by material of the extant genus *Empetrichthys* from deposits in Southern California (Webb et al. 2004). A molecular-based phylogeny indicates that the family originated 23 mya, which corresponds well to the Miocene fossils. The two subfamilies diverged between 11.5 and 16.8 mya (Doadrio and Dominguez 2004; Webb et al. 2004).

The family is thought to have originated in North America (Gilbert 1976; Burr and Mayden 1992). Increasing aridity during the Tertiary may have fragmented the once continuous range of the family, resulting in the divergence of the two subfamilies (Parenti 1981; Webb et al. 2004).

POECILIIDAE (69 SPECIES) The livebearers are primarily a Neotropical group with most of the diversity centered in Mexico, Central America, South America, and the West Indies, and with relatively few species in temperate North America (Parenti 1981; Rauchenberger 1988). There are no known fossils of this group in North America and no pre-Quaternary fossils known at all (Hedges 1996). This led Matthews (1998) to suggest a recent (< 1 mya) North American age for the family. However, recent molecular phylogenetic studies of *Poeciliopsis*, a large genus within the family

that occurs on the central Mexican Plateau, indicate that divergence within this genus occurred 6–18 mya (Mateos et al. 2002). As a consequence, the age of the family in North America must be at least Miocene. In further support of this, both morphological (Parenti 1981) and molecular phylogenies (Meyer and Lydeard 1993) show that the clade containing the Poeciliidae and the clade containing the Goodeidae are sister groups (i.e., derived from a common ancestor). Given that a molecular phylogeny (Webb et al. 2004) places the origin of the Goodeidae in the Miocene (ca. 23 mya), a similar age should apply to the Poeciliidae. North American poeciliids are most likely derived from Central American ancestors (Gilbert 1976; Burr and Mayden 1992; Lydeard et al. 1995).

CICHLIDAE (16 SPECIES) This large family has a broad Neotropical distribution occurring in Mexico, Central and South America, and the West Indies, with one species, the Rio Grande Cichlid (*Cichlasoma cyanoguttatum*), even reaching into the United States. Cichlids are poorly represented in the North American fossil record. In the Paleotropics, cichlids occur in Africa, Madagascar, and parts of southern Asia (India, Sri Lanka, Syria, and Iran) (Murray 2001a). There is a Miocene fossil of the modern genus *Cichlasoma* that was found in Haiti (Cavender 1986; Hedges 1996; Murray 2001a), and the oldest known cichlid fossil was found in an African deposit and dates from the middle Eocene (Murray 2001b). One view is that the separation between African and South American cichlids may postdate the formation of the Atlantic Ocean and that South American cichlids were derived from marine dispersal of cichlids from Africa, with molecular phylogeny suggesting a divergence time of 58–41 mya (Vences et al. 2001). However, a more recent review supports a vicariance hypothesis and thus requires an older age for divergence of New and Old World cichlids (Chakrabarty 2004).

The family as a whole is thought to have an early Tertiary origin, and movement from

South America to Central America perhaps occurred in the late Tertiary (ca. < 20 mya) (Myers 1966; Murray 2001a). Consequently, the origin of cichlids in North America (primarily Mexico) could have occurred as early as the middle Miocene (ca. 12–15 mya).

PERCIDAE (186 SPECIES) Percids are the second most speciose family of North American freshwater fishes. The family occurs in Europe, northern Asia, and eastern North America (Collette and Banarescu 1977; Berra 2001) and seems to represent a Laurasian clade (Wiley 1992; Carney and Dick 2001). The family likely originated in the early Tertiary (Paleocene; ca. 65 mya) when land connections existed between eastern North America and Europe (Figure 2.5D) (Wiley 1992; Carney and Dick 2001). However, fossil remains of percids are only dated to the Miocene (26 mya) (Carney and Dick 2001), and the earliest North American percid fossils are from Pleistocene deposits (ca. 2 mya) in areas now located in Texas and Oklahoma (genus *Perca*) and South Dakota (*Percina* and *Etheostoma*) (Cavender 1986). Given the currently available information, it is not possible to distinguish between a dispersal hypothesis, with percids evolving in Europe and then dispersing to North America via the North Atlantic connection, or a vicariance hypothesis, with percids evolving in Laurasia and then being separated by the formation of the North Atlantic Ocean (Carney and Dick 2001). However, given that fossils likely underestimate ages of percids, I have shown the Laurasian origin in Figure 2.4.

The first occurrence of darters in North America is likely far earlier than indicated by the Pleistocene fossils. A molecular phylogeny of the darter family (Percidae), with the rate of genetic change (i.e., the molecular clock) based on a fossil-calibrated phylogeny of centrarchids, shows the separation of darters from nondarter percids occurring 19.8 mya (Carlson et al. 2009). The 18.5 mya age of the most recent common ancestor to the darter genus *Nothonotus* provides further evidence of at least a Miocene origin of darters (Near and Keck 2005). Finally, a recent molecular phylogeny of logperches (genus *Percina*) showed that divergence began in the Pliocene (ca. 3–5 mya), although most speciation events in this group did occur within the Pleistocene (Near and Benard 2004).

FUNDULIDAE (34 SPECIES) Members of this group occur in North and Central America as well as Cuba; in North America all but two species occur east of the continental divide (Berra 2001). The oldest fossil evidence in North America dates from the middle Miocene (ca. 16 mya) and perhaps is of the modern genus *Plancterus* (Cavender 1986). North American fundulids apparently are derived from Central American ancestors (Briggs 1986, 1987) and, in support, phylogenies based on morphological and molecular data are consistent in placing the Central American family Profundulidae as basal to the Fundulidae (Parenti 1981; Wiley 1986; Bernardi 1997).

CYPRINODONTIDAE (35 SPECIES) Cyprinodontid fishes include both marine/estuarine and freshwater forms that are found primarily along coastal regions (with some notable exceptions) in North, Central, and South America, and the Mediterranean region including North Africa (Parenti 1981; Berra 2001). One view is that the distribution of the cyprinodontiform fishes suggests, in part, a reduced Pangean pattern (Figure 2.2), with members of the group absent from Australia and Antarctica (Parenti 1981). Correspondingly, the order Cyprinodontiformes likely existed at least from the late Triassic before the breakup of Pangea (Figure 2.5A), and New World cyprinodontids perhaps date from the early Tertiary (Parenti 1981). An alternative view places the origin of the group at a later time in the early Cretaceous when Africa and South America were only divided by a narrow saltwater passage (Figure 2.5B, C) (Briggs 1986).

The relatively great age of New World cyprinodontids is also supported by molecular studies of the amount of divergence between New and Old World species (Echelle and Echelle 1993). The North American cyprinodontids are likely derived from a marine ancestor (Gilbert 1976; Parker and Kornfield 1995). However, the hypothesis that lineages of inland species of *Cyprinodon* in North America have been derived independently from the widely distributed coastal species, although initially supported by a reduced data set of western species in the *Cyprinodon variegatus* complex (Echelle and Echelle 1992), has not been substantiated by a more complete study of the family (Echelle et al. 2005).

In spite of the suggested age of the order and of New World Cyprinodontidae, the fossil record for the family in North America is meager. The only known North American fossil, of the extinct species *Cyprinodon breviradius*, is from late Miocene/early Pliocene deposits near Death Valley, California (7 9 mya) (Miller 1981; Cavender 1986). However, there are recent phylogenies, based on water-soluble proteins (allozymes) and mitochondrial DNA, that provide times of divergence for North American genera and species (Echelle and Echelle 1992; Echelle et al. 2005). Molecular data suggest that modern New World genera of the Cyprinodontidae began diverging in the Miocene (7–9 mya)—dates that are earlier than those proposed by Miller (1981) based largely on geological inferences. Some species within the family are of much more recent origin, dating to less than one million years (Echelle et al. 2005).

COTTIDAE (30 SPECIES) In terms of the diversity of genera, the sculpins are primarily a marine family. However, the genus *Cottus* is well represented in North American fresh waters with 28 species, and there are at least two freshwater species of *Myoxocephalus*. The oldest fossils of the genus *Cottus* are from late Miocene (ca. 11 mya) deposits of North America in what is now Oregon (Linder 1970; Cavender 1986). Freshwater forms in both genera are derived from marine species (Gilbert 1976; Burr and Mayden 1992). In contrast to the Miocene age of *Cottus*, freshwater species of *Myoxocephalus* are more recent, likely invading freshwater habitats

in the early to middle Pleistocene around the beginning of the major continental glaciations (ca. 0.9 mya) (Kontula and Väinölä 2003).

ATHERINOPSIDAE (39 SPECIES) The New World silversides are primarily a marine family, although there are five genera that occur widely in freshwater habitats. Fossil silversides are known from Pliocene formations (ca. 4–5 mya)—one in what is now Arizona and the other from the Mesa Central of the Mexican Plateau (Cavender 1986). The Arizona fossils were most likely from a marine or brackish water habitat and have been assigned to the modern species *Colpichthys regis* (Todd 1976). The Mexican material also represented modern species of the genus *Menidia* (formerly placed in *Chirostoma*) (Barbour 1973; M. L. Smith 1981). Freshwater silversides are derived from marine ancestors, with perhaps several independent invasions of fresh water by species groups occupying the Mexican Plateau and those found in more northern regions of North America (Barbour 1973; Gilbert 1976; Burr and Mayden 1992).

SUMMARY

Fish evolution began in the early Paleozoic, perhaps 500–470 million years ago (mya). Modern bony fishes, the teleosts, appeared by the lower Mesozoic (230–206 mya), and by the middle Mesozoic (195 mya), representatives of most major groups of fishes were present. The distribution of marine and freshwater fishes worldwide, and the occurrence and distribution of North American freshwater fishes, have been strongly shaped by the movements of landmasses—plate tectonics.

The diverse fish fauna of North America came to occupy North America over a span of hundreds of millions of years, from as early as the late Paleozoic through the Pleistocene, and continuing into the present as populations respond to changing environmental conditions. Over two-thirds of the fauna, in terms of family origins, has occupied North America since the Paleogene (ca. 24 mya) or earlier, whereas ages of particular species can be much more recent. The minnows and darters, the two most speciose North American families, are also among the more recent (Miocene and Oligocene) arrivals.

Lineages of fishes in North America have various origins as well. Half of North American freshwater fishes have a marine origin, followed by those originating in North America (or in ancient landmasses of Pangea and Laurasia), Central and South America, and Eurasia. Because of the long and varied histories of fish lineages, contemporary assemblages of fishes should be viewed as being composed of suites of species with potentially widely differing histories, with adaptations that have likely been shaped to a greater or lesser extent by interactions in fish assemblages that were greatly different from those they are currently occupying.

SUPPLEMENTAL READING

Cavender, T. M. 1986. Review of the fossil history of North American freshwater fishes, 699–724. In *The zoogeography of North American freshwater fishes*. C. H. Hocutt and E. O. Wiley (eds.). John Wiley and Sons, New York, New York. An important reference to the fossil history of North American fishes.

Grande, L. 2001. An updated review of the fish faunas from the Green River Formation, the world's most productive freshwater Lagerstätten, 1–38. In *Eocene biodiversity: Unusual occurrences and rarely sampled habitats*. G. F. Gunnell (ed.). Kluwer Academic/Plenum Publishers, New York, New York. A comprehensive review of one of the most complete series of fossil fish faunas.

Smith, A. G., D. G. Smith, and B. M. Funnell. 1994. *Atlas of Mesozoic and Cenozoic coastlines*. Cambridge University Press, Cambridge, United Kingdom. An important reference for understanding continental margins during the Mesozoic and Cenozoic.

WEB SOURCES

Scotese, C. R. Paleomap Project. http://www.scotese.com/Earth.htm.

Paleogeography and Geologic Evolution of North America. http://jan.ucc.nau.edu/~rcb7/nam.html.

THREE

Reshaping North American Fish Faunas

THE ROLE OF LATE CENOZOIC CLIMATIC AND TECTONIC EVENTS

CONTENTS

Fish assemblages and populations are continually challenged by changes in their local and regional environments. These changes could be relatively minor, such as local shifts in stream habitats caused by alterations in pools or riffle structure, or changes in access to habitats caused by shifts in the distribution of large piscivores. More extreme changes might include annual shifts in water level and/or flow rates caused by variation in precipitation. On an even larger scale, changes could reflect long-term climatic shifts, such as the onset of the Pleistocene Ice Ages, or major tectonic events, such as the uplift of the Colorado Plateau in the late Cretaceous that resulted in major alterations of stream connections and drainage patterns (G. R. Smith et al. 2010). Examples of some of the large-scale events that characterized the middle to late Cenozoic and their impacts on fishes are the topic of this chapter.

TERTIARY AND QUATERNARY EVENTS

From the previous chapter it is apparent how North American fish assemblages have been shaped by large-scale geologic and climatic events, with some fish lineages, such as lampreys, having experienced events as far back as the late Paleozoic. Such climatic and geologic events have, over time, shaped the pattern of fish diversity that characterizes North America (Figure 1.5). In addition, fish assemblages, like other biotic assemblages, have undergone continual breakup and rearrangement (Jablonski and Sepkoski 1996). In this section the emphasis is on events occurring during the Cenozoic, particularly the late Tertiary and early Quaternary Periods, while still recognizing that many of the long-term climatic and geological impacts are part of continual processes shaping our planet.

Examples from Western North America

In the early Paleozoic (Cambrian), western regions of North America subsided and were covered by seas that lasted, to varying extents, into the Jurassic (Stokes 1986). The exposed continental margin of North America cut through portions of Alberta, Montana, and Utah, essentially following the Wasatch Line in Utah (Figure 3.1) (Stokes 1986; Aberhan 1999; Dickinson 2004). Furthermore, as the eastern margin of the Pacific Plate collided with the North American Plate, primarily by subduction (sliding beneath), terranes of largely oceanic origin were added to the western margin of North America. (A terrane is a discrete, fault-bounded crustal element that is added to a craton through plate movement—a craton is a continental nucleus.) Consequently, much of the extreme western margin of present-day North America is a collage of crustal fragments that have been tacked on to the North American Craton in a complex series of events extending from the Paleozoic through the Miocene (Schermer et al. 1984; Dickinson 2004)—a process termed the accretion of allochthonous terranes.

Middle to late Tertiary changes in landform and climate were extensive throughout North America, but were particularly so in the West. The subduction of the Pacific Plate resulted in orogeny (mountain building), including the formation of the Cascade and, much later during the middle Miocene, the Sierra Nevada ranges, as well as periods of intense volcanism (Schermer et al. 1984; Dickinson 2004). Because these processes have extended well into the Cenozoic, they are certainly recent enough to have impacted the flora and fauna of

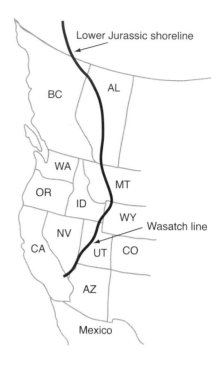

FIGURE 3.1. The approximate early Jurassic continental margin of North American. Based on Stokes (1986), Aberhan (1999), and Dickinson (2004).

western North America (Minckley et al. 1986); if not modern species, then certainly their evolutionary lineages.

In addition to the major geomorphic changes of mountain building and volcanism, the composition and distribution of Western fish assemblages have been shaped by a general climatic trend toward increasing aridity, resulting in the drying of large lakes and shrinking or loss of streams present during the Miocene and Pliocene, and the concomitant extinction of many populations (G. R. Smith 1978). The restriction or extirpation of some of the early components of the western fish fauna, such as mooneyes, smelts, pikes, sunfishes, and catfishes, was likely associated with habitat alterations brought on by uplift and climatic shifts that changed the low-gradient, meandering rivers of the Oligocene to higher-gradient streams flowing over diverse landforms, and with altered drainages that characterized the Miocene and later streams (Minckley et al. 1986).

The general trend of increasing aridity, coupled with aperiodic severe droughts lasting decades or even hundreds of years, prompted Matthews (1998) to suggest that western fishes must have suffered through periodic extirpations followed by long periods of recolonization. In fact, in reference to the development of the western North American fish fauna, Minckley et al. (1986) commented that "taxa that persist have dealt with far more spectacular geologic and climatic events than their counterparts in other parts of the Continent." In fact, fishes of western North America have suffered higher extinction rates compared to eastern fishes. As a result, many western species are relics of groups that were once more speciose but which have lost species through extinction over the past 1–3 million years (G. R. Smith et al. 2010).

Colorado Plateau

Distribution patterns of modern western fish faunas tend to correspond to the continental subplates formed from the accreted terranes, and drainages that extend into adjacent subplates tend to have faunas that have been derived from several sources (Minckley et al. 1986). For example, the Colorado Plateau (Figure 3.2) is both a tectonic and physiographic province that has remained internally stable. At the close of the Cretaceous, the region of the Plateau was near sea level; it then experienced approximately 2 km of uplift during the Cenozoic, especially during the Pliocene. The uplift occurred in two phases, with the second phase taking place only within the last 5 million years (Morgan and Swanberg 1985). The Colorado Plateau is drained by the Colorado River; hence, elevation changes of the Plateau have had major impacts on the directions of flow and drainage connectivity. The uplift resulted in the isolation of the Colorado Plateau as north-flowing streams from central Arizona were interrupted. Also, the uplift of the Wasatch Front, starting in the early Eocene, and the subsequent drop of the Great Basin in late Oligocene isolated the upper Colorado

River fauna from that of the Great Basin along its northwestern margin (Figure 3.2).

Origins of the upper Colorado River fish fauna (including watersheds of the Green and Colorado rivers) are ancient, likely having begun in the Oligocene and Miocene in streams draining the uplifted Rocky Mountains and flowing across the Colorado Plateau to interior basins in Colorado and New Mexico (the Miocene Bidahochi Basin) or northwestern Arizona (the Miocene Hualapai Basin) (Figure 3.2) (Oakey et al. 2004). A middle section of the Colorado River, currently comprising the Little Colorado, Virgin, and White rivers, drained to the southwest, while a third section, including the Gila River, was incorporated into the drainage after the retreat of the Bouse Miocene/Early Pliocene Embayment (Minckley et al. 1986)— an embayment that, while first thought to be a marine or estuarine extension of the Gulf of California, now appears to have been a series of lakes, with perhaps the lower being saline (Dillon and Ehlig 1993; Spencer and Patchett 1997; Roskowski et al. 2007). The upper and lower Colorado River systems were joined perhaps 10.6 to 3.3 mya, following headward erosion of streams of the middle and lower Colorado watersheds and through reoccupation and reversal of flow in older channels (Figure 3.2). Prior to this time, the upper Colorado River drainages flowed into a closed basin. It was not until the Pliocene that the Colorado River reached the Gulf of California (Minckley et al. 1986; Powell 2005).

The origins of the dominant components of the mainstem Colorado fish assemblage (Colorado Pikeminnow, *Ptychocheilus lucius*; Humpback Chub, *Gila cypha*; Roundtail Chub, *G. robusta*; Bonytail Chub, *G. elegans*; Speckled Dace, *Rhinichthys osculus*; Razorback Sucker, *Xyrauchen texanus*; Bluehead Sucker, *Catostomus discobolus*; and Flannelmouth Sucker, *C. latipinnis*) can be traced to these various geological events (Minckley et al. 1986; G. R. Smith et al. 2002). Most of the species have their closest relationships with populations in the north and west, but some also show relationships to the south.

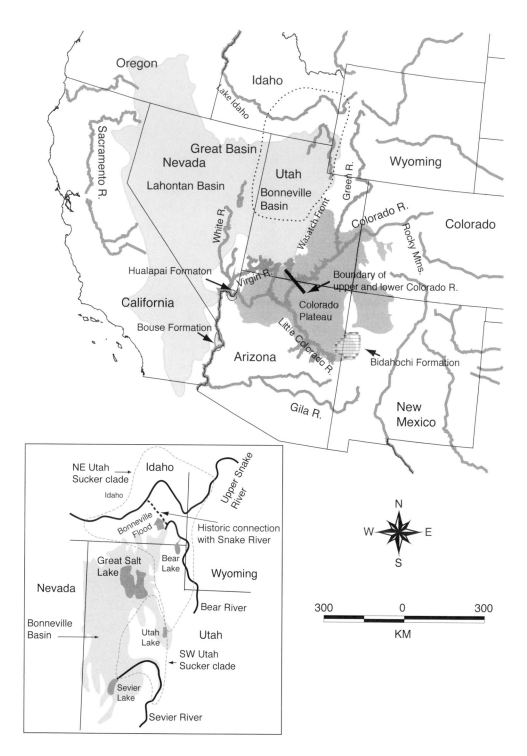

FIGURE 3.2. The Colorado Plateau (dark gray), Colorado River Drainage, and Great Basin (light gray), including other features mentioned in the text. The dotted line indicates the area covered by the map inset, which shows periodic connections of the Bonneville Basin with the upper Snake River. The Bonneville Basin (indicated by the light shading in the inset) reached its maximum extent during the late Pleistocene; dashed lines in the inset show boundaries of the two Utah sucker clades. Based on Minckley et al. (1986), Curry (1990), Spencer and Patchett (1997), Gross et al. (2001), Johnson (2002), Cook et al. (2006), and Desert Fishes Council (2012).

Colorado Pikeminnow, and Roundtail, Humpback, and Bonytail chubs, show relationships to the north and west, including the Sacramento-San Joaquin Basin in what is now California, but also to the Miocene Bidahochi Lake deposits to the southeast (Figure 3.2).

Although somewhat uncertain, Speckled Dace may have originated from populations in Tertiary Lake Idaho or ancestral Snake River drainages (Oakey et al. 2004) and thus would have relationships to the north and west. Speckled Dace are characteristic of higher-gradient, smaller streams and show extensive genetic structure among populations, including those in the Colorado River; populations occupying the upper, middle, and lower Colorado River form three distinct genetic groups (Oakey et al. 2004).

The Flannelmouth Sucker also shows relationships to the north and west, and the Bluehead Sucker shows relationships with forms in the Bonneville Basin to the west. Origins of the distinctive Razorback Sucker are less understood, but the divergence of the *Xyrauchen* lineage from that of *Deltistes* and *Chasmistes* likely occurred in the late Miocene, if not before, and suggests a relationship to the north or northwest (Miller and G. R. Smith 1981; Hoetker and Gobalet 1999).

The available evidence indicates that the Colorado River fish fauna is ancient and many of the changes in species composition predate Pleistocene events. Because of the age and action of climatic and tectonic events, faunal assembly of the main-channel Colorado River fish fauna likely occurred as a series of additions, separated in space and time, so that the modern fauna is a composite of species of different evolutionary origins and ages. However, it is also important to recognize that the post-Pleistocene history of the region is again characterized by physical changes. For instance, although the southwestern region has progressively become warmer and drier, the pattern is not one of continuous warming and drying but one of high variability in climatic patterns, including periodic severe droughts.

The impact of post-Pleistocene drought on western fishes is illustrated by work on genetic variability in Flannelmouth Sucker, one of the ancient, endemic species of the Colorado River. Genetic diversity in Flannelmouth Sucker is surprisingly limited for such an ancient species and is consistent with the hypothesis of a major, basin-wide, population crash during a known time (post-Pleistocene) of severe western drought, followed by a period of rapid repopulation growth and range expansion during a wetter period. Populations of Flannelmouth Sucker in the upper regions of the Colorado basin are the result of migration from refugia in the lower part of the system, generally within the last 10,000–11,000 years (Douglas et al. 2003).

Great Basin

The Great Basin comprises a large area of complex geology located northwest of the Colorado Plateau and including large areas of Nevada and Utah and parts of southeastern Oregon, southern Idaho, southwestern Wyoming, eastern California, and northern Mexico (Figure 3.2). As suggested by the term "Great," the Basin makes up almost 20% of the United States and constitutes the largest inland drainage in North America (Sigler and Sigler 1987). It includes more than 150 smaller drainage basins separated by approximately 160 regularly spaced mountain ranges forming the basin and range topography (G. R. Smith 1978; Sigler and Sigler 1987; Sada and Vinyard 2002). Two large sub-basins make up the Great Basin—the Bonneville Basin, occurring primarily in eastern Nevada and Utah, and the Lahontan Basin to the west, occurring mostly in Nevada and parts of eastern California. Most of the topography was formed in the last 20 million years, and many of the genera of fishes have occupied the area since the Pliocene (approximately 5 million years ago) and some since the Miocene (G. R. Smith 1981; Dowling et al. 2002; G. R. Smith et al. 2002). The fish fauna includes at least 102 species and subspecies in 15 genera of generally small-bodied forms, but the unique feature is the high level of endemism of both

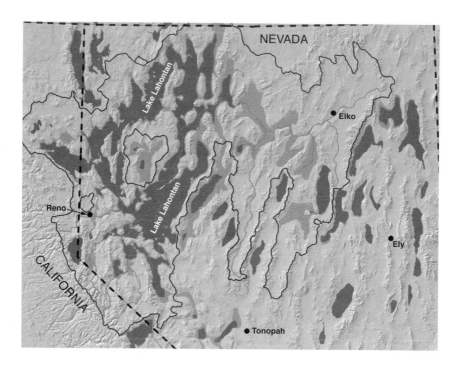

FIGURE 3.3. Pleistocene pluvial lakes and rivers of the western Great Basin in what is now Nevada. Dashed lines are state boundaries, dark gray areas show the maximum extent of late Pleistocene lakes, medium gray areas show the possible maximum extent of early- middle Pleistocene lakes, light gray lines show the modern drainages of the Lahontan Basin, and black lines indicate the late Pleistocene extent of the Lahontan Basin. Based on Reheis (1999). See also corresponding figure in color insert.

fishes and invertebrates (Sada and Vinyard 2002; G. R. Smith et al. 2010). At least 66% of the fish species are endemic to the Great Basin, most to specific drainages within the region (G. R. Smith 1978; Sada and Vinyard 2002). During the Pliocene and Pleistocene, repeated periods of high rainfall and changes in drainages due to volcanism resulted in a series of lakes, many quite large, in the Great Basin and created a very different environment compared to the modern-day desert—in fact, what is now Nevada was a land of abundant natural lakes (Figure 3.3). As shown by the green areas in Figure 3.3, some of the highest lake levels occurred in the early-middle Pleistocene, approximately 650,000 years ago (Reheis 1999). Two of the largest lakes were Lake Lahontan to the west and Lake Bonneville to the east. The Bonneville Salt Flats are part of the remains of Pleistocene Lake Bonneville, which is survived by the modern-day Great Salt Lake and two smaller freshwater

lakes, Bear and Utah, in the northeastern part of the Great Basin (Figs. 3.2 and 3.3) (Mock et al. 2006). The large Pleistocene lakes and interconnecting streams allowed aquatic organisms to colonize many of the lake basins; however, the subsequent increasing aridity during the late Pleistocene and post-Pleistocene, in part caused by the uplift of the Sierra Nevada Mountains, resulted in the isolation of the faunas, contributing to the high level of endemism and also to the frequent loss of species through extinction (Hubbs et al. 1974; Reheis 1999; G. R. Smith et al. 2002). Especially because of competition for the limited water in the now arid region, human impacts on the rate of extinction have also been particularly great (Sada and Vinyard 2002).

During the Pliocene and Pleistocene, the Bonneville Basin was connected at least twice to the upper Snake River via the upper Bear River (Figure 3.2) (Hart et al. 2004). In the late Pleistocene, when most of the upper Snake River was covered by glaciers, extensive lava flows blocked the connection, diverting the Bear River into the Bonneville Basin and contributing to a rise in water level of Lake Bonneville, at that time a freshwater lake. A second connection occurred 145,500 years later, when Lake Bonneville was at its high stand and a breach occurred along its northern shore. This resulted in a major erosive flood into the Snake River. Once lake levels dropped, the Bonneville Basin was again separated from the Snake River drainage, a situation enhanced by the increasing aridity of the region (Curry 1990; Hart et al. 2004; Mock et al. 2006).

As in other areas of North America, patterns of fish diversification are often more closely related to ancient drainage patterns than to modern-day patterns, and the Bonneville and Snake River basins provide excellent examples of this (G. R. Smith et al. 2002; Mock et al. 2006). For instance, mitochondrial and nuclear sequence data show a strong divergence among morphologically similar populations of the Utah Sucker (*Catostomus ardens*), a widespread endemic to the Bonneville Basin and the upper Snake River (Mock et al. 2006). The divergence reflects the ancient connection between the Bonneville Basin and the Snake River via the upper Bear River and divides Utah Sucker populations into a southwestern group of the Great Basin, centered around Utah Lake and the Sevier River, and a northeastern group in the Snake River and in the northeastern Bonneville Basin east of the Wasatch Mountains (Figure 3.2). The deep genetic divergence suggests that these two groups were separated 1.6–4.5 million years ago during the Pliocene or early Pleistocene. In addition, there is also genetic separation between the Sevier River populations and those in the Utah Lakes region, reflecting the post-Pleistocene isolation of these areas caused by increasing aridity. Surprisingly, the June Sucker (*Chasmistes liorus*), endemic to Utah Lake, shows little genetic differentiation from the Utah Sucker but strong morphological differentiation, suggesting strong recent selection for a more planktivorous lifestyle in contrast to the benthic feeding Utah Sucker.

The genetic separation between the northeastern Bonneville/lower Snake River and the southeastern Lake Bonneville is also reflected in other species, including the Leatherside Chub, which is now recognized as comprising two lineages—the Northern Leatherside Chub (*Lepidomeda copei*) and the Southern Leatherside Chub (*L. aliciae*) (J. B. Johnson et al. 2004). A similar pattern is shown by the Utah Chub (*Gila atraria*), which shows deep genetic divergence between a northeastern Bear Lake/Snake River clade and a southwestern Bonneville clade, which is again very like the Utah Sucker (Figure 3.2). Molecular evidence indicates that the division occurred sometime in the Pliocene or early Pleistocene (J. B. Johnson 2002). However, unlike the Utah Sucker, the genetic structure of the Utah Chub also shows a more recent connection between the Bonneville and Snake River basins that relates to the late Pleistocene Bonneville flood. In all examples, the impacts of the geological and climatic history of the region contribute greatly to understanding the ecology of the species and, in particular, to conservation efforts that might include transplanting populations (J. B. Johnson et al. 2004; Mock et al. 2006).

The fish fauna of the Great Basin colonized the region through numerous rivers and lakes present during various late Miocene, Pliocene, and Pleistocene pluvial periods. Some populations, such as Utah Sucker and Leatherside Chub, reflect the earlier Pliocene connections, in contrast to others, such as June Sucker, that show more recent responses to ecological opportunities. Since the Pleistocene, the Great Basin fish fauna has been progressively diminished and fragmented as aquatic habitats have dried and fishes have been isolated in small springs, spring runs, and the remaining lakes and streams (Sada and Vinyard 2002).

Examples from Northern and Eastern North America

Late Tertiary (Miocene and Pliocene) and early Quaternary geologic and climatic events also affected fish assemblages in northern, central, and eastern North America. However, in contrast to the high level of tectonic activity and volcanism of western North America, these regions of North America tended to be geologically more quiescent through most of the Pliocene but with major climatic impacts to fishes and other organisms caused by direct and indirect effects of late Tertiary and Quaternary (Pleistocene) glaciations. Glacial advances began in the Miocene and Pliocene of the late Tertiary, followed by numerous cold periods and concomitant glacial advances interspersed with temperate periods and their associated glacial retreats (Ehlers 1996).

Although there are many direct and indirect approaches to dating these cold and warm periods, one of the most fruitful approaches has been the use of ratios of various elements, and of isotopes of elements, that were incorporated into calcium-carbonate shells and skeletons of marine microorganisms and deposited in the stable environment of the deep sea (Lowe and Walker 1997). In particular, the ratio of two oxygen isotopes, ^{16}O and ^{18}O, found in the tests of Foraminifera has led to a much more precise understanding of the timing of cold and temperate periods. The ratio is known to vary with temperature, as the lighter isotope tends to accumulate in glacial ice during cold periods so that there is an enrichment of the heavier isotope in the deep sea (Lowe and Walker 1997).

Previously there were four major glacial advances recognized within the Quaternary for North America (Nebraskan, Kansan, Illinoian, and Wisconsinan); however, based primarily on information from isotopic ratios, the estimate of the number of glacial advances from the dawn of the Pleistocene, approximately 2 mya, is at least 18–20 for the entire planet and perhaps 13–18 major glacial advances in North America (Davis 1983; Ehlers 1996). The last major advance, the Wisconsinan, began perhaps 80,000 years ago (Ehlers 1996; Lowe and Walker 1997). Thicknesses reached by the ice sheets were impressive, reaching 90 m to several kilometers in some areas, and resulting in depressions of the land by 200–300 m (Lowe and Walker 1997; Lomolino et al. 2006). Even within major glacial advances, there was a strong pattern of major and minor variation. For instance, the Wisconsinan glaciation can be subdivided into three periods of advances, with the last advance, the late Wisconsinan, starting approximately 23,000–25,000 years ago (Ehlers 1996).

The Wisconsinan glaciation comprised two major ice sheets, the Cordilleran in northwestern North America and the Laurentide in eastern and northeastern North America, and one minor ice sheet, the Innuitian along the Arctic coastline (Figure 3.4). The development of the western Cordilleran ice sheet lagged behind that of the Laurentian and Innuitian ice sheets, with the latter two ice sheets reaching their maxima 20,000–24,000 years ago and remaining near maximum until 17,000 years ago. The Cordilleran did not attain its maximum extent until 14,500 years ago, followed by a rapid decline beginning around 12,000 years ago (Clague and James 2002; Dyke et al. 2002). The Laurentian ice sheet during the Wisconsinan glaciation extended across most of eastern Illinois, Indiana (except the south-central region), and most of Ohio, nearly to the present course of the Ohio River (Frye et al. 1965; Goldthwait et al. 1965; Wayne and Zumberge 1965; Clark et al. 1996; Ehlers 1996; Lowe and Walker 1997). Farther east, ice covered upper Pennsylvania and all of New York and New England (Muller 1965; Schafer and Hartshorn 1965). Except for montane glaciers, glacial penetration was less in western states, covering the upper half of most of Washington, Idaho, Montana, and all but the southwest corner of North Dakota (Figure 3.5) (Flint 1971). Higher elevations along the Rocky Mountains supported extensive glaciers as far south as New Mexico (Richmond 1965), and in California there were large glaciers in the Sierra Nevada range and even in the transverse ranges (the San Bernardino Mountains) of Southern California near Los Angeles (Owen et al. 2003).

FIGURE 3.4. The approximate locations of the Cordilleran, Laurentide, and In-nuitian ice sheets of the late Wisconsinan glaciation. Based on Clark et al. (1996), Ehlers (1996), Lowe and Walker (1997), and Dyke et al. (2002).

Advances of the glaciers had direct impacts on fishes as the landscape became covered with ice (Crossman and McAllister 1986; McPhail and Lindsey 1986; Matthews 1998). Ice dams caused changes in stream patterns and directions of flow and sometimes created large lakes. Because of the amount of water contained in the glaciers, sea level was lowered so that streams that now enter the sea separately may have been joined. Glacial scour altered the land, and the formation of terminal moraines created new lake habitats. Streams pouring off edges of the ice created plunge pools, and the melting of large blocks of ice formed kettle lakes. Several examples will help to illustrate general Pleistocene effects.

Changes in Drainage Patterns and Stream Connections in Eastern North America

Pleistocene events resulted in substantial changes to earlier drainages, although understanding the details of how glacial advances altered pre-Pleistocene drainage patterns is complex. The ongoing efforts to understand these events have involved geological research as well as biogeographic studies of fishes. An excellent example of drainage changes, as well as complexity, is the Central Highlands region of eastern North America, comprising the Eastern, Ozark, and Ouachita subregions (Figure 3.5A). Fishes endemic to the Ozark Highlands tend to show their closest relationships with fishes in the Ouachita Highlands, and fishes endemic to these two regions tend to have their closest relationships with fishes in the Eastern Highlands (Mayden 1985, 1987b, 1988). Such relationships among freshwater fishes strongly suggests that the three highland areas once shared common drainage connections, even though the three regions are now isolated by intervening lowlands.

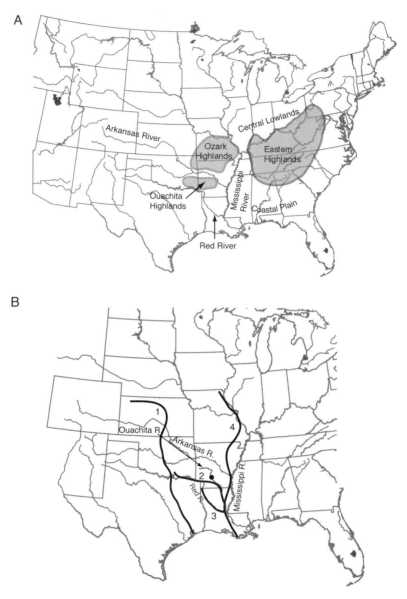

FIGURE 3.5. A. The Central Highlands region of eastern North America in relation to current drainage patterns. B. The formation of the modern Red River from Pre-Pleistocene drainages. Pre-Pleistocene drainages are shown in black: 1 = Plains Stream; 2 = Old Ouachita River; 3 = Old Red River; 4 = Old Mississippi River. The black dot shows the collecting site on the modern Ouachita River. Based on Mayden (1987a, 1987b, 1988).

The Highlands are remnants of an ancient topography that dates to the uplift of the Appalachian Mountains (Chapter 2; Wiley and Mayden 1985; Mayden 1988). The major vicariant events dividing the Central Highlands into eastern and western regions included the southward movement of Pleistocene glacial advances; in fact, the region of the central lowlands (Figure 3.5) was a highland area prior to the intrusion of massive ice sheets. The glacial advance was ultimately followed by the penetration of the lowland area connecting the Eastern and Interior Highlands by the Mississippi River, enlarged because of southward deflection and increased flows of streams that once drained into Hudson Bay (Missouri River) or the Laurentian stream system and the Atlantic Ocean (Ohio River) (Pflieger 1971; Mayden 1985; Wiley and Mayden 1985). The Interior Highlands area was separated into the Ozark and Ouachita highlands by the westward penetration and development of the Arkansas River (Mayden 1987b). Post-Pleistocene dispersal of fishes into some of the formerly glaciated regions from the unglaciated Central Highland areas was also important and adds to the complexity in understanding fish distributions (Berendzen et al. 2003; Near and Keck 2005).

Although the Highlands region is characterized by high fish diversity, the reasons for this diversity are still being debated. The Pleistocene dispersal hypothesis states that the Eastern Highlands represented a center of origin for lineages that subsequently dispersed along glacial fronts during the Pleistocene to streams of the Interior (Ozark and Ouachita) Highlands (Mayden 1987b; Strange and Burr 1997). As such, species in the Interior Highlands should be no older than the Pleistocene. Alternatively, the Central Highlands vicariance hypothesis (CHVH) predicts that the fauna diversified in a widespread and interconnected Highlands region during the Miocene and Pliocene and, after most speciation events had occurred, was fragmented by Pleistocene events into the Ozark and Ouachita Highlands west of the Mississippi River and the Eastern Highlands east of the Mississippi River (Figure 3.5) (Mayden 1988; Near and Keck 2005). Phylogeographic analyses using molecular data do show some support for predictions of the CHVH in divergence times of various lineages. The darter subgenera *Litocara* (genus *Etheostoma*) and *Odontopholis* (genus *Percina*) have species in the Ozark and Eastern Highlands, and both groups show deep divergences of species between the two regions that likely occurred in the Miocene (Strange and Burr 1997). Four species of the minnow genus *Erimystax*, which occur in the Ozark, Ouachita, Eastern Highlands, and adjoining areas, also indicate Miocene speciation events (Simons 2004), and divergence within the Hogsuckers (genus *Hypentelium*)

occurred prior to the Pleistocene (Berendzen et al. 2003). However, not all evidence supports Miocene or Pliocene ages of species. In a study of lineage divergences in the 20 species of the darter genus *Nothonotus*, times ranged from the Miocene (six events), Pliocene (four events), to the Pleistocene (eight events) (Near and Keck 2005). Divergences of subspecies of Studfish (*Fundulus catenatus*) occurred by dispersal or peripheral isolation in the late Pleistocene or later. Divergence of subspecies of Banded Sculpin (*Cottus carolinae*), perhaps by peripheral isolation, also occurred within the Pleistocene (Strange and Burr 1997), as did divergence within the Gilt Darter (*Percina evides*) (Near et al. 2001). Consequently, the rich fish fauna of the Central Highlands seems to be a product of both vicariant and dispersal events, facilitated by the region's great age and topographic diversity. The high fish diversity in many ways follows predictions of island biogeography theory and species-area relationships (Page 1983; Near and Keck 2005).

There are several consequences of ecological importance that are apparent from these events. First, much of the history of the faunas of the Highlands is pre-Pleistocene so that species or species groups have lineages dating to the middle or even early Cenozoic, and some groups thus have had the potential for long periods of interaction. Second, species have experienced major changes in range size (range being contracted during glacial advances), followed by expansion when habitats again became available as ice sheets retreated. Third, the faunas of present-day rivers may reflect species groups that originally occurred in separate drainages, so that species or lineages in a modern river may or may not share a long history.

For example, the modern fish fauna of the Red River and its tributaries (Figure 3.5B) is thought to comprise faunas from three distinct pre-Pleistocene river systems: the Plains Stream in the headwaters of the Red River, the Old Ouachita River (Little-Kiamichi-Ouachita system), and the Old (lower) Red River (Mayden 1985).

TABLE 3.1

Biogeographic Relationships of Species from a Sample of Fishes from the Ouachita
River, Arkansas, at the Confluence with the Little Missouri River (Ross, pers. observ.)

Taxa	Origin/ Pre-Pleistocene distribution	Source
Highland Stoneroller, *Campostoma spadiceum*	2	Mayden 1987a; Blum et al. 2008; Cashner et al. 2010
Blacktail Shiner, *Cyprinella venusta*	3	Mayden 1987a
Steelcolor Shiner, *Cyprinella whipplei*	1	Mayden 1987a
Redfin Shiner, *Lythrurus umbratilis*	4	Mayden 1987a
Bigeye Shiner, *Notropis boops*	1	Wiley and Mayden 1985; Mayden 1987a
Bullhead Minnow, *Pimephales vigilax*	4	Mayden 1987a
Mountain Madtom, *Noturus eleutherus*	2a	Mayden 1985, 1987a
Creole Darter, *Etheostoma collettei*	2a	Mayden 1985
Orangebelly Darter, *Etheostoma radiosum*	2a	Page 1983; Mayden 1985, 1987a
Speckled Darter, *Etheostoma stigmaeum*	3	Page 1983; Simon 1997
Redspot Darter, *Etheostoma artesiae*	3	Mayden 1985; Piller et al. 2001
Banded Darter, *Etheostoma zonale*	1	Page 1983; Mayden 1988
Channel Darter, *Percina copelandi*	1	Mayden 1987a

1 = Found in all Central Highlands (some with disjunct populations in central lowlands)

2 = Endemic to Ouachita Highlands

2a = Ouachita Highlands and various adjoining regions

3 = Widespread, primarily lowland species with sister species found in Central Highlands (i.e., cladogenesis likely before uplift of Central Highlands)

4 = Widespread but biogeographically noninformative species

The amalgamation of faunas is illustrated by examining the origins of 13 fish species taken in a single winter fish collection from a gravel bar on the Ouachita River, a tributary of the modern Red River (Figure 3.5B) (Table 3.1). Four species (Steelcolor Shiner, Bigeye Shiner, Banded Darter, and Channel Darter) are endemic, or largely so, to all three regions of the Central Highlands, and thus would have had the potential for interaction since the Pliocene or earlier. Four species (Highland Stoneroller, Mountain Madtom, Creole Darter, and Orangebelly Darter) are primarily restricted to the Ouachita Highlands and perhaps had a later origin compared to the previous four species. The remaining five species are widespread, generally lowland forms, some of which are sister species to forms occurring in the Central Highlands. This collection of fishes, comprising a few seine hauls along a single gravel bar, emphasizes that contemporary faunas can have different evolutionary origins, ecological histories, and ages of the taxa. The fish assemblage includes groups fragmented from a once intact Central Highlands fauna, some more recent taxa endemic to the Ouachita Highlands, and species derived from generally widespread, primarily lowland pre-Pleistocene taxa. Such separate origins have substantial consequences for the interpretation of factors like the coevolution of species' traits, which are treated in Chapter 13.

Changes in Drainage Patterns and Stream Connections in Northern and Northwestern North America

Farther north and west, portions of the Missouri River originally flowed northward into Hudson Bay, and the Bonneville Basin (discussed previously) was also likely once part of the Hudson Bay drainage during the late Miocene through connections via the Snake River (G. R. Smith 1981; Crossman and McAllister 1986). The past connections are reflected in the current fish faunas. For instance, the Bonneville Basin (located primarily in Utah) contains faunal elements from the north and northeast such as whitefishes, *Prosopium* spp.; suckers, *Catostomus* spp.; and the minnows *Richardsonius* and *Rhinichthys* (G. R. Smith 1981).

Southward Displacement

Beyond the area of direct glacial impact, cooling associated with the Pleistocene resulted in a general southward displacement of terrestrial plants and animals (Pflieger 1971; Whitehead 1973; Pielou 1991). In river systems that were oriented in a primarily north-south direction, such as the Mississippi River, fishes also responded to glacial advances and dropping temperatures by a general southward displacement (Cross 1970; G. R. Smith 1981; Cross et al. 1986). For instance, species that today have a primarily northeastern or north-central distribution, such as Redbelly Dace (*Chrosomus erythrogaster*), Northern Studfish (*Fundulus catenatus*), and Rainbow Darter (*Etheostoma caeruleum*) have disjunct populations as far south as Mississippi (Ross 2001), and Redbelly Dace and Creek Chub (*Semotilus atromaculatus*) have disjunct populations in northeastern New Mexico (Pflieger 1971).

AFTER THE ICE

Fishes that survived glacial advances did so in areas that remained ice free—the glacial refugia. As the ice retreated, fishes spread out from the refugia to colonize the newly available habitats. There were at least five major glacial refugia as well as various minor refugia that allowed the survival of organisms displaced by advancing ice. Refugia occurred in the Arctic as well as south of major glacial advances (Figure 3.6) (Flint 1971; Crossman and McAllister 1986; Stamford and Taylor 2004; Cox and Moore 2005). Minor refugia tended to occur along the boundary of the Laurentide and Cordilleran ice sheets or in coastal areas. Because of glacial refugia, repopulation of formerly glaciated habitats occurred both from the northwest (Beringia), as well as from the east, west, and south. Recolonization is a gradual process and is still ongoing so that formation of northern fish assemblages may be even more recent than within the last 10,000–12,000 years (Crossman and McAllister 1986; Lundberg et al. 2000). For example, species richness in formerly glaciated areas, as shown for Ontario, Canada, is related strongly to distance from glacial refugia and the time that recolonization corridors have been free of ice (Mandrak 1995).

In central North America, the majority of reintroductions to once glaciated areas occurred via the Mississippi Refugium (Figure 3.6), contributing species to north-central Canada, the Hudson Bay drainage, and the Arctic Archipelago (Mandrak and Crossman 1992; Matthews 1998). In the Canadian province of Ontario, which was totally covered by the Wisconsinan glacial advance, 77 out of 91 species, for which glacial refugia have been resolved, repopulated the area from the Mississippi Refugium (Mandrak and Crossman 1992). Over the larger area of the Hudson Bay drainage, the Mississippi Refugium again provided the greatest number of species (Crossman and McAllister 1986). For the Ontario fauna, 94% of the species for which refugia could be identified, survived the glacial advance in a single refugium (Mandrak and Crossman 1992). Whether assemblages tended to move as a group or as individuals is unknown, although recolonization likely occurred in waves of immigrants as passageways from various refugia became free of ice. For instance, of the 21 common species limited to the Great Lakes and Nelson River (located to the northwest and draining into Hudson Bay) watersheds,

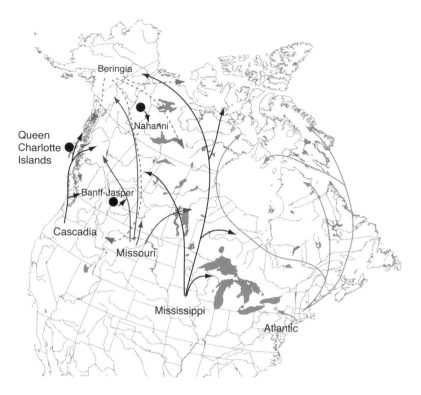

FIGURE 3.6. Glacial refugia during the Wisconsinan glacial advance and their contributions to repopulating formerly glaciated areas. Small refugia are indicated by closed circles. Lines with arrows show colonization routes; solid black lines show colonization from Cascadia, Nahanni, and Mississippi refugia; dashed gray lines show colonization from the Beringia Refugium; solid gray lines show colonization from the Atlantic Refugium. Based on data from McPhail and Lindsey (1970, 1986), Crossman and McAllister (1986), Mandrak and Crossman (1992), Matthews (1998), McCusker et al. (2000), C. T. Smith et al. (2001), and Stamford and Taylor (2004).

14 originated from the Mississippi Refugium, one species originated from both the Mississippi and Atlantic refugia, one species originated from the Atlantic Refugium, and one species originated from the Atlantic, Mississippi, and Missouri refugia (Figure 3.6) (Mandrak and Crossman 1992).

In western North America, four refugia, (Beringia, Cascadia [Pacific], Mississippi, and Missouri) contributed most to the formation of the northwestern Canada and Alaskan fish assemblages (McPhail and Lindsey 1970). The times of egress of fishes from these refugia differed because of the earlier retreat of ice from coastal refugia and from the Missouri Refugium of Great Plains compared to the Mississippi Refugium. What these examples suggest is that the fish assemblages in

formerly glaciated regions experienced a step-like increase in potential colonizers over time as passage from the various refugia became possible. In addition, as emphasized by Figure 3.6, regional faunas were established by colonizers from potentially a number of different refugia and thus have experienced different evolutionary histories and faunal associates.

The impact of postglacial dispersal is also illustrated by fishes occupying the Chehalis River valley, a small coastal drainage in western Washington that provided a refugium for lowland fishes of Puget Sound drainages and the Olympic Peninsula (McPhail 1967; McPhail and Taylor 1999). During the last advance of the Wisconsinan glaciation, the Puget Lobe of the Cordilleran ice sheet penetrated south to cover what is now Puget Sound (Figure 3.7)

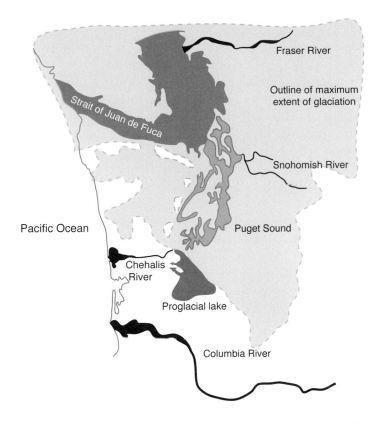

FIGURE 3.7. Modern and Pleistocene features of western Washington showing the location of the Chehalis River Refugium, modern-day Puget Sound (medium gray), the maximum southward penetration of the Puget Lobe of the Cordilleran ice sheet (light gray), early proglacial lakes, and other place names mentioned in the text. Based on McPhail (1967) and Porter and Swanson (1998).

(Porter and Swanson 1998). South of the ice sheet, the Chehalis River valley remained unglaciated over much of its area, as did the larger lower Columbia River farther south (McPhail 1967; Pielou 1991). Although early faunal exchange occurred between the Chehalis and Columbia rivers, during the middle to late Pleistocene these faunas remained distinct. Drainages north of the Chehalis River that now flow into Puget Sound were ice covered for approximately 900–1,000 years (Porter and Swanson 1998). The Puget Lobe reached its maximum southern extent 16,950 years ago and then began receding 16,850 years ago (Porter and Swanson 1998). As it began to recede, flow was to the south into the Chehalis River and thence to the Pacific Ocean. The Chehalis River fish fauna, comprising eight species of primary freshwater fishes, gradually dispersed northward as the ice withdrew, especially species such as the Longnose Dace (*Rhinichthys cataractae*) that are adapted to swiftly flowing water (McPhail 1967). The farthest northward penetration of fishes from the small Chehalis River Refugium was achieved by the Nooksack Dace (*Rhinichthys cataractae* ssp.) and the Salish Sucker (*Catostomus catostomus* ssp.), which reached the Fraser River system of southern British Columbia (Figure 3.7) (McPhail 1997; Pearson 2000; Hutchings and Festa-Bianchet 2009). As the ice sheet receded past what is now the mouth of the Snohomish River (Figure 3.7), sea water from the Strait of Juan de Fuca poured into the large proglacial lakes that had expanded to occupy the Puget Sound Basin, quickly changing the basin from fresh water to sea water and limiting further northward distribution of primary freshwater fishes (McPhail 1967).

Because glaciers are currently retreating, it is also possible to study postglacial colonization as it is now occurring. Cluster analysis of observed plant and animal taxa from newly developing streams in Glacier Bay, Alaska, results in three categories (Milner 1987). Newly emergent, meltwater streams support a limited biota consisting of algae and insects. Clearwater streams that are supported by runoff and snowmelt from the watershed have a greater diversity of insects compared to meltwater streams and also support a few Pink and Chum salmon (*Oncorhynchus gorbuscha* and *O. keta*). Clearwater streams fed from lakes, resulting in increased buffering of water quality, have extensive growth of algae and mosses, as well as a higher diversity of invertebrates and fishes.

SUMMARY

Biotic assemblages in general have undergone a continual cycle of breakup and rearrangement over geological time. As shown by this chapter, it is clear that fishes are no exception to this general pattern. Populations of freshwater fishes in most of North America have, through various means, been subjected to periodic fragmentation and restriction of their ranges. In some cases, as with typically lowland groups such as mooneyes and catfishes in western North America, their extirpation subsequent to the Oligocene was complete throughout the region. In other cases, range restriction and fragmentation was followed by population and range expansion, such as in Flannelmouth Sucker or in numerous species that recolonized northern North America following the retreat of the Pleistocene ice sheets. Recolonization of faunas most likely occurred as a mosaic, with specific faunal elements added over time from specific source regions. Especially in formerly glaciated regions of northern North America, recolonization of postglacial habitats occurred from often multiple refugia, greatly adding to the historical complexity of species and assemblages. Regions that remained free of ice and were otherwise less impacted by tectonic or climatic changes, such as the southeastern United States, support the greatest diversity of freshwater fishes.

The dynamic history of North American fish assemblages also carries an important conservation message. The goal of conservation of fishes should not only be to preserve species and assemblages in a "snapshot" of time but to conserve the full biodiversity of species and assemblages so that they have the potential to respond to natural (as well as anthropogenic) changes in their environments. This is clearly a challenging but critically important objective.

SUPPLEMENTAL READING

Pielou, E. C. 1991. *After the Ice Age, the return of life to glaciated North America*. University of Chicago Press, Illinois. An important general reference on the recolonization of formerly glaciated areas of North America.

Powell, J. L. 2005. *Grand Canyon, solving the earth's grandest puzzle*. Penguin Group, New York, New York. A fascinating account of the untangling of formation of the Grand Canyon, beginning with the work of John Wesley Powell.

Sada, D. W., and G. L. Vinyard. 2002. Anthropogenic changes in biogeography of Great Basin aquatic biota. *Smithsonian Contributions to the Earth Sciences* 33:277–93. Details the multiple ways that humans (including native hunters and gatherers, the first European settlers, and modern society) have interacted and impacted aquatic faunas in the Great Basin.

Smith, G. R., C. Badgley, T. P. Eiting, and P. S. Larson. 2010. Species diversity gradients in relation to geological history in North American freshwater fishes. *Evolutionary Ecology Research* 12: 693–726. A recent and thorough synthesis of factors shaping the North American freshwater fish fauna.

WEB SOURCES

Desert Fishes Council. 2012. Species tracking. http://www.desertfishes.org/?page_id=327.

Reheis, M. 1999. *Extent of Pleistocene Lakes in the western Great Basin: U.S. Geological Survey Miscellaneous Field Studies Map MF-2323*. U.S. Geological Survey, Denver, CO. http://geo-nsdi.er.usgs.gov/metadata/map-mf/2323/metadata.faq.html.

PART TWO

Formation, Maintenance, and Persistence of Local Populations and Assemblages

Chapters in Part 1 show how fish populations and assemblages are shaped by broadscale geological and climatic factors over broad temporal and spatial scales. The focus now turns to how local fish assemblages and populations are influenced by regional faunas, by the general nature of the landscape in which they occur, and by the interplay of temporal and spatial habitat heterogeneity (see the following figure). Instead of the mega- and macroscale domains (sensu, Delcourt and Delcourt 1988) that span temporal and spatial scales from 10,000 to billions of years and 100s to thousands of kilometers, the focus now is on meso- and microscale domains of 1–10,000 years and meters to tens of kilometers.

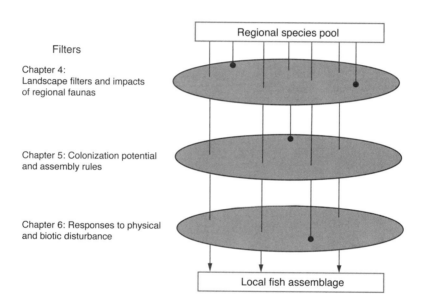

Examples of filters affecting the formation of local fish assemblages from a regional species pool and their associated chapters.

FOUR

Responses of Populations and Assemblages to Biotic and Physical Factors

CONTENTS

FISHES ARE CONFRONTED BY an environment that is complex and heterogeneous, with components of their habitat changing on multiple temporal and spatial scales. For instance, water temperature or turbidity might change on an hourly or daily frequency, altering the suitability of certain habitats; in contrast, substrata might change over a longer time period of days, months, years, or decades, and basic structures, such as shoreline characteristics in lakes or riffle–pool sequences in streams, might vary on a scale of many months to tens or hundreds of years.

Various approaches have been used to describe the dynamics of local habitats, albeit most have been based on terrestrial systems (Wu and Loucks 1995). While developed initially for estuarine systems, a useful approach to understanding the freshwater habitat mosaic is to view it holistically as an environment possessing both dynamic (short-term physical-chemical and biotic variability, prey fields, and predator occurrence) and static (long-term structural variability, sediment type, and shoreline context) components, each having its own influence on fishes (Peterson 2003; Peterson et al. 2007). The timing, positioning, and amount of overlap between the dynamic and the stationary components thus control the suitability of local habitats or patches.

The distinction between the terms "habitat" and "environment" has garnered considerable debate (e.g., Ryder and Kerr 1989). Considered in the static-dynamic dimensions of Peterson (2003), habitat comprises the "localized structured component that acts as a template" for organisms, whereas environment is "the sum of the biotic and abiotic surroundings, including habitat and other organisms." Static and dynamic features are, of course, not independent, and in the long run, dynamic features can influence "static" features. For instance, the bottom type in a stream, such as a coarse gravel substratum, is a function of the interactions of stream discharge, gradient, and bed materials. In the short term, changes in water depth, velocity, temperature, and quality across a gravel bar can alter the suitability of the coarse gravel substratum for particular fish species. For instance, in the Colorado River below Glen Canyon Dam, the hypolimnial release from the dam has greatly reduced ambient water temperature as well as turbidity. Annual water temperatures now range from 9 to 14° C and turbidity is very low; prior to the dam, summer water temperatures reached to near 30° C and the river carried extensive suspended sediments (Blinn and Poff 2005). As a consequence, native fishes such as Bonytail and Humpback chubs (*Gila elegans* and *G. cypha*) are now unable to spawn in what otherwise had been favorable main channel habitats and, in fact, are largely eliminated from the main channel (Minckley et al. 2003).

LANDSCAPE ECOLOGY

Aldo Leopold, one of the founders of the conservation movement in America, developed the concept of a land ethic, espoused, for instance, in the posthumously published *A Sand County Almanac* (Leopold 1949). The term "landscape ecology" was first used by a German scientist, C. Troll, in 1939 (Turner 1989), and the initial development of the field took place largely in Europe where it emphasized terrestrial patterns and humans as part of the landscape (Wiens 2002; Turner 2005). The field of landscape ecology thus deals with spatial patterns, the effect of temporal and spatial scales on how

organisms perceive and respond to patchiness within the environment, and linkages among the elements within the pattern.

During the period from 1940 through the 1970s, terrestrial and aquatic ecologists, by and large, continued to view their study systems as distinct units rather than as being part of an interconnected landscape, although they might infer landscape-level processes as being important in creating the ecological pattern they were studying (Turner 1989). For example, longitudinal changes in fish faunas (the patterns) were related to changes in stream sizes and gradients. The maturation of the field of landscape ecology occurred when ecologists began to study the effects of spatial patterns on ecological processes (Turner 1989). The development of the River Continuum Model, treated later in this chapter, incorporated the role of spatial patterns in watersheds on processes occurring at different points or patches in the watershed, such as changes in energy sources and functional groups of organisms, and thus represents the incorporation of a maturing view of landscape ecology.

On one level, all of us have an inherent feeling of what constitutes a landscape. A more formal, albeit terrestrially biased, definition of a landscape is "a kilometers-wide area where a cluster of interacting stands or ecosystems is repeated in similar form" (Forman and Godron 1981); it is important, however, to emphasize that landscapes can vary greatly in size. Although terrestrially focused landscape ecologists have tended to view streams and lakes as boundaries, water bodies have their own internal heterogeneity, and it is important to recognize aquatic landscapes as well as terrestrial ones (Wiens 2002). A definition more amenable to aquatic systems considers a landscape as "an area that is spatially heterogeneous in at least one factor of interest" (Turner 2005).

Terrestrial and aquatic landscapes are usually defined by the most obvious geomorphic, hydrologic, vegetational, or land-use features. Boundaries between adjacent landscapes are

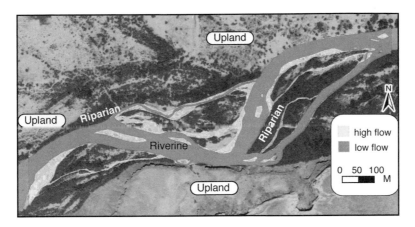

FIGURE 4.1. A section of the San Juan River in Utah showing the complex patterns and juxtaposition of riverine, riparian, and upland landscapes. Aquatic habitats expand during high flow, which can also restructure the streambed and adjoining wetlands. Land and river imagery provided by R. Bliesner, Keller-Bliesner Engineering. See also corresponding figure in color insert.

defined by distinct changes in spatial elements, such as a change from a riverine landscape, with patches defined by substratum and water mass characteristics, to a riparian landscape, with complex bank side vegetation, to a drier upland landscape (Figure 4.1). Landscapes are, however, interrelated so that impacts in upland and/or riparian landscapes can affect aquatic landscapes. For instance, fish species richness in the Aspen Parkland Ecoregion of Manitoba, Canada, was positively related to the quality of the terrestrial landscape within the catchment (Wilson and Xenopoulos 2008).

Landscapes themselves can be grouped into regions so that a region contains several to many landscapes and is a broad geographical area that may be ecologically diverse but that has a "common macroclimate and sphere of human activity and interest" (Forman 1995). Elements within a landscape include a background matrix, patches, and corridor, all of potentially varying shapes and sizes (Forman 1995). As perceived by organisms, what is termed a patch or a corridor can vary, so that patches for one species might become part of the surrounding matrix for another. However they are defined, the recognition that habitats are themselves embedded in an expanding hierarchy of other patches, and that the recognition of patches

by organisms is scale dependent, are perhaps two of the most important outcomes from the emerging field of landscape ecology (Kotliar and Wiens 1990; Grossman et al. 1995; Wiens 2002).

In small headwater streams, three important landscape attributes are (1) interactions at the interface between terrestrial and aquatic habitats, (2) how habitats are related spatially on a large scale, and (3) how refuges that provide relief from harsh environmental conditions are spread across the landscape (Schlosser 1995a). With relatively minor modifications, these attributes remain important for fishes in other freshwater habitats, including lakes and large rivers (see Part 5).

Patches

Landscapes are mosaics of patches, which are spatial units that differ from other such units in terms of function or appearance (Kotliar and Wiens 1990). How fishes perceive patches is controlled to a large extent by their size and life-history stage. The smallest scale at which an organism responds to patch structure is referred to as "grain," and the largest scale, equivalent to its lifetime home range, is referred to as "extent" (Kotliar and Wiens 1990). Grain

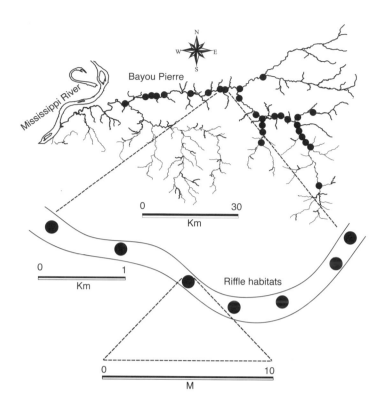

FIGURE 4.2. The distribution of Bayou Darters (*Nothonotus rubrum*) in Bayou Pierre, Mississippi, illustrating a three-level hierarchy of habitat patches. The closed circles in the top map show locations for Bayou Darters and the total range of circles thus defines the maximum extent. The bottom detailed map shows the distribution of riffles in one section of Bayou Pierre where each riffle would correspond to a habitat patch; within a single riffle (a size of approximately ≤ 10 m), Bayou Darters are selecting particular substratum sizes. Data from Ross et al. (1990, 1992b, 2001) and Slack et al. (2004).

and extent are organism-specific traits so that a benthic-oriented darter such as the Naked Sand Darter (*Ammocrypta beani*), which occurs primarily over clean sand (Ross 2001), might have a grain on the order of a millimeter or less as defined by the size of sand grains. In contrast, a water column minnow such as a Blacktail Shiner (*Cyprinella venusta*) might have a grain defined by water mass characteristics (e.g., depth, flow, and temperature) with a size on the order of centimeters or meters.

The distribution of patches in an organism's environment can be hierarchically nested between its grain and extent (Figure 4.2). For example, the Bayou Darter (*Nothonotus rubrum*) is endemic to Bayou Pierre, a tributary of the Mississippi River, where it is restricted to riffle habitats having a coarse, firm substratum (Ross et al. 1990, 1992b). The known range defines the maximum extent, even though actual extent on an individual basis is probably less. On a large scale, the extent of Bayou Darters constitutes a patch. At a finer scale, each riffle constitutes a habitat patch, and within each riffle there are smaller patches defined by the interplay of water depth, current velocity, and substratum size. For instance, Bayou Darters show strong selection for substratum particle sizes of 32–64 mm (Ross et al. 1992b), but during spawning, females select 1–2 mm diameter, coarse sand (Ross and Wilkins 1993). Consequently, the grain for Bayou Darters would seem to be approximately 1 mm.

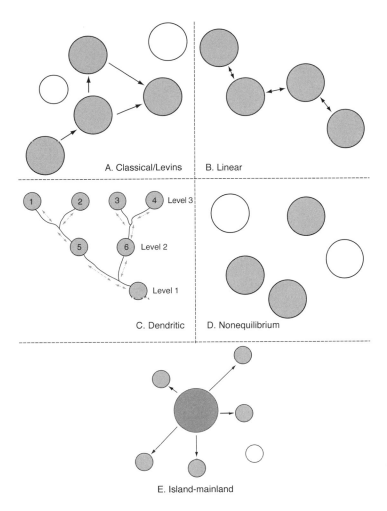

FIGURE 4.3. Various conceptual models of metapopulation structure. Shaded circles are occupied; open circles are unoccupied; arrows indicate directions of movement. Based on Harrison and Taylor (1997), Wiens (1997), Fagan (2002), and Farina (2006).

Metapopulations

Given that aquatic habitats can be viewed as temporal and spatial mosaics of varying suitability, it is not surprising that populations of a species are not distributed uniformly across the aquatic landscape but instead tend to be aggregated in areas offering the most suitable habitat components. This view of how populations are distributed in space has been formalized as the metapopulation concept—namely a metapopulation is "any assemblage of discrete local populations with migration among them" (Hanski and Gilpin 1997) (Figure 4.3).

As such, a metapopulation is "a set of populations that are interdependent over ecological time" (Harrison et al. 1988). Each local population or deme is subject to forces of local selection, including extinction, emigration, and immigration of individuals from other local populations. The balance of these forces determines the fate of local populations, the extent of sites occupied, and the overall size and genetic diversity of the metapopulation (Hanski and Gilpin 1997; Policansky and Magnuson 1998). The landscape concepts of patches and corridors and the ability of organisms to move between them are thus central to the metapopulation concept. (Movement is treated in Chapter 5.)

The term metapopulation (literally a population of populations) was first used by Richard Levins in 1970, although the mathematical description of a population of a single species comprising interconnected local populations appeared in 1969 (Hanski and Gilpin 1991). Levins (1970) described a landscape of occupied and vacant patches as a result of the colonization and extinction of local populations, with the overall collection of local populations comprising a metapopulation (Hanski and Gilpin 1991; Hanski and Simberloff 1997). This is often referred to now as the "classical" or "Levins style" metapopulation (Figure 4.3A) (Hanski and Simberloff 1997; Gotelli and Taylor 1999a). The concept has been readily, and often uncritically, applied to fish populations because the occurrence of habitat patches can result in spatial structuring of populations, such as linear or dendritic metapopulations in streams (Figure 4.3B, C) (Fagan 2002; Campbell Grant et al. 2007). However, without actual assessment of local colonizations and extinctions, rates of movement among local populations, and genetic structure, just because a species may comprise discrete local populations does not automatically indicate that it fits one of the metapopulation concepts (Hanski and Simberloff 1997; Gotelli and Taylor 1999a). For instance, a species existing as a number of discrete populations would, in the absence of movement, constitute a nonequilibrial population and would not be considered a metapopulation (Figure 4.3D).

Gotelli and Taylor (1999a) provide one of the few studies on freshwater fishes to rigorously test classical/Levins-style metapopulation predictions. They used a large data set of 46 fish species, censused 2–3 times per year for 11 years at 10 sites in the Cimarron River, Oklahoma. In the Levins-style metapopulation (Figure 4.3A), the probability of colonization should increase as the proportion of occupied sites increases because there would be more occupied sites from which dispersal could originate. Similarly,

the probability of extinction should decrease as the proportion of occupied sites increases because of the increased odds of rescue from another site (Gotelli and Taylor 1999a). Using the 11-year data set for each species, Gotelli and Taylor constructed a matrix showing the occurrence of each fish species at a census site in each year. The proportion of occupied sites was the number of sites occupied in a year, divided by the number of censused sites. The probability of extinction was the number of occupied sites in year (t) that were not occupied in year (t + 1), divided by the number of occupied sites in year (t). Finally, the probability of colonization was the number of vacant sites in year (t) that were occupied in year (t + 1), divided by the number of censused sites in year (t).

In contrast to the prediction of a Levins-style metapopulation, the probability of extinction overall was not related to the proportion of occupied sites, although 5 of the 36 species in the analysis did show a significant negative relationship and thus fit the prediction. Similarly, the probability of local colonization was not related overall to the proportion of occupied sites; at the level of individual species, only one species showed a significant positive relationship. Rather than site occupancy, the position in the river system was a more important predictor of colonization and extinction, with the probability of extinction increasing in upstream areas and the probability of colonization increasing in downstream areas. Although the Cimarron River fishes did not fit the predictions of the Levins-style metapopulation model, they did fit predictions of an island-mainland metapopulation model where local extinctions are independent of each other and colonizations occur from outside of the smaller patches (Figure 4.3E). An island-mainland model is appropriate when there is wide variance in the size of the local populations or high variation in patch quality (Harrison and Taylor 1997).

Stream fishes are often viewed as having linearly arranged metapopulations (Figure 4.3B), and in some instances this may be appropriate when species, such as the Bayou Darter in Mississippi, occur almost exclusively in main-channel habitats (Ross et al. 1992b). In many other instances, fishes occur in a broader range of stream sizes so that a dendritic model (Figure 4.3C) would better capture their population structure. As modeled by Fagan (2002), linear and dendritic models may differ in responses to perturbations, depending on how dispersal occurs, and also differ in their responses to fragmentation. Dendritic models tend to show more severe responses to fragmentation, and fragment sizes tend to be smaller and have greater variance. However, if dispersal is sufficient, and occurs both upstream and downstream, the increased connectivity afforded by dendritic versus linear systems increases opportunities for repopulation of extirpated patches, thus increasing the overall persistence of a metapopulation (Fagan 2002; Campbell Grant et al. 2007). Because the union of two streams (the nodes in a dendritic network) may provide increased habitat diversity and a concentration of other resources, such areas can be characterized by increased species diversity (Campbell Grant et al. 2007). Finally, the spatial geometries of disturbance and dispersal can be quite different in dendritic systems. For instance, recovery of lost headwater populations 1 and 2 (Figure 4.3C) would require recolonization from level-2 population 5, rather than from the physically closer (but not connected by water) level-3 populations 3 and 4.

Changes in patch quality can result in areas that vary in favorability to growth, survival, and reproduction of fishes. In the most extreme cases, favorable patches, where successful reproduction exceeds mortality and where emigration exceeds immigration, can supply individuals to patches that do not allow long-term survival and reproduction (i.e., mortality exceeds successful reproduction and recruitment to the population). Such pairs of sites are referred to as sources and sinks (Pulliam 1988; Farina 2006). For example, in a small Minnesota stream, Schlosser (1995a, b)

showed that Beaver dams functioned as source areas where most of the production of new individuals occurred. Stream sections between Beaver ponds tended to have lower retention and survival of fishes and thus operated as sinks. However, movement out of favorable source areas, which generally occurred during high-flow events, is also the mechanism whereby newly created Beaver ponds can be colonized by fishes, so there are potential advantages for movement out of areas with high population densities.

The distribution of Bayou Darters in Bayou Pierre of western Mississippi, with juvenile and adult fish occurring in discrete riffle patches (Figure 4.2) and not in intervening pools, suggested that Bayou Darters might comprise a linear metapopulation. Because a primary tenet of a metapopulation is movement among patches, Slack et al. (2004) first demonstrated that larval drift was occurring, such that larvae produced in an upstream patch could supply a downstream patch. Next, they tested the hypothesis that patches of riffle habitat lower in the stream system, where the substratum was softer and particle size generally smaller, functioned as sinks and that they were supplied by more upstream source areas (Ross et al. 2001; Slack et al. 2004). Although densities of Bayou Darters were greater in more upstream riffles, downstream riffles also supported Bayou Darters. The source-sink hypothesis was tested indirectly by comparing the age structure and somatic condition of the fish. In a sink, populations would be expected to have an altered age structure with fewer old individuals because of increased mortality. In fact, there were no differences in age structure between upstream and downstream riffles, with both supporting age-0 to age-3 fish in approximately the same proportions. Body condition also did not differ between upstream and downstream riffles. The data are consistent with a linear metapopulation model, but although apparent riffle quality is lower in downstream riffles, predictions of a source-sink hypothesis were not supported.

RELATING ASSEMBLAGES TO THE ENVIRONMENT • *Conceptual and Statistical Models*

Various conceptual and/or statistical models have been proposed relating the primary structure of assemblages (i.e., species presence and/or relative abundance), emergent assemblage structure (i.e., species richness, diversity, assemblage complexity, and trophic relationships), or ecological/life-history traits of species to environmental factors (Marsh-Matthews and Matthews 2000). Such models provide insight into how fish assemblages are formed and maintained. At the risk of oversimplification, these models can be placed into two groups: (1) conceptually based *a priori* approaches where general features of the habitat are used to make predictions of ecological traits of species or assemblages, or (2) data-based *a posteriori* approaches where species occurrences and/or abundances are related to habitat features, using some type of univariate or multivariate analysis. In contrast to the first group, which is based on a mechanistic understanding of how communities operate, this second approach is largely nonmechanistic.

A Priori Models

There are three principal conceptual models (e.g., a set of predictions arising from basic ecological principles) that are prevalent in the literature on lotic systems relating species traits or the emergent structure of assemblages to general environmental features (Goldstein and Meador 2004)—habitat templates (Southwood 1977), landscape filters (Poff 1997), and the river continuum concept (Vannote et al. 1980).

Habitat Template

In 1977 and 1988, Southwood suggested that the habitat is a template providing a predictive pattern for the evolutionary assembly of communities and life-history traits thereof, much like the periodic table of elements in chemistry. A key presumption is that the present-day ecological traits of the organisms will match the current ecological conditions, which, as previous chapters have suggested, is not always the case. Although very much aware of the importance of historical factors, Townsend and Hildrew (1994) set out to develop testable predictions of the habitat template model for species as well as assemblage traits. They used two axes, temporal habitat heterogeneity and spatial heterogeneity, in developing predictions of how species traits (e.g., reproductive type, age and size, parental care, movement, etc.) or assemblage traits (e.g., importance of biotic interactions) would respond to the habitat template (Figure 4.4). They viewed temporal heterogeneity as primarily a measure of the frequency of disturbance and habitat heterogeneity as primarily a measure of the availability of refugia. Thus as disturbance increases, the availability of refugia would become more important. Their model was developed with the Rhône River drainage, France, in mind as a testing arena, but the general constructs should apply to other systems, both lentic and lotic. For instance, considering the species trait of life span, the model would predict that life span should be short on the unstable side of the template (Figure 4.4) and long or short on the stable side (see also Chapter 9). Similarly, body size should be small on the unstable side and large or small on the stable side. In terms of assemblage characteristics, the more stable and complex habitats on the left side should lead to greater specialization in resource use, greater importance of biotic interactions, and increased potential for coevolution. The unstable and low complexity habitats on the right side should favor more generalists in resource use, as well as little potential for biotic interactions or coevolution.

Tests of the species-trait predictions of Townsend and Hildrew's habitat template model have been equivocal. For instance, an important prediction is that species traits such as body size and parental care should decrease in environments with low spatial heterogeneity and high temporal heterogeneity—the lower right side of the figure. Tests of these and other

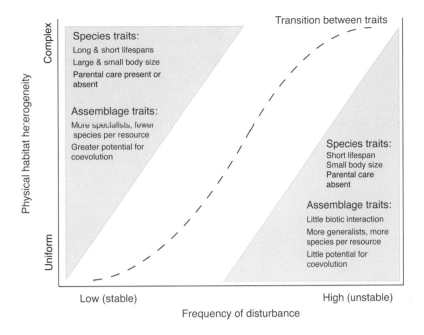

FIGURE 4.4. The habitat template model showing relationships of species or assemblage traits to the frequency of disturbance and complexity of the physical habitat. The large shaded triangles show areas of different predicted traits. The dashed line shows the transition points between traits on the upper left and lower right. The transition can shift to the left for short-lived species and to the right for long-lived species. Adapted from Resh et al. (1994) and Townsend and Hildrew (1994).

predictions based on the responses of 13 taxonomic groups of plants and animals occurring in the Rhône River drainage resulted in only mixed support, and support for fishes in this system was totally lacking (Resh et al. 1994). However, Resh et al. (1994) pointed out that the large preexisting data set used to test the predictions might have had methodological limitations that precluded a fair test of the model. Also, the occurrence of species in a habitat might reflect more of chance movement rather than actual habitat selection—resulting in a blurring of the match between species traits and the nature of the habitat.

Studies of northern U.S. midwestern streams (Poff and Allan 1995), and comparisons of functional convergence between European and eastern North American fish assemblages (Lamouroux et al. 2002), offer somewhat stronger support for the habitat template model—at least in terms of assemblage predictions. For instance, Poff and Allan (1995) found that two

predictions of the habitat template model—variable habitats should contain more resource generalists and nonvarying habitats should contain more specialists (cf. Figure 4.4)—were supported for stream fish assemblages. Hydrologic variables used by Poff and Allan (1995) included flow predictability and variation, base flow stability, and frequency of spates.

Landscape Filters

Recognizing that assemblages are the end products of "interacting multiple causes" at "multiple spatial and temporal scales," Poff (1997) proposed that functional attributes of species in assemblages are shaped by a hierarchical series of landscape filters that include both physicochemical and biotic factors. A particular community thus comprises species possessing the appropriate ecological "shapes" to have passed through the filters. The heuristic model presented earlier (see figure in Part 2) is similar to this approach. Key elements in

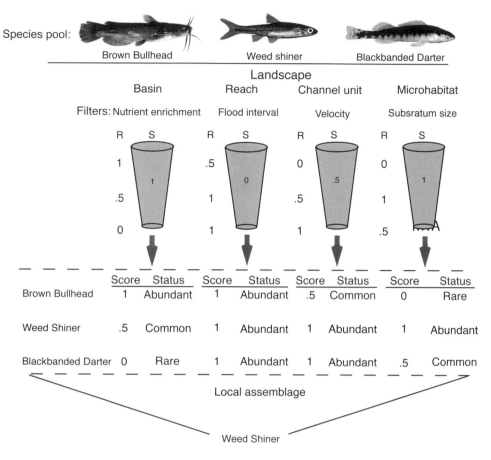

FIGURE 4.5. An example of the landscape filter model, based on four landscape scales and a source pool of three species, and illustrating how species traits interact with a hierarchical series of filters to determine species occurrences in assemblages. R = the resistance of each species to being removed by the filter; S = the strength of the filter relative to its ability to remove species. See text for additional explanation. Based on Poff (1997).

the operation of Poff's landscape filters are categorical niches, defined as discrete levels of species requirements along a given resource axis, and categorical filters, defined as the strength or resistance of a filter in allowing species to pass through it. The combination of categorical filter strength with the categorical niche determines the probabilities for a species to pass through a given filter (Figure 4.5). The example is based on four landscape scales, with one filter per each scale, and a source pool of three species. Two values are required to parameterize the model: a species-specific trait (the resistance of each species to being removed by the filter) and a landscape-level trait (the strength of the filter relative to its ability to remove species). Poff (1997)

arbitrarily chose three resistance categories: 1-strong resistance (not removed by any filter), 0.5-intermediate resistance (affected only by the strongest filter, which has a 50% chance of removing the species), and 0-weak resistance (100% chance of removal by strong filters and 50% chance of removal by intermediate filters). Similarly, Poff chose three filter strengths: 1-strong, 0.5-intermediate, and 0-weak.

To run the hypothetical model and determine the most likely species assemblage at the microhabitat level, I have provided likely species resistance values and filter strengths based on work done in the Pascagoula River drainage of southeastern Mississippi (Baker and Ross 1981; Ross and Baker 1983; Ross et al. 1987; Ross 2001). For instance, at the level of basin with the filter of nutrient enrichment, Brown Bullhead (*Ameiurus nebulosus*) would likely have a resistance of 1, allowing it to pass through the nutrient enrichment filter, which, because many species are strongly impacted by eutrophication, is also set at 1. At the basin level then, Brown Bullhead would have a score of 1 and be classed as abundant. At the reach level, the filter of flood interval has a strength of 0 (given the generally nonerosive floods of this region) and Brown Bullhead are assigned a resistance strength of 0.5. Therefore, at the reach level, Brown Bullhead would also have a score of 1 and be classed as abundant. At the channel-unit level, the filter of water velocity has a strength of 0.5 and Brown Bullhead have a resistance value of 0, given that they tend to occur in slow-velocity habitats, resulting in a score of 0.5 and being ranked as common. Finally, at the microhabitat level the filter of coarse substrate size has a strength of 1 (since bottom-inhabiting species generally respond strongly to particle size) and Brown Bullhead, which usually occur over fine substrata, have a resistance of 0, resulting in a score of 0 and a local abundance classed as rare. Poff (1997) cautioned that the model should not be used to predict absence because it does not incorporate all factors that might influence species presence or absence. Based on likely resistance levels and filter strengths as cited previously, two more species, Weed Shiner (*Notropis texanus*) and Blackbanded Darter (*Percina nigrofasciata*) can also be run through the four filter levels with the end result being a local assemblage most likely comprising Weed Shiner. Brown Bullhead are limited by the microhabitat filter of particle size, and Blackbanded Darter are limited by the basin filter of nutrient enrichment.

To expand the landscape model to real-life situations would require fairly extensive ecological information on the fish species potentially available at the basin-level species pool, along with knowledge of what filters are likely operating at each level of the hierarchy as well as their respective strengths. Although more research is needed to understand species susceptibility to various filters, for many regions of North America there is likely sufficient information to assess the relative resistance of species to various environmental filters as well as filters that are likely important at the different hierarchical levels. The information on composition of local assemblages that is provided by the mechanistic approach of landscape models is qualitative rather than quantitative. In other words, the models suggest relative abundance levels (i.e., abundant, common, or rare), but do not, and likely will not, provide quantitative measures of abundance (i.e., density).

Goldstein and Meador (2004) tested various predictions of the landscape model using their analyses of how fish species traits varied among different stream sizes. For the most part, their work supported the theoretical predictions from Poff's (1997) landscape model. For instance, their work supported the landscape model prediction that fish morphology would be best predicted by local factors such as hydraulic stress. Some differences between the predictions of the landscape model and the findings of Goldstein and Meador (2004) occurred with how reproductive and substratum-use traits varied within and among streams. Landscape predictions, as interpreted by Goldstein and Meador (2004), were that reproductive traits "will vary with flow and substrate variability" and that "substrate preferences are driven largely by microhabitat scale factors." In contrast, Goldstein and Meador (2004) found that both reproductive strategies and substratum use varied relative to stream size (i.e., among rather than within streams).

River Continuum Concept

The river continuum model (RCC) (Vannote et al. 1980) emphasizes continuity with gradual changes in species occurrences and functional groups from headwaters to downstream reaches (Figure 4.6). The model was conceived as an extension of the physical, geomorphic changes that occur longitudinally in rivers, with the idea that "over extended river reaches, biological communities should become established which

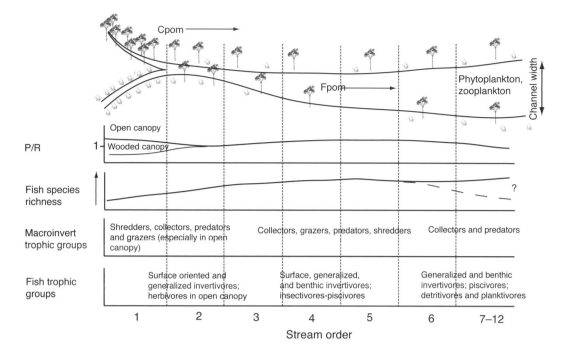

FIGURE 4.6. The relationship between stream size and the physical and biotic components, as proposed in the river continuum model. CPOM = coarse particulate organic material; FPOM = fine particulate organic material; P = primary production; R = respiration. Based on Vannote et al. (1980) and Paller (1994). See text for further explanation.

approach equilibrium with the dynamic physical conditions of the channel." In this sense the RCC is quite similar to the habitat template model.

Headwater streams, stream orders 1–3 (Box 4.1), are strongly influenced by the presence or absence of riparian vegetation. If riparian vegetation is well developed, or if the stream is otherwise strongly shaded by being in a deep canyon, energy input into the stream is largely derived from the surrounding terrestrial area (i.e., allochthonous input) rather than from within the stream (autochthonous input) so that instream measurements show greater respiration than production (P/R < 1). Such situations are common in many eastern and southeastern streams and montane western streams. As the stream increases in size, there is gradually more light penetration into the water and autochthonous production increases so that P/R > 1. In contrast, in headwater streams without well-developed riparian vegetation that could shade the stream, autochthonous production by

submerged vascular plants, periphyton, or algae is well developed so that P/R > 1. Such streams are typical of high elevations and latitudes and arid regions in general.

Medium-sized streams (orders 4–6) tend to be more open, have increased autochthonous production, and have greater diversity of insect functional groups (Figure 4.6). This general trend increases in large streams (order 6 and greater), except that increased turbidity can reduce light penetration and thus reduce photosynthesis (P/R ≤ 1). Medium and large streams have increased amounts of fine particulate organic materials (FPOM) produced by the upstream processing of coarse particulate organic materials (CPOM), such as leaves being processed by aquatic insects and bacteria.

Predictions of RCC about fish assemblage composition is at best limited to trophic groups and is based on food resources available in low-order, medium-order, and high-order streams. Vannote et al. (1980) also reasoned that low-order streams would tend to

BOX 4.1 • Stream Order

Streams can be categorized in various ways such as discharge, water depth, gradient, water quality, and branching pattern, to name but a few. The branching pattern of streams, or network analysis, has proven useful as a way to describe streams (Leopold et al. 1964; Leopold 1994). Pioneering work on network analysis of streams involved the concept of stream order, first proposed by Robert Horton (1945) and later modified by Strahler (1952, 1957). Using Strahler's modification of Horton's system, the smallest unbranched tributary is classified as first order. The union of two first-order streams results in a second-order stream, and the union of two second-order streams results in a third-order stream. To generalize, when two streams of equal rank join they form a segment of the next highest order. However, ordinal rank is not increased by the entrance of lower-order streams.

The decision as to what constitutes a first-order stream can be somewhat arbitrary and to a certain extent depends on the purpose of the study. For instance, a geomorphologist might be more interested in including small channels even though they are not perennial. Strahler (1952), in fact, suggested that first-order streams were wet-weather streams that were normally dry. Leopold et al. (1964) further refined this by suggesting that first-order streams are the smallest unbranched tributaries shown on a 1:24,000 scale topographic map. In contrast, a biologist might favor an approach where first-order streams are classified as the smallest perennial unbranched tributaries that have persisted long enough to contain plants and animals (Hynes 1970). Obviously the second approach requires field verification, so the methodology suggested by Leopold et al. (1964) tends to be the approach of choice for all but the smallest of watersheds. In addition, GIS technology now provides an automated means of determining stream order (Lu et al. 1996).

A useful property of stream order based on the Horton-Strahler system is that a number of physical properties of streams are correlated with stream order. For instance, discharge and drainage area tend to be positively correlated with stream order whereas gradient is negatively correlated (Knighton 1984). The number of stream lengths of each order are negatively related to order so that there tends to be 3–4 times more streams of order n-1 compared to order n. Comparing order n-1 and n, the former has sections that are about half as long and drain somewhat more than one-fifth of the area (Hynes 1970). Of course, stream order is a geographical-level variable and does not give information on microhabitat features such as pools, runs, or riffles. In addition, while it is a useful descriptor of the drainage network, streams of equivalent orders but from wet versus arid climates tend to be greatly different in size and biological attributes.

be cooler than high-order streams, although this could certainly vary depending on canopy cover and geographic region. In low-order streams, food resources for fishes would primarily be terrestrial invertebrates, with less importance of aquatic invertebrates or fishes as prey. In medium streams, food resources would expand to include more aquatic prey, both insects and fishes, and in large streams, the increased amount of autochthonous production could lead, in the absence of high turbidity, to the presence of planktivorous and herbivorous fishes. Although Vannote et al. (1980) said nothing about how invertivores might vary in abundance in high- versus medium-order streams, later work (e.g., Goldstein and Meador 2004) has inferred that RCC predicted a decline. Several early papers, not cited by Vannote et al. (1980), did provide information on changes in functional groups of fishes along a gradient of stream order, including a classic paper by Shelford (1911). Shelford's work on several small Michigan streams showed that headwaters of the streams were consistently occupied by several species, especially Creek Chub (*Semotilus atromaculatus*), and that fish species composition changed longitudinally in streams as a function of the physical changes in habitat. The headwater occurrence of Creek Chub (and in general the other three species within the genus) turns out to be a general pattern (Starret 1950; Kuehne 1962; Lotrich 1973; Ross 2001;

Boschung and Mayden 2004). As predicted by RCC, in very small headwater streams, Creek Chubs consume primarily allochthonous materials, including insects and even berries (Lotrich 1973; Moshenko and Gee 1973; Ross 2001).

As stream order increases, there is a general trend for both trophic groups and species richness to increase (e.g., Kuehne 1962; Sheldon 1968; Schlosser 1982, 1987; Paller 1994), although species richness may plateau or even decline in large streams (stream orders > 5) (Whiteside and McNatt 1972; Platts 1979; Fairchild et al. 1998). Changes in faunas are generally gradual and do not closely correspond with changes in stream order (Matthews 1986a). Also, the faunal characteristics of small streams that are tributary to lower reaches of large streams (termed adventitious streams) are different when compared to a similarly sized headwater stream (Gorman 1986). For instance, adventitious streams can have greater fish diversity than similarly sized headwater streams and, in fact, have faunas that are more similar to the main channel than to headwater streams (Thomas and Hayes 2006).

The RCC predictions of changes in functional groups, such as fewer insect predators and more planktivores and detritivores with increasing stream size, are generally supported (Goldstein and Meador 2004). For instance, in large eastern and southeastern rivers, including the Missouri and Mississippi rivers, fishes that primarily consume plankton include Paddlefish (*Polyodon spathula*) and Bigmouth Buffalo (*Ictiobus cyprinellus*) (Rosen and Hales 1981). Other species also consume phytoplankton or zooplankton, but from on or near the substratum, such as River Carpsucker (*Carpiodes carpio*) (Brezner 1958; Ross 2001). Detrital materials also contribute to the diet of Bigmouth Buffalo and its congeners, Smallmouth Buffalo (*I. bubalus*) and Black Buffalo (*I. niger*) (Walburg and Nelson 1966; McComish 1967; Ross 2001).

The RCC continues to be an important heuristic tool in understanding stream ecosystems. However, among its shortcomings, it did not treat a stream or river as a landscape such that the great variety and complexity of aquatic habitats was largely ignored (Fausch et al. 2002).

A Posteriori Models

Models that use large data sets on fish occurrence and environmental characteristics of lakes and streams in an attempt to find predictive suites of physical characters, or to identify fish assemblages characteristic of locations or environmental conditions, typify *a posteriori* approaches. Such studies depend on some form of multivariate statistical analysis (Box 4.2) so that their popular use has paralleled that of modern, high-speed computers, and especially the powerful personal computers that are now common. Although widespread use of such approaches is relatively recent, the underlying statistical techniques, such as factor analysis, were developed early in the twentieth century (Sokal and Rohlf 1995; Gotelli and Ellison 2004). The application of multivariate approaches in fish ecology also benefitted from the interest in numerical taxonomy during the same time period (Sneath and Sokal 1973). Many of the computer programs developed for numerical taxonomy, such as cluster analysis and ordination, were used as well in ecological applications.

Multivariate Statistics and Fish Assemblages

Pioneering multivariate studies relating habitat characteristics and fishes include G. R. Smith and Fisher (1970), dealing with the distribution patterns of fishes in Kansas, and Stevenson et al. (1974), who studied 53 species of western and central Oklahoma fishes from 27 drainage units. Both studies were based on factor analysis, which treats the variation and covariation of the original variables as a linear combination of underlying factors, with the number of factors usually being less than the original number of variables (Sokal and Rohlf 1995; Gotelli and Ellison 2004). Factor analysis thus is a means of reducing the number of variables (i.e., the factors replace the original variables) and in identifying possible causal factors that are behind the original correlations in the data set (Sokal and Rohlf 1995).

The Oklahoma study (Stevenson et al. 1974) included fish and environmental data from tributaries of the Arkansas, South Canadian, and Red river drainages. Species diversity is generally low in this environmentally harsh region, but numbers of individuals can be quite high (Matthews 1988). The analysis identified six factors defined by their responses to variation in environmental variables and the occurrence of fish species. Environmental variables, 13 in all, included average stream flow, annual precipitation, average winter temperature, number of days above freezing, elevation, and measures of water chemistry. For example, Factor 1, named the Ghost Shiner group because Ghost Shiner (*Notropis buchanani*) showed the highest loading (correlation) with this factor, also included Longnose Gar (*Lepisosteus osseus*), average flow, Freshwater Drum (*Aplodinotus grunniens*), Slenderhead Darter (*Percina phoxocephala*), runoff, and depth of salt deposits (only variables with high loadings on this factor are given). In other words, much of the variation among the seven listed variables could be captured by a derived variable, Factor 1. Species with high loadings on Factor 2 included the centrarchids—Largemouth Bass (*Micropterus salmoides*), Orangespotted Sunfish (*Lepomis humilis*), Bluegill (*L. macrochirus*), Longear Sunfish (*L. megalotis*), and White Crappie (*Pomoxis annularis*). Other species with high loadings were Suckermouth Minnow (*Phenacobius mirabilis*) and Western Mosquitofish (*Gambusia affinis*); high evaporation and low elevation were the associated environmental variables. This factor thus captures a lower gradient and warmer fauna dominated by centrarchids. Overall, the study showed the strong impact of climatic and habitat factors on the distribution of Great Plains fishes.

Studies such as these provide a way to reduce the number of variables to a more easily handled number and also offer an objective assessment of the linkages among species and of associations among species and their environments. As with current applications of multivariate techniques, the outcomes depend on

appropriate sampling designs, choosing appropriate spatial and temporal scales for analyses and meeting any assumptions (e.g., normality) of the statistical tests.

Another important feature of these studies is their predictive ability. For instance, the Oklahoma study of Stevenson et al. (1974) identified a group of fishes comprising Red River Pupfish (*Cyprinodon rubrofluviatilis*), Red River Shiner (*Notropis bairdi*), Speckled Chub (*Macrhybopsis aestivalis*), and Chub Shiner (*N. potteri*) that were positively related to indicators of natural brine. They surmised that, given increased salinity levels caused by oil and gas extraction, these species would show an expansion of their ranges, assuming that they had access to the new habitats. In support of this prediction, Red River Pupfish were introduced (perhaps by bait dealers) into a saline tributary (2.4 ppt) of the Cimarron River in northwestern Oklahoma. The pupfish are reproducing and appear to be established (McNeely et al. 2004).

Because different multivariate techniques have different strengths and weaknesses, more recent studies often combine several approaches. Rahel (1984) analyzed fish assemblages from 43 bog lakes in northern Wisconsin. Bog lakes are late successional-stage lakes in the transition from lakes to wetlands and are characterized by low pH, low oxygen levels, and generally low fish species diversity (in this case 20 species).

Fish species were grouped into assemblages using the multivariate technique of detrended correspondence analysis (DCA) on habitat distribution data (Box 4.2). Three assemblages were identified: the centrarchid assemblage consisting of bass and sunfish and associated species such as Northern Pike (*Esox lucius*), the cyprinid assemblage, and the Central Mudminnow (*Umbra limi*)-Yellow Perch (*Perca flavescens*) assemblage. The latter group, which occurred along with the other two assemblages, comprised a "core species group," to which others could be added in lakes with less harsh environments.

BOX 4.2 • Multivariate Statistics

In contrast to univariate models, such as linear regression where the response of the dependent variable is related to change in the independent variable or multiple linear regression where the response of the dependent variable is related to change in two or more independent variables, multivariate statistics deal with the simultaneous variation in two or more dependent variables (Manly 1986; Sokal and Rohlf 1995). As such, multivariate statistical techniques are ideally suited to dealing with the complexity of fish assemblages, both in comparing assemblages across space and/or time and for examining relationships among species and the physical and biological components of their environment. Multivariate statistics have become easy to use given the wide choice and availability of statistical software programs. However, this ease of use belies the underlying statistical complexity and the need for the user to have at least a conceptual understanding of what the statistical program is doing. In addition, such tests often assume that each of the variables has a certain structure, such as showing a normal distribution, that all the variables combined have a multivariate normal distribution, that variances among variables are homogeneous, or that the samples were collected randomly. Given the complexity and specialized nature of multivariate approaches, I have given only a brief introduction to some of the approaches that are referred to in the text. Principal components analysis, factor analysis, discriminant function analysis, correspondence analysis, and nonmetric multidimensional scaling can be considered ordination techniques because all are ways of ordering objects based on an array of variables or of ordering variables based on objects. Classification analysis provides another approach for recognizing patterns in multivariate data. Grouping objects (i.e., sites, individual organisms, or sampling units) based on measured variables (i.e., current speed, water depth, turbidity, or species composition) is termed Q analysis. For both ordination and classification, grouping measured variables based on the objects is termed R analysis because the grouping of variables by objects is based on correlation coefficients, such as Pearson's r (ter Braak 1995; Legendre and Legendre 1998).

ORDINATION TECHNIQUES
Principal Component Analysis

Principal component analysis (PCA) was first described by Karl Pearson in 1901, and in 1933 Harold Hotelling developed computational methods (Manly 1986; Gotelli and Ellison 2004). It is one of the simplest multivariate approaches and has been widely used in ecological studies, although more recently its use has been supplanted by other approaches (Ludwig and Reynolds 1988). The objective is to create linear combinations (components) of the original variables that are not correlated and which capture most of the variation. If successful, the original number of variables is replaced with fewer principal components, making interpretation of the data easier (Manly 1986). There is much to be gained in terms of data simplification if the original measured variables are highly correlated. Indeed, if the original variables are themselves not correlated (generally this would be rare in ecological studies), then the number of components would be the same as the number of original variables and nothing would be gained by the analysis (Manly 1986). Because PCA is sensitive to the magnitude of the original variables, they are usually standardized to means of zero and unit variances. In addition, because PCA is a linear model, its usefulness declines with data that are nonlinear.

Factor Analysis

Factor analysis was developed by Charles Spearman in 1904 for the purpose of measuring human intelligence (Gotelli and Ellison 2004). As with PCA, the goal is to reduce the original number of variables to fewer, noncorrelated, variables (factors). Unlike PCA where factors are linear combinations of variables, factor analysis assumes that the measured variables are a linear combination of underlying factors, with the number of factors usually being less than the original number of variables (Kim and Mueller 1978; Sokal and Rohlf 1995; Gotelli and Ellison 2004). Factor analysis is especially useful as an exploratory approach to identify possible causal factors behind the original correlations in the data set (Sokal and Rohlf 1995; Gotelli and Ellison 2004). Factor analysis can

(continued)

BOX 4.2 *(continued)*

use principal components as initial factors and, as with PCA, variables are first standardized to means of zero and unit variances (Manly 1986).

Discriminant Function Analysis

This approach is really a special case of factor analysis, where the goal is to extract factors (now referred to as discriminant functions) that best separate identifiable groups that are recognized prior to the analysis (Cooley and Lohnes 1971). Groups could be individuals of the same species or sex, or fish assemblages in the same latitude. The discriminant functions are linear combinations of variables that best separate the groups, and each function is uncorrelated with other functions. A useful feature of discriminant analysis is that once functions have been determined, they can be used to classify new data to groups (Gotelli and Ellison 2004).

Correspondence Analysis

Correspondence analysis (CA), or reciprocal averaging, is another approach used to elucidate group characteristics, such as species or functional groups, to habitat characteristics (ter Braak 1995). As is true of the other ordination techniques discussed here, CA requires the assumption that groups show unimodal distributions across the environmental variables. In contrast to the other approaches, CA does a simultaneous ordination of rows and columns to maximize the separation of the groups along each axis (Gotelli and Ellison 2004). Mathematical properties of CA, and other ordination techniques, result in compressing the extremes of an environmental gradient and accentuating the middle, resulting in what is variously referred to as the "horseshoe" or "arch" effect (Wartenberg et al. 1987; ter Braak 1995). Modifications to CA, collectively referred to as detrended correspondence analysis (DCA), were designed to deal with the distortion (ter Braak 1995). However, methods used to remove the curvature of scaling all have limitations (Wartenberg et al. 1987; Gotelli and Ellison 2004).

Nonmetric Multidimensional Scaling (NMDS)

Unlike the previous ordination techniques, which generally retain the original spacing of observations in multivariate space, NMDS is based on ranked distances. It can be used with any distance measure, and the goal of NMDS is to maximize distances of dissimilar objects and minimize distances of similar objects. It is particularly useful in ecological studies because it performs well with data containing many zero values and is robust to deviations from multinormality (Gotelli and Ellison 2004; Paavola et al. 2006).

CLASSIFICATION ANALYSIS

In contrast to ordination, the goal of which is to separate objects or variables along meaningful axes, classification analysis seeks to form discrete groupings. Cluster analysis, based on hierarchical methods, is the most commonly used form of classification analysis in ecological studies (Gotelli and Ellison 2004).

Cluster Analysis

Cluster analysis is particularly useful as an exploratory data tool by creating hierarchical groupings of objects by variables (Q analysis) or variables by objects (R analysis). Approaches to hierarchical cluster analysis are based on similarity or dissimilarity matrices and most commonly form groups by nearest-neighbor joining (Legendre and Legendre 1998; Gotelli and Ellison 2004).

USEFUL REFFRENCES

Gotelli, N. J., and A. M. Ellison. 2004. *A primer of ecological statistics*. Sinauer Associates. Sunderland, Massachusetts.

Jongman, R. H. G., C. J. F. ter Braak, and O. F. R. van Tongeren. 1995. *Data analysis in community and landscape ecology*. Cambridge University Press, New York.

Kim, J.-O., and C. W. Mueller. 1978. *Introduction to factor analysis*. Sage University Paper series on Quantitative Applications in the Social Sciences, series 07-013. Sage Publications, Beverly Hills, California.

Ludwig, J. A., and J. F. Reynolds. 1988. *Statistical ecology*. John Wiley and Sons, New York.

Manly, B. F. J. 1986. *Multivariate statistical methods, a primer*. Chapman and Hall, New York.

The lakes were grouped by their environmental characteristics using principal components analysis (PCA). Out of nine original variables, PCA was able to capture 71% of the environmental variation among the lakes with three derived variables (components) (Box 4.2). The first principal component reflected the influence of lake size and habitat diversity, with lakes having high correlations on PC-I being large and having well-developed, complex littoral zones (Figure 4.7). The second principal component was largely a measure of lake productivity and acidity, with lakes having high loadings on this axis characterized by higher pH, alkalinity, and conductivity values. The third axis (not shown in Figure 4.7) reflected lake depth and adjoining wetland development. Rahel then overlaid the distribution of fishes defined by the three DCA-identified assemblages on the ordination of lakes based on habitat characteristics (Figure 4.7). Centrarchid assemblages were characteristic of large, highly productive lakes, whereas the cyprinid assemblage tended to occur in smaller, less productive lakes, and the Central Mudminnow-Yellow Perch assemblage occurred in low productivity, highly acidic lakes. In terms of successional stages, the Central Mudminnow-Yellow Perch assemblage occupied late successional environments that were transitioning to wetlands (Figure 4.7), whereas the centrarchid assemblage was characteristic of lakes in an early successional stage. The successional pattern thus shows a change from high to low fish species diversity, as environmental conditions become more limiting.

The application of multivariate statistical approaches to understanding the distributions of fish species and to fish assemblage structure is now widespread and many of such studies are treated in other sections of this chapter and in other chapters. A more recent approach, again following the technological advances in computing power, applies point location data, such as from museum collections, and the information available in geographic information systems (GIS) to the prediction of potential species occurrences. Sometimes referred to as

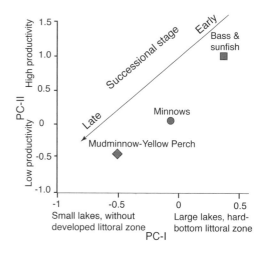

FIGURE 4.7. Principal components analysis of Wisconsin lakes on the basis of environmental characteristics. Data show mean factor scores of three fish assemblages, defined by detrended correspondence analysis, on PC-I and PC-II. See text for further explanation. Based on data from Rahel (1984).

"niche modeling," the approaches use information associated with actual species occurrences within the framework of a GIS to automatically generate a map of additional localities where the species is likely to occur. One of the first approaches used an iterative, artificial intelligence software package called GARP (Genetic Algorithm for Rule-Set Production) (Stockwell and Peters 1999; Peterson 2001). Genetic algorithms are useful in instances where the original data, generally museum records for species occurrences, and environmental data do not meet the assumptions of most multivariate statistics. Perhaps a chief distinction between these niche models and the multivariate approaches discussed earlier is that the former generally incorporate more environmental data (often 30 or more data layers), focusing especially on topographic and climate data that can be placed in a GIS mapping system. Another obvious difference is that such studies generally are done remotely, without actual fieldwork other than the initial fish collections. As such, niche modeling links the information of museum data on species occurrence with the power of modern GIS systems, but is generally limited to data that

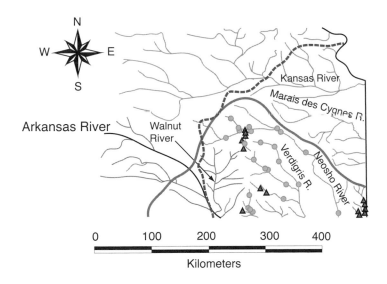

FIGURE 4.8. The prediction of the Kansas distribution of Bluntface Shiner (*Cyprinella camura*) using a niche model (GARP). The solid gray line surrounds the known distribution of Bluntface Shiner in Kansas; the dashed line shows the predicted distribution. Circles show the locality data used to build the model; triangles show data used to test the model predictions. Based on McNyset (2005). See text for additional explanation.

are available remotely in electronic databases. However, as long as environmental data can be provided that are suitable for GIS, there is really no limit on what could be included (McNyset 2005). Detailed information on local habitat use, such as focal-point water velocities, substratum selection, or vertical water column position are typically not included, nor are the influences on species local occurrences caused by interactions with other species. Thus niche modeling does not include local dimensions of the realized niche of species but instead is more analogous to the fundamental niche (Hutchinson 1957a). Wiley et al. (2003) properly refer to these as "partial niche models."

Predictive capabilities of GARP models, as with other multivariate models (e.g., Ross et al. 1987), can be tested by using only a random subset of the species occurrence data to build the models. The withheld set of occurrence data can then be plotted on the map along with the model predictions. Of course another way to test model output is to "ground truth" the information with follow-up field studies. By either approach, GARP models have generally proven quite

successful in predicting species occurrences for North American birds (Peterson 2001), marine fishes (Wiley et al. 2003), Mexican freshwater fishes (Domínguez-Domínguez et al. 2006), and Kansas freshwater fishes (McNyset 2005), among others. For example, McNyset (2005) was able to closely predict the distribution of 12 species of fishes in Kansas using GARP modeling.

Errors in niche modeling include overprediction (including habitats where the species does not occur) and omission (not including known habitats). For Bluntface Shiner (*Cyprinella camura*), one of the species studied by McNyset (2005) in Kansas, the omission error was calculated to be 17%, although essentially all streams occupied by Bluntface Shiner were included. Overprediction errors are much more problematical because their actual verification would require knowing the true range of the species at the pixel level of the distribution map (in which case running a niche model would be somewhat superfluous). Although no training or test data for Bluntface Shiner were from Walnut River (Figure 4.8), they are known to occur there (McNyset 2005). However, GARP modeling also predicted that Bluntface Shiner would occur in Kansas tributaries of the Kansas and Marais des Cygnes rivers—drainages where they are not known to occur. In this situation, it could be that the species actually occurs in these areas but the areas have not been adequately sampled, that the species once occurred in these drainages but has been eliminated, or, as seems to be true in this instance (Miller and Robison 2004), Bluntface Shiner have never occurred in these drainages, perhaps because of geographic or ecological barriers.

Niche models are also useful in the context of predicting ranges, or locations for refugia, of rare, threatened, or endangered species, as well as determining the probability of invasions by alien species. For instance, Chen et al. (2007) successfully modeled the North American distribution of two large minnows native to eastern Asia, Silver Carp (*Hypophthalmichthys molitrix*) and Bighead Carp (*H. nobilis*) that were originally brought into North America for aquaculture purposes. First Chen et al. (2007) constructed niche models using data associated with the natural distribution of these fishes in China, and then projected this information onto North American streams showing where invasions were likely to occur. Such information is useful to resource managers to be alert to the occurrence of nonnative species. Presently, the field of niche modeling is extremely active and includes development and application of new algorithms, such as Maxent, that show different strengths and weaknesses when compared to GARP (A. T. Peterson et al. 2007).

One or Several Models?

No single model of how fish assemblages are formed or identified would likely ever be sufficient. Each approach illustrated here, as well as the many others available in the literature, offer different insights into assemblage formation and spatial or temporal occurrence. *A priori* models provide an opportunity to infer what functional groups or species traits are likely to occur in an area. These approaches can be particularly powerful in efforts to understand how faunas might change as a consequence of anthropogenic factors such as eutrophication, desertification, construction of impoundments, increasing aridity, or elevated temperatures. Although *a priori* models are useful because of their generality, in most cases they are limited in their ability to predict the occurrence and/ or abundance of a particular species. *A posteriori* models tend to be more restricted in their application, such as to a particular stream or geographic region, with the notable exception of recent advances in niche modeling using GARP or similar approaches. What *a posteriori* models might lack in generality they gain in their ability to relate the occurrence of particular species, or suites of species, to particular physical factors and to provide an objective means of defining recurring groups of species (i.e., fish assemblages).

LOCAL VERSUS REGIONAL EFFECTS ON ASSEMBLAGES

Any fish assemblage is the outcome of a myriad of factors operating over a broad range of spatial and temporal scales. Depending on the strength of the factors, the structure of some assemblages might be determined more by local temporal or spatial effects, whereas others by historical or broadscale factors. Focusing on spatial dimensions, one recent approach compares the relative importance of local to regional effects. Local effects on fish assemblages include the nature of the local habitat—especially how it varies spatially and temporally. Regional effects could encompass the regional species pool, as well as landscape features such as climate, elevation, and geographic location.

Much of the role of regional or landscape factors in shaping fish assemblages has been discussed earlier in this chapter. In this section I focus on what has been a primary issue in the study of local versus regional factors, the relationship of the species richness of the local fauna to that of the regional fauna. The principal question is whether local faunas are saturated

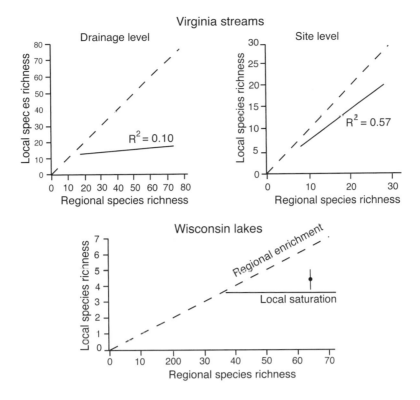

FIGURE 4.9. The relationship between native fish diversity of local assemblages to regional fish diversity in Virginia streams at the drainage and local scales (based on Angermeier and Winston 1998) and Wisconsin lakes (based on Tonn et al. 1990). Dashed lines indicate a hypothetical direct relationship between regional and local diversity; for Virginia streams, solid lines indicate actual relationships between regional and local diversity. The closed circle and vertical line indicate the mean and 95% confidence interval of local species richness for Wisconsin lakes. Used from Ross and Matthews (in press) with permission from Johns Hopkins University Press.

with species and thus resistant to the addition of new elements, or whether local faunas will increase in species richness as a function of regional species richness (Ricklefs 1987; Cornell and Lawton 1992). In other words, the question is to what degree local fish assemblages are determined by processes acting within the local area versus processes operating at a regional level. Asymptotic relationships between local and regional species diversity (i.e., saturation) suggest the primacy of local control over assemblages; in contrast, if the relationship remains linear (i.e., unsaturated), then local processes would be subordinate to the effect of the regional species pool (Ricklefs 1987). Although it seems clear to contrast local versus regional effects, local assemblages obviously contribute to the

regional species pool thus creating a "chicken and egg" problem (Cornell and Lawton 1992).

A study on stream fish assemblages in Virginia showed that regional diversity generally explained more of the variation in local native fish assemblages than did local variables, but local variables also showed some, albeit reduced, explanatory power, with the most important being habitat complexity (Angermeier and Winston 1998). Graphs of local versus regional species richness, although not reaching asymptotes, had low slopes when comparisons were on large spatial scales, such as the drainage level, and all intercepts with the y-axis were significantly greater than zero, both suggesting a tendency for an asymptotic relationship (Figure 4.9). However, when analyzed at a local scale such as site, diversity was strongly related to regional diversity, indicating that the local sites were not saturated with species. The number of introduced species in a local area was also positively related to the regional number of introduced species and, in contrast to native species, showed no evidence of saturation. In addition, the number of native fish species did not influence the number of nonnative species, suggesting that high native-fish diversity does not preclude invasion by nonnative fishes.

The strong influence of regional compared to local factors has also been shown for lakes. Jackson and Harvey (1989) demonstrated that fish faunas of watersheds within the Laurentian Great Lakes showed the effect of large-scale regional processes reflective of postglacial dispersal or climate but were much less related to measures of environmental similarity (e.g., lake depth, area, and pH), although such factors likely have some role in affecting species composition. In a study comparing small lakes in Wisconsin and Finland, Tonn et al. (1990) showed that species richness in individual lakes was related to regional species richness but that local richness reached an asymptote, suggesting that individual lake faunas became saturated with species (Figure 4.9). However, regional factors alone could not explain local species composition because biotic factors, particularly the presence of large predators, also influenced species composition. Tonn and Magnuson (1982) also showed the effects of predator composition, as well as lake morphometry and winter oxygen levels, on the structure of fish assemblages in small Wisconsin lakes.

These studies all suggest a general, but highly variable, link between regional and local species richness. In contrast, in a study of fishes of the Interior Highland region (Ozark and Ouachita mountains of Arkansas, Oklahoma, Missouri, and Kansas) Matthews and Robison (1998) found that regional (river basin) species richness accounted only marginally for species richness at local sites if all species were considered, and that the regional-local species

richness relationship was nonexistent within the minnow or darter families. Overall, within all levels of basin (regional) richness, there was great variation in numbers of species in local assemblages. They attributed the lack of strong regional effects to local physical factors at sites within basins, or to within-basin zoogeographic chance in movement or distributions of species. In an analysis of midwestern stream fishes at 65 sites in 13 drainages from Nebraska and Iowa south to Texas, local factors had more effect on species richness than did the overall size of the regional species pool (Marsh-Matthews and Matthews 2000). However, in contrast to emergent assemblage properties (i.e., species richness), primary assemblage structure (i.e., the occurrence of particular species) was strongly influenced by broad geographic factors, primarily latitude, reflective of the fact that many species have restricted north-south distributions (Conner and Suttkus 1986; Cross et al. 1986).

Regional and historic filters, as emphasized by Tonn et al. (1990), clearly can have a major influence on local assemblages and in some cases, especially southeastern streams and northern lakes, the richness of local fish assemblages is strongly affected by regional diversity. Species composition also can be influenced by large-scale factors such as latitude or divisions between major river basins. However, not all assemblages show a relationship between regional and local diversity, as evidenced by harsh midwestern streams and speciose upland streams. In addition, the scale of the study influences the outcome—namely, how small is the local area and how broad is the regional area.

SUMMARY

Fish populations and assemblages are shaped broadly by landscape features and by the mosaic pattern of patches within a landscape. The scale on which fishes perceive patches varies among taxa and life-history stage. Many fish populations likely comprise linear or dendritic metapopulations, although rigorous tests of metapopulation structure of North American freshwater fishes are extremely uncommon and one of the most rigorous studies supports an island-mainland metapopulation model.

Statistical models relating fish species and assemblages to environmental factors are increasingly widespread and generally fall into two groups. *A priori* models attempt to predict assemblage characteristics and, less often, species occurrence from general environmental features. Three common *a priori* models are the habitat template, the river continuum, and the landscape filter, and all have some predictive successes. *A posteriori* models use multivariate statistical techniques to find suites of environmental variables (both physical and biotic) that best predict the occurrence or abundance of species, assemblages, or functional groups. Such models rely on the approximately simultaneous collection of fishes and potentially predictive variables. However, a new class of multivariate models, "niche models," uses museum collections of fishes in concert with independently collected digitized environmental data sets suitable for GIS programs. Because the various models have different strengths and weaknesses, the use of several models is often appropriate, with the choice based on the research questions and the nature of the data.

Fish assemblages are a product of local conditions and regional factors, including the regional fish fauna. Studies of both lentic and lotic fish assemblages suggest that the relative influence of local versus regional factors varies, although there is support for the influence of regional factors in both, as well as the importance of local factors, including the presence or absence of predators and habitat complexity. The importance of local versus regional effects seems to be heightened in harsher systems. However, as with population models, the number of studies of local versus regional effects on fish assemblages is quite limited, especially in terms of representation of different geographical regions.

SUPPLEMENTAL READING

Gotelli, N. J., and A. M. Ellison. 2004. *A primer of ecological statistics.* Sinauer Associates. Sunderland, Massachusetts. A useful source for understanding ecological statistics.

Hanski, I., and M. E. Gilpin. 1991. Metapopulation dynamics: Brief history and conceptual domain. *Biological Journal of the Linnean Society* 42:3–16. Provides a background of the metapopulation concept.

Leopold, A. 1949. *A sand county almanac.* Oxford University Press, New York, New York. A "must read" for students interested in conservation.

Turner, M. G. 2005. Landscape ecology: What is the state of the science. *Annual Reviews of Ecology and Systematics* 36:319–44. A good source for current views of landscape ecology.

Wiley, E. O., K. M. McNyset, A. T. Peterson, C. R. Robins, and A. M. Stewart. 2003. Niche modeling and geographic range predictions in the marine environment using a machine-learning algorithm. *Oceanography* 16:120–27. Provides a good background and examples of niche modeling.

FIVE

The Formation and Maintenance of Populations and Assemblages

CONTENTS

ECOLOGISTS STUDYING FISH populations and assemblages most often are dealing with their subjects already formed, at least in the sense of ecological time. In Part 1, the examples of the colonization of new habitats that were opened up following glacial retreat showed how species might be added to such systems. Obviously, opportunities to study natural assemblages in the early stages of their formation are rare. However, opportunities do occur when assemblages repopulate after being eliminated due to loss of habitat during droughts or because of intense flooding that can cause local or regional extirpation of fishes.

This chapter deals with issues of how species are added to assemblages, both from the standpoint of how preexisting species might alter the outcome and from the standpoint of how adept fishes are in moving about their environment (see figure in Part 2). In thinking about how fish assemblages might be formed or reconstituted, a key question is how species are added to an assemblage, and whether there are "rules," other than having a suitable habitat, that govern the order and kind of species additions. Are fish assemblages built solely on the basis of random colonization opportunities by the component species or does the presence of a certain species, or groups of species, change the probability of successful establishment by yet other species? The resolution of the question could be approached at a larger scale to include both historical (i.e., lineage) and dispersal components, although much of the ecological literature has focused on a finer scale that emphasizes dispersal (but see Gorman 1992; Winston 1995). As natural communities are increasingly besieged by nonnative taxa, the answer to the questions of how species are added to assemblages has also taken on new importance in the efforts to develop models predicting the vulnerability of native communities to invasion by nonnative taxa (see also Chapter 15).

Understanding the formation and maintenance of assemblages also requires an understanding of the vagility of fishes. Without some degree of movement, fishes would be unable to colonize new habitats or to recolonize habitats from which they have been eliminated because of biotic or physical impacts. Movement ability varies widely among fish taxa and within a given taxon by life-history stage, physiological state, and simply by individual differences.

COLONIZATION OF ASSEMBLAGES

The Search for Assembly Rules

The question of whether or not there are ecological "rules" governing how species are added to communities has intrigued ecologists for decades. Indeed, long before that, Plato provided the first attempt in the fourth century BC when he considered how human communities were assembled (Belyea and Lancaster 1999). A seminal paper by Diamond (1975) on the addition of bird species to islands initiated a debate that continues in some measure to the present. Diamond hypothesized that species composing a community are "selected and coadjusted in their niches and abundances, so as to fit with each other and to resist invaders." A more formal definition of assembly rules was proposed by Belyea and Lancaster (1999) in an effort to make the distinction between underlying

mechanisms, rules related to the mechanisms, and the resultant patterns resulting from the rules. In their definition, assembly rules are "general and mechanistic, and operate within the case-specific constraints imposed by colonization sequence and environment."

For several decades now since Diamond's (1975) publication, debate has centered on issues of whether communities are structured as opposed to random. Such debate is, in part, because Diamond did not first clearly demonstrate that the bird communities he studied were actually structured rather than random collections of species before he proposed specific rules for assembly (Weiher and Keddy 1995). Later studies focused on distinguishing between structured versus random assemblages by testing observed patterns of species occurrences against null models of species assembly, resulting in the demonstration that at least some of the patterns that ecologists used to invoke assembly rules could be attributed to chance (Connor and Simberloff 1979). The use of null models increased the rigor of studies on community structure and assembly rules, although philosophical differences in constructing null models, such as the appropriate species pool for building the model, have continued to foster debate (see Gotelli and Graves 1996; Box 5.1).

A challenging issue is one of choosing a preinteraction species pool—one that has not already been subjected to long periods of interaction (such as competition) that would have caused the removal or stifling of speciation of certain species or lineages (Colwell and Winkler 1984). As a consequence of sampling a postinteraction species pool, the strength of the effect being measured is consistently underestimated. The issue of sampling a postinteraction species pool is termed the "Narcissus Effect" (Colwell and Winkler 1984), named after the Greek god Narcissus who, according to one version of the story, when gazing into a pool saw only his own image and could not discern the depth of the pool.

Clearly there are challenges to providing rigorous tests of community assembly rules. One

BOX 5.1 • Null Models in Community Ecology

The experimental tools that are conventionally available in community ecology are field, laboratory, and so-called natural experiments (Gotelli and Graves 1996). Well-designed experiments require both tests and controls so that the effect of a potential factor can be evaluated. However, ecological communities, or even a subset such as fish assemblages, are notoriously complex, making large-scale manipulations or laboratory experiments impractical from the standpoint of logistics, time, or money. For instance, assessing the impact of Species X on the colonization probability of Species Y could require eliminating Species X from randomly chosen sites, introducing Species Y, and then comparing the colonization success of Species Y between sites with and without Species X.

Coupled with the ethical or legal issues of manipulating systems that might already be at risk, the potential of being able to "manipulate" communities statistically is certainly an attractive alternative (Connor and Simberloff 1986). In the case of Species X and Y, a null model might be constructed such that the probability of the two species co-occurring in the absence of any interaction could be calculated and then compared with the actual co-occurrence. Simply put, a null model "is an attempt to generate the distribution of values for the variable of interest in the absence of a putative causal process" (Connor and Simberloff 1986). However, the challenge is to design an appropriate null model for the situation being tested! The literature on null models, relative to testing forces involved in shaping communities, is a litany of proposed methods followed by criticisms thereof. (For a few examples see Connor and Simberloff 1986; Brown et al. 2000, 2002; Stone et al. 2000; and Fox 2001.)

The use of null models in addressing the question of community assembly has focused on the interactions of species with potential sites and the interactions among species (Gotelli and Graves 1996). As outlined by Gotelli and Graves (1996), all models that have been used to test the hypotheses generated by assembly rules have identifiable shortcomings. There are key questions to consider: (1) what species to include in the species pool from which potential assemblages are randomly, or otherwise, extracted; (2) rather than species, should some other category, such as functional groups (i.e., a collection of species all of which have a similar ecological function) be used; (3) what metric is most appropriate to determine whether or not assemblage organization follows a random pattern; and (4) what is the most appropriate way to design a null model (being that the possibilities are large)? Obviously, null models, although potentially very useful, are not without their own set of issues. The use of null models in ecological applications is treated in detail by Gotelli and Graves (1996).

approach that has been taken is to consider predictions that would be met by assemblages if, indeed, assembly rules were important. These have primarily focused on the random versus structured dichotomy. This approach was taken by Gotelli and McCabe (2002), who conducted a meta-analysis of 96 published studies of species presence-absence matrices, comprising a wide range of taxa and geographic locations, to determine the generality of Diamond's (1975) model of assembly rules. Their results showed that most plant and animal assemblages do follow predictions of Diamond's assembly rules in that there are fewer species combinations than expected by chance (i.e., communities are nonrandom), a result also shown by community assembly studies using laboratory microecosystems (Drake et al. 1993). In an effort to move studies of community assembly forward, Weiher and Keddy (1995) argued that the issue of whether there are nonrandom communities has long been resolved in the affirmative and that it is time to move on to newer questions. Although fish assemblages treated by Gotelli and McCabe (2002) showed less evidence for structuring by competition than did other groups of organisms, the low sample size (n = 3) precluded any meaningful generalities.

Application to Fish Assemblages

There have been relatively few studies that specifically addressed the reality of assembly rules governing fish assemblages or included fish assemblages in more general analyses. The studies of fish assemblages that have evaluated the predictions derived from community assembly rules have had mixed results in terms of finding support.

Matthews (1982) provided an early attempt to determine if assembly rules (sensu, Diamond 1975) could be applied to the species-rich Ozark stream fish assemblages of eastern North America. The study focused on six small streams occupied by a suite of 13 largely insectivorous species in the minnow, silverside, and topminnow families. All of the species had overlapping ranges even though each species did not occur in all of the streams. One of the hypotheses tested was that pairs of the common/abundant species were associated randomly among the six small streams. The alternative hypothesis, that species occurred only in certain combinations, would support the idea of assembly rules. The outcome of this statistical experiment showed that the occurrence of species pairs based on the random assortment of species into watersheds (keeping both the number of watersheds in which a species was common/abundant and the number of species per watershed equal to observed values) was not different from the observed values. From this, Matthews (1982) concluded that no other process, other than random assortment, was required to explain the pattern of mutual abundance of the small insectivorous fishes in watersheds.

Although this simple statistical experiment is appealing, a follow-up paper (Biehl and Matthews 1984) used more sophisticated statistical approaches to reexamine the results of Matthews (1982). Relative to the previous conclusion, they found that both the way the simulations were computed as well as the way the observed and expected values were compared (chi-square test) may have affected the outcome of the study (i.e., reducing the confidence in the conclusions). Subsequent studies have built on Matthews's work; not surprisingly, the following two examples are from biologists who did

their doctoral studies under the direction of W. J. Matthews.

Winston (1995) tested predictions coming from Diamond's assembly rules with assemblages of phoxinin cyprinids (i.e., all minnow genera present in the study area except the leuciscin, *Notemigonus*) inhabiting streams of the Red River drainage in Texas, Oklahoma, and Arkansas. He found that species pairs that were most similar morphologically, and thus most likely to experience greater competition, co-occurred significantly less often than expected by chance. A potential problem with this outcome is that the absence of species' co-occurrences could be due to historical as well as ecological reasons. For instance, closely related species could have arisen through allopatric speciation and thus have different spatial distributions by virtue of being adapted to slightly different environments (Winston 1995).

In a second test, Winston (1995) was able to rule out the historical explanation by showing that species pairs that were evolutionarily most closely related (based on a cladistic analysis) did not show co-occurrence patterns that differed from random patterns. Consequently, he argued that interspecific competition was the "most likely mechanism causing the pattern." These results support a prediction of Diamond's assembly rules—namely that certain species combinations do not occur in nature. However, the actual mechanism(s) responsible for the nonrandom pattern of species occurrence in this study are unknown.

In a study of benthic fishes living in small headwater streams of northeastern Oklahoma, Taylor (1996) also showed that species occurrence patterns can be affected by other resident species. Taylor compared the actual species area relationship with one predicted by a null model and showed that actual species richness relative to the area of habitat declined much more steeply than expected for smaller habitats. This suggested that some factor in addition to area was influencing the number of species. By following this regional analysis with manipulative field experiments, Taylor (1996) was able to show that asymmetrical interference competition between Banded Sculpin (*Cottus carolinae*) and Orangethroat Darter (*Etheostoma spectabile*) was responsible in part for the difference between observed and expected species densities (see also Chapter 11).

Thus there is evidence that freshwater fish assemblages can be structured, as well as evidence that competition between species pairs is one of the mechanisms causing such structuring (see also Chapter 11). However, in general, studies addressing factors governing the formation of fish assemblages have been surprisingly rare and have been restricted to only one general region, the Great Plains of North America.

A Return to Historical Effects

The concept of community assembly rules, as it has developed, focuses primarily on current ecological interactions. As I have already emphasized in earlier chapters (see Chapter 2; Figure 2.1), contemporary species interactions must be viewed in the context of history. For instance, Gorman (1992) analyzed assemblage formation in several Ozark upland streams and argued that general patterns of habitat use and ecological interactions were "ancient and not the result of an ongoing process." Consequently, the occurrence of fish species on a macrogeographic level (i.e., drainage) was best predicted by historical, biogeographic factors. In contrast, ecological processes best predicted abundance patterns in local assemblages. Thus community assembly was largely determined by historical factors, whereas contemporary ecological processes affected patterns of species' abundances but not occurrences. The contrast with the studies by Winston (1995) and Taylor (1996) is perhaps one of scale—at what spatial level do contemporary interactions control not only abundances but also presence or absence of species?

Species Characteristics and Assemblage Formation

Are species characteristics such as body size, geographic range, trophic position, and population size related to their ability to colonize habitats?

In a long-term study of fishes in the Cimarron River in Oklahoma, Gotelli and Taylor (1999b) found that colonization potential of native fish species at 10 study sites was, not surprisingly, correlated with species' abundance, but that geographic range size and body size were not. The species pool considered by Gotelli and Taylor (1999b) was based on the known native fish fauna of the Cimarron River from the Oklahoma study sites, so the concern of adaptation to the regional environment was not an issue.

Other studies have examined the colonization potential of species that are not native to the general region of the study systems. Information on these situations can provide general insight into factors important in the addition of species to assemblages (see also Chapter 15).

THE MATCH BETWEEN SPECIES TOLERANCES AND THE PHYSICAL HABITAT Moyle and Light (1996a, b) developed a set of empirically based predictions regarding colonization success in fishes. Although the predictions were largely developed to understand invasions of nonnative species, most are also applicable to colonization potential of native species with the caveat that they were developed for western (California) fish assemblages. A principal conclusion of Moyle and Light's work in California streams and estuaries was that "the most important factor determining the success of an invading fish species is the match between the invader and the hydrologic regime."

Colonization in Some Systems but Not Others

In general, most studies support the view that all aquatic systems are potentially invasible, although systems that have been highly altered tend to be more prone to invasions (Moyle and Light 1996a). Numerous authors have suggested that species-rich assemblages, such as those found in the southeastern United States, are more resistant to invasion by nonnative taxa when compared with species-poor systems, although there are exceptions to this (e.g., Ross 1991; Lodge 1993). However, given the condition that where there are species-rich systems, and that these systems have gained species richness both by *in situ* speciation and colonization, it stands to reason that species with appropriate physiological tolerances and ecological characteristics can enter and become established.

Colonization by Some Taxa and Not Others: The Role of Trophic Position

Resource use, and especially trophic position, are thought to be important in predicting the invasion success of nonnative fishes. In California streams, successful invaders of generally unmodified aquatic systems tend to be at one extreme or the other of the trophic pyramid—either top predators or detritivores/omnivores. In both cases, these are trophic levels that tend not to be food limited (Moyle and Light 1996b). A more recent study incorporating stable isotope analysis (see Chapter 11 for details on stable isotope analysis) tested the Moyle and Light hypothesis, focusing on the base of the food web (Gido and Franssen 2007). In fact, the trophic position of nonnative fishes tended to be lower than native fishes in four of the five study rivers, although the differences were small and not significant at the 0.05 level. Because their study included five streams located from New Mexico to Kansas, they were also able to compare the trophic position of four species in streams where they were native and in those where they were nonnative. In this case, three of the four species showed significantly lower trophic positions in assemblages where they were nonnative relative to where they were native. Thus this study is at least consistent with the hypothesis that low trophic level facilitates colonization.

The San Juan River, a major tributary of the lower Colorado River, has been changed by impoundments and flow modifications and has a fish fauna that includes both native and nonnative species (Gido and Propst 1999). The food web of small-bodied fishes from secondary channels and from the margin of the main channel includes three trophic levels, as determined from stable isotope analysis ($\delta^{15}N$ ratios): a primarily detrital base, a primary consumer level dominated by midges (Chironomidae), and a secondary consumer level that includes all fishes (Gido et al. 2006). Fishes from the secondary channels showed higher food overlap than those in the main channel, indicating convergence on the same food source. Also, fishes found in swifter water in the main channel, principally Speckled Dace (*Rhinichthys osculus*) and Channel Catfish (*Ictalurus punctatus*), had slightly elevated trophic levels. Although all fishes were grouped as secondary consumers, the more common native fishes tended to feed at a higher trophic level (mean native = 3.18) than most of the nonnative fishes (mean nonnative = 3.03), a finding at least consistent with part of the Moyle and Light hypothesis (Figure 5.1).

The Moyle and Light hypothesis, predicting the invasion of nonnative fishes at the highest or lowest trophic levels, has also been tested for fishes in the Gila River system in southwest New Mexico (Pilger et al. 2010). This portion of the Gila River has received relatively low human impacts but has been subject to widespread introduction of nonnative fishes, which potentially put continued persistence of native fishes at risk. The study was based on both stable isotope ratios and gut analyses and showed that native and nonnative fishes fed at multiple trophic levels. Large-bodied native fishes, Adult Sonora Sucker (*Catostomus insignis*) and Desert Sucker (*Catostomus clarkii*), fed at low trophic levels as adults, but at a higher trophic level as juveniles. In contrast, large-bodied nonnative fishes, such as Rainbow (*Oncorhynchus mykiss*) and Brown (*Salmo trutta*) trout, Smallmouth Bass (*Micropterus dolomieu*), and Yellow Bullhead (*Ameiurus natalis*) fed at lower levels as juveniles and then at higher trophic levels as adults. Native small-bodied fishes such as Longfin Dace (*Agosia chrysogaster*), Spikedace (*Meda fulgida*), Speckled Dace, and Loach Minnow (*Tiaroga cobitis*), were primarily insectivorous; the Headwater Chub (*Gila nigra*) was the only native piscivorous fish but also consumed large amounts of algae. Overall, nonnative fishes foraged at higher trophic levels on fishes and large, predaceous aquatic invertebrates, compared to native fishes. The introduction of nonnative

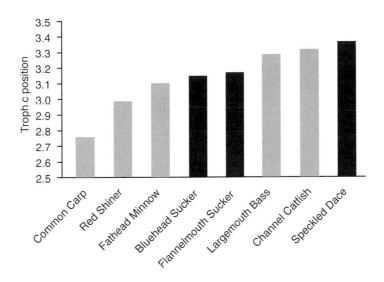

FIGURE 5.1. Trophic position, as determined from stable isotope analysis (δ^{15}N ratios), for native (black bars) and nonnative (gray bars) fishes in the San Juan River. Based on Gido et al. (2006).

fishes has extended the maximum food chain length in the upper Gila River fish assemblages to higher trophic levels, although long-term survival of at least some of the native fishes is in question as a consequence (Pilger et al. 2010).

Colonization Models

Strange and Foin (2001) developed a model to predict the invasion success of six fish species (including native and nonindigenous) to coolwater (trout zone) streams, the Sagehen and Martis creeks, located on the eastern slope of the Sierra Nevada mountains near Lake Tahoe. The model incorporated life-history responses to biotic factors of competition and predation and the physical factor of stream discharge. Predictions of the model were tested against a long-term data set of relative abundance of fishes for Martis Creek from 1979 to 1994 and Sagehen Creek from 1952 to 1961. Model predictions generally captured major trends in species' relative abundances for Martis Creek and showed excellent fit overall to data for Sagehen Creek, although some species showed better fit than others (Figure 5.2). Key points from this study of a six-species system are that invasion sequence and biotic interactions (competition and predation) were all important, but that the relative importance of these factors

varied based on physical conditions. In particular, the sequence of floods and droughts had a major impact on the relative abundance and potential for invasion of each species. One of the lessons from studies such as this, which incorporate the impact of physical habitat variation, is that the formation of fish assemblages is largely nondeterministic—that is, the final assemblage cannot be totally predicted by knowing the species pool available for colonization. As stated by Strange and Foin (2001), "assembly from the same species pool is likely to result in multiple or alternate states as the physical regime alters rates of biotic processes."

Another set of studies used the extensive database on successful and unsuccessful introductions of fishes in California to develop models predicting characteristics of successful colonizers (Marchetti et al. 2004a, b). In terms of becoming established, the best predictive model included physical characteristics of the invading species (body size and physiological tolerance), demographic characteristics (area of population), niche characteristics (trophic position), and behavioral characteristics (parental care of young). Of these, greater physiological tolerance and the degree of parental care seemed particularly important in aiding establishment.

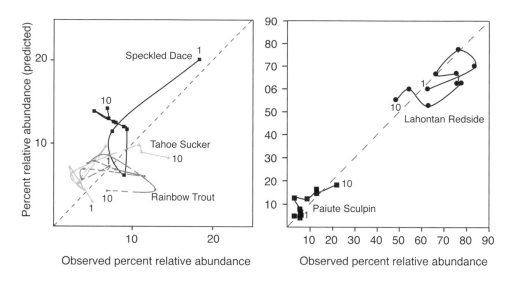

FIGURE 5.2. Relative abundances of five of the six species from Sagehen Creek, California, based on model predictions and observed values over a 10-year period. Numerals indicate starting and ending years; the dashed line shows a 1:1 relationship. Based on data from Strange and Foin (2001).

MOVEMENT AND ASSEMBLAGE FORMATION

Freshwater fishes, with few exceptions, are naturally limited in their access to new habitats by suitable aquatic connectivity. Thus understanding how fish assemblages are formed also requires understanding the movement capability of potential colonizers found in regional species pools. To a certain extent, this tempers what one can learn about community assembly by examining nonnative taxa, given that such taxa often move more through human intervention than by natural processes (see also Chapter 15).

The terms "dispersal" and "migration" are often used to describe the movement of organisms outside of their home range. Briefly, dispersal has most often been used to refer to a general outward spreading of individuals away from each other, such as from a starting group. Consequently, it is a group definition because it is based on the movement of one individual relative to the movement of one or more other individuals (Begon et al. 1996). Dingle (1996) recommended the term "ranging" in place of dispersal. Ranging refers to individual behavior and is "the departure from the current habitat patch (emigration), the seeking of a new patch,

and the occupation of the first available and suitable habitat patch discovered (immigration)." In older works, the definition of migration was constrained to include movement to and from an area or areas. For instance, Harden Jones (1968) defined migration as "a class of movement which impels migrants to return to the region from which they have migrated." If an organism did not return, however far or periodic its travels, then the term migration was deemed inappropriate. A recent trend is to recognize migration as a persistent movement of individuals of a species from one habitat to another so that the location of each life-history stage is optimized in terms of needed resources (Begon et al. 1996; Dingle 1996). Migration usually involves distinct pre- and postmigratory behaviors and includes physiological changes, such as the reallocation of energy to support long-distance movement (Dingle 1996).

In the previous section, the match between a habitat and a potential colonizer was an important factor in understanding the establishment of nonindigenous species. However, the match between an organism and a potential habitat means nothing if the organism lacks access to the habitat. For instance, a study of the ranging ability of various aquatic taxa, including taxa from bacteria to fishes, hypothesized that the patterns of occurrence in lakes for species with weaker movement abilities would be better predicted by regional patterns of connectivity, whereas the occurrence of taxa with greater movement abilities would be better predicted by local conditions (Beisner et al. 2006). Not surprisingly, the patterns of occurrence of freshwater fishes were best predicted by the spatial distribution of lakes and patterns of aquatic connectivity. In contrast, for taxa with high movement ability, such as bacteria, the local environment was a better predictor of occurrence than measures related to aquatic connectivity.

Movement of some sort is integral to a wide range of fish behavior, from finding food or mates, to avoiding or reducing predation risk, to defending feeding or breeding territories, to long distance, often annular, movements frequently associated with breeding. In the context of this chapter, the categories of interest include those that result in an organism moving outside of its home range such that new habitats could be colonized or former habitats recolonized. Other aspects of movement are treated in subsequent chapters (especially Chapters 9 and 14). Evidence for the movement abilities of fishes comes from a variety of sources including observational studies on the colonization of new habitats and from studies using some sort of tagging procedure.

Movement Inferred from Colonization Studies

Colonization of newly available habitats by fishes has been studied on a variety of temporal and spatial scales, including those studies examining movement of fishes into newly available habitats following the retreat of the Wisconsinan ice sheets, the recolonization of habitats after amelioration of harsh conditions, such as complete dewatering of a stream reach, or the response of fishes, usually over a short time period, to newly created stream habitats.

During the Wisconsinan glacial advance, the Laurentide ice sheet covered eastern Canada including Ontario (Chapter 3; Figure 3.4) so that there were no ice-free habitats available for freshwater fishes (Mandrak and Crossman 1992). As the ice began retreating, large meltwater lakes and connecting rivers formed along the southern boundary of the ice sheet. Lake Algonquin was one such meltwater lake that covered parts of modern-day Lake Huron, Lake Superior, and Lake Michigan approximately 12,000 years ago (Hinch et al. 1991; Mandrak and Crossman 1992) (Figure 5.3). The area inundated by Lake Algonquin in Ontario also included the future basins of a number of smaller lakes and, as Lake Algonquin subsided, its fish fauna would have had access to these smaller emerging lakes (Hinch et al. 1991). In addition, fishes from Lake Algonquin also colonized lakes at higher elevations that were not directly inundated by the glacial meltwater—in this case by moving up the lakes' outlet streams that flowed into the discharge of Lake Algonquin (Figure 5.3).

Prairie streams of the central United States, on the edge of the Great Plains and Osage Plains, are often subjected to periods of intense drought, such that streams may be reduced to a few isolated pools or totally dewatered (Matthews 1988). In spite of such extreme conditions as total drying of long sections of streambed, the fish faunas of these streams tend to be surprisingly resilient and fairly quickly recolonize reaches after stream flow resumes (Matthews 1987, 1988). The fish fauna of Brier Creek, an Oklahoma tributary to the Red River arm of Lake Texoma, has been particularly well studied since the late 1960s and provides an example of the ability of fishes to recolonize newly watered habitats (C. L. Smith and Powell 1971; Ross et al. 1985; Matthews 1987; Matthews et al. 1988).

In 1980, the region of southern Oklahoma was hit by a severe drought and high temperatures, resulting in the dewatering of large stream sections and even heat death of one of the common species (Orangethroat Darter) (Matthews et al. 1982; Ross et al. 1985). In spite of such extreme conditions, by the following

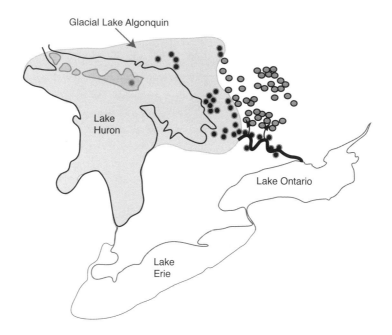

FIGURE 5.3. Dispersal from glacial meltwater lakes: Glacial Lake Algonquin in Ontario, Canada, and modern lakes in Canada and the United States. The black circles indicate lakes within the borders of Glacial Lake Algonquin or along the glacial outlet and colonized as Lake Algonquin subsided; the gray circles indicate higher elevation lakes colonized by fishes moving from the outlet of Lake Algonquin up through streams draining the lakes, as indicated by the arrows. Based on Hinch et al. (1991).

year the fish assemblage of Brier Creek was essentially unchanged from previous years (Ross et al. 1985), indicating that fishes had recolonized habitats once stream flow resumed. Drought returned in the summer of 1982, and a 1.5 km section of the stream in the headwaters (station 2, Figure 5.4) was totally dry by autumn and remained so until heavy rains in March of 1983 (Matthews 1987). Once flow resumed, monthly collections showed that the fish fauna was rapidly rebuilt by species colonizing from refugia outside of the dewatered reach, followed by spawning of some of the colonizers; the fish fauna of this area of Brier Creek typically includes only 9–10 species, so colonization essentially reestablished the entire fauna in approximately four months (Ross et al. 1985; Matthews et al. 1988). The favorable stream conditions were, however, short lived. Flow ceased in July, and by September the section was reduced to one pool with no surviving fish (Figure 5.4) (Matthews 1987).

A long-term study to evaluate the impact on trout populations of new habitats created by the placement of log weirs in high-elevation streams in the Rocky Mountains included 500-m reaches of six streams and involved four species and three genera of salmonids (Brook Trout, *Salvelinus fontinalis*; Cutthroat Trout, *Oncorhynchus clarkii*; Brown Trout; and Rainbow Trout) (Gowan and Fausch 1996). Newly available habitats created by the weirs were primarily colonized by trout that were coming from outside of the 500-m study reaches; however, once the trout were in the new habitats they tended to remain there. Marked fish within the study reaches tended to be sedentary, but the presence of unmarked fish in the new habitats is a strong indication of the ability for long distance movement.

Movement Inferred from Tagging Studies

Clearly, the previous examples indicate that fishes are capable of moving through connecting

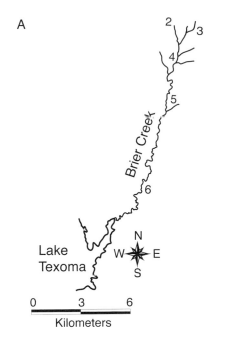

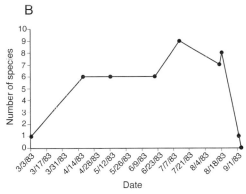

FIGURE 5.4. A. Brier Creek, a tributary to the Red River arm of Lake Texoma in Oklahoma. Numbers show sampling stations. Map based on C. L. Smith and Powell (1971) and Ross et al. (1985). B. Recolonization of a dewatered section of Brier Creek, Oklahoma, following resumption of stream flow. Drying occurred in a 1.5 km reach at station two. Severe drought dewatered the stream again by September. Data from Matthews (1987).

waterways, but for a long time there have been questions of how far fishes routinely or periodically move and what proportion of a population is involved in movement. In 1959, Shelby Gerking summarized what was then known about movement in both marine and freshwater fishes. He identified 34 fish species (24 freshwater taxa; 10 marine taxa) characterized by restricted movement that was not associated with spawning. Following the classic work by Gerking, a view referred to as the "restricted movement paradigm" developed among fish ecologists: namely, that freshwater fishes for the most part had local populations and that extensive movement was uncommon. However, Gerking was careful to qualify his conclusions. First, he emphasized the importance of experimental design on the outcome of movement studies. Even if fishes tend to remain primarily in a home range, if the resampling area is small relative to the home range of the fish (keeping in mind that the home range is not known), then there is a bias toward finding few fish and concluding that long-range movement is common. Conversely, if the resampling area is large relative to the size of the unknown home range, then there is a bias toward finding many fish and concluding that fishes have restricted movement. Gerking (1959) was also careful to point out that restricted movement behavior was not universal among any given population and said, "That stray fish occur has never been doubted, and their importance in repopulation of decimated areas and the distribution of the species is not questioned." In fact, Funk (1955), as cited by Gerking, proposed that lotic freshwater fishes had populations composed of a mobile group and a sedentary group (often referred to as "movers" and "stayers"). Funk (1955) also pointed out the problem of bias toward finding limited movement for fishes when sampling was concentrated in "limited areas near release sites."

Movement of freshwater fishes is generally studied by marking individuals in some way and then attempting to recapture them later (termed mark-recapture). As pointed out by Funk, sampling is almost always greater near the point of release compared to sampling at great distances from the release, especially if there are many potential routes of fish movement (as in lakes or in streams with numerous tributaries). As a consequence, sampling effort is unequal over distance from the point of release, so that short movements tend to be recorded more often than long movements. The change in sampling intensity with increasing distance from the point of release is referred to as "distance weighting" (Porter and Dooley 1993; Albanese et al. 2003). Thus unless the study compares recoveries against the probability of recapture (i.e., capture probability decreases as fishes disperse outward from the point of release) or, preferably, is designed to alleviate the issue of distance weighting, there would be strong bias for interpreting recovery data as supporting limited fish movement (Box 5.2).

In spite of the cautionary words by Gerking (1959) about the problem of experimental design and the presence of fish straying, and the suggestion by Funk (1955) about sedentary and mobile groups, the restricted movement paradigm became entrenched in the literature. One of the first to take issue with the restricted movement paradigm, specifically in regard to salmonids, was Gowan et al. (1994), who pointed out that although most tagging studies captured the majority of fish near the point of release, in 78% of the salmonid studies that they reviewed, over half of the fishes were never seen again after being marked. Whether these fish represented mortalities or fish that simply moved much greater distances is the crux to understanding the level of movement of fishes. Gowan et al. (1994) also suggested that the mobile fraction of fish populations had been downplayed through the use of such deprecating terms as "strays" and argued that more attention needed to be given to the experimental design of fish movement studies and to the underlying mechanisms involved in fish movement. The greater realization of the often high degrees of movement shown by freshwater fishes has had important consequences for the better understanding and management of fish populations (Fausch et al. 2002).

A study of fish movement in a small Ouachita Highlands stream (Arkansas) involved four species of stream fishes (Creek Chub, *Semotilus atromaculatus*; Blackspotted

Topminnow, *Fundulus olivaceus*; Green Sunfish, *Lepomis cyanellus*; and Longear Sunfish, *L. megalotis*) (Smithson and Johnston 1999). The study area was 500 m long and consisted of 10 pools, and the possibility of movement of fishes outside of the study area was determined to be unlikely. Most fishes were recaptured in the same pool where they were initially marked; however, there were differences among the species. Compared to the other three species, Blackspotted Topminnow moved significantly greater distances. For all species, there were some individuals that moved greater distances than others, although there were no apparent morphological correlates associated with greater movement and, in fact, the same individuals switched between static and mobile behaviors over the course of the study. This suggests that individual fish periodically engage in exploratory travel, perhaps assessing habitat quality in areas outside of their home pool.

Field and theoretical approaches were used by Skalski and Gilliam (2000) to examine characteristics of movement of primarily four species (Bluehead Chub, *Nocomis leptocephalus*; Creek Chub; Redbreast Sunfish, *Lepomis auritus*; and Rosyside Dace, *Clinostomus funduloides*) in a small southeastern stream. In contrast to Smithson and Johnston (1999), species showed only weak differences in their degree of movement. However, within a species there was evidence for both "movers" and "stayers." In addition, the distribution of distances moved tended to be leptokurtic, with more short and long movements and fewer intermediate movements compared to a normal distribution. Relationships between the propensity to move and morphological characteristics were complex and varied among species. Within Bluehead Chub, the probability of movement increased with size for individuals that had slow growth but decreased with size for those having fast growth. In Creek Chubs the probability of movement increased with body size but was not related to growth rate. There was no relationship between body size and the probability

BOX 5.2 • Distance Weighting in Studies of Fish Movement

An appropriate experimental design is critical for assessing fish movement using mark-recapture approaches. This is especially so because the likelihood of capturing a marked fish declines with distance from the point of release, leading to the risk of underestimating longer movements. A robust experimental design is an important issue even with essentially linear stream systems—additional complexity of the aquatic system (e.g., tributary streams, lakes with numerous coves, etc.) further increases the challenge of obtaining reliable data. Following Rodríguez (2002), in quantitative terms, the density of marked fish multiplied by meters away from the region of their release, n(x), is given by

$$n(x) = N_o s \, \pi f(x), \quad \text{(Equation 5.1)}$$

where N_o is the number of fish originally marked, s is the probability of their surviving to the sampling period, π is the catchability of the fish, and f(x) describes the decline in density as a function of the distance from the release area (referred to as a dispersal function). The key point that this equation makes is that the number of recaptures at any given location must be evaluated relative to the probability of recapture.

Albanese et al. (2003) evaluated the impact of distance weighting on the assessment of movement by three species of southeastern stream fishes. He used a modeling study to illustrate the impact that increasing the number of 50 m sampling sections would have on distance weighting. Fish were considered marked in ten 50 m sections. The modeling approach showed how the zone of uniform sampling (i.e., sampling at or near 100%) changed as the sampling area was increased. If sampling only occurred within the 500 m marking section, then the impact of distance weighting was extreme. The proportion of total possible movements sampled (PS_d) was only 100% for sampling within ± 50 m (i.e., the zone of uniform sampling) and declined sharply for movement distances > ± 50 m. Clearly, a study design that only included sampling within the same stream reach used to mark fish would be strongly biased toward short-term movements. With a 1,000 m sampling effort, the PS_d values were 100% for fish movements within ± 250 m, and with a 2,000 m sampling effort, PS_d values are 100% for fish movements within ± 750 m. Both of these are much more robust designs in terms of understanding movement. However, even with the 2,000 m

sampling effort, any fish movements greater than ± 1,200 m would have been undetected.

In the following figure, based on Albanese et al. (2003), the modeling study shows the proportion of total possible movements (PS_d) that would be detected for three different sampling designs: one in which the sampling reaches were the same as the marking reaches (500 m), one in which the sampling reaches (1,000 m) were twice that of the 500 m marking reach, and one in which the sampling reaches (2,000 m) were four times that of the 500 m marking reach. Solid lines show PS_d values; dashed lines show the sampling lengths. Movement can be either upstream (+) or downstream (–) from the marking section.

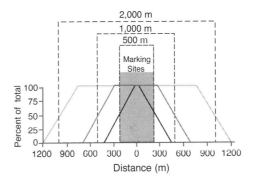

The effect of the size of the resampling areas (indicated by dashed lines) on the ability to detect marked fishes, relative to the distances that the fishes move. As determined by a modeling study, the largest resampling area (2,000 m) can detect 100% of fish movements up to 750 m upstream or downstream from the marking site (indicated by shading). The smallest sampling area (500 m and the same as the marking area) could only detect 100% of fishes that moved less than approximately 50 m. Based on Albanese et al. (2003).

In theory, the probabilities of capture from distance weighting could be used to adjust the observed captures of fish (see Albanese et al. 2003). In actuality, outside of the zone of uniform sampling, the numbers of fish captured were so low that adjustments generally were not possible—adjustments are not possible if no fish are captured! The take-home message is that an understanding of distance weighting using PS_d values is most useful for the *a priori* design of the sampling study. The *a posteriori* application of correction factors generally cannot correct for a poor study design.

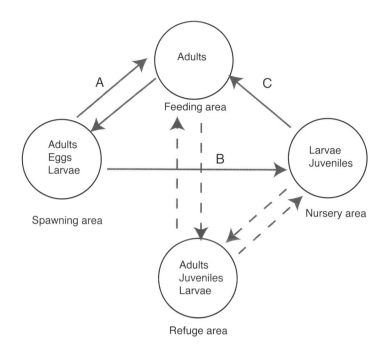

FIGURE 5.5. Patterns of migration in fishes. The circles enclose life-history stages using a particular resource. Solid lines indicate regular movement; dashed lines indicate aperiodic movement to refuge areas when there are harsh environmental conditions. Distances (A, B, C) between circles can vary greatly from a few meters to hundreds or thousands of kilometers. Adapted from Harden Jones (1968) and Schlosser (1995).

of movement in Redbreast Sunfish or Rosyside Dace. Correcting the movement data to account for distance weighting had little effect. However, this does not negate the strong potential effect of distance weighting but emphasizes the importance of an appropriate sampling design that, *a priori*, deals with the problem of distance weighting—a conclusion also reached by Albanese et al. (2003) (Box 5.2).

A general pattern that emerges from these studies of fish movement is that although the majority of fishes are often sedentary, there is often another, albeit smaller, group that undertakes much more extensive movements. Although use of the terms "movers" and "stayers" is descriptively appealing, several studies, including Smithson and Johnston (1999), have shown that the same individuals may switch from static to mobile behavior. In other words, fishes are continually responding to their local environment relative to their physiological needs, and the variation in when and how much

they move reflects this. Hence a "mover" one day might be a "stayer" the next.

Ontogeny and Movement

During their life cycle, freshwater fishes are faced with many challenges, including feeding, growth, predator avoidance, and reproduction. In some species and/or habitats, these activities may occur over a small spatial scale on the order of meters, whereas in others, tens or hundreds of kilometers may separate these and other activities (Figure 5.5). An important point that Figure 5.5 illustrates is that movement can occur at any life-history stage. For egg and larval stages, movement is generally passive and occurs via transport by water currents. The downstream drift of fish eggs and/or larvae is widespread, but by no means universal, among freshwater fish taxa (Gale and Mohr 1978; Brown and Armstrong 1985; Pavlov 1994). Entry into the drift by larval fishes can be due to turbulence that dislodges larvae (termed catastrophic drift) or because of active choice, as larvae swim up off the bottom. As described by Pavlov (1994), once in the water column, drift can be passive (most common for early larval stages), drifting downstream but not oriented to the direction of flow. Drift can also be active, where fish are actively swimming downstream, or active-passive, where fish show orientation to the current but only weak swimming ability. Several examples of species from different regions and habitat types in North America serve to illustrate these patterns.

MOVEMENT AT THE FERTILIZED EGG STAGE
Various minnow species in Great Plains streams have adapted to life in large, turbid rivers with shifting sand substrata by having a semibuoyant egg stage. This is apparently an adaptation to the unpredictable summer flows, since eggs are released during high-discharge periods (Platania and Altenbach 1998; Dudley and Platania 2007), and perhaps to avoid the risk of suffocation by sediment accumulation that would threaten eggs deposited directly on a shifting substratum. In the Rio Grande drainage (including the Rio Grande and Pecos rivers), four native species in three genera form a reproductive guild of broadcast spawners with semibuoyant eggs. The species are the Rio Grande Silvery Minnow, *Hybognathus amarus*; Speckled Chub, *Macrhybopsis aestivalis*; Rio Grande Shiner, *Notropis jemezanus*; and Pecos Bluntnose Shiner, *N. simus pecosensis* (Platania and Altenbach 1998). Two additional taxa that were endemic to the Rio Grande (Phantom Shiner, *Notropis orca*; Rio Grande Bluntnose Shiner, *Notropis simus simus*) were also likely members of this guild; unfortunately these taxa are now extinct (Bestgen and Platania 1990; Platania and Altenbach 1998).

Embryonic and early larval development of fishes in this guild occurs as they drift downstream with river flow. The distances required for the egg and early larval development to occur are impressive—during the time of passive transport of the eggs, they could travel some 144 km. The newly hatched protolarvae, which also remain in the water column, could be carried an additional 216 km depending on water temperature (which controls the rate of development) and current speed (Platania and Altenbach 1998). Clearly, this guild of Rio Grande fishes shows that the eggs and early larval stages of certain species have the ability to move great distances. Upstream movements of juveniles and adults counter the downstream movements of the eggs and protolarvae. In relation to Figure 5.5, the distance (A) from adult habitat to spawning habitat can be quite short. In contrast, the distance (B) from the spawning habitat (open water) to the nursery habitat (downstream in shallow, slow water along the shoreline) can be a hundred or more kilometers. This occurs following the protolarval stage. Later larval stages and juveniles inhabit the shallow, warm, and productive river margins as they gradually move upstream. This upstream distance (C) is also on the same scale as the distance from the spawning habitat (B). This life-history pattern, while demonstrating the ability for long distance movement of early life-history stages and adults, is obviously very susceptible to man-made barriers in rivers and to flow modifications (Winston et al. 1991; Dodds et al. 2004; Dudley and Platania 2007). The semibuoyant eggs require at least some current speed to remain in suspension. Consequently, if they enter the slack water of a reservoir they tend to sink and die from suffocation. The high concentration of nonnative predators in reservoirs is also generally lethal to drifting eggs and larvae (Dudley and Platania 2007). Dams also preclude the upstream return movement of juveniles and adults. As a consequence, reproductive output of most of the breeding adults of silvery minnows seems to be lost as developing embryos and larvae drift into impoundments (Alò and Turner 2005); it is no surprise that most species of this guild are extirpated or have their ranges greatly reduced and have required federal listing as threatened or endangered.

MOVEMENT AT THE LARVAL STAGE In contrast to the previous examples of fishes with semibuoyant eggs, most freshwater fishes have eggs that are demersal and adhesive, remaining attached to bottom materials such as gravel, sand, wood, or other solid materials prior to hatching. However, larvae of many species do have a free-swimming stage and can enter the water column of streams. The diverse arrays of invertebrate and vertebrate organisms that are carried in the water column are collectively referred to as drift. The drift of larval fishes can occur at any larval stage but is most prevalent at the earliest (protolarvae) and the intermediate (mesolarvae) larval stages and can result in downstream movement on the scale of meters to hundreds of kilometers. Relative to Figure 5.5, distances of larval drift (B) and movement of juveniles to adult feeding areas (C) are generally roughly equivalent to distances moved by adults to the spawning ground (A). The duration of larval drift varies widely among taxa, from very long periods of drift in various species of minnows, as discussed previously, to very short time periods in the drift, as in certain species of darters (Slack et al. 1998). The density of drifting fish larvae can be impressive. In the Smith River, a coastal river of northern California, White and Harvey (2003) found that some 2.5 billion sculpin (*Cottus*) embryos and larvae move down the river to the estuary over a four-month period.

Movement via larval drift is common in many North American fish families. Of the 15 families that make up 90% of North American species (Chapter 1; Table 1.1), larval drift or drifting by early juvenile stages commonly occurs in 10 families (Table 5.1). Exceptions to this are livebearing fishes (families Poeciliidae and Goodeidae), fishes in the families Cyprinodontidae and Fundulidae where larvae tend to remain on or near the bottom (Foster 1967), and fishes where young are often guarded in a confined nesting cavity, such as madtom catfishes (*Noturus*), or closely guarded by parents (family Cichlidae).

The numerically dominant cyprinids are widely represented in drift throughout North American streams, including the western and central United States (Robinson et al. 1998),

northern rivers in Canada and the United States (Muth and Schmulbach 1984), and southeastern streams (Gallagher and Conner 1980; Slack et al. 1998). As shown in the previous section for species with drifting eggs and larvae, distances traveled by larvae can exceed several hundred kilometers. In the Colorado River drainage, estimates of movement of native cyprinid larvae range from approximately 9 km for Humpback Chub (*Gila cypha*) and Speckled Dace in the Little Colorado River (Robinson et al. 1998), to over 200 km for Colorado Pikeminnow (*Ptychocheilus lucius*) in the Green River (Tyus and Haines 1991).

For percids, larvae of the commercially and recreationally important Walleye (*Sander vitreus*) and Yellow Perch (*Perca flavescens*) exhibit substantial drift with greatest abundances occurring at night (Gale and Mohr 1978; Corbett and Powles 1986; Johnston et al. 1995). The occurrence in drift samples of darter species in the genus *Etheostoma* has been documented in several studies, with peak abundances typically at night from 2100 to 0300 h (Gale and Mohr 1978; Lathrop 1982; Brown and Armstrong 1985; Paller 1987). Even Bayou Darters (*Nothonotus rubrum*), members of a genus that inhabits swift water and coarse substrata, show downstream drift of at least several hundred meters. This is at least far enough for them to travel between patches of riffle habitats that juveniles and adults of this species selectively occupy (Slack et al. 2004). Larval drift also occurs within the darter genus *Percina*. For instance, larval Snail Darters (*Percina tanasi*) show downstream transport of up to several kilometers, followed by return upstream movement of juveniles and adults (Kuehne and Barbour 1983). As shown earlier, this type of life cycle that includes downstream drift followed by upstream movement of juveniles and adults is greatly at risk from man-made barriers to movement. Indeed, the population of Snail Darters in the Little Tennessee River was extirpated by the infamous Tellico Dam (Ono et al. 1983). More recently, additional populations of Snail Darters have been discovered in several

TABLE 5.1

Prevalence of Larval and Early Juvenile Drift in Numerically Dominant Freshwater Fish Families of North America

Family	Common name	Occurrence in drift	Source
Cyprinidae	Minnows	Common in most genera and species	Gale and Mohr 1978; Gallagher and Conner 1980; Floyd et al. 1984; Muth and Schmulbach 1984; Brown and Armstrong 1985; Peterson and VanderKooy 1995; Gadomski and Barfoot 1998; Robinson et al. 1998; Slack et al. 1998; Marchetti and Moyle 2000; Harvey et al. 2002; Johnson and McKenna 2007
Percidae	Perches	Common in most genera and species	Gale and Mohr 1978; Floyd et al. 1984; Corbett and Powles 1986; Johnston et al. 1995; Peterson and VanderKooy 1995; Marchetti and Moyle 2000; Slack et al. 2004; Johnson and McKenna 2007
Catostomidae	Suckers	Common in all genera but not in all species	Gallagher and Conner 1980; Muth and Schmulbach 1984; Corbett and Powles 1986; Kay et al. 1994; Brown and Armstrong 1995; Johnston et al. 1995; Peterson and VanderKooy 1995; Kennedy and Vinyard 1997; Robinson et al. 1998; Slack et al. 1998; Marchetti and Moyle 2000; Johnson and McKenna 2007
Poeciliidae	Livebearers	Rare—poeciliids are livebearers	Brown and Armstrong 1985; Peterson and VanderKooy 1995; Marchetti and Moyle 2000
Ictaluridae	North American catfishes	Common in most genera; rare in *Noturus*	Floyd et al. 1984; Muth and Schmulbach 1984; Brown and Armstrong 1985; Holland-Bartels et al. 1990; Slack et al. 1998; Marchetti and Moyle 2000
Goodeidae	Goodeids	Rare—goodeids are livebearers	Miller 2005
Atherinopsidae	New World silversides	Common in *Labidesthes*; rare in *Menidia*	Frietsche et al. 1979; Peterson and VanderKooy 1995; Slack et al. 1998; Marchetti and Moyle 2000
Salmonidae	Trouts and salmons	Present in *Coregonus*, *Prosopium*, *Thymallus*, and perhaps *Stenodus*; uncommon or of short duration in *Oncorhynchus*, *Salvelinus*, and *Salmo* and varies among species	Scott and Crossman 1973; Randall et al. 1987; Crisp 1988; Thorpe 1988; Winnell and Jude 1991; Pavlov 1994; Moyle 2002
Cyprinodontidae	Pupfishes	Uncommon but larvae are free-swimming after hatching	Hardy 1978
Fundulidae	Topminnows	Uncommon	Peterson and VanderKooy 1995; Robinson et al. 1998

(continued)

TABLE 5.1 *(continued)*

Family	Common name	Occurrence in drift	Source
Centrarchidae	Sunfishes	Common in all genera	Floyd et al. 1984; Muth and Schmulbach 1984; Peterson and VanderKooy 1995; Marchetti and Moyle 2000
Cottidae	Sculpins	Common in *Cottus*	Gadomski and Barfoot 1998; Marchetti and Moyle 2000; Harvey et al. 2002; White and Harvey 2003
Petromyzon-tidae	Lampreys	Common in all genera	Gadomski and Barfoot 1998; Harvey et al. 2002; White and Harvey 2003
Cichlidae	Cichlids	Uncommon—Neotropical cichlids are brood guarders	Winemiller et al. 1997; Miller and Norris 2005
Clupeidae	Herrings	Common in all genera	Muth and Schmulbach 1984; Peterson and VanderKooy 1995; Gadomski and Barfoot 1998

Tennessee River tributaries, although of the nine known populations, six are considered marginal (Williams et al. 1989).

Catostomids, the third most speciose North American fish family, are also well represented in larval drift. Historically, many western catostomids drifted long distances downstream from spawning areas, subsequently followed by upstream spawning movements of adults. In the Little Colorado River, Bluehead (*Catostomus discobolus*) and Flannelmouth (*C. latipinnis*) sucker larvae drifted at least 9 km downstream from the spawning area (Robinson et al. 1998). Another Colorado River endemic, the Razorback Sucker (*Xyrauchen texanus*) has even more extensive movements as larvae. Because of the close linkage of adult and larval distances, Razorback Suckers, along with other examples of larval drift, are included in the following section on adult movement.

MOVEMENT AT THE ADULT STAGE In the Great Lakes of North America, Lake Sturgeon (*Acipenser fulvescens*) adults make spawning migrations out of the lake habitats and into tributary rivers. For example, in Lake Superior, adult sturgeon travel upstream for spawning at a single riffle in the Sturgeon River, Michigan, a distance of 69 km (Auer and Baker 2002). After hatching, larvae drift downstream at least 45 km and in some cases 61 km. In reference to Figure 5.5, the distance from adult feeding area to the spawning area (A) is 69 km; the nursery area is essentially the lower 10 km of river habitat downstream of the spawning site, and so this distance (B) is approximately 59–69 km; the other distance (C) would include the movement within Lake Superior to juvenile and adult feeding grounds.

Most suckers (family Catostomidae) also exhibit seasonal spawning migrations in which they move upstream into small tributaries from larger streams or lakes. Depending on the species, distances moved vary greatly, and movement may occur in groups or in larger schools (Curry and Spacie 1984) (see also Chapter 14). Adults make the return downstream migration after spawning, whereas newly hatched larvae are passively transported downstream by water flow, with larval numbers often increasing in surface waters at night (Gale and Mohr 1978).

A striking example of extensive travel to spawning habitats by a catostomid is provided by the Razorback Sucker, a species endemic to the Colorado River system. In the Green River and its tributaries, movements of Razorback Suckers are bounded upstream by the Flaming Gorge Dam and downstream by Lake Powell and the Glen Canyon Dam. Within this reach

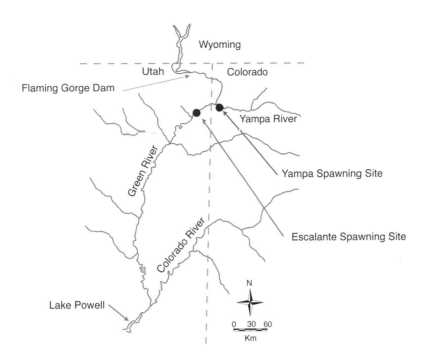

FIGURE 5.6. Spawning movements in fishes as illustrated by the Razorback Sucker, (*Xyrauchen texanus*), an endemic catostomid in the Colorado River drainage. Spawning locations of Razorback Sucker are indicated by black dots. Based on Tyus and Karp (1990), Modde et al. (1996), and Modde and Irving (1998).

there are two spawning sites known for Razorback Sucker—one in the Yampa River upstream from its confluence with the Green River, and one in the Green River downstream of the Yampa River (Tyus and Karp 1990; Modde et al. 1996; Figure 5.6). Fish in breeding condition may travel at least 30–106 km to reach the spawning sites, followed by equivalent downstream movement (Tyus and Karp 1990). After hatching from demersal, adhesive eggs, larvae drift downstream into nursery habitat (historically provided by large backwaters) (Modde et al. 2001). Another Colorado River endemic, the Colorado Pikeminnow, makes equally impressive long distance spawning movements (Tyus and McAda 1984).

The most impressive long-distance movements occur in fishes that travel between salt and fresh water for purposes of spawning (diadromy) (see also Chapter 9). In fishes that spawn in fresh water and then spend part of their life in the sea where they feed (anadromy), one-way distances traveled can be hundreds or even thousands of kilometers. For instance, Chinook Salmon (*Oncorhynchus tshawytscha*) travel almost 2,000 km as the spawning adults move from the Pacific Ocean upstream to spawning sites in the Yukon River (corresponding to distance A in Figure 5.5) (Scott and Crossman 1973). Post-yolk-sac fish (fry) as well as parr (young salmonids during the first year or two of life) and smolts (older juveniles ready to return to the sea) make the return journey downstream and then out to sea (distances B and C in Figure 5.5). Once in the open ocean where they are actively feeding, Pacific salmon may travel over thousands of kilometers during that time (usually 1–6 years) they spend at sea (Healey and Groot 1987; Thorpe 1988; Walter et al. 1997).

SUMMARY

The process of forming fish assemblages, although complex, involves characteristics of the environment, characteristics of the fish species in the regional species pool, and characteristics of the fishes and other biota in the local environment. Fish assemblages tend to be structured rather than random groupings of species, although random processes may at times be important. Also, very few studies representing even fewer geographical regions have rigorously addressed the issue of structure in freshwater fish assemblages. A major factor seems to be the fit of the potential colonizer with the environmental features of the new habitat. Following this, trophic position (low or high rather than intermediate), and if there is parental care of young, are important attributes of successful colonizers.

Fishes show the ability to move long distances, and depending on the species, movement may occur at any life-history stage. Even within fish populations that are relatively sedentary, individuals may make periodic or aperiodic movements, most likely in response to assessing resource availability or the risk of predation. Although the terms movers and stayers have been used to describe the differences in movement among individuals in a population, data seem to indicate that the same individual can shift between the two states. Hence the terms apply more to the state of an individual rather than differences among individuals. As long as there are periodic water connections and sufficient time, the well-developed capability for movement in most fish taxa allows fishes to colonize new areas or to enter preexisting assemblages.

SUPPLEMENTAL READING

Albanese, B. W., P. L. Angermeier, and C. Gowan. 2003. Designing mark-recapture studies to reduce effects of distance weighting on movement distance distributions of stream fishes. *Transactions of the American Fisheries Society* 132:925–39. Explores the role of changing capture probabilities with distance from the release point in estimates of fish movement.

Belyea, L. R., and J. Lancaster. 1999. Assembly rules within a contingent ecology. *Oikos* 86:402–16. An overview of the literature on community assembly.

Fausch, K. D., C. E. Torgersen, C. V. Baxter, and H. W. Li. 2002. Landscapes to riverscapes: Bridging the gap between research and conservation of stream fishes. *BioScience* 52:483–98. Emphasizes the need to view streams and rivers as complex landscapes and the importance of access to these varied habitats by fishes.

Gowan, C., M. K. Young, K. D. Fausch, and S. C. Riley. 1994. Restricted movement in resident stream salmonids: A paradigm lost? *Canadian Journal of Fisheries and Aquatic Sciences* 51:2626–37. A reanalysis and counter argument to the restricted movement paradigm.

Mandrak, N. E., and E. J. Crossman. 1992. Postglacial dispersal of freshwater fishes into Ontario. *Canadian Journal of Zoology* 70:2247–59. An important paper on the glacial refugial origins of Ontario freshwater fishes.

Taylor, C. M. 1996. Abundance and distribution within a guild of benthic stream fishes: Local processes and regional patterns. *Freshwater Biology* 36:385–96. Uses field collections and manipulative studies to test predictions of hypotheses regarding the abundance and distribution of fishes.

SIX

Persistence of Fish Assemblages in Space and Time

CONTENTS

THE FIRST TWO CHAPTERS in Part 2 examined how fish species and assemblages are affected by broadscale landscape features, how various models relate assemblages to the environmental variables, how fish assemblages are formed, and the role that movement plays over different life-history stages in allowing fishes to access new habitats and to move among habitats so that their fitness is maximized. This chapter focuses primarily on the temporal and spatial dynamics of fish assemblages, or how fish populations and assemblages cope with relatively short-term physical and biotic challenges.

Understanding the type, frequency, and magnitude of variability in fish assemblages is important for several reasons (e.g., Grossman et al. 1990; Matthews 1998). First, assessing the impact of anthropogenic environmental changes requires knowing the background level of natural variation in assemblages. Second, the degree to which assemblages are resistant to changes over space and time is related to the strengths of control mechanisms operative within the assemblage. Although assemblages generally are structured and not random collections of species from a regional species pool (Chapter 5), once established, they may be acted upon by external or internal processes. With some exceptions (Strong 1983), assemblages showing high variation in species composition and abundances may primarily be governed by external, stochastic (i.e., random) processes such as floods, droughts, or other major events. These events can control such processes as species persistence, colonizations, or even extinctions. In communities with strong stochastic influences, the importance of biotic interactions (i.e., competition or predation) in affecting community structure is considered to be lessened because of the frequent changes in species composition. In contrast, assemblages that show little variation may be controlled primarily by deterministic processes, such that the characteristics of the environment result in a particular suite of species (e.g., the landscape filters described in Chapter 4). In assemblages that show little variation in species composition, the possibility of well-developed biotic interactions is considered to be greater (Grossman et al. 1982; Lepori and Malmqvist 2009). Importantly, processes controlling communities should not be viewed in an either-or situation. Stochastic and deterministic processes can act hierarchically (i.e., stochastic processes influence the species on which deterministic processes act). The relative importance of stochastic versus deterministic controls varies with disturbance levels, although not necessarily monotonically (Lepori and Malmqvist 2009).

RESPONSES TO ENVIRONMENTAL PERTURBATIONS

Types of Perturbations

Natural perturbations have shaped the evolution of fish populations and, in the case of severe events, have resulted in the local extirpation of populations or the total extinction of species. For instance, large-scale Cenozoic climatic changes resulted in the extinction of numerous western North American fishes at the end of the Miocene and also the early Pleistocene (G. R. Smith 1981). Natural perturbations include droughts, floods, fires within the watershed, climatic changes, and biotic changes (such as the addition or loss of a predator). Human-induced changes might include chemical spills or piscicide applications; changes in land use, such as mining, agriculture, or timber harvesting resulting in flooding, increased water temperature, nutrient or herbicide runoff, and erosion; major barriers to fish movement as a result of dams or water diversions; stream channelization; and the introduction of nonindigenous species.

One way to view both natural and human-caused disturbances is by their extent. Events that persist longer than the life spans of the species in an assemblage and impact large spatial areas are referred to as press disturbances, in contrast to pulse disturbances, which are of short duration and are generally point source or brief hydrologic events (Bender et al. 1984; Detenbeck et al. 1992). Based on Detenbeck et al. (1992), press disturbances would include impacts of channelization, large-scale habitat alterations, timber harvesting, mining, and changes in nutrient input; pulse disturbances would include floods, chemical spills, droughts, nonchemical removal of biota, and localized construction activity.

Determining what amount of environmental variation actually represents a disturbance or perturbation (since the terms generally are used interchangeably although there are exceptions; e.g., Pickett and White 1985) to aquatic organisms is also challenging—especially for terrestrial, hominid biologists! Natural variations in physical conditions, even some viewed as "a disturbance," are generally beneficial in the long-term to the well-being of aquatic systems. This would include changes in stream flow (including flow into lakes, ponds, and reservoirs), turbidity, temperature, ice cover, or insolation. For instance, without periodic high, scouring flows in streams, streambed complexity (Mount 1995) and complexity of riverine food webs (Wootton et al. 1996; Power et al. 2008) can be greatly reduced, resulting in population declines or loss of fish species. Likewise, the annual or semiannual turnover in many lakes results in redistribution of nutrients to surface waters and oxygenation of bottom waters (Wetzel 2001).

The recognition of the value of periodic disturbance in ecological communities in the 1970s and 1980s led to models of how periodic disturbance fostered increased species diversity. This corresponded with the recognition that most communities probably did not exist at some sort of steady state or equilibrium (Levin and Paine 1974; Sousa 1984). The intermediate disturbance hypothesis (Levin and Paine 1974; Connell 1978) predicts that the greatest species richness would occur at some intermediate level (intensity and/or frequency) of disturbance. The logic is basically that intermediate levels of disturbances provide sufficient time for species to colonize affected patches of habitat yet keep the habitat from being dominated by only a few species (Connell 1978; Sousa 1984). In a similar way, the dynamic equilibrium model (Huston 1979) predicts that diversity of communities is the outcome of two processes—the rate of population growth of competing species, balanced against the frequency of population reductions, caused by various types of disturbances. In contrast to disturbance functioning by mediating competitive interactions between species, the role of intermediate disturbance in a study of stream macroinvertebrates was due to the removal of more sensitive species, so that invertebrate communities converged to a core group of species moderately resistant to disturbance (Lepori and Malmqvist 2009).

What constitutes a disturbance also changes over ecological and evolutionary time and among taxa. Viewed in the evolutionary context of species and assemblages, a force that once was a major disturbance might be less so today given the strong selection for populations to withstand environmental change (Sousa 1984). In ecological time, as elaborated later in the chapter, life-history stages of a species differ in their abilities to respond to disturbances, just as species differ in their responses. Furthermore, the seasonal timing of disturbance can affect the level of impact on fish populations and, because of the wide variation in body size and mobility among fish species, disturbance must be viewed relative to the spatial and temporal dynamics of species (Pickett and White 1985).

Efforts to define disturbance have taken two main approaches. One approach defines a disturbance by its magnitude, whereas the other defines a disturbance by the population, species, or community responses to it or to its impact on the physical environment (Resh et al. 1988; Matthews 1998). In the former case, a sudden change in water temperature or stream flow that exceeded some arbitrary value, say ± two standard deviations, would be judged as a disturbance, whereas a change of less than ± two standard deviations would not. In the latter case, if there were no apparent biological or physical response to what would seem to be a disturbance, such as a major flood event, then the event would not be considered a disturbance. The latter approach has generally been preferred (e.g., Resh et al. 1988), and a useful working definition of a disturbance proposed by White and Pickett (1985) and used by other authors (e.g., Resh et al. 1988; Yount and Niemi 1990) is "any relatively discrete event in time that disrupts ecosystem, community, or population structure, and that changes resources, availability of substratum, or the physical environment." As such, a disturbance is "the primary event, or *cause*, from which certain effects follow" (Yount and Niemi 1990).

The Metric

Responses to environmental change can basically be measured by the presence or absence of species, irrespective of the actual numbers or relative abundance of individuals. This qualitative measure is referred to as persistence, in contrast to stability, which is based on abundance measures (Connell and Sousa 1983). Quantitative measures include relative abundances, or actual numbers or densities of the component species. The choice of metric has a strong influence on the detection of change, or lack thereof, in fish populations (Rahel et al. 1984; Yant et al. 1984; Matthews et al. 1988; Grossman et al. 1990; Rahel 1990; Matthews 1998). For instance, presence-absence, ranks in abundance, relative abundance measures, and actual numbers of individuals form a transformation series of increasing sensitivity to change. That numbers of individuals of a given species show the greatest variation is not surprising, especially because most long-term studies of fish assemblages employ sampling techniques that are not designed to provide rigorous quantitative data on population sizes (Matthews 1998).

Spatial and Temporal Scales

In addition to the appropriate metric, the spatial and temporal scales over which a measurement is made also affect the outcome. To assess stability, the temporal scale must encompass at least one full turnover in the assemblage (Connell and Sousa 1983); if it does not, then what is really being measured is simply the impact of long-lived organisms on the local community. This point can have a major impact on apparent regional differences in responses of fish assemblages to environmental change. Some southwestern fish assemblages, such as in the San Juan River of the Colorado River drainage, consist primarily of species like Flannelmouth (*Catostomus latipinnis*), Bluehead (*C. discobolus*), and Razorback (*Xyrauchen texanus*) suckers; Roundtail Chub (*Gila robusta*); Speckled Dace (*Rhinichthys osculus*); and Colorado Pikeminnow (*Ptychocheilus lucius*) (Tyus et al. 1982; Propst and Gido 2004). With the exception of the short-lived (ca. 3 years) Speckled Dace, these San Juan River species commonly live more than 20 years, and in the case of Colorado Pikeminnow and Razorback Sucker, over 40 years (John 1964; McCarthy and Minckley 1987; Scoppettone 1988; Lanigan and Tyus 1989; Osmundson et al. 1997). In contrast, southeastern fish assemblages, such as in Black Creek of the Pascagoula River drainage, Mississippi (Baker and Ross 1981; Ross et al. 1987), are composed primarily of small minnows, topminnows,

darters, and sunfishes, most of which have life spans of only 1–5 years (Ross 2001). A study of 4–5 years would essentially capture one complete assemblage turnover for the Black Creek fishes, whereas an equivalent study in the San Juan River would need to extend to 20 years or more to achieve the same result. In probably the majority of studies, the temporal scale is defined more by the duration of funding or graduate student tenure (both commonly on the order of 1–5 years) than by consideration of the life history of the fishes—with some notable exceptions

The spatial scale of a study also has a major impact on the ultimate outcome (Connell and Sousa 1983; Rahel 1990). If the spatial scale does not include the normal population bounds of the component species (see Chapter 5), then it is likely that any measure will record extensive changes in assemblage structure. In contrast, a large study area might include a number of subpopulations comprising various metapopulations of the component species (see Chapter 4), so that variation or loss of taxa in one area is damped out by their survival in another. Connell and Sousa (1983) suggest that the spatial scale should correspond to the least area that is necessary for the recruitment of adults through successful reproduction, survival, and growth of young. Recalling the types and extent of movement in Chapter 5, this guideline would result in widely differing spatial scales, depending on the species and region. However, again with a few notable exceptions, the spatial scale of most studies is somewhat arbitrary or driven by sampling logistics or cost. Thus not only the analytical scale, as illustrated previously, but also the temporal and spatial scales of the analysis have strong effects on the outcome and, not surprisingly, because of the interactions between metrics and scales, there are conflicting views on the nature of change in fish populations and assemblages.

Assessing Assemblage Change

The stability and persistence of assemblages should be investigated on multiple levels within the hierarchical framework (Rahel 1990). Separating actual changes in species presence or absence from artifacts of sampling also is a pervasive and significant problem (Magnuson et al. 1994). Preferably, the goal of current research should not be to determine whether local assemblages change or not—virtually all local assemblages undergo change as individuals are added or removed (due to natality and mortality, or movements). Instead, as the temporal extent of data increases (several to many samples over a period of years to decades), and as change is measured on multiple spatial scales, it becomes of more interest to ask how much a local fish assemblage, or distribution of fish species within a watershed, have changed during various intervals (Ross and Matthews, in press). Also, an important issue is whether changes largely are driven by major events such as extensive droughts or floods, with little change during times that lack apparent disturbance factors. In other words, are the changes that can be observed in fish assemblages over long periods of time related more to gradual changes, or to dramatic events that may (or may not) leave their mark for years or more?

Dealing with Environmental Change

Fish populations can deal with environmental or biotic stressors in three primary ways. First, populations might lack means of dealing with perturbations and be eliminated from a region altogether. Second, individuals in a population might show resistance by withstanding such challenges through morphological, physiological, or behavioral adaptations, such as refuge-seeking behavior that would overall increase their tolerance to environmental perturbations. Finally, populations might emigrate from, or perish in, the stressed habitat but recover following a perturbation by return immigration of the displaced individuals or colonization by individuals from other populations, such that pre- and postdisturbance assemblage structures are the same or similar. This approach to dealing with environmental change was termed

adjustment stability by Connell and Sousa (1983) and resilience by Dodds et al. (2004). Connell and Sousa (1983) considered that this response included two components: amplitude and elasticity. Amplitude is a measure of how far a population or assemblage can be displaced from its predisturbance state and still return; elasticity, drawing further on the analogy with a rubber band, is a measure of how quickly populations or assemblages can return to a predisturbance condition.

Resistance

It is not surprising that fish assemblages occurring in geographic regions prone to extreme climatic conditions are generally persistent in the face of such environmental challenges, given that these faunas have been under strong, long-term selection to deal with such conditions, and intolerant species would have been extirpated (C. L. Smith and Powell 1971). Examples of fish faunas inhabiting areas prone to disturbance include faunas of Great Plains streams. Fishes in these streams are subjected to periods of low flow and even dewatering as well as to intense periods of flooding. Such conditions likely occurred before widespread human-caused changes in the nineteenth and twentieth centuries, although the amount of silt has probably increased as a consequence of changes in land use (Matthews 1988).

In response to this environment, some fishes have evolved increased tolerances to low dissolved oxygen levels and high temperatures. Comparisons of physiological tolerances among minnow species from more benign upland streams in Arkansas with those inhabiting apparently harsher Great Plains streams in central and western Oklahoma generally showed that minnows from harsh environments were more resistant (Matthews 1987). As a group, Matthews showed that the minnows from the harsh streams had significantly greater tolerance to high water temperatures compared to fishes from the relatively benign streams, although one prairie minnow, the Emerald Shiner (*Notropis atherinoides*), had a

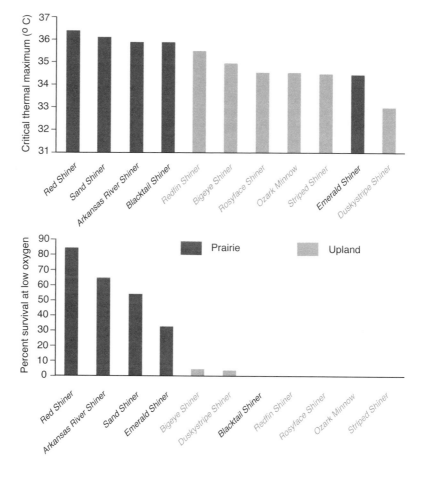

FIGURE 6.1. Contrasts in physiological resistance of minnows from relatively harsh versus benign environments. Harsh environments include prairie streams from central and western Oklahoma; benign environments include streams from upland regions of Arkansas. Oxygen tolerance was measured as the percentage of fish surviving 8.5–10 h at low oxygen levels (0.2–0.9 ppm dissolved oxygen). Based on data from Matthews (1987).

critical thermal maximum (CTM) more in line with the upland fishes (Figure 6.1). However, Emerald Shiners, along with three other plains minnows, showed better survival at low oxygen levels compared to upland fishes or Blacktail Shiner (Figure 6.1).

Pupfishes (genus *Cyprinodon*) occur widely in fresh and brackish water habitats in Mexico (Miller 2005), in coastal brackish water areas along the Gulf of Mexico and Atlantic coasts (Johnson 1980; Nordlie 2003), and in desert regions of the southwestern United States (Naiman and Soltz 1981). In all regions, *Cyprinodon* species show high resistance to temperature extremes (Feldmeth 1981; Bennett and

Beitinger 1997; Nordlie 2003). For instance, the Sheepshead Minnow (*Cyprinodon variegatus*) of the Atlantic and Gulf coasts can survive temperatures from a low of –1.8° C to a high of 43° C, the widest temperature range of any of the over 200 estuarine/salt marsh fishes reported by Nordlie (2003). Populations of cyprinodont fishes inhabiting the Death Valley region of Nevada and California also have wide temperature tolerances, being able to withstand temperatures of < 1° C to 40–44° C (Brown and Feldmeth 1971; Soltz and Naiman 1978; Feldmeth 1981).

Refuge-seeking behavior also allows fishes to resist adverse conditions in their environment and can have a role in surviving floods and

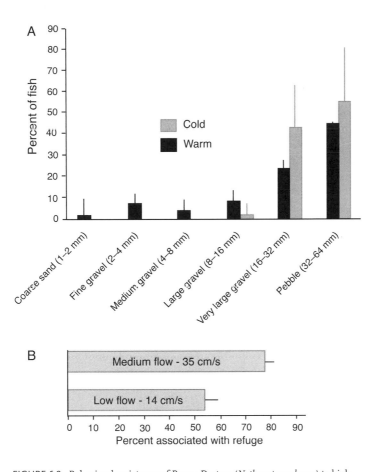

FIGURE 6.2. Behavioral resistance of Bayou Darters (*Nothonotus rubrum*) to high winter stream flow in western Mississippi.
A. At cold temperatures (7.5–11.0° C) in a laboratory stream with a current speed averaging 31 cms⁻¹, Bayou Darters shift their habitat selection to patches with larger particle sizes compared to habitat selection at warm temperatures (22° C).
B. Even in habitats with large substrata, Bayou Darters select patches with larger refuges (bricks) at moderate versus low current speeds. In both figures, bars are 95% confidence intervals. Based on data from Ross et al. (1992b).

droughts. Fishes in a variety of regions increase their resistance to downstream displacement during floods, especially during the winter when lowered water temperature limits their swimming ability, by actively selecting habitats with large structure such as woody debris, rocks, or other large and relatively immovable structures. Cutthroat Trout (*Oncorhynchus clarkii*) in a tributary of the Smith River in California showed twice the site fidelity in pools with large woody debris compared to those without (Harvey et al. 1999). During a winter flood event, trout in pools with large woody debris tended to remain in those pools in contrast to

the greater movement shown by trout in less complex habitats. Fishes in an arid eastern Oregon stream also were more resistant to floods in structurally complex habitats compared to simple habitats (Pearsons et al. 1992). In the southeastern United States, Bayou Darters (*Nothonotus rubrum*), a species endemic to the Bayou Pierre system of Mississippi, responded to winter water temperatures (7.5–11° C) in an artificial stream by shifting their distribution to habitat patches with larger particle sizes of coarse gravel and pebbles (Figure 6.2A). In addition, under winter conditions, as flow increased from 14 to 35 cms⁻¹, Bayou Darters

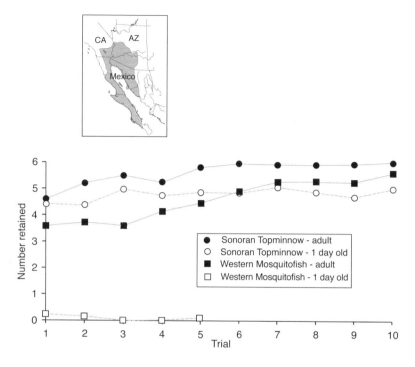

FIGURE 6.3. Behavioral resistance of adult and one-day-old Sonoran Topminnow (*Poeciliopsis occidentalis*) to displacement by floods, compared to the nonnative Western Mosquitofish (*Gambusia affinis*). Data show mean number retained in a laboratory stream system for each of ten, 60 s trials with a 60 s rest period between each trial; six fish of each species were used in each trial. One-day-old Western Mosquitofish only survived for five replicates. Based on data from Meffe (1984). The shaded area of the map inset shows the approximate boundaries of the Sonoran Desert.

selected habitat patches with large refuges to current (in this case bricks 14 × 7 × 7.5 cm) over habitats with just pebbles (Figure 6.2B).

Sonoran Topminnow (*Poeciliopsis occidentalis*), a small poeciliid native to streams, marshes, and springs in the Sonoran Desert, has evolved behavioral responses to flash floods that commonly occur in the region, in contrast to the morphologically similar Western Mosquitofish (*Gambusia affinis*), which is not native to the region (Figure 6.3). In habitats that are not periodically disturbed, Western Mosquitofish can reduce or eliminate Sonoran Topminnow, probably through predation on the young (Meffe 1984). Using field observations in a tributary of the Santa Cruz River in southern Arizona, combined with laboratory experiments, Meffe (1984) showed that Sonoran Topminnow of all life-history stages responded to floods by rapidly moving to shoreline eddies

and remaining there until high flows receded. In contrast, Western Mosquitofish responded more slowly and in a less organized manner to flooding and tended to move back out into high flows sooner, thus exposing themselves to downstream displacement. As a consequence, Sonoran Topminnow showed stronger resistance to downstream displacement by flooding in contrast to the nonnative Western Mosquitofish, although both species showed improvements in flood resistance through repeated exposure (Figure 6.3). By testing one-day-old fish, Meffe also demonstrated that the behavioral response to floods is innate in Sonoran Topminnow; one-day-old Western Mosquitofish were almost all displaced. Western Mosquitofish were also completely displaced from Sabino Creek, another southern Arizona stream, by a record winter flood, whereas a native minnow (Gila Chub, *Gila intermedia*) and a nonnative centrarchid (Green Sunfish, *Lepomis cyanellus*) were not (Dudley and Matter 1999).

Morphology, in concert with appropriate behavior, can also provide resistance to harsh environments. For example, most all fishes that obtain oxygen from the water will move closer to the water's surface as oxygen levels are depleted—the response is termed aquatic surface respiration (ASR) (Kramer 1987). However, most fishes, because of jaw morphology and head shape, must spend additional energy through body or fin movements to maintain the extreme body angle required for ASR and cannot survive severe subsurface oxygen depletion for extended periods of time (Lewis 1970). Fishes such as livebearers and topminnows have flattened heads and superior mouths, morphologies that are particularly adapted to ASR, and are able to use the highly oxygenated surface film while remaining in a nearly horizontal (< 10°) position. By exploiting the surface film, these fishes can survive extended periods in water that is otherwise low in oxygen (Lewis 1970; Kramer 1987; Timmerman and Chapman 2004). More recent work indicates that ASR might be most important as an immediate response measure to low oxygen levels. For instance, Sailfin Mollies (*Poecilia latipinna*), as well as a variety of other fish groups, are able to gradually increase oxygen capacity of the blood when chronically subjected to an oxygen-poor environment (Timmerman and Chapman 2004). This occurs through an increase in red blood cells and through increased hemoglobin concentration.

Resilience

Fish populations are often faced with environmental perturbations that are too severe for one or all life-history stages to resist through morphological, physiological, or behavioral mechanisms. Such perturbations might include extreme floods, drying of essential habitats such as feeding or spawning areas, changes in water quality, or total drying of an aquatic habitat. Although the initial effect can be the total or partial loss of species making up an assemblage or the lack of successful reproduction, the ability to recover once environmental conditions become more favorable is described by resilience.

Resilience to major environmental perturbations is provided through the ability of fish populations and assemblages to repopulate an area once conditions improve. This may occur through the return of displaced individuals and through the often-accelerated production of new individuals (Ross et al. 1985; Matthews 1986b; Fausch and Bramblett 1991). The extent of resilience is influenced by the size of the affected habitat and by the size and proximity of refuges where fishes can survive. Watershed geometry (e.g., Chapter 4; Figure 4.3) plays an important role in resiliency (Grant et al. 2007).

In southern Oklahoma, fishes in Brier Creek (see Chapter 5; Figure 5.4) recolonized a dewatered section of stream within four months once flow resumed. Recolonization to this pulse disturbance was initially by movement of fish out of isolated pool refugia, followed by spawning. In a southeastern study, Albanese et al. (2009) removed adult and juvenile fishes from 416 m and 426 m reaches of two small streams, Middle Creek and Dicks Creek, located in the James River drainage of Virginia (Figure 6.4). They followed recolonization for approximately one year along a 130 m reach in Dicks Creek, and two years along a 126 m reach in Middle Creek. The reaches were located in the middle of each of the two removal sections (Figure 6.4). The larger Dicks Creek site involved 19 species, whereas the smaller Middle Creek site involved 6 species. The fish fauna in both streams was dominated by minnows, which make up 85% of the fish in Dicks Creek and 92% in Middle Creek.

The resilience of individual species studied by Albanese et al. (2009), measured by their rates of recovery to the simulated pulse disturbance, varied widely; rates of recovery also differed between the two streams. Mountain Redbelly Dace (*Chrosomus oreas*) rapidly recolonized, reaching over 60% of the original population size within one month in Dicks Creek but

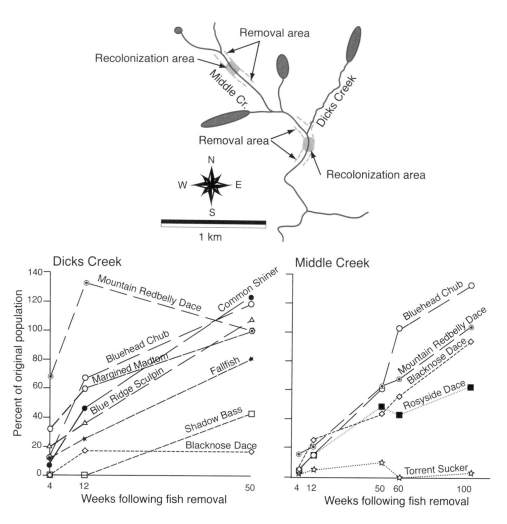

FIGURE 6.4. Varying levels of resilience of two southeastern fish assemblages as demonstrated by recolonization of experimentally defaunated areas. The map shows the study area, which was located in the James River drainage of Virginia; ovals are impoundments. Recovery was followed for approximately one year in Dicks Creek and two years in Middle Creek. Based on data from Albanese et al. (2009).

less than 20% in Middle Creek (Figure 6.4). At the end of one year, Mountain Redbelly Dace populations had fully recovered in Dicks Creek, but required an additional year for full recovery in Middle Creek. After one year, five of the eight censused species in Dicks Creek attained 80% or greater recovery; Shadow Bass (*Ambloplites ariommus*) and Blacknose Dace (*Rhinichthys atratulus*) showed much lower resilience. In Middle Creek, three of the five censused species reached 90% or greater recovery after two years, whereas Torrent Sucker (*Thoburnia rhothoeca*) did not recover and Rosyside Dace

(*Clinostomus funduloides*) only reached 60% of the original population size. Even though species varied significantly in their resilience to the defaunation, the fish assemblages, as measured by pre- and postremoval similarity, appeared resilient. This occurred because abundant species remained abundant, and more abundant species have a greater effect on faunal similarity measures that include relative abundance (Matthews 1998). As stressed by Albanese et al. (2009), this has important conservation implications. Measures that focus only on the assemblage level could overlook the loss of rare species following natural or human-caused perturbations.

In western North America in the Willamette River drainage of western Oregon, Lamberti et al. (1991) followed recovery of Cutthroat Trout in Quartz Creek, a high-gradient stream that had suffered a pulse disturbance in the form of a catastrophic debris flow. The debris flow had severe impacts on the physical and biotic characteristics of a 500 m stream reach. Physical changes included loss of woody debris, loss of canopy cover, a reworking of channel sediments, and an overall simplification of the channel, resulting in reduced hydraulic retention. Chlorophyll α was low immediately after the debris flow, but the newly opened canopy and the reduction in grazing by macroinvertebrates later resulted in a doubling of chlorophyll α compared to a control reach. Macroinvertebrate density initially showed high variation, followed by recovery after one year to densities shown in the control reach, and recovery of species richness after about two years. Cutthroat Trout, the only fish species in Quartz Creek, were initially extirpated. Resilience, as measured by percent recovery to predebris flow conditions, increased rapidly after one year; overall recovery of Cutthroat Trout, which required three years, initially began by immigration of juvenile fish (age-1+) into the disturbed area, followed in the second and third years by enhanced recruitment of fry.

Although responding to press disturbance, salmonids in a Canadian stream required at least a year to successfully recolonize a rewatered section of the Bridge River after a long period of no or greatly reduced water flow. The Bridge River, a British Columbia tributary of the Fraser River, was impounded in 1963 and most of the captured flow (annual mean discharge of 100 m^3s^{-1}) was redirected into another watershed for hydropower production (see also Chapter 14). As a consequence, there was no flow in a 4 km section below the dam, after which groundwater and small tributaries resulted in a small flow of 0.7 m^3s^{-1} for the next 11 km before being substantially augmented by a large tributary (Decker et al. 2008; Bradford et al. 2011). Once the flow was restored in the 4 km reach, the flow was only at a level of 2–5 m^3s^{-1}, 2–5% of the original annual mean discharge (as the channel was regraded to accommodate the reduced flow), but did result in some positive responses. There was rapid recolonization of periphyton and aquatic insects, but colonization by fishes was much slower. Juvenile salmonids (primarily Steelhead and Rainbow Trout, *Oncorhynchus mykiss*; Coho Salmon, *O. kisutch*; and Chinook Salmon, *O. tshawytscha*) did not move upstream even though an invertebrate prey base was available within three months. Instead, colonization was primarily the result of upstream movement of adult anadromous fishes that successfully spawned in the restored habitat. Coho and Chinook salmon spawned in the fall of 2000 and Steelhead spawned the following year. By one year after the resumption of flow, populations of age-0 Rainbow Trout and juvenile Coho and Chinook salmon in the rewetted area were equivalent to downstream populations in the continuously wetted site.

Resilience is also shown by life-history responses of fishes, such as the timing and duration of reproductive cycles or the length of the reproductive life span. The Splittail (*Pogonichthys macrolepidotus*), a cyprinid endemic to the Sacramento-San Joaquin Estuary in west-central California, requires inundated floodplains for successful reproduction (Sommer et al. 1997). Submerged terrestrial vegetation on inundated floodplains is used as a feeding area by prespawning adults, as a spawning substratum, and as a larval nursery area. Low-flow years result in substantial reductions in the production of age-0 fish, in contrast to large increases of age-0 fish during wet years. Because the adults have a reproductive life span of three or more years, as well as a high fecundity, the populations are moderately resilient to periodic drought years that limit successful reproduction (Sommer et al. 1997).

Fishes in a small Ontario, Canada, lake also demonstrate resilience to harsh conditions through survival of long-lived adults. The lake was acidified by the addition of sulfuric acid for eight years and then recovery studied for

13 years as part of a large investigation on the effects of acid precipitation (Mills et al. 1987). Three of the five species studied by Mills et al. (1987) (Lake Trout, *Salvelinus namaycush*; Pearl Dace, *Margariscus margarita*; White Sucker, *Catostomus commersonii*) survived the acidification but were not able to successfully reproduce as the pH level dropped. Once acidification stopped and the pH began to gradually rise, recruitment of all three species gradually resumed. Two other species, Fathead Minnow (*Pimephales promelas*) and Slimy Sculpin (*Cottus cognatus*), were extirpated from the lake during the acidification period, but only Fathead Minnow successfully recolonized from a nearby lake during the 13 years following acidification. Population levels of fishes, especially the top predator, Lake Trout, had not reached preacidification levels by the end of the 13-year study of recovery—likely a reflection of the still recovering prey base.

Fishes also may show resilience to unfavorable spawning conditions by having extended reproductive seasons. The Longnose Shiner (*Notropis longirostris*), a cyprinid found in small, upland streams in the southeastern United States, has a short life span of only 1–2 years. However, resilience to poor spawning conditions is achieved by having a protracted spawning season that begins in February and can extend into October (Heins and Clemmer 1976; Ross 2001). Similar patterns of extended spawning seasons in association with short life spans are shown for numerous other southeastern minnows such as Red Shiner (*Cyprinella lutrensis*), Blacktail Shiner (*C. venusta*), and Weed Shiner (*Notropis texanus*) (Ross 2001).

In summary, resilience in fishes can be achieved by movement of adults or juveniles back into a previously disturbed area. Resilience to poor spawning conditions or unfavorable conditions for larval/juvenile survival occurs through elevated longevity of adults so that they can wait out poor years. On an annual basis, short-lived fishes show resilience to poor spawning conditions by having extended reproductive seasons. Overall, there is considerable variation

in the resilience of fish species and fish assemblages to perturbations. Variation occurs across multiple levels including the nature, timing, and severity of the disturbance; the type and location of the aquatic system; species characteristics; and life-history stage (Schlosser 1985; Detenbeck et al. 1992; Albanese et al. 2009).

Levels of Persistence and Stability in Lotic Systems

Considering a wide range of studies, lotic systems tend to show moderate to high levels of persistence and low to moderate levels of stability, with the degree of stability influenced by the metric used to test it. In a survey of 49 primarily North American stream sites that had been subjected to various types of disturbance, Detenbeck et al. (1992) found full or near-complete recovery within two years. Analysis of 25 long-term studies (≥ 2 years; median = 11 years; range 2–45 years) designed or amenable to testing assemblage persistence and stability, and including from 3 to 95 species, showed 76% high persistence and 52% high stability in at least one type of measure (Table 6.1). Environmental harshness, especially if the harshness was related to anthropogenic impacts, had a strong effect on assemblage persistence and stability (Figure 6.5A). In systems judged to have low stress, 100% of the assemblages were persistent and 80% stable. In contrast, for systems judged to have moderate or high stress, only 22% were persistent and 11% stable. In systems with obvious human disturbance, only 14% of the assemblages were considered persistent or stable (Figure 6.5B). However, the sample size is too limited to separate the impacts of human versus natural disturbances; of the 10 studies having moderate to high disturbance, only three were disturbed by nonhuman impacts. The data in Table 6.1 are also biased by geographical region; most of the studies were at lower latitudes (mean latitude = 35.7°; range = 31°–42°) and 84% were done east of the continental divide. However, recall the challenges of assessing assemblage persistence and stability in western fish faunas with long-lived species.

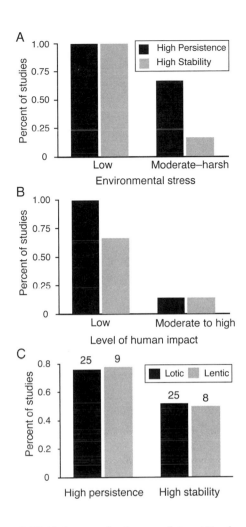

FIGURE 6.5. Impacts of environmental stress (A) and the level of human disturbance (B) on the degree of persistence and stability of lotic fish assemblages, and (C) a comparison of assemblage persistence and stability in lotic and lentic systems. Numbers above bars show sample sizes. Based on data from Tables 6.1 and 6.2.

There is also a bias in stream size as only two studies dealt with large rivers.

Examples of Persistence and Stability in Lotic Systems

Brier Creek, an Oklahoma tributary of the Red River (now inundated by Lake Texoma), is routinely subjected to extreme conditions and has been particularly well studied. In spite of extreme conditions, including total dewatering of some stream reaches, the fish fauna over an 18-year period showed strong persistence on a stream-wide basis, in that abundant species continued to remain abundant and rare species remained rare, with only a few exceptions. Stability of the Brier Creek fish fauna showed greater variation, as measured by indices of similarity of the sampled fish fauna among years. The fish fauna at individual collection sites (i.e., at the local assemblage level) showed less persistence and stability.

The timing of perturbations can have a major influence on the resultant impacts to aquatic organisms. If flooding in Brier Creek occurs when fish are spawning, there are severe impacts on larval survival. For instance, Harvey (1987) showed that larval cyprinids and centrarchids that were less than 10 mm TL were displaced downstream and killed by a major flood event.

Long-term data also exist for Piney Creek, a permanent upland stream in the Ozarks (Ross et al. 1985; Matthews 1986b; Matthews et al. 1988) that offers a more benign habitat (i.e., no dewatering and less temperature variation). Not surprisingly, Piney Creek fishes also showed high persistence; however, in contrast to Brier Creek, the fish fauna in Piney Creek also showed greater faunal stability, both overall and at the assemblage level. Piney Creek had a severe flood in 1982; however, immediately after the flood there were no major changes in rank abundance of the 10 most abundant species (Matthews 1986b). Less common species did show changes in abundance, so local assemblages were altered immediately postflood. Eight months after the flood, the overall fish fauna and the fauna at individual collecting stations had essentially recovered to preflood conditions, leading Matthews (1986b) to conclude that the Piney Creek fish fauna showed stability and persistence across years and across a range of flow conditions.

Although some studies have shown that fish assemblages rebound rather quickly from flooding, as discussed previously, and as documented also by Taylor et al. (1996) for mainstem and tributary sites in the upper Red River system of Oklahoma, other studies indicate that floods or droughts acted to change or reset

TABLE 6.1

Long-Term (≥ 2 years) Studies of North American Stream Fish Assemblages Organized
from Low to High Levels of Stress and from Low to High Latitudes

Site	Habitat[a]	Temporal scale (years)	Spatial scale (km)[b]	Latitude[c] (DD)	Longitude[c] (DD)	Stations	Potential stressor and stress level	No. spec analy
Black Creek, MS	Medium stream	9	14.2	31.2	89.5	5	Annual overbank flooding (low stress)	25
Pearl River, MS	Large river	16	42	31.5	90.1	8	Periodic flooding; upstream impoundments (generally low stress)	28
Undisturbed streams, Savannah River Site, SC	Small stream	2	1.35	33.4	82	9	Annual variation in flow (low stress)	15[d]
Kiamichi River, OK	Medium stream	6	106	34.5	95.4	6	Annual high flows (low stress)	10
Ball Creek, NC (upstream site)	Small stream	4	0.03	35.1	83.4	1	Spring flooding; late spring to fall drought (generally low stress)	3
Ball Creek, NC (downstream site)	Small stream	4	0.03	35.1	83.4	1	Spring flooding; late spring to fall drought (generally low stress)	4
Coweeta Creek, NC	Medium stream	4	0.03	35.1	83.4	1	Spring flooding; late spring to fall drought (generally low stress)	5
Coweeta Creek, NC	Medium stream	10	0.04	35.1	83.4	1	Annual variation in flow, including droughts (generally low stress)	16
Piney Creek, AR	Medium stream	15	27	36.1	92.1	5	Periodic flooding (low stress)	10
Martis Creek, CA	Small stream	11	2.9	39.3	120.1	4	Periodic flooding; severe spring flood in 1983; nonnative predators (generally low stress)	5

Author's conclusions	Human disturbance	Low stress	High persis-tence	High stability	Source
...emporal persistence; temporal stability based on ...umbers of individuals and rank-order data	N	Y	Y	Y	Ross et al. 1987
...emporal persistence overall; temporal stability ...verall based on similarity analyses; high varia-...on in numbers of individuals (CV = 1.03)	N	Y	Y	Y	Gunning and Suttkus 1991; data analyzed by Matthews 1998
...emporal persistence overall; temporal stability ...ased on rank-order data and similarity analyses; ...noderate stability based on CV of 0.44	N	Y	Y	Y	Paller 2002
...emporal persistence of common species overall; ...emporal stability overall based on rank-order ...ata and similarity analyses; stability at 3 indi-...idual stations and instability at 3 others based ...n rank-order data.	N	Y	Y	Y	Matthews et al. 1988
...emporal persistence of resident species; moder-...te temporal stability of resident species based on ...ctual abundances (mean CV = 0.53)	N	Y	Y	Y	Freeman et al. 1988
...emporal persistence of resident species; tempo-...al stability of resident species based on relative ...bundance; moderate to low temporal stability ...ased on actual abundances (mean CV = 0.75)	N	Y	Y	Y	Freeman et al. 1988
...emporal persistence of resident species; tempo-...ral stability of resident species based on relative ...abundance; moderate to low temporal stability ...based on actual abundances (mean CV = 0.62)	N	Y	Y	Y	Freeman et al. 1988
Temporal persistence of common species (mean C_j =0.79; range = 0.67–1.0); temporal stability altered by drought (predrought, drought, and postdrought assemblages distinct)	N	Y	Y	N	Grossman et al. 1998; additional analysis by STR
Temporal persistence overall; temporal stability overall based on rank-order data and similarity analyses; temporal stability at individual stations based on rank-order data	N	Y	Y	Y	Ross et al. 1985; Matthews et al. 1988
Temporal persistence overall; low temporal stability based on relative-abundance data (spe-cies abundances changed dramatically after 1983 flood)	N	Y	Y	N	Strange et al. 1992

(continued)

TABLE 6.1 *(continued)*

Site	Habitat[a]	Temporal scale (years)	Spatial scale (km)[b]	Latitude[c] (DD)	Longitude[c] (DD)	Stations	Potential stressor and stress level	No. of species analyzed
Otter Creek, IN	Medium stream	12	0.12	39.5	87.4	1	Upstream mill dam; no other major impacts (low stress)	18
French Creek, NY	Small to medium stream	42	43	42.1	79.6	9	Normal seasonal variation in flow and temperature (low stress)	41
Sagehen Creek, CA	Small stream	10	21	34.4	120.2	11	Periodic flooding; severe winters; no major human disturbances (low to moderate stress)	8
Bogue Chitto River, LA	Medium stream	27	165	31	90.2	7	Land use changes including increases in human population, dairy farming, cattle ranching, gravel mining, road construction, and silviculture (moderate stress)	95
Little Uchee Creek, AL	Small stream	35	4	32.5	85.3	2	Increase in pine monoculture; 69% human population increase in region; 39% decline in annual flow; flashier runoff (moderate stress)	12
Wacoochee Creek, AL	Small stream	35	11	32.6	85.1	4	Increase in pine monoculture; 69% human population increase in region; flashier runoff (moderate stress)	20

Author's conclusions	Human disturbance	Low stress	High persis-tence	High stability	Source
emporal persistence (mean C_j = 0.80); low to oderate stability (PSI = 0.47); low stability sed on numbers of individuals (CV = 1.37)	N	Y	Y	N	Whitaker 1976; Grossman et al. 1982; data reanalyzed by Matthews 1998
emporal persistence overall; temporal stability sed on species abundances	N	Y	Y	Y	Hansen and Ramm 1994
emporal persistence; moderate temporal stabil-y based on rank-order data; low temporal stabil-y based on changes in standing crop	N	Y	Y	N	Gard and Flittner 1974
ow to moderate temporal persistence (C_j = 66–4%); temporal stability low (27 year comparison) high (11- and 16-year comparisons)	Y	N	N	N	Stewart et al. 2005; additional analysis by STR
ow to moderate temporal persistence with rare pecies eliminated (mean C_j = 0.57; range 0.22–.0); moderate temporal stability (mean I_m = 0.71; ange 0.22–0.96)	Y	N	N	N	Johnston and Maceina 2009; additional analysis by STR
Low to moderate temporal persistence with rare species eliminated (mean C_j = 0.27; range 0.14–0.50); low temporal stability (mean I_m = 0.53; range 0.24–0.88)	Y	N	N	N	Johnston and Maceina 2009; additional analysis by STR

(continued)

TABLE 6.1 *(continued)*

Site	Habitat[a]	Temporal scale (years)	Spatial scale (km)[b]	Latitude[c] (DD)	Longitude[c] (DD)	Stations	Potential stressor and stress level	No. spec analy
Halawakee Creek, AL	Small stream	35	8	32.7	85.3	2	Increase in pine monoculture; 69% human population increase in region; flashier runoff (moderate stress)	15
Martis Creek, CA	Small stream	5	2.9	39.3	120.1	4	Periodic flooding; nonnative predators (moderate stress)	7
Blue River, KS	Large river	45	350	39.6	96.6	14	Reservoir construction; introduction of nonnative species (moderate stress)	29
Wabash River, IN	Large river	25	147	39.8	87.4	29	Dam construction; positive and negative changes in water quality; urbanization; periodic flooding (moderate stress)	75
Cedar Fork Creek, OH	Medium stream	10	0.27	40.6	82.6	1	Annual flooding (moderate stress)	30
Brier Creek, OK	Small stream	18	15	33.9	96.8	5	Flash flooding and drought (moderate to high stress)	10
Aravaipa Creek, AZ	Medium stream	15	41	37.6	104	3	Flash flooding and drought (moderate to high stress)	7
Disturbed streams, Savannah River Site, SC	Small stream	2	1.2	33.4	82	8	Postthermal discharge; periodic anoxic discharge; toxic chemicals (high stress)	14[d]
Purgatoire River tributaries, CO	Small streams; some intermittent	5	41	37.4	103.8	5	Flash flooding and drought (high stress)	11

SOURCE: Modified from Matthews (1998) and Ross and Matthews (in press).

a. Relative stream size is based on stream orders when available (small stream: orders 1–3; medium stream: orders 4–6; large river: orders

b. Spatial scale, if not stated, approximated from map of study area

c. Latitude and longitude are from the approximate midpoint of the study area

d. Average over all sites

Author's conclusions	Human disturbance	Low stress	High persis- tence	High stability	Source
...ow temporal persistence with rare species ...iminated (mean C_j = 0.33; range 0.22–0.40); ...ow temporal stability (mean I_m = 0.53; range ...36–0.71)	Y	N	N	N	Johnston and Ma- ceina 2009; additional analysis by STR
...emporal persistence overall; temporal stability ...ased on rank-order data; number and biomass ...ata showed high variation (last sample in 1983)	N	Y	Y	Y	Moyle and Vondracek 1985
...ow to moderate persistence (mean C_j = 0.41; ...ange 0.2–0.54) temporal stability based on rela- ...ive abundances	Y	N	N	Y	Gido et al. 2002
...Moderate temporal persistence overall; low ...emporal stability overall based on Bray-Curtis ...imilarity; similarity decreased with greater time ...between samples to approximately 0.25; low ...similarity at individual stations based on multi- ...variate measures using abundances	Y	N	Y	N	Pyron et al. 2006
Temporal persistence; temporal stability indi- ...cated by consistency in rank-order data; high ...variation in numbers of individuals	N	Y	Y	Y	Meffe and Berra 1988
Temporal persistence overall; low temporal stabil- ...ity overall based on rank-order data and similarity ...analysis (I_m = 0.40)	N	N	Y	N	Ross et al. 1985; Mat- thews et al. 1988
Temporal persistence of species overall; temporal ...stability based on rank-order data; actual num- ...bers fluctuated extensively	N	N	Y	Y	Meffe and Minckley 1987
Low temporal persistence; low temporal stability ...overall based on rank-order data and similarity ...analyses; mean CV based on actual abundance ...= 0.59	Y	N	N	N	Paller 2002
Temporal persistence at 4 of 5 sites (5th site ...had intermittent flow); low temporal and spatial ...stability due primarily to variation in numbers of ...rare species; stability greater in sites with deep ...pools than with only shallow riffles	N	N	Y	N	Fausch and Bramblett 1991

C_j = Jaccard coefficient

CV = coefficient of variation

PSI = proportional similarity index

I_m = Morisita's Index of Similarity

assemblage structure. Matthews and Marsh-Matthews (unpublished data) have recently found that two severe droughts resulted in a substantial change in the Brier Creek fauna, which did not recover to its former state until 3–4 years postdrought. Another example of how fish assemblages are affected by perturbations emphasizes the significance of timing of the event. In Coweeta Creek, North Carolina, a severe drought resulted in three distinct assemblages over a 10-year period corresponding to predrought, drought, and postdrought conditions (Table 6.1) (Grossman and Ratajczak 1998; Grossman et al. 1998).

In the Sierra Nevada mountains of California, a severe spring flood in Martis Creek shifted the assemblage from being dominated by native, spring spawning species, to domination by the nonindigenous, fall spawning Brown Trout (*Salmo trutta*) (Table 6.1) (Strange et al. 1992). The importance of timing of floods relative to life history is also shown by responses of a northwestern fish assemblage in the John Day drainage, Oregon. Fishes that spawned in late spring and summer, such as Speckled Dace and Bridgelip Sucker (*Catostomus columbianus*), showed high losses of young-of-year individuals to summer flooding, whereas early spring spawners, such as Rainbow Trout, were more susceptible to spring flooding when the developing embryos were still in the redds (gravel nests) (Pearsons et al. 1992). In addition, losses of fishes due to flooding were greater in stream sections with low habitat complexity, leading Pearsons et al. (1992) to suggest that complex habitats may act as sources of individuals for the colonization of structurally simple habitats.

Levels of Persistence and Stability in Lentic Systems

There are relatively few studies with suitable data for assessing the temporal persistence and stability of lakes and reservoirs (Table 6.2). However, in an analysis of nine long-term studies (median duration = 15 years; range 11–72) of lentic fish assemblages (eight natural lakes and one impoundment), the levels of persistence

and stability were essentially the same as those for lotic systems (Figure 6.5C; Table 6.2). Reduced persistence or stability of assemblages tends to occur in altered habitats and/or in habitats impacted by nonnative plants or animals. In contrast to lotic systems, all but one of the lentic studies had suffered moderate or major human impacts, primarily through commercial fishing pressure, the introduction of nonnative plants and animals, and overall urbanization within the watershed.

Examples of Persistence and Stability in Lentic Systems

Numerous lakes exhibit evidence of negative impacts on persistence and stability. Lake Michigan has received considerable study because of heavy fishing pressure and the introduction of nonindigenous species such as Sea Lamprey (*Petromyzon marinus*), Alewife (*Alosa pseudoharengus*), Rainbow Smelt (*Osmerus mordax*), Coho Salmon, Chinook Salmon, Rainbow Trout, Brown Trout (*Salmo trutta*), and Brook Trout (*Salvelinus fontinalis*). Many native species such as Lake Trout, Bloater (*Coregonus hoyi*), and Cisco (*C. artedi*), have shown substantial declines and/or increases as numbers of nonindigenous fishes have fluctuated (see also Chapter 15). Although overfishing, Sea Lamprey predation on large fishes, and competitive interactions all contributed to the decline of indigenous fish species, another factor has been predation on early life-history stages (Stewart et al. 1981; Eck and Wells 1987; Miller et al. 1989). Major shifts in species composition of a small Michigan lake after the loss and then reintroduction of Largemouth Bass (*Micropterus salmoides*) have also been observed (Mittelbach et al. 1995). In Lake Mendota, Wisconsin, which has been impacted by extensive shoreline urbanization and introduction of nonnative vegetation, the fish fauna showed both low temporal persistence as well as low stability (Lyons 1989)

Lentic systems showing greater persistence and stability were generally, but not always, less impacted by habitat alteration or introduction of nonnative species (Table 6.2). In six small Michigan lakes, changes in the fish assemblages were generally low over a four-year period, as determined from a measure of community heterogeneity (based on the average percent dissimilarity over all possible pairs of seine sites within lakes) (Benson and Magnuson 1992). Somewhat surprisingly, heterogeneity among seine hauls within a site (33 m of shoreline) was of the same magnitude as heterogeneity among sites, thus suggesting that fishes were responding to small-scale patchiness of the environment.

In a 15-year study of two depauperate Arctic lakes, Johnson (1994) examined long-term stability of Arctic Char (*Salvelinus alpinus*) populations. Arctic Char was the sole species in one lake and one of two species in the other. After an initial period of moderate (one lake) and high (the other lake) fishing exploitation, the lakes were allowed to recover for several years. In both lakes, age and size structure of Arctic Char returned to the original condition, indicating population stability.

Using long-term data, Gido et al. (2000) examined stability and persistence of a pelagic reservoir fish assemblage over a 43-year period in Lake Texoma, a large impoundment on the Oklahoma-Texas border. Except for the introduction of two species within the study period, Striped Bass (*Morone saxatilis*) and Threadfin Shad (*Dorosoma petenense*), the fauna showed persistence and stability as determined from the rank order of species. Numbers of individuals of each species showed greater fluctuations, with coefficients of variation of the 11 most abundant species ranging from 11–108% over the years.

Persistence and Stability Summary

At the scale of the entire fish assemblage and across a wide range of systems, persistence is fairly common and stability somewhat less so, although both can be impacted by the timing, type, and magnitude of perturbations. Persistence and stability in lentic and lotic systems are generally reduced following severe human disturbances (see also Schlosser 1982; Matthews 1998), and such disturbances occur in both types of systems, although they are more common in lentic (89%) compared to lotic studies (28%). At the population level, rare species, and especially those with limited vagility, recover more slowly or not at all from perturbations, as shown by Albanese et al. (2009) for fishes in the James River drainage, Virginia. Although rare species have less of an effect on most measures of assemblage similarity, there is a greater risk of losing such species either from local habitats or system-wide. Because of this, although responses of assemblages to perturbations indicate generally high stability, postimpact fish assemblages (even if judged highly similar to preimpact assemblages by most measures of assemblage structure) might differ in the loss of rare species.

Although there are fewer lentic compared to lotic studies, those in lentic systems tend to compare longer time intervals (median 15 versus 11 years) and half as many species (mean 10, range 1–20, versus mean 21, range 3–95). Compared to lotic studies, lentic studies were also generally at higher latitudes (mean 49, range 34–64, versus mean 36, range 31–42)—regions that typically have lower fish species diversity. Thus our understanding of persistence and stability in streams, reservoirs, and lakes is incomplete because of relatively few studies and biases in geographic location, species richness, and length of comparisons.

Persistence and Stability of Local Associations

The previous sections dealt with levels of change in species assemblages over time periods of two or more years and over moderate to broad spatial scales. Much less is known about how close contacts of species in associations change over time—for instance how long do multispecies groups remain and do they remain together long enough for reciprocal evolutionary responses (coevolution) to occur? It can be difficult to detect association patterns in species of mobile animals. Individuals found in the same sample may or may not have been in close

TABLE 6.2

Long-Term (≥ 2 years) Studies of North American Lake and Reservoir Fish Assemblages
Organized from Low to High Levels of Perceived Stress and from Low to High Latitudes

Site	Habitat[a]	Temporal scale (years)	Surface area (ha)	Lat.[b] (DD)	Lon.[b] (DD)	Productivity	Potential stressor and (categorization of stres
Small Wisconsin Lakes (7)	Lake	11	1–1,608	46.0	89.6	O–M	Natural variation (generally low)
Little Nauyuk Lake, NW Territory, Canada	Lake	15	44.6	68.4	107.7	O	Initial heavy fishing expl tion followed by cessatio exploitation (low)
Gavia Lake, NW Territory, Canada	Lake	15	17.4	68.4	107.7	O	Light to moderate fishin exploitation followed by (sation of exploitation (lo
Lake Texoma	Reservoir	43	36,000	33.9	96.8	M	Natural and human-caus variation in inflow; summer hypoxia; turbid inflo introduction of nonnativ fishes (moderate)
Chequamegon Bay, Lake Superior; shallow water	Lake	24	16,000	47.7	90.8	M	Changes in water quality introduction of nonnativ species (moderate)
Chequamegon Bay, Lake Superior; deep water	Lake	24	16,000	47.7	90.8	M	Changes in water quality; introduction of nonnativ species (moderate)
Lake Mendota, WI	Lake	72	3,980	43.1	89.4	E	Shoreline urbanization; introduction of nonnativ aquatic plants (moderate high)
Wintergreen Lake, MI	Lake	15	15	42.4	85.4	E	High nutrient inflow; loss native predator due to wi terkills in 2 years followe by reintroduction of nativ predator (high)
Lake Michigan	Lake	12	5,775,673	43.2	87.1	M	Commercial fishing; fish stocking; introduction of nonnative species (high)

SOURCE: Modified from Ross and Matthews (in press).

a. Lakes are natural. Reservoirs are impoundments.

b. Latitude and longitude are from the approximate midpoint of the study area.

No. of species analyzed	Author's conclusions	Human disturbance	Low stress	High persistence	High stability	Source
–22	Moderate temporal persistence (species turnover 0–1.1% per year)	N	Y	Y	N/A	Magnuson et al. 1994
1	Temporal persistence; temporal stability based on age structure	Y	Y	Y	Y	Johnson 1994
1	Temporal persistence; temporal stability based on age structure	Y	Y	Y	Y	Johnson 1994
11	Temporal persistence; temporal stability based on rank-order data and similarity analyses; moderate stability based on abundance measures (mean CV = 0.58)	Y	N	Y	Y	Gido et al. 2000
12	Temporal persistence (mean C_j = 0.80); temporal stability based on multivariate analyses of abundances	Y	N	Y	Y	Hoff and Bronte 1999
8	Temporal persistence (mean C_j = 0.83); low temporal stability overall (3 distinct assemblages over time)	Y	N	Y	N	Hoff and Bronte 1999
20	Low temporal persistence (8 species extirpated); low temporal stability based on relative abundance	Y	N	N	N	Lyons 1989
8	Moderate temporal persistence (2 species extirpated by winterkill); low temporal stability based on pre- and postwinterkill assemblages and pre- and postpredator introduction assemblages; high temporal stability based on recovery to initial conditions following native predator reintroduction	Y	N	N	N	Mittelbach et al. 1995
18	Temporal persistence; low temporal stability based on changes in relative abundance, fishery landings, and catch-per-unit effort	Y	N	Y	N	Eck and Wells 1987

C_j = Jaccard coefficient

CV = coefficient of variation

E = eutrophic

M = mesotrophic

O = oligotrophic

PSI = proportional similarity index

enough contact to have had direct interactions with each other. The capture of individuals of two species in a sample may not equate to their direct interaction because most survey methods for fishes, or other mobile vertebrates, have fuzzy boundaries and may sample different microhabitats (Ross and Matthews, in press). Also, fishes found together in a relatively long reach of stream (e.g., Marsh-Matthews and Matthews 2002) may, especially in highly structured habitats, occupy different pools or riffles and thus do not encounter each other daily. For mobile animals distributed across a heterogeneous landscape, consistent spatial associations among species could exist because some species (or some life stages within a species) select similar microhabitats totally independent of each other (see also Chapter 13) (Chapman and Chapman 1993; Grossman et al. 1998; Wilson 1999).

Matthews and Marsh-Matthews (2006a) provided one of the clearest studies of the longevity of multispecies associations. Based on 19 snorkeling surveys taken over 22 years, they examined persistence of associations of eight taxonomic species (including minnows, a top-minnow, sunfishes, and black basses) and 11 "ecospecies" (with the sunfishes separated into piscivorous adults versus insectivorous juveniles) across 14 adjacent pools within a kilometer reach of Brier Creek, Oklahoma (Figure 6.6). For each completed survey of the 14 pools (Figure 6.6A), species associations were compared by constructing a triangular similarity matrix of species pairs based on relative abundances (Figure 6.6B). Next, the strength of species associations over time was determined by sequentially comparing the 18 matrices using the Mantel test, a statistical procedure for comparing the correlation between matrices (Legendre and Legendre 1998). Concordance (based on Z-scores provided by the Mantel test) declined as the interval between samples increased (Figure 6.6D), so that associations within a year were largely concordant, but associations across years within a season were concordant only in late summer. Overall, species associations were

concordant for approximately half of the 18 intervals between snorkeling surveys. Associations were not typically changed by events like floods and droughts, but the second of two very severe droughts in three years coincided with distinct changes in associations of species or ecospecies. The empirical evidence of Matthews and Marsh-Matthews (2006a) suggests that although some smaller species groups (pairs, triads, or foursomes) might remain consistently in direct contact across years or even decades, there was little evidence that whole assemblage, multispecies associations were constant, or that selection pressures due to such multispecies groups would be consistent across such periods of time.

Persistence, Stability, and Control of Fish Assemblages

As pointed out at the beginning of the chapter, an important reason for trying to understand the levels of persistence and stability in fish assemblages is that assemblages that show high persistence, and particularly those with high stability, may be more influenced by biotic interactions (and have greater likelihood of deterministic control) than by abiotic factors (with greater likelihood of stochastic control). Clearly, the level of physical disturbance can influence the position of a community along a gradient of deterministic to stochastic control, as suggested by various authors, with biotic interactions likely more important in communities with low levels of disturbance and abiotic factors more important in communities with high levels of disturbance (e.g., Grossman et al. 1982; Peckarsky 1983). This is not to say that all assemblages showing stability are deterministically controlled or that all assemblages lacking stability are stochastically controlled. For instance, an assemblage with strong deterministic control based on competitive interactions among species could show a lack of stability because of differential time lags in the effects of species interactions (Strong 1983).

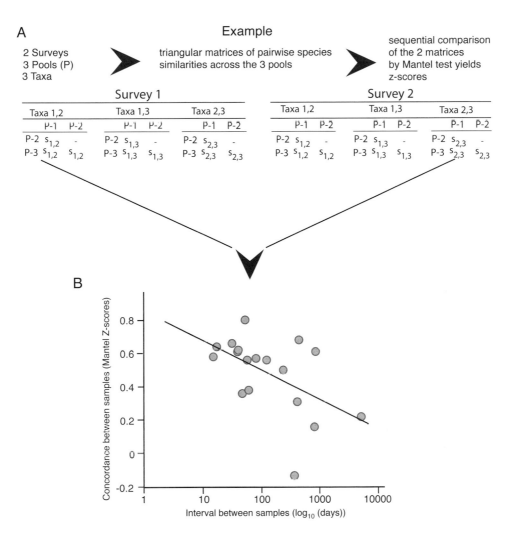

FIGURE 6.6 Persistence of associations among fish taxa in Brier Creek, Oklahoma, based on 19 snorkeling surveys taken over 22 years.

A. A simple example of similarity matrices based on two surveys with three taxa and three pools, where $s_{x,y}$ is the similarity of taxa x and y between two pools.

B. Actual data showing the strength of the associations between consecutive samples relative to the time between the surveys; the solid line is the regression line based on the array of Z-scores. Based on Matthews and Marsh-Matthews (2006a).

One way of addressing the question of the degree of deterministic control of fish assemblages would be to follow an assemblage over time in the absence of major disturbance. However, such systems are uncommon in nature, except for some isolated springs. An alternative approach would be to use a seminatural, artificial stream system. If all streams offer essentially the same environments, then deterministic control should result in high similarity among fish assemblages. Matthews and Marsh-Matthews (2006b) stocked seven outdoor artificial streams with identical numbers and kinds of species and then followed assemblage composition for 388 days. Even though the streams were as identical as possible, differences did develop over time in the extent of algal cover and in the level of predation. Different levels of predation were caused by the differential survival of sunfishes among pools. Somewhat surprisingly, even in the absence of natural disturbances, the assemblages diverged significantly in composition, so that the ultimate structure of any of the experimental assemblages "could not be predicted from its initial structure." In other words, the study did not support predictions of strong deterministic control.

SUMMARY

Fish assemblages change over both ecological and evolutionary time scales. Assemblages controlled primarily by random processes (i.e., primacy of stochastic control) have greater variation in species composition and abundances, compared to those influenced primarily by nonrandom processes (i.e., primacy of deterministic control). Disturbances include any event that disrupts a community or a particular assemblage in some way. Measures of assemblage responses to disturbance include those of persistence (i.e., the presence or absence) of fish species or by stability, a measure that includes the kinds of species and their abundances. Responses of fishes to disturbance can be in the form of resistance to environmental stressors or through resilience—a measure of the ability of populations and assemblages to recover following the disturbance.

Studies of persistence and stability of fish assemblages are challenged because of greatly varying turnover rates in assemblages from different parts of North America. In addition, metrics used to determine stability form

a hierarchical series of increasing sensitivity to change, such that the choice of analysis also influences the outcome. Furthermore, studies should be cognizant of biases. For example, studies on lotic systems are biased toward low-latitude, southeastern regions, in contrast to those on lentic systems, which are biased toward higher latitudes.

SUPPLEMENTAL READING

Albanese, B. W., P. L. Angermeier, and J. T. Peterson. 2009. Does mobility explain variation in colonization and population recovery among stream fishes? *Freshwater Biology* 54:1444–60. Shows the variation in resilience, through recolonization ability, across streams and species.

Grossman, G. D., J. F. Dowd, and M. Crawford. 1990. Assemblage stability in stream fishes: a review. *Environmental Management* 14:661–71. A review of data on fish assemblage stability using coefficients of variation.

Grossman, G. D., P. B. Moyle, and J. O. Whitaker, Jr. 1982. Stochasticity in structural and functional characteristics of an Indiana stream fish assemblage: A test of community theory. *The American Naturalist* 120:423–54. An important paper that stimulated much research and discussion on the relative stability of fish assemblages.

Matthews, W. J., R. C. Cashner, and F. P. Gelwick. 1988. Stability and persistence of fish faunas and assemblages in three midwestern streams. *Copeia* 1988:945–55. Comparisons of fish assemblage persistence and stability using resemblance measures.

Matthews, W. J., and E. Marsh-Matthews. 2006. Temporal changes in replicated stream fish assemblages: Predictable or not? *Freshwater Biology* 51:1605–22. A test of stability of nonperturbed fish assemblages using an outdoor, experimental stream system.

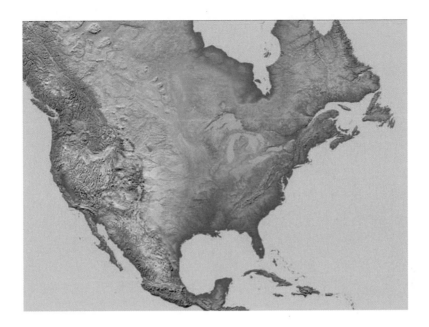

FIGURE 1.4. North American topography illustrating the tectonically active regions of the west; the erosional areas to the east of the western mountains; and the lower, depositional areas primarily in eastern North America. Colors indicate topographic elevation, from white at the highest elevations grading through gray, tan, yellow, and green for progressively lower elevations. The image of North America was generated with data from the Shuttle Radar Topography Mission (SRTM). Courtesy of NASA/JPL-Caltech.

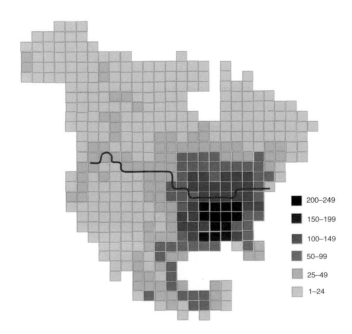

FIGURE 1.5. The number of fish species in equal-sized grids across North America, and the limit of Pleistocene glaciations (solid blue line). Reproduced with permission from G. R. Smith.

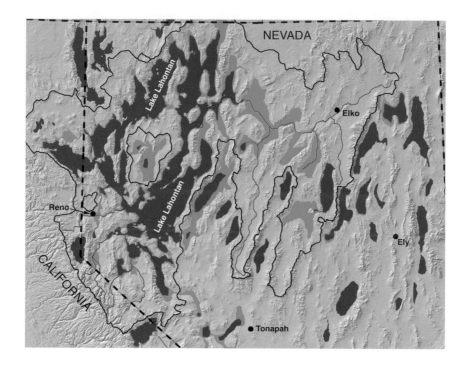

FIGURE 3.3. Pleistocene pluvial lakes and rivers of the western Great Basin in what is now Nevada. Dashed lines are state boundaries, blue areas show the maximum extent of late Pleistocene lakes, green areas show the possible maximum extent of early-middle Pleistocene lakes, red lines show the modern drainages of the Lahontan Basin, and black lines indicate the late Pleistocene extent of the Lahontan Basin. Based on Reheis (1999).

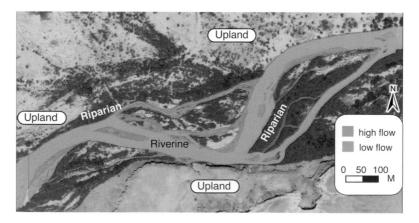

FIGURE 4.1. A section of the San Juan River in Utah showing the complex patterns and juxtaposition of riverine, riparian, and upland landscapes. Aquatic habitats expand during high flow, which can also restructure the streambed and adjoining wetlands. Land and river imagery provided by R. Bliesner, Keller-Bliesner Engineering.

FIGURE 8.2. Elements of jaw protrusion in derived chondrosteans, the Acipenseriformes, showing the freedom of the upper jaw from the braincase. Illustrated by a
21-day posthatch White Sturgeon (*Acipenser transmontanus*). Specimen preparation
and photograph courtesy of Katie May Laumann.

FIGURE 10.3. Male nuptial coloration in the Redspot Darter (*Etheostoma artesiae*).

FIGURE 10.4. Egg mimics on the dorsal fin (arrow) of the Fantail Darter (*Etheostoma flabellare*).

FIGURE 12.8. Apparent rarity through general resemblance, a form of crypsis in the Frecklebelly Madtom (*Noturus munitus*). Based on Armbruster and Page (1996).

FIGURE 12.9. Razorback Sucker (*Xyrauchen texanus*). The nuchal hump in this species and in the Humpback Chub (*Gila cypha*) most likely evolved as a defense against Colorado Pikeminnow (*Ptychocheilus lucius*), the large gape-limited predator of the Colorado River system. Illustration by Howard Brandenburg, courtesy of the New Mexico Game and Fish Commission.

FIGURE 12.10. Examples of posterior fin or body spots in Flier (*Centrarchus macropterus*).

FIGURE 13.3. Pebble nest of a Bluehead Chub (*Nocomis leptocephalus*) from a small stream in Mississippi. The nest is 66 cm × 40 cm. Photo courtesy of Mollie Cashner.

FIGURE 13.4. Pacific salmon migrations import large quantities of marine-derived nutrients into freshwater ecosystems. Picture shows decomposing head of a Chum Salmon (*Oncorhynchus keta*) along the shore of the Skokomish River, Washington. Photograph courtesy of Peter Bisson.

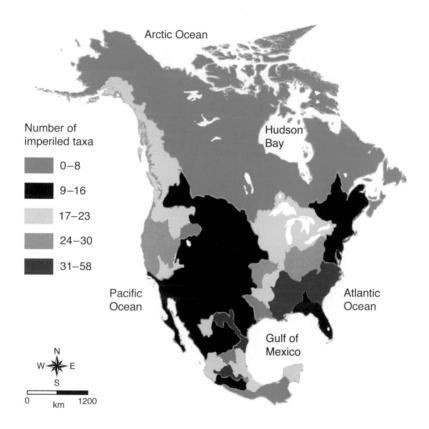

PART 5, FIGURE 1. The distribution of imperilment (extinct, endangered, threatened, or vulnerable) of North American freshwater fishes. Degrees of imperilment are indicated for ecoregions. Redrawn from Jelks et al. (2008) with permission of the American Fisheries Society.

FIGURE 14.9. The nature-like passage stream around a weir on the San Juan River near Farmington, New Mexico. Photo courtesy of James Morel.

FIGURE 15.4. A small spring in Ash Meadows, Nevada. The right side of the picture shows recent efforts to clear emergent and shoreline vegetation.

FIGURE 15.5. A sandy shoreline along the northern shore of Lake Michigan, near the
Straits of Mackinac.

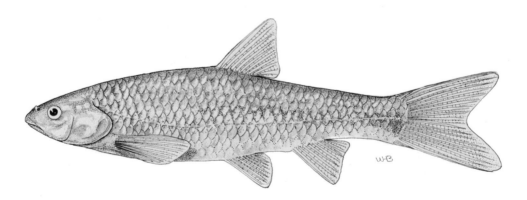

FIGURE 15.8. The Rio Grande Silvery Minnow (*Hybognathus amarus*). Picture courtesy of Howard Brandenburg,
courtesy of the New Mexico Game and Fish Commission.

PART THREE

Form and Function

The previous parts dealt with how fishes respond to their environments and to each other at small and large spatial and temporal scales. However, this has been largely a "phenomenological" approach (sensu, Koehl 1996), where species and individuals have been treated as "black boxes" that perform certain functions. The chapters in this part take more of a "mechanistic" approach in exploring the interactions of morphology, ecology, and evolution, and the resultant impacts on fish populations and assemblages.

SEVEN

Morphology and Functional Ecology of the Fins and Axial Skeleton

CONTENTS

VERTEBRATE EVOLUTION BEGAN in an aquatic environment in the early Paleozoic (500+ mya), followed by the evolution of tetrapods and then the evolution of terrestriality in the middle Devonian (390 mya) (Clack 2002; Nelson 2006). The aquatic and terrestrial environments occupied by vertebrate organisms offer their own sets of challenges and opportunities. For instance, unlike air, water is incompressible for all practical purposes and has much greater viscosity (the resistance of a fluid to deformation because of internal friction). Viscosity becomes increasingly significant as body size decreases and so is an especially important issue for larval stages of fishes (Webb and Weihs 1986). Because the viscosity and density of water are much greater than in air, movement in water must overcome greater drag compared to terrestrial vertebrates moving over land or flying. As a consequence, aquatic organisms, other than those where speed is not an issue, have streamlined body shapes to reduce the energy requirements of locomotion. Also, volume for volume, oxygen content in water is about a thirtieth of that in air (Kramer 1987), and obtaining oxygen from water is additionally challenging by the need to move a viscous medium across respiratory surfaces. Compared to movement on land, the lack of a solid surface to push against reduces the resultant force, although water is a much more efficient medium to push against compared to air. In contrast to terrestrial vertebrates, because their density is close to that of water, aquatic vertebrates gain all or a majority of their bodily support from water rather than having to invest in a skeletal system that can carry the weight of the body. In addition, little energy is required to move vertically. In a now-classic study, Schmidt-Nielsen (1972) provided a way of comparing some of the costs and benefits of movement in water, air, and on land. He determined that the net energetic cost of powering 1 gram of vertebrate over 1 km relative to body size was lowest for swimming, intermediate for flying, and greatest for running. The disciplines of fish biomechanics and hydrodynamics are presently very active, due in part to new technologies allowing the precise quantification of water flow patterns around swimming fishes (Lauder and Tytell 2006). This chapter explores the interaction of morphological evolution in fishes with their success in various freshwater habitats.

Basics of Fish Propulsion

The body of a fish essentially consists of a compression resistant notochord or vertebral column, surrounded by lateral musculature, and wrapped in a complex arrangement of connective tissue and skin (Danos et al. 2008). In contrast to terrestrial locomotion, where the limbs involved in locomotion must also support the body, fishes can use a variety of mechanisms for locomotion, both independently and in concert, and can employ a variety of control surfaces such as scales, body projections, and fins to affect their posture and position in the water column (Webb 1994, 2006).

Forces to Overcome

To achieve forward motion, the force generated by a swimming fish must equal (constant swimming speed) or exceed (acceleration) the resistance to movement caused by drag (Webb 1975; Blake 1983a). The two components of drag are friction drag and pressure drag, both of which can best be understood by boundary-layer theory. Water moving across the body of a fish, either by the fish moving through water or holding position in flowing water, has a gradient in relative velocity that increases from 0, where water molecules are in contact with the fish, to that of the free-stream velocity, the velocity of the undisturbed water at some distance from the fish (Blake 1983a). The region between the free-stream velocity and the velocity at the fish is referred to as the boundary layer (Figure 7.1). Flow in the boundary layer can be laminar, resulting in low friction drag, or turbulent, where the resultant eddies form a thicker boundary layer compared to laminar flow and overall friction drag is increased (Webb 1975; Blake 1983a). The change from laminar to turbulent flow is predicted by the Reynolds number, a hydrodynamic measure calculated as

$$R_e = \frac{LU}{v}, \qquad \text{(Equation 7.1)}$$

where L = fish length; U = speed; and v = the kinematic viscosity of water, which is approximately 0.01 cm^2s^{-1} (Webb 1975; Purcell 1977).

Friction drag arises from the viscosity of water in the boundary layer. The greater the surface area of the body, the greater the friction drag. Friction drag also increases exponentially with swimming speed. For laminar flow, the exponent is 1.5, rising to 1.8 for turbulent flow in the boundary layer (Alexander 1967a). Pressure drag is caused by eddies generated along and behind the body by the separation of the boundary layer from the body of the fish (Figure 7.1A). The farther back that boundary separation occurs, the lower the underpressure and the size of the wake. Because turbulent boundary layers separate farther back than laminar boundary layers (Figure 7.1B), the pressure drag resulting from separation of a laminar boundary

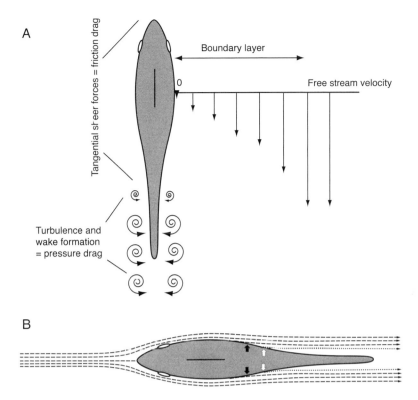

A

Tangential sheer forces = friction drag

Boundary layer

Free stream velocity

0

Turbulence and
wake formation
= pressure drag

B

FIGURE 7.1. Flow separation around a fish holding position in flowing water.
A. Flow lines, friction and pressure drag, and the boundary layer at a point tangential to the
body. The relative thickness of the boundary layer is greatly exaggerated. The length of the
arrows indicates the relative velocity of water, ranging from zero in contact with the body of
the fish to the free-stream velocity indicated by the arrows of identical length on the right.
B. Changes in flow separation from the body in laminar (dashed lines, black arrows) and
turbulent (dotted lines, white arrows) flow. Based on Webb (1975) and Blake (1983a).

layer is higher than that for a turbulent one
(Blake 1983a). Streamlining also reduces boundary layer separation and thus lowers pressure
drag. Other things being equal, pressure drag
increases at approximately the square of velocity (Alexander 1967c). Because of how friction
and pressure drag are formed, a body shape that
reduces friction drag has the opposite effect on
pressure drag. Friction drag is related to surface area, so a body shape that minimizes the
surface-to-volume ratio, such as a sphere, would
have the lowest friction drag. Among freshwater
fishes, a more globular shape, such as shown by
some sunfishes, would have lower friction drag
but a higher pressure drag in contrast to a more
elongate, streamlined fish such as a trout, which
would have higher friction drag but a lower pressure drag (Alexander 1967c).

Generated Forces

Water flowing over the body and fins of a fish
can generate lift because the shapes are acting
as hydrofoils—such lift is often referred to as
dynamic lift. Bernoulli's equation predicts that
pressure will decrease as the velocity of fluid
increases across a surface, so lift for a hydrofoil occurs when flow across the upper surface
exceeds that of the lower, resulting in a pressure
differential (Webb 1975). Such conditions occur
when the angle incidence (α) of the hydrofoil
increases from zero (Figure 7.2). The lift generated by a hydrofoil acts normal to the drag force
and increases with the angle of incidence up to
a point where flow lines begin to separate from
the hydrofoil (usually about 15°), resulting in a
sudden increase in pressure drag and a sudden
decrease in lift so that a stall occurs. Because

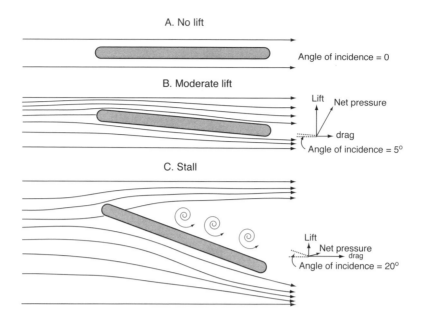

FIGURE 7.2. Flow lines, lift, drag, and the resultant pressure force at three angles of incidence (α) of a hydrofoil. Drag is parallel to the axis of flow (or motion) while lift is normal to the axis of flow or motion. Based on Webb (1975) and Blake (1983a).

the amount of lift generated by turbulent flow is greater than for laminar flows, as a consequence of later separation of flow lines as described previously, higher values of lift occur at higher Reynolds numbers (Webb 1975; Blake 1983a).

Freshwater fishes occupy a wide range of habitats with a correspondingly high range of current speeds and degrees of turbulence. To maintain hydrodynamic stability, change posture, initiate changes in course, or change location, fishes must control translational and rotational forces. Translational forces refer to movement of a body from one point in space to another without rotation and occur in three planes: surge, slip, and heave (Figure 7.3). Surge refers to movement forward or backward, slip refers to sideways movement, and heave refers to movement up or down. Rotational forces refer to movement around the center of mass and occur along three axes: yaw, pitch, and roll (Figure 7.3). Yaw describes the rotation about the center of mass from side to side, pitch is the rotation up or down, and roll is the rotation along the horizontal axis of the body. Some actions do not result in a change of rotational or

translational state because they result in keeping the body in the same location (e.g., hovering) (Alexander 1967c; Webb 2006).

Body Shape, Fin Location, and Maneuverability

Control and maneuverability during hovering or active movement are related closely to fin placement relative to the center of mass, the control of fin rays and fin area by muscles, and swimming speed (Alexander 1967c; Webb 2006). Four zones are recognized relating to fin placement and function (Figure 7.4): (1) an anterior body zone of rudders and lift surfaces positioned anterior to the center of mass that are important in translational forces; (2) a zone of keels located at the center of mass that are particularly important in controlling roll; (3) a zone of stabilizers located immediately posterior to the center of mass and important in controlling yaw, pitch, and roll; and (4) a zone of locomotion and rudders located well posterior to the center of mass that is again important in translational forces (Aleev 1969; Gosline 1971). Anterior control surfaces (zone 1) can include pectoral fins, the head, or the anterior part of

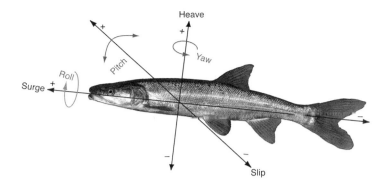

FIGURE 7.3. Terms used in describing translational (black font and arrows) and rotational (gray font and arrows) changes in state about the center of mass in fishes. Photograph of Colorado Pikeminnow (*Ptychocheilus lucius*) courtesy of Tom Kennedy. Based on Alexander (1967a) and Webb (2006).

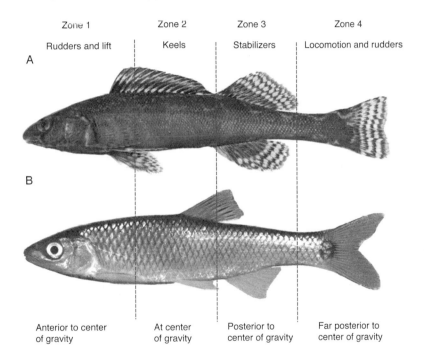

Zone 1	Zone 2	Zone 3	Zone 4
Rudders and lift	Keels	Stabilizers	Locomotion and rudders

A

B

Anterior to center of gravity	At center of gravity	Posterior to center of gravity	Far posterior to center of gravity

FIGURE 7.4. Potential fin functions relative to the center of gravity in (A) higher teleosts illustrated by the Freckled Darter (*Percina lenticula*), and (B) lower teleosts illustrated by the Blacktail Shiner (*Cyprinella venusta*). Based on Aleev (1969) and Gosline (1971).

the spinous dorsal fin, with the head particularly important in turning motion in elongate body shapes (Webb 2006). A fin, such as the spinous dorsal in zone 1, acts to deflect the fish away from its forward course, but during rapid forward progress in a straight line, it is advantageous for it to be folded down, which also helps to reduce drag. Pectoral fins can also be furled during high swimming speeds (Webb 2006). A single dorsal fin located over the center of mass (zone 2) serves as keel but does not stabilize or deflect the forward course of the body. Many lower teleosts, such as herrings, minnows, suckers, catfish, and trout (groups in the Clupeomorpha, Ostariophysi, and Protacanthopterygii; Figure 7.5), have dorsal fins in

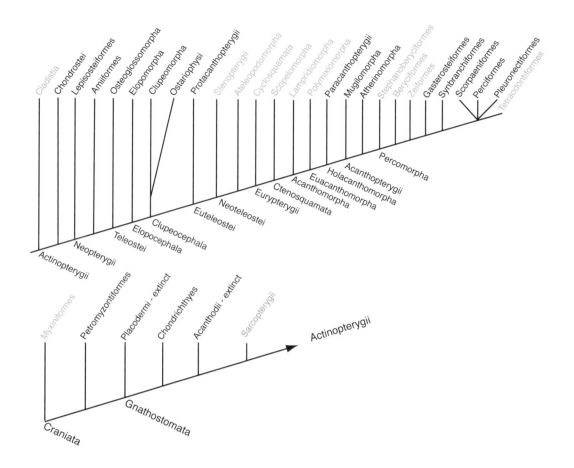

FIGURE 7.5. Major levels of fish evolution. Names at the base of the cladogram define inclusive groups (e.g., Osteoglossomorpha to Tetraodontiformes are included within the Teleostei). Names at the ends of branches refer to particular lineages. Black text identifies groups that have, or had, representation in North American freshwater habitats. The Sarcopterygii includes lobefin fishes as well as tetrapods. Based on Nelson (2006).

this general position or in a position slightly posterior to the center of mass where the fin can also function as a stabilizer (rudder) or aid in propulsion (Figure 7.4B) (Aleev 1969; Gosline 1971). In higher teleosts, such as Moronidae, Centrarchidae, and Percidae (groups in the Acanthomorpha; Figure 7.4A), the dorsal fin consists of two parts, the more anterior spinous dorsal fin and the more posterior soft dorsal fin. The spines can be raised or lowered depending on need. It is important to remember, however, that fins can serve multiple purposes, including camouflage, communication, and in the case of spines, defense.

Many freshwater fishes achieve static lift (= buoyant lift) by having air bladders or low-density fatty inclusions within the body cavity so

that the mass of water displaced approaches the mass of the fish (Gee 1983). However, because the vertebral column bounds the upper extent of the abdominal cavity, low-density inclusions result in the center of buoyancy being beneath the center of mass (Eidietis et al. 2003). The difference between the center of mass and the center of buoyancy is termed the metacentric height, and a negative value, typical of most fishes, results in a rolling torque (a reason why a recently dead or an incapacitated fish turns belly up). Fishes must use behavioral changes, such as resting on the bottom or leaning against structures, or fin movements, to compensate for this inherent instability. To a certain extent, this rolling torque likely was reduced by the location of the swimbladder dorsal to the gut in actinopterygians compared to the ventral position of the lung (the precursor to the swimbladder) in early bony fishes (the lobefin fishes within the Sarcopterygii) such as lungfishes (Lauder and Liem 1983; Webb 2002). Some actinopterygians also have a more anterior location of gas volume such that the pitching torque generated by the mass of the head skeleton is reduced (Webb 2006).

Types of Locomotion

Fish swimming modes can be divided into those involving the body and caudal fin (BCF) and those using various combinations of paired or median fins for locomotion (MPF) (Blake 2004). BCF locomotion is undulatory, involving alternate waves of contractions on either side of the body, because of sequential innervation of lateral body muscles (serial myomeres) that are three-dimensionally folded and divided into blocks by connective tissue (myosepta) (Danos et al. 2008). Furthermore, BCF swimming can be subdivided into steady, continuous swimming versus unsteady, transient (burst and coast) swimming (Blake 2004). Burst-and-coast propulsion occurs in many pelagic and nektonic fishes, and in fishes with streamlined bodies, it can provide considerable energy savings per distance traveled in contrast to steady swimming (Blake 1983b).

BCF swimming typically is categorized into three to five modes: anguilliform, subcarangiform, carangiform, thunniform, and ostraciiform (Breder 1926; Webb 1975; Lindsey 1978). The modes are named after exemplar species and characterized by increasing concentration of the propulsive force in the caudal fin, although they do not imply phylogenetic relationships (Webb 1975; Blake 2004). The ostraciiform mode has a complete, or nearly complete, absence of body undulation with all propulsive power generated by oscillation of the caudal fin. Because the ostraciiform swimming mode is exemplified by tropical marine boxfishes, marine electric rays, and tropical African freshwater elephant fishes (Lindsey 1978; Helfman et al. 2009), and is not represented by

any North American freshwater fish group, it will not be discussed further.

The remaining four modes were originally defined by perceived differences in swimming based on morphology and not on hydrodynamic analyses and, among other things, overlooked the three-dimensional geometry of the body during swimming. Recent research indicates that two-dimensional views of dorsal midline profiles of anguilliform, subcarangiform, carangiform, and thunniform modes are essentially indistinguishable, at least during certain swimming speeds. Because of this, the traditional modes of BCF swimming in fishes are not always representative of hydrodynamic differences and lack a functional basis (Blake 2004; Lauder and Tytell 2006). Current research suggests that thunniform and carangiform modes are quite similar in most, although not all, features. Because of the high similarity between the carangiform and thunniform modes (Blake 2004), and because I know of no North American freshwater fish using a thunniform swimming mode, it is not treated further. The remaining three BCF modes are not distinct in all attributes and are grouped differently based on different functional and morphological criteria, including propulsive wavelengths, wake patterns, tendon lengths, and red muscle activity (Table 7.1) (Lauder and Tytell 2006; Danos et al. 2008). Thus, although useful as general shorthand descriptors of BCF swimming, the taxon-named swimming modes are not totally distinct but share various features.

ANGUILLIFORM BCF LOCOMOTION In anguilliform swimming, which is ontogenetically and phylogenetically the basal mode of BCF swimming in ray-finned fishes, the Actinopterygii (Figure 7.5), the entire body is employed to generate thrust through a series of waves moving from head to tail (Gosline 1971). In contrast to early studies indicating that large amplitude undulations occurred all along the body over a range of swimming speeds, recent work indicates that body waves have increasing amplitude posteriorly, thus increasing water displacement

TABLE 7.1

Similarities and Differences among Commonly Recognized Modes of Body and Caudal Fin (BCF) Locomotion

Trait	Anguilliform	Subcarangiform	Carangiform
Length of active, ipsilateral red-muscle blocks	Short block	Intermediate	Long block
Length-specific propulsive wavelength	Low	Intermediate	High
Wake momentum	Lateral	Primarily downstream	Primarily downstream
Length-specific body amplitude	Greater anteriorly	Smaller anteriorly	Smaller anteriorly
Lateral myoseptal tendon	Short	Short	Long
Epineural myoseptal tendon	Present	Present	Absent
Epipleural myoseptal tendon	Present	Present	Absent
Three-dimensional myoseptal shape	Same	Same	Different
Firing duration of red muscles	Short burst	Short burst	Long burst

SOURCE: Based primarily on Danos et al. (2008) with additional information from Lauder and Tytell (2006).

toward the tail, and that the anterior body region only shows strong undulation during acceleration and not during steady swimming (Müller et al. 2001; Lauder and Tytell 2006). Fishes using anguilliform swimming are elongate and flexible, such as freshwater eels, lampreys, some catfishes, and the larvae of most fishes (Blake 1983a). In contrast to nonanguilliform swimming, anguilliform swimmers are also generally adept at backward locomotion (Webb 2006).

Anguilliform swimming, at least as shown by eels, does differ from other swimming modes in several ways (Table 7.1). Red muscle activation tends to occur in short blocks ipsilaterally, in contrast to long blocks in the carangiform mode and intermediate blocks in the subcarangiform mode (Danos et al. 2008). One of the original descriptors of swimming modes, the propulsive wavelength adjusted for body length, is still useful, being short in anguilliform swimming, intermediate in subcarangiform modes, and high in carangiform modes (Tytell and Lauder 2004; Danos et al. 2008). Even though it tends to increase posteriorly, wave amplitude is also somewhat greater anteriorly

in anguilliform swimming, in contrast to the other modes that are highly similar in this regard (Lauder and Tytell 2006). Wake form differs in anguilliform swimmers, with wakes having lateral momentum but not substantial downstream flow momentum (the momentum opposite the line of thrust of the body), in contrast to other swimming modes. The difference most likely is caused by the absence of a distinct caudal fin structure in eels in contrast to fishes having caudal fins that are distinct from the body (Lauder and Tytell 2006). In five other features, anguilliform and subcarangiform modes do not differ (Table 7.1). These include four features of the myosepta (the sheets of connective tissue separating blocks of myomeres and onto which muscle fibers insert) involving the lateral myoseptal tendon length, the presence of epineural (located on the dorsal surface of the vertebral centrum) and epipleural (located above the abdominal ribs) tendons, and the shape of the myosepta; the fifth similarity is in the firing duration of red muscle fibers (Danos et al. 2008). Red muscle fibers are oxidative and used in slow, prolonged swimming; as such, they are highly vascularized and contain abundant myoglobin, a red oxygen-binding pigment characteristic of muscle (Syme 2006).

LARVAL FISHES AND ANGUILLIFORM LOCO-MOTION During their larval period, the majority of all North American freshwater fishes use anguilliform locomotion in the sense of generating more than one complete propulsive wavelength within the length of the body (Webb and Weihs 1986). Anguilliform swimming in larvae occurs because the musculature and axial skeleton are not sufficiently developed to use lift-based subcarangiform or carangiform modes, both of which would place greater compressive force on the axial skeleton and require more muscular power. In addition, because of their small size and speed, larval fishes operate in an environment dominated by viscous rather than inertial forces so that any cessation of swimming movement stops forward progress—there is no coasting in the absence of inertial forces. The balance between viscous and inertial forces is determined by the Reynolds number (R_e), the same equation described previously for prediction of laminar versus turbulent flow in a boundary layer. $R_e < 1$ indicates a totally viscous environment and $R_e > 1,000$ indicates a totally inertial environment; at intermediate values both forces are represented (Lauder and Tytell 2006), but for values of R_e below 300–450, viscous forces predominate over inertial forces (Webb and Weihs 1986; Fuiman 2002). It is difficult for us to really imagine life at low Reynolds numbers. Purcell (1977), in discussing swimming in microorganisms and the impact of the primacy of viscous over inertial forces, said: "If you are at [sic] very low Reynolds number, what you are doing at the moment is entirely determined by the forces that are exerted on you *at that moment*, and by nothing in the past."

Once the yolk sac is absorbed, larval fishes generally swim at 1–3 body lengths per second (Fuiman 2002). Thus R_e of a 5-mm larval fish would be 25–75, at the lower end of the intermediate range, and subject to viscosity effects. In a viscosity-dominated environment, pushing against the water by an elongate body is more effective than using caudal fin propulsion (Webb and Weihs 1986), but because of the unimportance of inertial forces, larvae must swim continuously to move. (Recall that the law of inertia, or Newton's first law, states that a particle at rest or moving in a straight line with a constant velocity will continue to do so, provided the particle is not subject to an unbalanced force.) As soon as the larvae stop actively swimming, they come to a halt (Purcell 1977; Blake 1983a). As fishes increase in size, the importance of inertial forces increase relative to viscous forces so that once R_e reaches 300–450, they can employ an energy-saving burst-and-glide approach to locomotion (Fuiman 2002).

NON-ANGUILLIFORM BCF LOCOMOTION Increased posterior localization of body undulation and power and the development of distinct caudal fins characterize the traditional modes of subcarangiform and carangiform locomotion (Table 7.1). Subcarangiform swimming occurs in the majority of nonlarval North American freshwater fishes, including salmonids, cyprinids, catostomids, centrarchids, and percids; however, specific studies on swimming are limited to relatively few species of salmonids, cyprinids, and centrarchids (Blake 1983a; Lauder and Tytell 2006). Subcarangiform fishes typically have fairly flexible but low-aspect-ratio caudal fins, such as the fins of many minnows, suckers, catfishes, sunfishes, and darters (Figure 7.4). Aspect ratio expresses the amount of lift generated by a hydrofoil, with lift increasing with aspect ratio, and is determined by the square of fin span divided by fin area. Examples of carangiform swimmers within North American freshwater fishes are less common but potentially include herring and shad (family Clupeidae). Carangiform swimming also is likely approached by two large cyprinid fishes endemic to the Colorado River system, Humpback Chub (*Gila cypha*) and Bonytail Chub (*G. elegans*); both have high-aspect-ratio caudal fins and narrow caudal peduncles, although there are no supporting biomechanical or

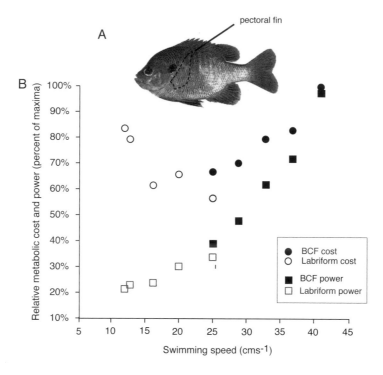

FIGURE 7.6. A. Bluegill (*Lepomis macrochirus*), with the dotted line showing the outline of the pectoral fin used in labriform locomotion. B. Gait change and relative metabolic power and cost as a function of swimming speed in Bluegill (mean length 19.5 cm). The gait transition occurs at approximately 1.3 body lengths per second. Based on data from Kendall et al. (2007).

hydrodynamic studies on these species. Consequently, in terms of BCF swimming, the vast majority of North American freshwater fishes occupy the anguilliform-subcarangiform range of swimming modes.

MPF LOCOMOTION Similarly to categorization of BCF swimming modes, Breder (1926) recognized six undulatory modes of MPF locomotion and one oscillatory mode, based principally on the median or paired fins involved, the general appearance of the waveforms, and the length of the fin relative to body length, but not on functional aspects of fin kinematics (Blake 1980, 1983a). In a simplified system, Blake (1983a; 2004) recognized the distinction between undulatory and oscillatory modes and divided undulatory modes into two groups based on fin kinematics. Type I includes fishes using fins with high amplitude, low frequency, and long wavelengths, such as dorsal fin locomotion in

the Bowfin (*Amia calva*). Amiiform locomotion is also advantageous in allowing backward as well as forward locomotion (Webb 2006). Type II includes fishes using fins with low amplitude, high frequency, and short wavelengths, such as pipefishes (Syngnathidae), a primarily marine group but with some species, such as the Gulf Pipefish (*Syngnathus scovelli*), entering into fresh water. Because thrust is achieved most efficiently by accelerating a large mass of water to a low velocity rather than the reverse, type I locomotion is more efficient than type II (Blake 1983a).

Oscillatory fin locomotion, also referred to as labriform swimming, is shown by fishes using pectoral fins for locomotion, such as mudminnows (*Umbra* spp.), sticklebacks (Gasterosteidae), and certain centrarchids (*Lepomis* and *Pomoxis*) (Figure 7.6; Drucker and Lauder 2000; Walker 2004; Jones et al. 2007). Fishes using MPF locomotion are common in complex habitats such as weedy ponds, lake margins, or streams with abundant submerged or emergent vegetation or woody debris and are adept at backward as well as forward locomotion (Webb 2006).

Gaits, Maneuverability, and Specialization

Swimming fishes can accommodate a wide range of speeds and maneuverability, with different speeds referred to as "gaits" or "gears." Not all fishes or all life stages have all gaits, and potentially different structures may be employed at different gaits (Webb 1994, 2006; Blake 2004). Slow swimming in juvenile and adult fishes can occur via median and paired fins (MPF), which are gradually supplemented and then replaced by BCF swimming as speed increases or when increased acceleration is needed (Webb 1994). For instance, Bluegill (*Lepomis macrochirus*) switch from labriform, pectoral swimming, to undulatory body swimming as speed and power requirements increase (Figure 7.6) (Kendall et al. 2007). Initially, the cost of locomotion decreases while power output increases until approaching a speed of 25 cms^{-1}, corresponding to about 1.3 body lengths per second. Bluegill can achieve swimming speeds above 25 cms^{-1} by involving increased muscle mass, albeit by increased cost, by changing from labriform to BCF locomotion. MPF locomotion also affords a high degree of maneuverability at a low metabolic cost (Kendall et al. 2007).

Gait changes also can involve different muscle groups, obviously in the change from MPF to BCF swimming, but also within BCF swimming. At low speeds, oxidative red muscle, which is aerobic, supplies the power. Red muscle is slow contracting but capable of maintaining cruising speeds without fatigue. As speed increases, fast oxidative glycolytic (pink) and/or fast glycolytic (white) muscle fibers are added (Webb 1994). White fibers have greater power output than red or pink muscle, are good for sprints and fast starts, but have shorter times to fatigue. White and red fibers do not overlap in use in less derived teleosts, including bowfin, gars, hiodontids, and clupeids; in more derived teleosts having multiple innervation of white fibers, graded responses are possible with overlap in firing of red and white muscles at intermediate swimming speeds (Webb 1994).

Loss of Gaits and Specialization in Water-Column Fishes

Fishes that primarily use BCF cruising or sprinting gaits are adapted to exploit widely distributed food resources and are well represented in the marine environment but less common in freshwater habitats (Webb 1994). Fishes having cruising or sprinting gaits are characterized by high-aspect-ratio caudal fins; relatively stiff fins; large anterior body masses; and rigid, streamlined bodies (Webb 1984). Streamlined body shapes are defined as having fineness ratios >2, where fineness ratio is body length divided by maximum diameter (Blake 1983a). Fishes with specialized accelerator gaits, such as ambush predators, have enhanced fast-start gaits, resting on the bottom or hovering in the water column when not actively pursuing prey. Fishes with specialized accelerator gaits entrain a large amount of water along the body to assure maximum thrust during fast starts. As such, they are characterized by having a large caudal area and a low aspect ratio caudal fin, a deep caudal peduncle, a relatively flexible body, and with the dorsal and anal fins located or extended posteriorly. Fishes using accelerator gaits are adapted to take locally abundant, evasive prey (Webb 1984). Two variants of accelerator gaits are characterized by (1) resistance minimizers, such as esocids, which have a large caudal area and elongated, circular anterior body shape; and (2) thrust maximizers, such as cottids, which have a dorso-ventrally flattened body and large fin depth along most of the body length as well as enlarged pectoral fins that also contribute to fast starts (Webb 1984). Esocids, a group of fishes represented in North America since at least the Cretaceous Period of the late Mesozoic (Chapter 2), have the highest acceleration rates recorded to date (Webb 1994). As a trade-off, MPF gaits usually are poorly developed in accelerators (Webb 1994).

Maneuverers are found in complex structured habitats, such as macrophyte beds, but also occur in open water where their maneuvering ability allows them to position their body to capture small prey items. These fishes, such as sunfishes, have large, flexible MPF and short, deep bodies (Gosline 1971). Not surprisingly, MPF gaits, which excel in low-speed swimming, are emphasized in maneuverers, whereas station-holding gaits and BCF sprinting, cruising, and fast-start gaits may be suppressed (Webb 1994). Prey used by maneuverers tend to be relatively nonevasive (Webb 1984).

Loss of Gaits and Specialization in Substratum Fishes

Burrowers are characterized by having elongate bodies with a loss or reduction of projecting appendages, as seen in larval lampreys (ammocoetes) and in American Eels (*Anguilla rostrata*). Specialized burrowers are slow swimmers and show the suppression of fast-start, cruising, or sprinting gaits. Some burrowers might use burst-and-coast swimming to compensate for high-speed swimming costs (Webb 1994). Fishes having flow-refuging gaits live on the bottom of swiftly flowing streams and take advantage of the reduced current flow in the boundary layer. Many stream fishes rely on frictional contact with the substratum to minimize the swimming energy required to maintain position. Examples include the cottid form with low lift and high drag (Webb 2006). Flow-refuging fishes offset drag by having structures to increase friction with the stream bottom or to create negative lift, especially with enlarged pectoral fins as in catostomids, salmonids, darters, and cottids. Such frictional forces are not high, but because of similarities in densities of fish and water, they can be important. As current speed increases, many lotic fishes, such as Longnose Dace (*Rhinichthys cataractae*), Logperch (*Percina caprodes*), and River Darter (*Percina shumardi*) increase their frictional contact with the substratum by releasing gases from their swimbladders and thus increasing their density (Gee 1983). High drag–low lift cottiform fishes suppress MPF and BCF sustainable gaits (Webb 1994).

In addition to morphological and physiological features associated with flow-refuging gaits, a number of North American freshwater fishes have distinctive behaviors that are initiated as flow increases and fishes risk downstream displacement. Oral grasping, literally grabbing a stationary object with the mouth and hanging on to remain stationary in flowing water, has been documented in 10 genera and 19 species of minnows, but these observations have been in a laboratory stream environment and not in nature (Table 7.2). Oral suction for the purpose of remaining stationary, closely related to oral grasping, is known for two species of suckers, and the Bluehead Sucker (*Catostomus discobolus*) has been observed using oral suction on rocks to maintain position in a natural stream (Table 7.2). Studies investigating whether there is oral grasping and suction in minnows and suckers for this purpose, rather than for feeding, have also included species in the families Characidae, Ictaluridae, Fundulidae, Poeciliidae, Atherinopsidae, Centrarchidae, Percidae, and Cichlidae, none of which showed this kind of oral grasping or suction (Leavy and Bonner 2009).

Another flow-refuging behavior is referred to as "parr posture" (Arnold et al. 1991). In this behavior, fishes orient into the current, supporting themselves on the tips of their pectoral fins, with the center of mass located in the middle of a triangle formed by the anterior support of the two pectoral fins and the posterior support of the pelvic fins or body (Webb 2006). This behavior allows fishes such as cyprinids (Ward et al. 2003), salmonids (Arnold et al. 1991), and cottids (Webb et al. 1996) to maintain position in a current, where drifting prey are likely available, at a low metabolic cost. At increased current speeds, fishes lower themselves onto the bottom, lower the dorsal fin, and use the downward force of water on the angled pectoral fins to maintain position. A similar behavior, sometimes accompanied by arching

TABLE 7.2

Oral Grasping and Oral Suction as Station-Holding Behaviors in North American Freshwater Fishes

Common name	Scientific name	Grasping (G); Suction (S)	Lab (L); Field (F)	Source
Bluntface Shiner	*Cyprinella camura*	G	L	Adams et al. 2003
Red Shiner	*Cyprinella lutrensis*	G	L	Leavy and Bonner 2009
Blacktail Shiner	*Cyprinella venusta*	G	L	Adams et al. 2003; Leavy and Bonner 2009
Roundnose Minnow	*Dionda episcopa*	G	L	Leavy and Bonner 2009
Mississippi Silvery Minnow	*Hybognathus nuchalis*	G	L	Adams et al. 2003
Redfin Shiner	*Lythrurus umbratilis*	G	L	Leavy and Bonner 2009
Silver Chub	*Macrhybopsis storeriana*	G	L	Adams et al. 2003
Golden Shiner	*Notemigonus crysoleucas*	G	L	Adams et al. 2000
Emerald Shiner	*Notropis atherinoides*	G	L	Adams et al. 2003
River Shiner	*Notropis blennius*	G	L	Adams et al. 2003
Longnose Shiner	*Notropis longirostris*	G	L	Adams et al. 2003
Taillight Shiner	*Notropis maculatus*	G	L	Adams et al. 2003
Sand Shiner	*Notropis stramineus*	G	L	Leavy and Bonner 2009
Weed Shiner	*Notropis texanus*	G	L	Adams et al. 2003
Topeka Shiner	*Notropis topeka*	G	L	Adams et al. 2000
Channel Shiner	*Notropis wickliffi*	G	L	Adams et al. 2003
Bluntnose Minnow	*Pimephales notatus*	G	L	Adams et al. 2003
Splittail	*Pogonichthys macrolepidotus*	G	L	Young and Cech 1996
Creek Chub	*Semotilus atromaculatus*	G	L	Leavy and Bonner 2009
Desert Sucker	*Catostomus clarkii*	S	L	Ward et al. 2003
Bluehead Sucker	*Catostomus discobolus*	S	L, F	Ward et al. 2003

of the back as current flow increases, has been documented in darters (*Etheostoma* and *Percina*) (Matthews 1985a).

Another pattern of station holding is to minimize drag by having a flattened (depressed) body form such as flounders, soles, and rays. This pattern is common in areas where there is insufficient surface roughness to increase friction. However, space for organ systems results in a "blister-shaped" dorsal profile and relatively large surface area, both of which result in increased lift (Webb 2006). When the body is appressed to the substratum, with the marginal fins lowered to be in contact with the bottom, flow over the eyed side results in lift because of a pressure reduction relative to the stagnation pressure (the pressure in the absence of flow) on the blind side. As flow increases, pressure on the blind side can be reduced by beating of the posterior parts of the dorsal and anal fins, resulting in a substantial, posteriorly directed flow under the body. Because pressure is reduced as flow increases, this behavior helps to equalize pressure between the eyed and blind sides, thus reducing the amount of lift and providing increased station-holding ability (Arnold and Weihs 1978; Webb 1989). Flatfishes also can bury into a soft substratum (Brainerd et al. 1997), a behavior likely important in crypsis as well as for reducing the risk of downstream displacement from currents. The Hogchoker (Achiridae, *Trinectes maculatus*), a small flatfish that occurs in coastal waters and rivers along the Atlantic and Gulf coasts (Burgess 1980), typically is found over nonvegetated, mud or sandy areas in rivers and illustrates this pattern (Ross 2001).

Finally, many fishes are gait generalists, using MPF and BCF gaits, both with well-developed red and white muscles. Generalists, such as Largemouth Bass (*Micropterus salmoides*), typically have good hovering and station-holding gaits, as well as fast-start, sprint, burst-and-coast, and cruising gaits (Webb 1994). In addition to Black Basses (genus *Micropterus*), generalists include many groups that are represented in North American fresh water, such as salmonids, cyprinids, and ictalurids (Webb 1984). Burst-and-coast swimming can occur in a variety of gaits besides generalists.

EVOLUTIONARY TRENDS IN FORM AND FUNCTION

Anguilliform locomotion, which is used by jawless fishes such as lampreys and more derived freshwater eels, does not require as much resistance to compression as other types of BCF locomotion. Consequently, the evolution of nonanguilliform BCF swimming occurred in concert with selection for a vertebral column capable of resisting longitudinal compression by opposing muscles. The vertebral column consists of a series of rigid blocks separated by joints so that bending is permitted but not compression. High flexibility that is important to anguilliform locomotion becomes a hindrance in subcarangiform and carangiform locomotion, and one way to reduce flexibility is to decrease the number of vertebrae. For instance, American Eels have more than 100 vertebrae, and lampreys (Petromyzontidae) normally have between 46 and 74 myomeres (retaining a notochord and lacking

ossified vertebral elements) (Scott and Crossman 1973; Ross 2001). In contrast, the number of vertebrae has been reduced to no more than 24 in more derived teleosts (Gosline 1971).

Placoderms, an extinct group of early fishes that are the sister group to all other gnathostomes (Figure 7.5), dominated the benthic environment of Devonian seas, with some groups likely entering freshwater habitats. These were generally heavily armored fishes with strongly dorso-ventrally compressed bodies and long tails (Moy-Thomas and Miles 1971; Janvier 1997b). Placoderms likely used BCF locomotion, but because heavy body armor would enhance station holding but detract from acceleration, a fast-start gait was likely absent (Webb 1994). The Acanthodii, another now extinct group of Paleozoic fishes that lived from the late Ordovician to early Permian periods, were mostly pelagic (Nelson 2006). Although initially marine, by the Devonian they primarily occurred in freshwater habitats. Fossil Acanthodii have been recovered from Laurasian shale deposits, including some in what is now eastern North America (Bardack and Zangerl 1968). Because they were primarily pelagic and lacked heavy armor, they likely were the first fishes to possess a fast-start gait (Webb 1994). In general, the performance of Paleozoic fishes was probably not high compared to modern fishes. Paleozoic fishes lacked vertebral strengthening and median fins, and appendages generally lacked mobility so that they had more of a role in trim than in propulsion (Webb 1994). An important aspect in the evolution of increasing maneuverability in fishes has been the reduction in the size of the fin base, thus increasing the extent of possible fin movement (Webb 2006).

Beginning in the Carboniferous, major advances in fish locomotion and control occurred within the Actinopterygii, the ray-finned fishes (Figure 7.5). MPF gaits likely arose when fins, which were initially used for trim, acquired independent movement. Over time, actinopterygian gaits improved with increases in strength of axial and appendicular skeletons and increased flexibility of paired fins (Webb 1994).

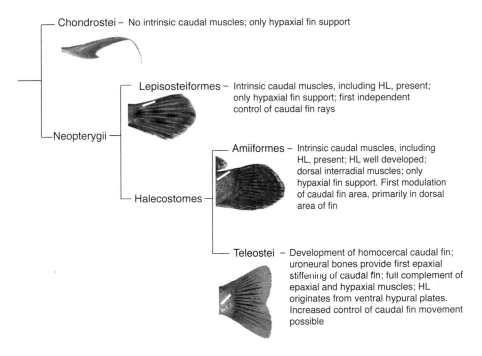

Chondrostei – No intrinsic caudal muscles; only hypaxial fin support

Lepisosteiformes – Intrinsic caudal muscles, including HL, present; only hypaxial fin support; first independent control of caudal fin rays

Neopterygii

Amiiformes – Intrinsic caudal muscles, including HL, present; HL well developed; dorsal interradial muscles; only hypaxial fin support. First modulation of caudal fin area, primarily in dorsal area of fin

Halecostomes

Teleostei – Development of homocercal caudal fin; uroneural bones provide first epaxial stiffening of caudal fin; full complement of epaxial and hypaxial muscles; HL originates from ventral hypural plates. Increased control of caudal fin movement possible

FIGURE 7.7. Patterns in the evolution of caudal fin shape and control in the ray-finned fishes (Actinopterygii). The heavy white line indicates the approximate position of the hypochordal longitudinalis muscles (HL), which are involved in stiffening the upper lobe of the caudal fin. Based on Lauder (1989).

Caudal fins of many teleosts are capable of extensive changes in shape, providing fine control over forces generated during swimming (Lauder and Tytell 2006), and such changes are possible through the evolution of the caudal skeleton and associated musculature.

The primary support of the teleostean caudal fin is the terminal portion of the upturned vertebral column. Extending below and behind the upturned vertebral column are flat bony plates (the hypural bones) that support the fin rays (Box 7.1). The development of caudal fin locomotion resulted in most of the thrust being concentrated on one or several vertebrae at the base of the tail, and the evolution of nonanguilliform BCF swimming has been accompanied by a reduction in the number of vertebrae in the upturned part of the caudal fin to a maximum of two, and fusion of vertebrae within the caudal fin to increase resistance to compression (Gosline 1971, 1997). In the Chondrostei, a group of early ray-finned fishes that includes living forms such as sturgeon and paddlefish, the notochord and

vertebral elements extend well into the upper fin lobe, and fin rays and other supporting elements are connected to the ventral side of the vertebral column (i.e., hypaxial in position) (Figure 7.7). Because the caudal fin lacks intrinsic muscles, the ability to change fin shape or stiffness is limited only to effects of trunk musculature on the caudal fin (Lauder 1989). In gars (Lepisosteiformes), the caudal fin is still supported only by hypaxial elements, but there are muscles within the caudal fin, including the hypochordal longitudinalis muscle (HL), that allow the beginning of independent control of caudal fin rays (Figure 7.7). The HL originates on caudal vertebrae, or on the ventral hypural bones in teleosts, and extends posteriorly to insert on the caudal fin rays. It functions first in stiffening the upper caudal lobe so that it is less inclined to the horizontal during swimming (Lauder 1989).

Bowfin (Amiiformes) also have only hypaxial fin support but show more development of intrinsic caudal musculature, including the first occurrence, in the upper part of the caudal

Major advances in fish locomotion are linked to the evolution of the caudal fin. The following figure illustrates key elements of a teleostean caudal skeleton as shown by a mesolarval River Carpsucker (*Carpiodes carpio*) from the Rio Grande. The hypural bones are the broad plates that provide support for the fin rays. The upturned portion of the vertebral column is generally greater in larval fishes, becoming reduced and differentiated during development into preural (proximal) and ural (distal) centra.

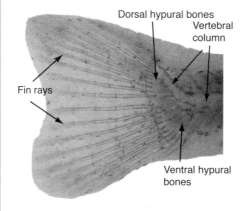

Caudal fin morphology of a larval River Carpsucker (*Carpiodes carpio*). Photo courtesy of Trevor Krabbenhoft.

compared to less derived groups. In modern teleosts such as *Lepomis*, tail thrust depends to great extent on caudal muscle activity, particularly the HL. At low to moderate constant swimming speeds, the HL is not contracted so that the upper lobe of the caudal fin provides less thrust than the lower lobe. However, during rapid acceleration, the HL is active and serves to stiffen the upper caudal lobe so that the caudal fin generates much more symmetrical thrust (Lauder 1989).

In lower teleosts, the pectoral fins are low on the body, the pelvic fins are posteroventral, and the pectoral fin base tends to be more horizontal (Figure 7.4B). In the more derived spiny-rayed teleosts (the Acanthomorpha) (Figure 7.5), the pelvic fins are anterior and ventrolateral rather than posterior, and the pectoral fins are located higher on the body, with the fin base shifted to a more vertical position (Figs. 7.4A, 7.6A) (Rosen 1982). The increased body depth and expansion of the median fins in certain groups, such as sunfishes, results in the placement of MPF propulsors around the center of mass. Such placement likely provides increased stability for hovering, slow swimming, and maneuvering (Webb 2006). However, the pattern in lower teleosts provides increased resistance to rolling, with the ventral position of pectoral and pelvic fins allowing increased stability through contact with the bottom. For instance, in a study of rolling stability in Creek Chub (*Semotilus atromaculatus*), Largemouth Bass, and Bluegill, Creek Chub showed significantly greater resistance to rolling compared to the two centrarchids (Eidietis et al. 2003). Creek Chub were particularly adept at gaining support from the ground using parr posturing, as described in the previous section. Furthermore, recent studies of pectoral fin function in salmonids have dispelled the idea that the more ventrally placed pectoral fin position of lower teleosts limits its functional importance. The salmonid pectoral fin shows a wide range of kinematic function in hovering, swimming, and braking. In slow-speed turning using pectoral fins, salmonids achieved the same angular velocities as Bluegill (a higher teleost) performing the same turning motion (Drucker

fin, of interradialis muscles, which go between adjacent caudal rays. The HL is increasingly differentiated and developed. Because of the development of intrinsic caudal musculature, the amiiform lineage shows the first ability to modulate caudal fin shape (Lauder 1989).

The development of the homocercal caudal fin occurs in teleosts (Figure 7.7). For the first time, the caudal fin is supported by epaxial as well as hypaxial elements. The epaxial support is provided by elongated neural arches, termed uroneurals, from the caudal vertebrae. In addition, there is a full complement of intrinsic caudal musculature and the HL originates from ventral hypural bones rather than the caudal vertebrae (Lauder 1989). These changes allow for much greater control over fin shape and stiffness and Lauder 2003). Thus changes in fin position and body morphology within the teleosts should be viewed not as one approach being superior to another, but instead as different solutions to the common problems of survival and reproduction (Rosen 1982). Certainly, the lower teleosts such as the Ostariophysi (Figure 7.5) are among the most speciose of all fish groups with close to 8,000 recognized species (Nelson 2006).

NATURAL SELECTION, PHENOTYPIC PLASTICITY, BODY FORM, AND FUNCTION

North American freshwater fishes exhibit a wide diversity of body forms, with variation in forms partitioned among orders and families, but also among and within species. As I have shown in previous sections of this chapter, ecological functions of varying morphologies, such as caudal fin shape and body shape, can be related to function, although some relationships have been more thoroughly studied than others. For instance, minnows having more elongate dorsal fins, wide caudal fin spans, and streamlined bodies showed greater swimming speed, as tested in a laboratory swim tunnel, and minnows with greater swimming speeds occur in habitats having higher current velocities (Leavy and Bonner 2009).

In general, fishes with shorter, deeper bodies tend to be associated with structurally complex habitats compared to those with longer, shallower bodies. A number of freshwater fish groups show polymorphisms in body form that are likely related to specific habitats (Baker et al. 2005). Examples include the endemic fish species of Lake Waccamaw, sunfishes of northeastern lakes, and sticklebacks from lakes of British Columbia and Alaska, with all the examples representing divergences since the middle to late Pleistocene.

Lake Waccamaw

Lake Waccamaw, North Carolina, a large (3,618 ha) Carolina bay lake of late Pleistocene origin, has a known fish fauna of at least 41 species, including three endemic forms (Shute et al. 1981; Menhinick 1991), with all colonization of the lake from adjoining streams (Hubbs and Raney 1946). The lake is generally clear and offers only sparse cover and a clean sand bottom (Krabbenhoft et al. 2009). The three endemic fishes in the lake, Waccamaw Silverside (*Menidia extensa*; Atheriniformes), Waccamaw Killifish (*Fundulus waccamensis*; Cyprinodontiformes), and Waccamaw Darter (*Etheostoma perlongum*; Perciformes) are sister species to the stream-dwelling Inland Silverside (*Menidia beryllina*), Banded Killifish (*Fundulus diaphanus*), and Tessellated Darter (*Etheostoma olmstedi*), respectively (Hubbs and Raney 1946; Krabbenhoft et al. 2009). Although representing three highly divergent lineages, the three endemic fish species have evolved similar shape differences from their stream ancestors, all having longer and shallower caudal peduncles and more slender bodies in comparison with their ancestral stream forms; the Waccamaw Silverside and Waccamaw Killifish also have significantly shorter and more slender heads than their stream ancestors (Figure 7.8). The local streams show much greater physical heterogeneity compared to the lake. Lake forms are likely adapted to a less complex habitat and more sustained swimming in open water in contrast to stream forms; even the benthic darter shows morphological shifts toward greater use of the water column (Krabbenhoft et al. 2009; Krabbenhoft 2010, pers. comm.).

Sticklebacks

As the Cordilleran ice sheet began its retreat from northwestern North America over the last 12,000–20,000 years (see Chapter 3), with the concomitant rebound of the land, newly formed lakes were independently colonized by the marine anadromous form of Threespine Stickleback (*Gasterosteus aculeatus*). This has provided a replicated system for studying stickleback evolution (McPhail 1994; S. A. Foster et al. 2003; Foster and Baker 2004). Arising from

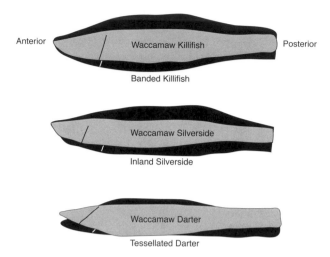

FIGURE 7.8. Outlines of body shapes of the three endemic species from Lake Waccamaw, North Carolina (gray), and their closest stream relatives (black). Shapes are drawn from deformation grids of mean body shape for the six taxa, with differences in shape magnified three times. The diagonal lines (white = stream, black = lake) indicate the approximate posterior margin of the gill cover. Note the more dorsally oriented head in the Waccamaw Darter compared to the Tessellated Darter. Based on Krabbenhoft et al. (2009).

one or perhaps two original founder populations (Taylor and McPhail 1999; Wund et al. 2008), Threespine Sticklebacks found in lakes of British Columbia and Alaska often show distinct body forms associated with different lake habitats. The divergent body shapes are most likely related to variation in selection between highly structured, benthic habitats versus open water (limnetic) habitats. Benthic forms tend to be larger bodied, have fewer and shorter gill rakers, wider mouth gapes, and greater body depths in contrast to limnetic forms, which also tend to have larger, protruding eyes (Bentzen and McPhail 1984; Schluter and McPhail 1992; Rundle et al. 2000; Baker et al. 2005). Allopatric benthic and limnetic ecotypes occur in thousands of lakes, where developmental plasticity at least plays a role in shaping morphology; sympatric benthic and limnetic species pairs have evolved in at least six lakes; and partially differentiated benthic and limnetic ecotypes are known from two lakes (Schluter 2001; Baker et al. 2005; Wund et al. 2008).

Sunfishes

A similar pattern of benthic and pelagic forms is apparent with sunfishes (*Lepomis* spp.), with colonization of the lakes likely occurring following the beginning retreat of the Laurentide ice sheet approximately 17,000 years ago (Chapter 3). In the Adirondack region of the northeastern United States, Bluegill and Pumpkinseed (*Lepomis gibbosus*) often occur together in lakes, with Bluegill occupying the water column and feeding primarily on planktonic prey, and Pumpkinseed occupying the benthic, littoral zone and feeding mostly on benthic invertebrates (Robinson et al. 1993, 2000). However, in lakes that were never colonized by Bluegill, the normal bottom-associated Pumpkinseed may also have a planktonic form that occurs in the water column and feeds primarily on planktonic prey such as cladocerans. Morphological divergence between the open- and shallow-water forms varies among lakes but is generally well developed (Robinson et al. 2000). Morphologically, the planktonic form of Pumpkinseed tends to approach the morphology of typical Bluegill, with shallower and longer bodies and longer pectoral fins. Robinson et al. (1993) suggest that the most likely explanation for the development of the planktonic form of Pumpkinseed is the exploitation of planktonic prey made available by the absence of Bluegill, although Robinson et al. (2000) showed that diffuse competition from other planktivorous fishes can also reduce the extent of morphological divergence.

Trade-Offs in Form and Function

Throughout this chapter the diversity of body forms in North American freshwater fishes, related to habitat use and behavior, suggests that some shapes are more efficient (e.g., in loco-motion, predator avoidance, or prey capture) in some habitats compared to others. The examples of changes in body form of fishes in structurally simple versus complex habitats indicate that natural selection favors certain forms over others in particular habitats. However, there are few studies that have directly measured trade-offs in efficiency of different body forms in one type of habitat versus another. Indeed, well-designed studies of trade-offs of specializations in different habitats on any animal group have been relatively rare (Futuyma and Moreno 1988).

Trade-offs in foraging of sympatric benthic and limnetic "species" of Threespine Stickle-backs were investigated by Bentzen and McPhail (1984), who used laboratory-reared progeny of benthic and limnetic forms from Enos Lake, Vancouver Island, British Columbia. Benthic fish consumed larger prey relative to body size and were twice as efficient in foraging on a detrital substratum compared to limnetic fish. Both male and female benthic forms averaged 0.44 prey per feeding strike in contrast to male limnetic fish who averaged only 0.2 prey per strike. (Female limnetic forms would not feed on benthic prey.) In contrast, limnetic fish showed greater success in capturing small zooplankton, perhaps because of their larger, more protruding eyes. At the end of a five-day experiment where sticklebacks were introduced into mesh enclosures suspended in the lake and allowed access to a natural array of plankton, limnetic fish averaged 103 prey in the stomach, in contrast to approximately two prey in stomachs of benthic fish.

In a similar study, Robinson (2000) used benthic and limnetic forms of Threespine Stickleback to first ask whether morphological differences had a genetic basis and then to ask if the two forms differed in foraging efficiency between benthic and limnetic habitats. Fish were taken from limnetic and from shallow water, benthic habitats in Cranby Lake, Texada Island, British Columbia, which contains one of two known populations polytypic for body shape. The morphologies differ along a continuum from benthic to limnetic shapes (Baker et al. 2005). To determine if there was a genetic basis for the polymorphism, Robinson (2000) used laboratory-reared progeny from 30 crosses of benthic × benthic and 30 crosses of limnetic × limnetic morphotypes. Progeny of the benthic and limnetic fish retained their shape differences, even after being reared in a common environment, thus indicating at least a partial genetic basis for the differences. Second, laboratory-reared progeny were tested for feeding rate (number of prey consumed per unit time) and feeding efficiency (number of bites per prey item). The hypothesis that there would be trade-offs in foraging relative to morphology was supported. Limnetic forms were superior to benthic forms in the capture of planktonic prey, and benthic forms were superior to limnetic forms in capturing benthic prey—even though the shape differences were only on the order of 4–10%. When feeding on plankton, on average, benthic forms were less efficient, taking approximately twice the number of bites per prey item and 0.8 fewer prey per minute compared to limnetic forms. When feeding on benthic prey, limnetic forms had lowered efficiency, averaging 1.4 more bites per prey and 0.22 fewer prey per minute than benthic forms. The studies by Bentzen and McPhail (1984) and Robinson (2000) clearly illustrate that relatively minor shape differences, as well as other morphological correlates such as gill-raker length and number, mouth size, and eye shape, are significantly related to functional correlates—in these cases feeding rates and efficiencies.

Another, rather stringent test of functional trade-offs in Pumpkinseed focused on fish from Paradox Lake, New York. Fish in this lake are unimodal in the distribution of body shape, and shape differences are so subtle that comparison of the extremes of the distribution are difficult to separate by eye, instead requiring a multivariate shape analysis (Robinson et al. 1996; Robinson and Wilson 1996). Body shapes of fish from the pelagic habitat are more fusiform compared to those from the benthic habitat. Pumpkinseed were collected from pelagic and benthic habitats, and for each habitat, 10 fish were selected from the extremes of the morphological shape distribution (i.e., for the pelagic habitat, the 10 least fusiform and the 10 most fusiform body shapes; Robinson et al. 1996). For fish from each habitat, Robinson et al. determined body condition, measured as the percentage of lipids in muscle tissue. They also determined the relative growth rate, expressed as the standard length at a given age, with age determined from scale annuli. In fish from the pelagic habitat, those most extreme for a limnetic (fusiform) shape had higher condition factors and faster growth than those least extreme in limnetic shape. Fish from the benthic habitat did not show significant relationships in growth or condition factor nor in the multivariate index of body shape. In spite of this, the results are important because they illustrate functional differences in a case where measured morphological differences are very slight (Robinson et al. 1996).

These examples of trade-offs in function from different body forms illustrate how apparently minor changes in shape can have functional consequences. However, the linkage of form with function can be challenging because such relationships often are not linear, they can be dependent upon a particular context, and different structures sometimes can have equivalent functions. Because a large suite of morphological characters might govern a particular function, the choice of what is chosen to measure becomes critical to the outcome (Koehl 1996). For instance, in the previous example with Pumpkinseed, the studied relationship of body shape of pelagic fish to apparent foraging success might also be impacted by such things as gill-raker shape and morphology, head dimensions, and eye position, which were not studied.

DOES MORPHOLOGY PREDICT ECOLOGY?

This section takes a somewhat different approach to the relationship between morphology and function and considers whether ecological functions such as habitat and food use can be accurately predicted from morphology. To be able to do so would allow basic ecological questions of overlap in resource use, niche breadth, and species packing in assemblages (see Chapter 11) to be answered by studies of fish shapes, such as from museum specimens, rather than primarily through actual field or laboratory observations. The latter approach rapidly becomes logistically challenging as the number of potential resource dimensions and the number of species increase. In addition, short-term studies of resource use can be misleading because of natural variation, whereas morphology tends to average out such short-term environmental variation and, consequently, might be more useful as an indicator of overall ecological patterns (Ricklefs and Travis 1980; Douglas 1987). The underlying premise, then, is that morphological "space" can be directly mapped onto ecological "space," where morphological and ecological spaces represent multidimensional hyperspace as determined via multivariate statistical analyses (see Chapter 4) (Strauss 1987). If there were not a strong relationship between morphology and ecology, then fish assemblages might not be structured or constrained by physical adaptations but perhaps by other processes (Douglas and Matthews 1992).

For a long time, morphological studies have been used to infer ecological relationships of fishes as well as other organisms (e.g., fishes: Keast and Webb 1966, Gatz 1979a, b; lizards: Collette 1961; birds: Van Valen 1965; bats: Findley 1976, with Findley [1976] being one of the earliest studies to use a multivariate, morphometric approach). The history of ecomorphological studies shows a mix of approaches and assumptions, with studies attempting to verify morphological and ecological relationships (generally rare) interspersed with those assuming strong relationships between morphology and ecology, and then using morphological studies to address ecological questions (generally more common).

Tests of the Ecomorphological Hypothesis

Studies of relationships between morphology and ecology show a progression of approaches, from providing only verbal descriptions of morphological and ecological relationships (e.g., Keast and Webb 1966), to using basic correlation analysis and limited use of factor analysis to summarize morphological differences (Gatz 1979b), to using sophisticated multivariate analyses capable of statistically testing the relationships between morphology and ecology (Douglas and Matthews 1992). In a series of papers, Gatz (1979a, 1979b, 1981) provided the first studies of freshwater fishes in which the relationships between morphology and ecology were tested statistically. He used 56 morphological features that included superficial body shape, pigmentation, mouth form, fin form and placement, gill-raker number and shape, swimbladder length and volume, digestive tract morphology, and brain structure. Ecological characters focused on broad-based dietary categories determined from stomach content analysis, with only limited information on habitat. Using information from the literature, for each morphological character he developed assumed functional predictions. The predictions were then tested against other morphological data, or with the ecological data. For instance, orientation of the mouth was assumed to indicate the location, relative to the fish, where feeding occurred. Hence a ventral mouth should indicate primarily benthic prey in the gut. Tests against prey categories known to occur on the surface versus those near or on the bottom showed general support for the morphological prediction. Using this approach, Gatz found that dietary hypotheses of species based on morphological features, even among morphologically similar species, were supported 90% of the time by actual dietary studies. Although this work provided important information, the prey categories were quite broad and ecological features were not, with a few exceptions, tested directly against morphological features but instead against hypotheses of function developed from the morphological characters.

Building on the studies by Gatz, Felley (1984) developed two suites of morphological characters that were then tested against detailed habitat information on a group of 43 minnow species from locations primarily in Oklahoma. The first morphological character group (a priori characters) was assumed to be related to habitat use, and characters were chosen primarily based on Gatz (1979a). The second group (a posteriori characters) was determined by statistically testing associations between morphology and habitat use by factor analysis and then choosing morphological measurements that were significantly related. These were then tested against habitat use in a different group of fishes than those used to identify what morphological features were important; however, the two groups differed in generic composition rather than being a randomly chosen group. No characters in the a priori group successfully predicted habitat use, illustrating the challenges in making generalizations of habitat use from one group of species to another. Among the a posteriori characters, the location of food source and the position in the water column were successfully predicted from morphology. Benthic species had morphologies associated with greater maneuverability, such as deeper bodies, dorsal fins near the center of mass, and longer fins, in contrast to more open-water species—an outcome similar to the previously mentioned studies on benthic and limnetic forms of fishes. The nature of the food source, measured as the relative use of organic detritus, also was predicted successfully from morphology. Species using larger amounts of detritus had longer intestines, greater amounts of pigmentation in the body cavity, and reduced size of the cerebellum. Overall, the Felley's study, with both successes and failures in the prediction of habitat and food use from morphology, suggests caution in choosing certain aspects of an organism's morphology and trying to relate it to function. As Felley (1984) cautioned, "We must consider the whole of a species' morphology to properly identify its place in its ecosystem."

One of the first studies to use multivariate analyses to rigorously test the association between morphology and ecology focused on an assemblage of southeastern fishes (Douglas and Matthews 1992). The study was based on 65 collections at 11 sites in the Roanoke River drainage of Virginia along a gradient from headwaters to fifth order streams. The analysis included most major groups of North American freshwater fishes—minnows, darters, suckers, madtom catfishes, and sculpins. Ecological data included food habits, based on analysis of stomach contents to major taxonomic levels, and microhabitat use (current speed, substratum size, and water depth). The shape data were based on 34 morphological characters and were analyzed in three different ways, resulting in three shape matrices. For each data set (food, habitat, and morphology), species were grouped by unweighted pair-group with arithmetic mean (UPGMA) cluster analysis based on a measure of dissimilarity (i.e., a distance measure) (see Chapter 4), and the matrices were compared using the Mantel test. The Mantel test determines whether one matrix is correlated with another by testing against the null hypothesis of randomly arranged elements in the second matrix. The randomization is repeated many times and if the observed value of the correlation of the two matrices falls within the cluster of correlations based on the randomization, the two matrices are not related; if the observed value falls above or below the correlations based on randomization, then the two matrices are either positively or negatively correlated (Manly 1986).

The results showed that only the food use, morphological shape, and fish taxonomy matrices were significantly correlated. Although food use and body shape covaried, beyond the family level, taxonomy explained most of the variation in both the shape and ecology data sets. Consequently, the relationship between body shape and food use was completely overshadowed by the relationship between body shape and taxonomy. Because of the strong presence of phylogenetic history in structuring resource use

of fish assemblages, morphological data may not be very useful in describing the structure of an assemblage any further than is already described by taxonomic data. Douglas and Matthews concluded "that taxonomy and trophic ecology are inextricably intertwined, and that resource use in the assemblage reflects [a] phylogenetic artifact." To control for this, they repeated their analyses for a single family, the Cyprinidae. In this case, body shape was not significantly related to food use but was significantly related to habitat use, suggesting a greater importance of resource separation of freshwater fishes along habitat compared to food dimensions (Douglas and Matthews 1992). Freshwater fishes certainly differ along habitat dimensions; however, in a review of resource partitioning in fishes, Ross (1986) showed that both marine and freshwater fishes tended to separate more along food than habitat resources, and freshwater fish assemblages separated about equally along trophic and habitat dimensions. More likely, the difference in the relationship of morphology to food use in cyprinids was an artifact of the level of prey identification (see Chapter 11).

In a more limited study (eight cyprinids, three centrarchids, and four percids), Wood and Bain (1995) showed that body shape, expressed in multivariate space, was significantly correlated with habitat use (also expressed in multivariate space) in minnows and darters but not sunfishes. The lack of an association between sunfish morphology and habitat use was perhaps because of the few sunfish species, the choice of morphological measurements, or high variability in one of the morphological measurements.

Studies Assuming Validity of the Ecomorphological Hypothesis

Working on the assumption of a close relationship between morphology and ecology, Strauss (1987) used morphological analyses to compare species packing among seven fish assemblages from North and South America. He found that the separation of species in morphological space was not related to the number of species in an assemblage, so that species differences in morphology (and assumed ecology) remained similar across a broad range of fish assemblages. At the same time, Douglas (1987) found that minnow species were more tightly packed in morphological space in contrast to species of sunfishes. He also showed that as communities become more diverse, the amount of morphological space increases, suggesting that increased species packing is not because of smaller morphological distance between species. However, the increase in morphological space was essentially because of the increase in taxonomic diversity—morphological similarity was highly correlated with taxonomic similarity, which is not surprising given that taxonomies are largely based on morphology. Because of the confounding factor of phylogenetic relationships, Douglas (1987) concluded that comparisons of morphological space are more meaningful within than among major lineages.

On a much broader geographical scale, Winemiller (1991) compared packing in morphological space (based on 30 morphological characters) among 10 lowland fish assemblages from five major geographical regions, including North, Central, and South America and Africa. Key research questions included whether fish assemblages in faunistically richer areas showed greater morphological diversification and whether species in such areas showed greater or lesser similarity in morphology in contrast to assemblages from high latitude, species-poor areas. Tropical fish assemblages showed greater morphological diversity, which, assuming a close association between morphology and ecology, suggests greater niche diversity in such areas. However, species packing, inferred from morphological distance among species, did not differ among the 10 assemblages.

SUMMARY

Fishes primarily use their body and caudal fins (BCF) or median and paired fins (MPF)
in locomotion, in some cases switching from MPF to BCF as the demand for speed relative to maneuverability increases. Within BCF locomotion, freshwater fishes vary in the amount of the propulsive wavelength that is generated within the body, ranging from anguilliform to carangiform as the caudal fin is increasingly involved and body undulations are reduced. However, almost all fishes use anguilliform locomotion during the larval period. In addition to providing propulsion, fins are used in a variety of ways to control stability, position in the water column, turning, and station holding. A major trend in the evolution of fishes has been the increased control of fin shape and function, especially with the caudal fin. Higher and lower teleosts differ in fin number and position, but differences generally are related to equally successful, albeit different, ways of dealing with environmental challenges. Fin positions, body shapes, and mode of swimming are generally related to water currents, habitat complexity, and prey use. Fishes in structurally complex habitats tend to have larger fins and deeper bodies in contrast to fishes in open-water habitats, and even slight differences in body form provide habitat-specific advantages in feeding. Although morphology is strongly related to ecological function, direct predictions of ecological function from morphology have proven difficult to make. In part this reflects the multiple roles that structures might have and the diverse ways that adaptation has responded to selective pressures within habitats.

SUPPLEMENTAL READING

Blake, R. W. 2004. Review paper: Fish functional design and swimming performance. *Journal of Fish Biology* 65:1193–222. An overview of functional morphology and fish locomotion.

Douglas, M. E., and W. J. Matthews. 1992. Does morphology predict ecology? Hypothesis testing within a freshwater stream fish assemblage. *Oikos* 65:213–24. One of the first multivariate analyses of the relationship between ecology and morphology.

Lauder, G. V., and E. D. Tytell. 2006. Hydrodynamics of undulatory propulsion, 425–68. In

Fish Biomechanics. Fish Physiology. Vol. 23. R. E. Shadwick, and G. V. Lauder (eds.). Elsevier Academic Press, San Diego, California. Explains the relationship of water currents and thrust generated from undulatory locomotion in fishes.

Webb, P. W. 2006. Stability and maneuverability, 281–332. In *Fish Biomechanics. Fish Physiology.* Vol. 23. R. E. Shadwick, and G. V. Lauder (eds.). Elsevier Academic Press, San Diego, California. Explains the relationship between the degree of maneuverability in fishes and their hydrodynamic stability.

WEB SOURCES

The Tree of Life Web Project. http://tolweb.org.

EIGHT

Form and Function in the Feeding of Fishes

CONTENTS

THE PREVIOUS CHAPTER FOCUSED on the overall body shape of fishes, functional aspects of swimming or position holding, and the ways in which body form and function are related to habitat. This chapter looks at some basic elements of the head skeleton of North American freshwater fishes and how bone and muscle shapes and positions relate to modes of feeding. Ray-finned fishes, the Actinopterygii (see Figure 7.5), display tremendous diversity in structural and behavioral aspects of prey capture—a not surprising finding given the great diversity of the group. In fact, diversity of trophic morphology (i.e., teeth, jaws, suspensorium, etc.) is a hallmark of

actinopterygian radiation. What fishes eat is most closely (morphologically) related to their dentition, but jaw construction and jaw movements strongly influence how they eat (Gosline 1987). Although the size of prey consumed by fishes is generally related to body size, with the general exception of suspension feeders, the prey size–body size relationship is affected by morphological and energetic constraints (the cost-to-benefit ratio of a prey item). Gape limitation generally refers to prey dimensions relative to the size of the oral jaw opening, and predators that are constrained in the size of prey they consume by the size of their mouths are said to be gape limited. However, the opening of the oral jaws is only the first of three "filters" encountered by prey on their way into the esophagus of a predator. For instance, in Largemouth Bass (*Micropterus salmoides*) the oral jaw opening is larger than the passage through the pharyngeal jaws at the back of the mouth. (In a 200 mm SL fish the pharyngeal gape is only 55% of the oral gape.) Prey size can be further limited by the throat diameter, which is constrained by the cleithral bones of the pectoral girdle. To some extent the size limitations of the pharyngeal gape and throat gape are offset by flexibility and deformability of the prey and/or by crushing or tearing actions of the pharyngeal teeth (Wainwright and Richard 1995). The study of functional morphology of fishes, both in terms of locomotion (Chapter 7) and feeding, has benefitted extensively from technological developments (Box 8.1).

EVOLUTIONARY TRENDS IN TROPHIC MORPHOLOGY

The structures involved in the feeding of fishes (upper and lower jaws, jaw suspension, mouth

and throat cavities, and associated musculature) show a progression in complexity and degrees of freedom of movement over evolutionary time (Schaeffer and Rosen 1961; Lauder 1982). These changes are associated with greatly expanded modes of foraging, especially as shown by teleosts. Bony fishes have a highly kinetic skull containing some 60 skeletal units that are controlled by approximately 80 groups of muscles (Winterbottom 1974; Sanford and Wainwright 2002).

Tracing the patterns of evolution in jaw function is challenging because of the tremendous diversity of ray-finned fishes, and because more derived groups in each major lineage frequently differ in jaw function and morphology from more ancestral groups in the same lineage and may independently converge on the same or similar patterns. Ancestral patterns for each group are presented first in Table 8.1, followed by patterns in more derived forms. In early Paleozoic ray-finned fishes (subclass Chondrostei), the bones of the upper jaws (the premaxilla and maxilla) were fused to other dermal bones of the skull and only the lower jaws (mandibles) moved

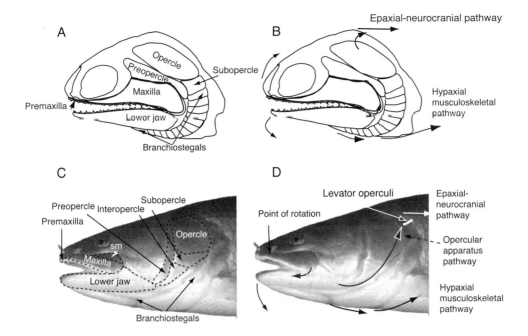

FIGURE 8.1. Trends in the evolution of jaw mobility in preteleostean, ray-finned fishes (Actinopterygii).
A. Jaw elements of early ray-finned fishes, illustrated by the late Devonian chondrostean, *Limnomis*. Drawing based on Daeschler (2000).
B. Muscle forces and rotational movements involved in the two pathways of jaw opening in *Limnomis*.
C. Jaw elements of Bowfin (*Amia calva*), an early neopterygian; dotted lines highlight outlines of some of the bones; sm = supramaxilla.
D. Muscle forces and rotational movements involved in the three pathways of jaw opening in Bowfin. Rotation of the maxilla was the first level of upper jaw mobility. Muscle forces and rotational elements are based on Lauder (1980, 1982).

(Figure 8.1A, B; Table 8.1). The bones of the cheek and throat (opercles, preopercles, subopercles, and branchiostegal rays) also had limited mobility so that lateral expansion of the buccal cavity was minimal (Schaeffer and Rosen 1961; Lauder 1982). Mouth opening was achieved by two pathways: (1) elevation of the head caused by contraction of the dorsal (epaxial) muscles—the epaxial-neurocranial pathway, and (2) depression of the lower jaw from contraction of throat musculature (sternohyoideus muscle) and contraction of hypaxial musculature connecting to the lower jaw via the pectoral girdle to the ventro-medial hyoid series (the hypaxial musculoskeletal pathway) (Figure 8.1B). Although most fossils of early ray-finned fishes occur in marine habitats, several genera are recorded from fresh water, including *Limnomis* and *Cuneognathus* (Friedman and Blom 2006). *Limnomis*, one of the earliest of North American bony fishes

(Figure 8.1A), is known from the Late Devonian Catskill Formation in Pennsylvania. It apparently lived in shallow, low-velocity habitats, such as floodplains and oxbow lakes, where it fed on soft-bodied invertebrates. It is considered a pioneer in the exploitation of North American freshwater ecosystems (Daeschler 2000).

The jaw mobility of living species of the subclass Chondrostei (Sturgeons and Paddlefishes, order Acipenseriformes) differs greatly from Paleozoic and early Mesozoic forms. With the exception of the highly specialized Paddlefish (*Polyodon spathula*), all sturgeons and the primitive Chinese Paddlefish (*Psephurus gladius*) have the upper jaw free from the braincase (neurocranium) (Figure 8.2). The resultant mobility allows jaw projection and is related to active foraging on invertebrates and fishes (Bemis et al. 1997; Findeis 1997). In addition, the role of head lift in opening the mouth, important in

TABLE 8.1

Trends in Jaw Form and Function in Ray-Finned Fishes, the Actinopterygii

Group	Characteristics
Chondrostei	Two muscle couplings for lower jaw opening; premaxilla and maxilla fused to skull and in series; maxilla toothed (Lauder 1982; Rosen 1982; Westneat 2004)
Neopterygii	
Amiiformes	Three muscle couplings for lower jaw opening; maxilla free to rotate; suction feeding (Lauder 1982; Rosen 1982; Westneat 2004)
Teleostei	Premaxilla and maxilla primitively in series (Rosen 1982)
Osteoglossomorpha	Premaxilla and maxilla in series; small alveolar process on premaxilla overlaps anterior head of maxilla; premaxilla fixed to skull or with limited movement (Rosen 1982; Nelson 2006)
Elopomorpha	Premaxilla and maxilla in series; small alveolar process on premaxilla overlaps anterior head of maxilla; premaxilla with limited movement (Schaeffer and Rosen 1961; Rosen 1982)
Ostarioclupeomorpha (Clupeomorpha and Ostariophysi)	Premaxilla partially or fully free to rotate; premaxilla and maxilla primitively in series but tandem in more derived forms; ascending process on premaxilla and jaw protrusion in many Ostariophysi but not Clupeomorpha (Harrington 1955; Schaeffer and Rosen 1961; Motta 1984; Nelson 2006)
Euteleostei	
Protacanthopterygii	Premaxilla secondarily fused to skull; premaxilla and maxilla usually in series (exceptions in some Southern Hemisphere smelts) (Lauder 1982; Nelson 2006)
Neoteleostei	
Paracanthopterygii	Jaws in tandem with maxilla excluded from gape; small ascending process on premaxilla; some groups with protrusile jaws (Schaeffer and Rosen 1961; Rosen 1982; Nelson 2006)
Acanthopterygii	Unique mechanisms of jaw protrusion; premaxilla and maxilla in tandem with maxilla excluded from gape; well-developed ascending premaxillary process (Schaeffer and Rosen 1961; Rosen 1982; Motta 1984)

ancestral Chondrostei (Figure 8.1A,B), is lost in modern acipenseriform fishes. Instead, modern acipenseriforms are unique in having an enlarged mandibular arch muscle involved with jaw opening and projection (Bemis et al. 1997).

At the early neopterygian level (see Chapter 7; Figure 7.5), represented by Bowfin (*Amia calva*), the upper jaw shows the first level of mobility (with the exception of the more derived chondrosteans mentioned previously). The maxilla is detached posteriorly from the cheek bones and has a pivot point anteriorly where it abuts the premaxilla so that the posterior part of the bone swings forward during jaw opening and acts to prevent fluid inflow (therefore, loss of suction) through the sides of the mouth (Figure 8.1C) (Lauder 1980, 1982; Rosen 1982). The epaxial-neurocranial pathway is retained for elevation of the head. In addition, opening of the lower jaw is accomplished through two independent biomechanical pathways (Figure 8.1D). The hypaxial musculoskeletal pathway is essentially the same as shown in early actinopterygians through the hypaxial muscle contraction via the pectoral girdle, sternohyoideus, and hyoid bones. A newer opercular apparatus pathway involves a muscle (levator operculi) causing dorsal rotation of the opercular bones (opercle, subopercle,

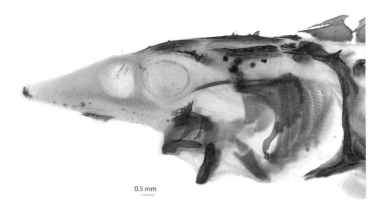

0.5 mm

FIGURE 8.2. Elements of jaw protrusion in derived chondrosteans, the Acipenseriformes, showing the freedom of the upper jaw from the braincase. Illustrated by a 21-day posthatch White Sturgeon (*Acipenser transmontanus*). Specimen preparation and photograph courtesy of Katie May Laumann. See also corresponding figure in color insert.

and interopercle) via a ligament connecting the interopercle to the posterodorsal edge of the lower jaw (Figure 8.1D). Two independent pathways for depression of the lower jaw allow for much greater flexibility in the timing of depression of jaw elements and in fluid movement through the oral cavity; these two pathways are retained in most all Neopterygii (Lauder 1982).

Teleosts (Table 8.1; Chapter 7; Figure 7.5) show the next stage of jaw kineticism with the development of a posterior process on the premaxilla, the alveolar process (Figure 8.3B). Although present, this process remains weakly developed in the most basal teleostean lineages such as the Bony Tongues (Hiodontidae) (Figure 8.3A). The alveolar process overlaps with the anterior arm of the maxilla, and the maxilla and premaxilla swing forward as a unit when the lower jaw is opened. The premaxilla is hinged anteriorly with the ethmoid region of the skull (Rosen 1982). Exceptions occur in generalized predators where the force of the strike on the premaxilla and powerful closing of the lower jaw dictate little or no mobility of the premaxillary bone (Schaeffer and Rosen 1961; Lauder 1982; Westneat 2004). In the Osteoglossomorpha, represented in North American freshwater by Mooneye and Goldeye (genus *Hiodon*), the ancestral condition is largely maintained, with bones of the upper jaw in series and with limited premaxillary movement (Figure 8.3A). The Elopomorpha, represented in freshwater by American Eel (*Anguilla rostrata*), show a similar pattern (Table 8.1). Basal members of the Ostarioclupeomorpha lineage, which include herrings

and shads (Clupeomorpha) and minnows, suckers, and catfishes (Ostariophysi), maintain the ancestral teleostean plan, as do more derived clupeomorphs such as Threadfin Shad (*Dorosoma petenense*) (Figure 8.3B). However, several groups of ostariophysans, including Cypriniformes (minnows and suckers), have developed highly protrusile jaws, have the premaxilla and maxilla in tandem, and have the maxilla excluded from the gape (Motta 1984; Table 8.1; Figure 8.3C). Trouts, salmons, pikes, and pickerels (Protacanthopterygii) retain the basic teleostean pattern of a premaxilla with limited movement, maxillary rotation during jaw opening, and the premaxilla and maxilla in series, although fusion of the premaxilla to the skull is considered a secondary change (Lauder 1982; Table 8.1). The two lineages of advanced teleosts, the Paracanthopterygii and the Acanthopterygii, both have the premaxilla and maxilla in tandem and the maxilla excluded from the gape (Figure 8.3D). The Paracanthopterygii, represented in North American freshwater by Trout-Perches (Percopsidae), Pirate Perches (Aphredoderidae), Cavefishes (Amblyopsidae), and Burbot (Gadidae), only have a small ascending process on the premaxillae (Figure 8.3E), and only some groups have protrusile jaws (Table 8.1). Acanthopterygians, which are represented by close to 15,000 species worldwide, have a toothless maxilla that is excluded from the gape, a premaxilla with well-developed ascending and alveolar processes, highly protrusile jaws, and show a variety of different mechanisms for jaw protrusion (Table 8.1; Figure 8.3E) (Lauder 1982).

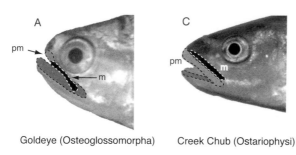

Goldeye (Osteoglossomorpha)　　Creek Chub (Ostariophysi)

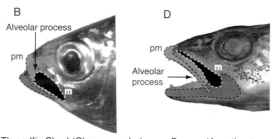

Threadfin Shad (Clupeomorpha)　　Sauger (Acanthopterygii)

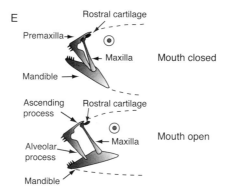

FIGURE 8.3. Trends in mobility of upper jaw elements in teleosts.
A. Goldeye (*Hiodon alosoides*), a lower teleost.
B. Threadfin Shad (*Dorosoma petenense*), a lower teleost.
C. Creek Chub (*Semotilus atromaculatus*), a lower teleost.
D. Sauger (*Sander canadensis*), an advanced teleost. Names in parentheses indicate higher-level groups referred to in the text; pm = premaxillary bone; m = maxillary bone.
E. A schematic diagram of upper jaw bones of an advanced percomorph showing the configuration of jaw bones during mouth opening. Photo of Goldeye by J. Baker; the drawing is based on Alexander (1967b) and Ferry-Graham et al. (2008).

Jaw Protrusion

Approximately 50% of bony fishes (the Teleostei) have protrusile jaws, meaning that components of the upper and sometimes lower jaws move forward during the feeding strike, and the evolution of jaw protrusion is associated with the wide radiation of teleostean feeding mechanisms (Motta 1984; Gosline 1987; Holzman et al. 2008a). Jaw protrusion also occurs in some living chondrosteans (sturgeons and Chinese Paddlefish). Mechanisms of jaw protrusion have arisen independently in different groups of fishes and may incorporate various approaches (Table 8.1) (Motta 1984; Ferry-Graham et al. 2008; Holzman et al. 2008a).

An initial development in the evolution of protrusile jaws was the freeing of the rostral cartilage from the premaxilla and the presence of a moderate ascending process on the premaxilla; development of a long premaxillary ascending process was a later adaptation. In teleosts with a long ascending process of the premaxilla (Figure 8.3E), jaws are protruded as the ascending process slides along the rostral cartilage cushion at the front of the skull; however, a long ascending process does not always indicate extreme jaw protrusion—some fishes (primarily marine groups such as wrasses) with a long ascending process have only limited protrusion and instead the ascending process is involved in stabilizing the jaw (Gosline 1987). As mobility increased, the rostral cartilage became free from the rostrum so that it could also move forward. Jaws with a short ascending process on the premaxilla tend to rotate about the front of the skull as the mouth is opened and have limited protrusion.

Protrusion can be independent of opening and closing of the mouth, at least to a certain degree.

Even though they evolved independently, jaw protrusion in the ostariophysans and in the percomorphs have many similarities. In both groups, the upper jaw, formed from the left and right premaxillary bones, is projected forward and generally upward during feeding as the ascending processes of the premaxillae slide along the rostral cartilage and the ventral ends of the alveolar processes lap over the lower jaw and close off the sides of the gape, creating a more rounded mouth opening (Figure 8.3E) (Alexander 1966, 1967b; Ferry-Graham et al. 2008). A rounded mouth is advantageous in terms of sucking in prey, although the reduced mouth opening limits the size of prey that can be consumed (Carroll et al. 2004). In contrast, the killifishes and their allies (Cyprinodontiformes; Atherinomorpha) have the alveolar process of the premaxilla tightly connected by strong ligaments to the lower jaw so that it cannot swing forward during mouth opening and occlude the jaw opening (Ferry-Graham et al. 2008). Because of this connection, the mouth tends to move forward and downward during protrusion. Some occlusion along the sides of the gape can occur by the lip membrane, but in most cases the open jaw has more of a "beak-like" rather than a rounded appearance.

Jaw protrusion in North American freshwater fishes seems to occur by at least four primary means (Table 8.2): (1) protrusion caused by depression of the lower jaw through various ligamentous connections, with the maxilla often serving as a supporting strut (Type A); (2) protrusion directly as a result of maxillary twisting (Type B); (3) a combination of mandibular depression and maxillary twisting (Types A + B); and (4) protrusion primarily by neurocranial elevation (Type C) (Alexander 1967a, b, c; Gosline 1981; Motta 1984; Hernandez et al. 2008). The most widespread mechanism, represented by five orders and seven families, is protrusion of the premaxillae caused by mandibular depression (Table 8.2). The number of taxa achieving protrusion of the premaxillae via twisting of the

maxillae, although once considered to be high (Alexander 1967b), has apparently been overestimated (Motta 1984). Among families represented in North American fresh water, this mechanism is only documented in mullet (Mugilidae) (Table 8.2). Basal members of the order Cyprinodontiformes, such as the genera *Kryptolebias* (formerly the genus *Rivulus*) and *Fundulus*, employ both maxillary twisting and mandibular depression (Hernandez et al. 2008, 2009). Finally, jaw protrusion that involves the tilting back of the head (neurocranial elevation) is perhaps more common than previously thought (Motta 1984). Examples occur among advanced percomorphs such as the Centrarchidae and African cichlids (jaw protrusion in New World cichlids has apparently not been studied) (Table 8.2). In addition, secondary loss of jaw protrusion has occurred repeatedly through loss or reduction of premaxillary movement. Examples of North American genera that show loss of jaw protrusion include the darter genus *Percina* and the cavefish genus *Amblyopsis* (Gosline 1981, 1987).

Advantages of Jaw Protrusion

Jaw protrusion potentially provides a variety of advantages, as suggested by Alexander (1967b), Lauder and Liem (1981), and others. Several of the suggested advantages of jaw protrusion are explored further in the following section titled "Modes of Prey Capture." One of the most commonly cited advantages is that the jaws accelerate ahead of the body as a fish moves toward a potential prey. For instance, in Largemouth Bass jaw protrusion equals about 10% of head length, and a 10-cm Largemouth Bass gains 27 cms^{-1}, equivalent to 87% of average attack velocity, by protruding the mouth; a 25-cm Largemouth Bass has a higher protrusion velocity of 39 cms^{-1}, but because of the overall faster approach speed related to larger body size, this is equivalent to 50% of average attack velocity (Nyberg 1971). Other things being equal, the benefit of protrusion for a gain in approach velocity is greater in small versus large fishes.

A second suggestion involves increased hydrodynamic efficiency for suction feeding

TABLE 8.2

Basic Mechanisms of Jaw Protrusion in North American Freshwater Fishes
Families are listed if protrusion mechanism has been described in any member genus

Mechanism	Order	Family	Studied genera (source)
Type A (protrusion by mandibular depression and not maxillary twisting)	Cypriniformes	Cyprinidae	*Abramis** *Cyprinus**
		Catostomidae	*Catostomus*
	Gasterosteiformes	Gasterosteidae	*Gasterosteus*
	Atheriniformes	Atherinopsidae	*Menidia*
	Cyprinodontiformes	Poeciliidae	*Xiphophorus* (Hernandez et al. 2009)
			Gambusia (Hernandez et al. 2008, 2009)
			Heterandria (Hernandez et al. 2008, 2009)
	Perciformes	Centrarchidae	*Lepomis* (Gosline 1981)
		Percidae	*Perca*
Type B (protrusion by maxillary twisting)	Mugiliformes	Mugilidae	*Mugil*
	Gasterosteiformes	Gasterosteidae	*Gasterosteus*? described by Alexander (1967b); questioned by Anker (1974)
	Perciformes	Percidae	*Perca*? described by Alexander (1967b); questioned by Liem and Osse (1975)
Type A + B (protrusion by mandibular depression and maxillary twisting)	Cyprinodontiformes	Fundulidae	*Fundulus* (Hernandez et al. 2008)
		Rivulidae	*Kryptolebias* (formerly Rivulus; Hernandez et al. 2008)
Type C (protrusion by neurocranial elevation and mandibular depression)	Perciformes	Centrarchidae	*Micropterus*
			Lepomis (Gillis and Lauder 1995)
		Cichlidae	*Petrotilapia**
			*Eretmodus**
			*Spathodus**

SOURCE: Compiled from Motta (1984) unless otherwise listed.

*Genera not native to North America

obtained by having a rounder mouth opening. Protrusion may also help suction efficiency by further expanding the buccal cavity. However, a round mouth opening is not limited to fishes with protrusible jaws because many primitive teleosts and even a preteleost such as Bowfin, all of which lack jaw protrusion, also have rather circular mouth openings as a result of the forward rotational movements of the lower part of the maxillae (Lauder 1979, 1980).

A third suggested advantage is that of increasing the volume of the mouth and pharyngeal cavities—most fishes with protrusible jaws can keep the premaxillae protruded while the mouth is partially closed. The advantage results from the greater speed of closing the mouth and the greater volume of water that can be sucked into the mouth without being blown out again (and thus potentially expelling the prey) as the mouth closes (Alexander 1967b).

A fourth suggested advantage of jaw protrusion is that fishes feeding on benthic prey can remain closer to horizontal when feeding. When the mouth of a typical percomorph is closed, the lower jaw slopes upward from the point of articulation (Figure 8.3E). If the upper jaw is not protruded, the open mouth will point somewhat upwards, so that the fish would have to approach prey at a steeper angle. In addition to providing a better body orientation for predator avoidance, jaw protrusion could also increase the ability to grasp prey (Alexander 1967b).

MODES OF PREY CAPTURE

There are two major stages in the acquisition of food by fishes, namely prey capture (often involving the jaws) and prey processing (often involving the teeth in the back of the mouth—the pharyngeal teeth) (Liem 1980a; Wainwright and Bellwood 2002). Fishes employ a range of jaw and throat musculature and bones to capture prey, and in many cases a single species might use several different modes depending on the size and location of a prey item. The three primary approaches of prey capture are manipulation, ram, and suction (Liem 1980a, b; Wainwright et al. 2007). Prey manipulation involves a diversity of jaw movements designed to clip off parts of prey; rasp tissue from prey (as in lamprey feeding); bite prey with strong jaws; scrape material from the substratum; or grip prey, or parts thereof, and pull them from hiding. Ram feeding involves forward velocity to overtake mobile prey and bring them into the jaws. Suction feeding is based on the creation of a negative pressure gradient so that a discrete water mass containing a prey item is pulled into the mouth (Liem 1980a, b; Wainwright et al. 2007). Often these approaches are used in concert, such as manipulation and suction, or ram and suction.

Suction feeding, which likely originated with jawed vertebrates, represents the primitive yet dominant feeding mode in bony fishes and is used by all bony fishes at some stage in their lives (Liem 1980a; Lauder and Shaffer 1993; Wainwright et al. 2007). Suction is maximized by a relatively small mouth opening (primarily as a result of lateral occlusion of the gape by the ventral arms of the premaxillae and maxillae), well-developed hyoid muscles to rapidly lower the floor of the mouth, and development of other cranial elements to quickly expand the mouth and throat cavities so that a rapid pressure gradient develops (Wainwright and Day 2007; Hernandez et al. 2008). In contrast, effective ram feeding occurs in fishes with large mouths and strong adductor muscles to forcibly close the jaws, and also strong bones in the jaws and head to effectively deal with the stresses of jaw closing (Hernandez et al. 2008). Manipulation, especially exemplified in various cichlid species, involves mobile jaws, modified tooth shapes, and changes in development of jaw musculature.

Although it is convenient to categorize fishes into various feeding modes, doing so obscures the often high level of phenotypic or genetic variability in feeding modes within a single taxon. Some species show developmental plasticity in feeding behaviors and feeding morphology in response to the type of food or its location. For instance, Western Mosquitofish (*Gambusia affinis*) that fed on attached or free prey developed differences in head shape. Those raised on attached food developed shorter, wider heads; anteriorly shifted eyes; lower snout positions; and a sloping caudal peduncle compared to fish raised on free prey. Differences were largely due to developmental plasticity rather than differential mortality. Free prey were attacked using suction whereas attached prey were obtained by biting and scraping (Ruehl and DeWitt 2005).

Suction Feeding

Suction feeding normally involves four phases: preparation, expansion, compression, and recovery (Gibb and Ferry-Graham 2005). During preparation, the buccal cavity is compressed, followed by a rapid expansion, which brings water and prey into the mouth with little or no mastication, and prey are swallowed by action of the pharyngeal jaws (modified gill-arch elements) (Wainwright 2006). Expansion is followed by a slow compression phase where water is expelled through the gill openings (and rarely through the mouth), and a recovery phase where jaw elements return to a relaxed, prefeeding condition. Suction feeding has been studied in various groups, including the Cyprinidae (Alexander 1969), but especially in the Centrarchidae.

There are functional trade-offs in suction feeding between fishes that have relatively slow ram speeds, such as Sunfishes, and those that have high ram speeds such as trout and salmon, Black Basses, and Bowfin. Although both groups employ suction and ram feeding, in the latter group suction is considerably reduced relative to ram (Carroll et al. 2004). Fishes with slow attack speeds should have larger pressure differentials between buccal and opercular cavities and greater forces of suction compared to fishes with generally higher attack speeds (Lauder 1986). This has been supported by recent studies comparing Bluegill (*Lepomis macrochirus*) (low attack speed) and Largemouth Bass (high attack speed) that show higher fluid speeds toward the mouth and greater pressure gradients in Bluegill (Higham et al. 2006a, b).

In Bluegill (size 14–18 cm SL) feeding on earthworms or tethered ghost shrimp, the initial strike takes place close to the prey at 0.6–2.3 cm or 0.04–0.15 body lengths (Gillis and Lauder 1995; Holzman et al. 2008b). The actual strike may involve a ram component (forward velocity), jaw protrusion, cranial elevation, and suction (Ferry-Graham et al. 2003). Fluid speeds vary with the rapidity of mouth opening, which is faster for mobile prey, and average approximately 30 cms^{-1} with maximum values of over 250 cms^{-1} measured at one-half of the peak gape size in front of the mouth (Day et al. 2005; Higham et al. 2005, 2006b). Absolute fluid speeds, and thus overall ability of suction feeding, also increase with body size, although induced flows scale with measures of body size so there is no overall change in how buccal expansion translates to water motion (Holzman et al. 2008c). Because larger fish also have higher swimming speeds, the combined effect of higher speed and suction results in increased closing speed.

The mushroom-shaped suction plume is highly localized, with the highest suction velocity at the mouth opening and declining to only 5% of the maximum suction velocity at a distance of only one mouth diameter away (Day et al. 2005).

Bluegill often combine ram and suction feeding. As ram speeds increase, the water being drawn into the mouth becomes more focused in front of the mouth (drawn in from all directions in stationary fish) and the shape of the ingested water volume becomes more elongate (Figure 8.4A, B) (Higham et al. 2005). From the perspective of the prey, the combination of suction and ram results in a faster closing time between predator and prey because the water mass containing the prey is moving toward the fish at the same time the fish is moving toward the prey. As long as the ratio of ram speed to fluid speed caused by suction stays within 0–20%, there is apparently no trade-off between ram speed and fluid speed. Bluegills usually decelerate during prey capture, perhaps to reduce ram speeds to ≤ 20% of fluid speed so that no hydrodynamic trade-offs occur, and also to allow finer control of movements, which is required because of the increased degree of focusing at moderate ram speeds compared to no ram speed (Higham et al. 2005).

Successful suction feeding requires timing the maximum pressure gradient so that it centers on the volume of water with the prey item and has sufficient pressure generation to overcome prey escape behaviors of swimming or attaching to the substratum (Holzman et al. 2007; Wainwright and Day 2007). A critical aspect of suction feeding is that the flow of water into the mouth is essentially unidirectional so that the prey item is carried into the mouth and then into the throat while the water exits through the opercular opening or gill slits. Unidirectional flow is achieved by a wave of muscle activity that moves from anterior to posterior, resulting in a sequence of closely timed events. First, the lower jaw is depressed, followed by the dorsal rotation of the neurocranium. This is followed by the floor of the mouth (hyoid region) being depressed and finally by the maximal expansion of the opercular region. This pattern

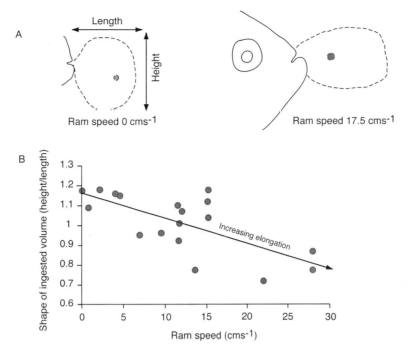

FIGURE 8.4. Shape of the ingested water volume relative to ram speed in suction-feeding Bluegill.
A. The outline (dashed line) of the ingested water volume in stationary and swimming fish. Closed circle indicates the midpoint of the prey.
B. The height to length ratio of the ingested water volume versus ram speed. Based on Higham et al. (2005).

of movement for suction feeding on nonelusive prey is conserved across a wide range of teleosts, having been observed in fishes in the orders Cypriniformes, Cyprinodontiformes, Perciformes, and Pleuronectiformes (Gibb and Ferry-Graham 2005).

In the case of elusive behavior by the prey, ram speed allows a predator to approach closer to the prey and thus overcome the escape force as well as increase the overall closing speed between predator and prey (Higham et al. 2005; Wainwright and Day 2007). An additional aspect for suction feeders is to minimize clues to mobile prey so that any avoidance behavior is delayed. Bluegills are able to reduce hydrodynamic cues to prey, such as bow waves or other swimming disturbances, by coordinating aspects of the strike. The strike is initiated prior to the bow wave reaching the prey so that by the time a prey has hydrodynamic cues, forward jaw protrusion and strong suction already have

the prey moving into the mouth. The result is that Bluegill, at least in terms of hydrodynamic cues, are essentially "ambush predators" (Holzman and Wainwright 2009). In general, low ram speeds would be expected where prey perceive their predators by hydrodynamic clues, whereas high ram speeds would be expected for predators feeding on visually oriented prey (Holzman and Wainwright 2009).

The relative ability for suction feeding varies among other sunfishes, other centrarchids, and, of course, among other kinds of fishes. At least within the Centrarchidae, morphological features related to the force of suction feeding can be expressed in a simple model (Figure 8.5) (Carroll et al. 2004). Because suction involves, in part, the dorsal rotation of the skull, suction increases relative to the cross-sectional area of the epaxial muscles and as L_i (the distance between the centroid of force generated by epaxial muscles and the rotational fulcrum, located approximately at a point between the posttemporal bone of the skull and the supracleithrum of the pectoral girdle) increases. Suction decreases as the projected buccal area increases and as L_2 (the distance from the fulcrum and the center of force of the buccal area) increases. Other things being equal, the force of suction in feeding should be greater in deep-bodied fishes with correspondingly large epaxial muscle areas and small buccal areas and be lower in more streamlined fishes with reduced cross-sectional areas of epaxial muscles and large buccal areas. Within the genus *Lepomis*, Spotted Sunfish (*L. punctatus*), which also have a small buccal area, short L_2, large epaxial muscles, and large L_1, are similar to Bluegill in generating high degrees of suction. In contrast, Redear Sunfish (*L. microlophus*), which are specialized to feed on benthic, hard-bodied prey such as mollusks, have relatively smaller epaxial muscle area, smaller L_1, greater buccal area, and larger L_2 compared to Spotted Sunfish and Bluegill (Carroll et al. 2004). Other species with deep bodies and large epaxial muscle area, such as flatfishes, also are suction specialists (Gibb and Ferry-Graham 2005).

There are trade-offs in morphological specializations required for suction. Greater suction requires smaller mouths, smaller buccal areas, and deeper bodies (to accommodate the epaxial muscles), and so reduces the maximum size of prey and, because of reduced streamlining (see Chapter 7), the maximum approach velocities (Carroll et al. 2004); however, deeper bodies increase maneuverability. The trade-offs in body form relative to suction ability, as predicted from the model (Carroll et al. 2004; Wainwright et al. 2007), are illustrated by five species of southeastern centrarchids (Figure 8.6). The deeper-bodied species of sunfish show greater reliance on suction compared to ram feeding (high approach velocity), in contrast to the more streamlined Largemouth Bass; however, ram feeders may experience a lower overall strike accuracy (measured as how centered the prey is in the ingested water mass) (Higham et al. 2006a). Largemouth Bass, and other species of Black Basses (genus *Micropterus*) are

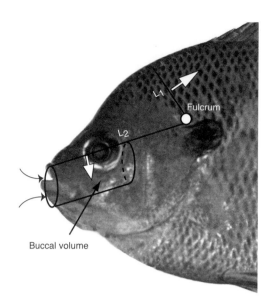

FIGURE 8.5. A schematic of relative forces generated during suction feeding. The schematic model of suction feeding (based on Carroll et al. 2004 and Wainwright et al. 2007) is overlain on the head of a Bluegill. Locations are approximate and are only to show relative locations of forces. The fulcrum is located at a point between the posttemporal bone of the skull and the supracleithral bone of the pectoral girdle. L_1 is the distance from the fulcrum to the centroid of force generated by epaxial musculature; L_2 is the distance from the fulcrum to the centroid of force exerted on the buccal area. The force generated by suction increases as the epaxial muscle area and L_1 increase and decreases as the projected buccal area and L_2 increase.

specialized piscivores, even though the specialization for piscivory limits morphological diversity that might include other feeding modes (Collar et al. 2009). Successful feeding on fish prey is related to a larger body and mouth size, greater buccal volume, reduced jaw protrusion, and reduced size of the pharyngeal jaws so as not to inhibit passage of prey into the esophagus (Carroll and Wainwright 2009; Collar et al. 2009). Despite differences in the role of suction versus ram feeding in centrarchids, the pattern of muscle activity involved in creating suction has been conserved across the group (Wainwright and Lauder 1992).

Largemouth Bass, fed on Goldfish (*Carassius auratus*), initiated the strike sequence at one body length removed from the Goldfish by a rapid approach and an explosive expansion of

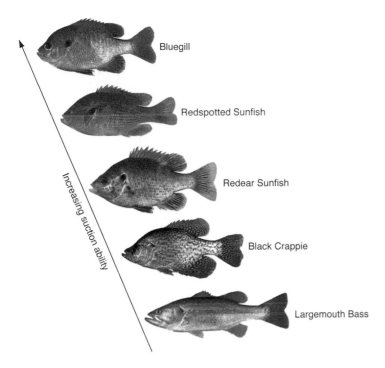

Increasing suction ability

Bluegill

Redspotted Sunfish

Redear Sunfish

Black Crappie

Largemouth Bass

FIGURE 8.6. Relative suction ability in five species of southeastern centrarchids. Based on model predictions from Carroll et al. (2004) and Wainwright et al. (2007). The Redspotted Sunfish is shown instead of its close relative the Spotted Sunfish, which was used in the two studies.

the mouth and throat (Svanbäck et al. 2002). However, strike initiation distances vary with prey kind and mobility. For Largemouth Bass feeding on tethered Ghost Shrimp (*Palaemonetes* spp.), strike distances were much smaller (1.25 cm or 0.07 body length) compared to fish feeding on Goldfish (Holzman et al. 2008b). Approach velocities increase proportionately with fish size, but times required to open and close the mouth (the gape cycle time) are fastest for small compared to large bass and can range from 37 ms in a 33 mm SL bass to 80 ms in a 201 mm SL Largemouth Bass (Richard and Wainwright 1995). Strikes that are the fastest show higher synchrony of the initiation of muscle firing, the fastest time to peak gape and forward rotation of the jaws, and result in the highest pressure gradient (Svanbäck et al. 2002). A typical strike requires the simultaneous elevation of the head, depression of the hyoid apparatus (the bones supporting the jaw joint), and depression of the lower jaw. Lower jaw depression causes the anterior rotation of the maxillae and protrusion of the premaxillae (Figure 8.3E). Jaw closing can occur with the mouth cavity still expanded and typically happens when the

prey is between the upper and lower jaws by adduction of the lower jaw and lowering of the head (Richard and Wainwright 1995).

Suspension Feeding

Various freshwater fishes, including some in the families Cyprinidae and Clupeidae, employ suction to move water containing organic detritus or small food items such as phytoplankton or zooplankton into their mouths. Although initially it was thought that most fishes removed particles from the water by direct sieving action of their gill rakers, removal methods turn out to be much more varied (Figure 8.7) (Rubenstein and Koehl 1977). When fluid flow is primarily perpendicular to the gill rakers, the rakers can function as dead-end filters or sticky filters; when flow is essentially parallel to the gill rakers, they can function as crossflow filters, which act to concentrate prey at the back of the throat (Rubenstein and Koehl 1977; Brainerd 2001; Sanderson et al. 2001). Although suspension feeding occurs in at least 60 species of marine and freshwater fishes, actual mechanisms of particle retention are known for only a few (Goodrich et al. 2000).

A. Dead-end and hydrosol filtration

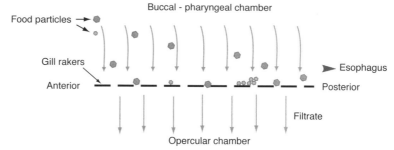

B. Crossfilter filtration

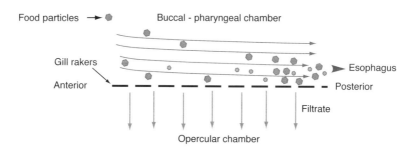

FIGURE 8.7. Modes of filtering in suspension-feeding fishes.
A. Dead-end and hydrosol filtration. In dead-end filtration (shown by large particles), food particles are retained that are larger than the pore spaces of the gill-raker sieve. In hydrosol filtration (shown by small particles) food particles are retained by sticking to mucus on the gill rakers or over the pore spaces.
B. Crossflow filtration, showing the increasing concentration of food in the posterior part of the oral chamber as water progressively exits past the gill rakers. Based on Brainerd (2001) and Sanderson et al. (1996, 2001).

DEAD-END FILTERS Dead-end filtration involves fluid flow perpendicular to the filter, with fluid (the filtrate) passing through the gill-raker filter while leaving a build-up of food particles on the surface of the gill rakers (Figure 8.7A). All particles that exceed the pore size of the sieve are retained and smaller particles pass through with the fluid (Rubenstein and Koehl 1977). However, as food particles are trapped on the filter, they act to change properties of the filter (pores become clogged and pore sizes decrease) and to increase flow resistance, consequently requiring some mechanism to transfer the food from the gill rakers to the esophagus. A European minnow, the Common Bream (*Abramis brama*) uses gill rakers as a type of dead-end filter to capture zooplankton. However, rather than trapping prey in the pore spaces between adjacent gill rakers, prey are collected in channels formed by adjacent transverse ridges on the gill arches and by gill rakers on adjacent arches (Hoogenboezem et al. 1991). Using muscles that originate on the gill arch and insert on the expanded bases of the gill rakers, Common Bream can also reduce channel widths, and thus effective pore sizes, relative to the size of prey being consumed (Hoogenboezem et al. 1993). Common Carp (*Cyprinus carpio*), a nonindigenous minnow of European origin that is widespread in North America, also possesses muscles that would allow reduced channel widths; whether similar mechanisms are used among native North American minnows to increase suspension-feeding ability is unknown (Van den Berg et al. 1994).

HYDROSOL (STICKY) FILTERS Hydrosol filters act similarly to dead-end filters, except that they possess a sticky mucus that serves to trap particles smaller than the actual pore sizes of the filter (Figure 8.7A). Suspension feeding using hydrosol filtration has not been documented among native North American freshwater fishes but is known for the Nile Tilapia (*Oreochromis niloticus*), a fish widely used in North America for aquaculture. Nile Tilapia apparently adjust the level of mucus secretion from the gill rakers relative to the particle sizes being ingested. For large particles, mucus secretion is minimal so that gill rakers function as a dead-end filter. However, for small particles such as bacteria, phytoplankton, and diatoms, there is active mucus secretion, often forming strands of mucus extending off of the gill rakers, and thus functioning as a hydrosol filter (Sanderson et al. 1996).

CROSSFLOW FILTERS Crossflow filtration involves the fluid moving parallel to the gill-raker surface and is the same technique used by industry to clarify beverages such as beer or wine (Brainerd 2001). During crossflow filtration, the food particles flow parallel to the surface of the gill rakers while water gradually exits through the gill rakers, resulting in a progressively concentrated slurry of food as it moves to the back of the throat or to the roof of the oral chamber (Figure 8.7B). Unlike a dead-end sieve, the gap widths between the gill rakers do not act as thresholds to prey-size retention. Also, unlike hydrosol or dead-end sieves, flow of liquid along the gills can be relatively fast, with measured values of approximately 55 cms^{-1} as determined for Gizzard Shad (*Dorosoma cepedianum*) and Goldfish (Sanderson et al. 2001).

Gizzard Shad and two nonnative minnows, Goldfish and Common Carp (*Cyprinus carpio*), use suction to pass suspensions of food and water rapidly from anterior to posterior. Gill rakers are not mucus covered and particles rarely stick to the surface of the gill rakers, indicating that these fishes use crossflow filtration to concentrate food particles and directly pass the concentrated slurry of food particles into the esophagus (Drenner et al. 1982b; Sanderson et al. 2001; Callan and Sanderson 2003).

Sacramento Blackfish (*Orthodon microlepidotus*), a minnow native to the San Joaquin-Sacramento River system in California, is an example of a species concentrating food particles on the roof of the mouth via crossflow filtration. During feeding, Sacramento Blackfish move water containing planktonic algae and zooplankton into the mouth using pulses of suction (Johnson and Vinyard 1987). Nonselective feeding occurs in individuals larger than 200 mm SL, but smaller fish use sight to direct feeding actions at individual prey or groups of prey (Johnson and Vinyard 1987). As the suspension of water and food items moves through channels between the rows of gill rakers, water gradually exits and the concentrated slurry of food particles is carried dorsally to the roof of the pharynx where food particles become trapped in the mucus-covered palatal organ (Sanderson et al. 1991, 1998). Suspension feeding involves movements similar, albeit exaggerated, to normal respiration. As such, it is no surprise that crossflow filtration and particle retention occur at a reduced level in Sacramento Blackfish undergoing normal respiration (Sanderson and Cech 1995).

Ram Feeding

Ram and suction feeding can be seen as two ends of a continuum of feeding tactics, as shown previously for Largemouth Bass and Bluegill. In ram feeding, fishes primarily close the distance to their prey by forward movement rather than using muscles to pull the water mass containing the prey toward them. Fishes that employ ram often have elongate jaws and feed on active prey such as fishes, either by stalking the prey or by capturing prey from ambush such as gars, pike and pickerel. Ram feeding also occurs in large planktivores, such as Paddlefish (*Polyodon spathula*), that entrain plankton in their gill rakers as they swim through the water with their mouths open (Porter and Motta 2004).

Ambush Predators

Fishes such as gars (genera *Lepisosteus* and *Atractosteus*) and pike and pickerel (family Esocidae) frequently employ ambush feeding. In a typical feeding sequence, based on a study of Florida Gar (*Lepisosteus platyrhincus*), feeding actions take place from anterior to posterior and involve expansive, compressive, and recovery phases (Porter and Motta 2004). Initially, Florida Gar slowly move alongside their prey, and the actual strike is sideways and initiated by strong lateral bending. Strike distances are small, and in the case of Florida Gar, average only 11.7% of the gar's total body length. Strikes occur quickly and rapidly, averaging only 32 ms in duration and speeds of 3.6 body lengths per second. The sequence of events involves the mouth opening beginning with the strike initiation, depression of the hyoid area, achievement of maximum gape, jaw closing, and hyoid elevation. Once prey are held by the jaws there are short, quick lateral and forward movements to manipulate the prey before it is moved to the throat and swallowed (Porter and Motta 2004).

The attack approach in pike and pickerel, based on Tiger Muskellunge (a hybrid between Northern Pike, *Esox lucius*, and Muskellunge, *E. masquinongy*), Chain Pickerel (*E. niger*), and Northern Pike, differ from that in Florida Gar in that *Esox* attack moving forward and not sideways to their prey (Webb and Skadsen 1980; Harper and Blake 1991). However, other aspects are similar. The attack generally begins with a slow approach toward prey using the median and paired fins (MPF) (Webb and Skadsen 1980), and the actual strike starts either with the body straight and then quickly bending into an S-shaped posture and then bending into an opposite S-shape as propulsion continues (termed pattern A) or starting with the body already in an S-shape (termed pattern B) (Webb and Skadsen 1980). The S-start provides for maximum straight-line acceleration toward the prey. In both Tiger Muskellunge and Chain Pickerel, type A starts occurred at a greater distance (22% and 37% body length, respectively) compared to type B starts (7.8% and 14% body length, respectively)

(Webb and Skadsen 1980; Rand and Lauder 1981). Strikes occurred in well less than a second, but compared to Florida Gar, the strike duration was longer and generally less than 200 ms for Tiger Muskellunge (Webb and Skadsen 1980), between 84 and 189 ms for Northern Pike (Harper and Blake 1991), and between 51 and 98 ms for Chain Pickerel (Rand and Lauder 1981). Acceleration rates, from first movement to reaching the prey, were 2.4–4.0 body lengths per second in Chain Pickerel (Rand and Lauder 1981) and 3.7–4.0 body lengths per second in Northern Pike (Harper and Blake 1991), and thus essentially the same as for Florida Gar.

Suspension Feeding

Suspension feeding occurs both by active suction, discussed in the section on suction feeding, and by ram. During ram suspension feeding, fishes move through the water with their mouths open and the trunk musculature basically takes the place of buccal and opercular muscles in terms of gill ventilation and suspension feeding (Burggren and Bemis 1992). Although ram suspension feeders are relatively common in the marine environment, with such examples as herrings, anchovies, mackerels, Manta Rays (family Mobulidae), Basking Sharks (*Cetorhinus maximus*), and Whale Sharks (*Rhincodon typus*) (Sanderson and Wassersug 1993; Helfman et al. 2009), this approach is infrequent in freshwater fishes. Paddlefish, which are adapted to large rivers and their associated floodplain lakes with their often abundant plankton densities, swim constantly soon after hatching (Burggren and Bemis 1992). Both juvenile and adult Paddlefish (*Polyodon spathula*) rely on ram ventilation of the gills and use ram suspension feeding (Figure 8.8). During feeding, Paddlefish increase their swimming speed 1.6 times above that used for gill ventilation. Flow in the mouth cavity can be 19 cms^{-1} or about 60% of the swimming velocity during ram suspension feeding. Although initial prey capture is based on ram and not suction, the second phase of suspension feeding requires suction to increase the flow velocity in the

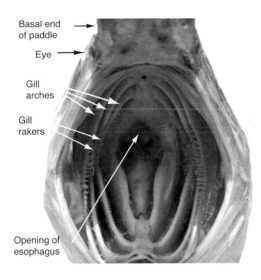

Basal end of paddle

Eye

Gill arches

Gill rakers

Opening of esophagus

FIGURE 8.8. Orientation of the gill arches and gill rakers of Paddlefish (*Polyodon spathula*), a ram suspension-feeding fish. Photo by W. T. Slack and S. G. George.

mouth above the swimming speed. Referred to as a "hydrodynamic tongue," this stream of water helps to move prey from the gill rakers into the esophagus (Sanderson et al. 1994). In adult Paddlefish, the gill rakers function as a dead-end filter. The sieve formed by the gill rakers has an average "mesh size" of approximately 0.06–0.09 mm and prey can be retained down to an average width of 0.10 mm (Rosen and Hales 1981). Although the gill rakers are well developed in adult Paddlefish, juvenile Paddlefish lack well-developed gill rakers until they reach approximately 100 mm TL. Juveniles also show increased feeding selectivity compared to adults, leading to the suggestion that juvenile Paddlefish do not filter feed (Rosen and Hales 1981). However, behavioral studies show that Paddlefish between 60 and 130 mm TL will use ram filter feeding (Burggren and Bemis 1992), indicating that gill rakers might function in some other way, as discussed previously in the section titled "Suction and Suspension Feeding."

Manipulation

Manipulation constitutes the third main category of prey capture, as defined by Liem (1980a). It involves use of teeth of the upper and lower jaws and includes a range of approaches including biting, rotational feeding, picking, and scraping.

Biting

Most simply, biting refers to the forceful contact of the oral jaws with the prey and can include feeding on entire prey items, as demonstrated by many piscivorous and insectivorous fishes, removing pieces from a prey item, such as scale eating in various topminnows and pupfish (Able 1976), or tearing off chunks of a prey, as shown by shaking and rotational feeding of American Eels (Helfman and Clark 1986). Biting fishes generally possess well-developed oral jaws and have strong adductor muscles to allow forceful jaw closing (Alfaro et al. 2001). As emphasized earlier, major categories of feeding modes are not necessarily exclusive. Consequently, fishes that employ biting might also use ram feeding to close the distance to their prey. Compared to suction feeding, biting is an evolutionarily derived behavior in bony fishes, and muscle activity sequences differ between fishes that use suction, such as Bluegill and Largemouth Bass, and fishes that use little or no suction. Suction requires a short onset time of muscle activity to generate the highest force, whereas in biting there is less constraint on the timing of muscle firing, and increased jaw strength and muscle mass tend to result in longer closing times. In addition, the muscles, such as the sternohyoideus, that pull down and back on the hyoid region to help expand the oral cavity in suction feeding are able to take on other roles in biting (Alfaro et al. 2001; Horn and Ferry-Graham 2006).

Fishes that consume parts of other fishes, such as chunks of flesh, scales, or ectoparasites, although more common in tropical freshwater or marine habitats, are also represented by North American freshwater fishes (see also Chapter 12). Species showing cleaning behavior actively remove ectoparasites (such as trematodes and leeches), mucus, diseased tissue, or unwanted food particles from other fishes, either of the same or different species (Losey 1972). Sheepshead Minnow (*Cyprinodon variegatus*), Striped Killifish (*Fundulus majalis*), Rainwater Killifish (*Lucania parva*), and Diamond Killifish (*Adinia xenica*) all either show posturing that would invite cleaning activity from other individuals or exhibit actual cleaning activity themselves (Hastings and Yerger 1971; Able 1976).

Predators that bite or tear pieces out of their prey overcome the limitations of gape size encountered by those that swallow their prey whole. American Eels employ two behaviors, shaking and rotational feeding, to tear pieces from prey that are too large to consume whole by suction feeding (Helfman and Winkelman 1991). Jaw teeth in American Eels are numerous, small, and sharp (setiform teeth) and, while not providing the ability to directly bite off pieces of prey, they are effective at grasping a prey so that other behaviors can be effective in tearing off pieces. Small pieces are removed by grasping the prey and shaking it—an approach common in a number of different fish groups. Rotational (spinning) behavior is employed to remove larger chunks but this behavior is uncommon among fishes (Helfman and Clark 1986). During rotational feeding, the body is extended and spins rapidly along its longitudinal axis. In fact, American Eels can achieve up to 14 rotations per second, approximately three times faster than achieved by human figure skaters (Helfman and Clark 1986). Compared to suction feeding on entire prey items or using shaking to remove pieces of prey, rotational feeding requires more energy and is likely employed in the wild only when American Eels come across large prey that cannot be consumed by other behaviors (Helfman and Winkelman 1991).

Picking and Scraping

Fishes with jaws specialized for picking and scraping employ fine motor control of the jaws to obtain individual food items from the substratum or water column, while being able to avoid or reduce the intake of nonnutritive items

(Horn and Ferry-Graham 2006). The effectiveness of a picking-based feeding mechanism requires finely controlled, "forceps-like" movements of the upper and lower jaws (Motta 1988). Pickers are also characterized by jaws in which the force application has been shifted anteriorly and a biomechanical coupling that allows upper and lower jaw movements to be synchronized. This morphological transformation appears to be associated with functional specialization for algal scraping and grazing.

Fishes in the order Cyprinodontiformes, which includes North American freshwater representatives of the families Aplocheilidae (rivulines), Fundulidae (topminnows), Poeciliidae (livebearers), Goodeidae (goodeids), and Cyprinodontidae (pupfishes), often employ a picking type of feeding mechanism (Ferry-Graham et al. 2008; Hernandez et al. 2008). Jaw protrusion in derived compared to basal cyprinodontiform fishes (Table 8.2) or to the atheriniform fishes (the sister group to the Cyprinodontiformes) is more specialized. It is in part controlled by the increasing complexity and greater amounts of cellular tissue of a ligament connecting the premaxilla and the mandible (the premaxillomandibular ligament) and by changes in insertion of a branch of the major jaw-closing musculature, the adductor mandibulae, so that it inserts on the premaxilla in addition to the mandible (Figure 8.9) (Hernandez et al. 2009). These changes allow greater coordination of the upper and lower jaws and increase the extent of jaw protrusion in derived versus basal cyprinodontiform fishes. For example, the derived Mexican Molly (*Poecilia sphenops*) has a divided premaxillomandibular ligament and a more complex branching of the adductor mandibulae, which inserts on the premaxilla in addition to the maxilla and mandible. This is in contrast to the more basal livebearer, the Western Mosquitofish, and the basal killifish, the Redface Topminnow (*Fundulus rubrifrons*; formerly *F. cingulatus*), whose branches of the adductor mandibulae complex only insert on the maxilla and the mandible (Figs. 8.9 and 8.10) (Hernandez et al. 2008, 2009).

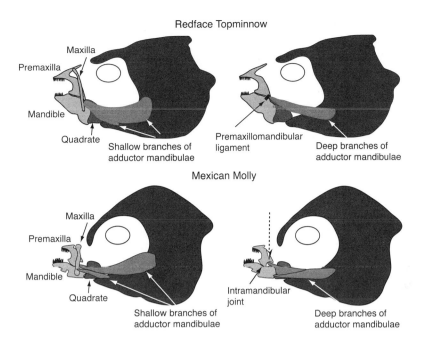

FIGURE 8.9. Upper and lower jaw bones and insertions of the adductor mandibulae complex in the Redface Topminnow (*Fundulus rubrifrons*) and the Mexican Molly (*Poecilia sphenops*). The left panels show the shallow branches of the adductor mandibulae complex; the right panels show the deep branches of the adductor mandibulae complex. For the Redface Topminnow, the heavy black line shows the location of the premaxillomandibular ligament (present but not shown in the Mexican Molly). For the Mexican Molly, the dashed arrow shows the branch of the deep adductor mandibulae inserting on the premaxilla and the black arrow shows the location of the intramandibular joint. Based on Gibb et al. (2008) and Hernandez et al. (2008, 2009).

The increased independence of control over the premaxilla in the Mexican Molly may provide for greater dexterity that is needed for grazing on epiphytic algae through repeated scraping motions (Hernandez et al. 2008; Gibb et al. 2008). In addition, the lineage including *Heterandria*, *Xiphophorus*, and *Poecilia* is characterized by increased mobility of lower jaw shape as a result of mandibular bending allowed by a joint between the dentary and the angular/articular bones (Figure 8.9) (Gibb et al. 2008). Consequently, evolutionary changes in jaw morphology of cyprinodontiform fishes have been in the direction of greater jaw dexterity during feeding events. During the early evolution of cyprinodontiforms, increased dexterity and precision, granted by the ligament that directly connects the upper jaw to the lower jaw, may have enhanced the ability of basal species to select individual prey items from the substratum or water column using picking-based prey capture behavior. During the later evolution of this clade, the direct control of upper jaw movements during retraction, granted by the insertion of a branch of the adductor mandibulae complex on the premaxilla, enhanced the ability of derived species to remove encrusting material using a nipping or scraping-based feeding behavior; lower jaw bending apparently further increased the efficiency of algal scraping (Gibb et al. 2008; Hernandez et al. 2008, 2009). Of the species included in Figure 8.10, all microcarnivores except for *Floridichthys* have basic conical teeth, whereas tooth morphologies of omnivores include weakly spatulate, spatulate, or tricuspid shapes, and tooth shapes of grazers include recurved spatulate or tricuspid teeth.

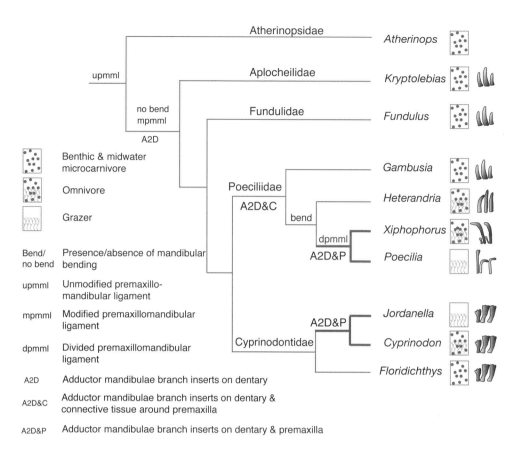

FIGURE 8.10. Morphological, feeding, and phylogenetic relationships of selected fishes in the order Cyprinodontiformes and the sister group Atheriniformes. Lineages possessing modified adductor mandibulae muscles are also shown in bold. Drawings on the far right show general premaxillary or dentary tooth shapes; rectangles show trophic levels. Based on Hernandez et al. (2008, 2009), Gibb et al. (2008), with additional ecological and morphological information from Parenti (1981) and Ross (2001).

Prey Processing

Manipulation, postcapture processing such as mastication, and swallowing of prey involve the pharyngeal jaws, which are formed from upper and lower elements of the pharyngeal arches, associated bones, and ligaments, and from processes of other bones (Gosline 1973; Lauder 1982, 1983a, 1983b; Lauder and Wainwright 1992). The level of prey processing varies greatly among fishes. In suction-feeding fishes, prey tend to be swallowed with minimal postcapture processing. In contrast, fishes such as the Cyprinidae totally lack true teeth in the oral jaws, and all food processing is achieved by the toothed pharyngeal jaws. (*Danionella dracula*, a recently described minnow from northern

Myanmar [Burma], does have tooth-like extensions from the upper and lower jaw bones [Britz et al. 2009].) In more derived fishes, teeth in the oral jaws and the pharyngeal teeth can be well developed (Sibbing 1991a). The pharyngeal jaws are used to remove hard outer structures of prey or otherwise reduce the size of the prey so that they can pass into the esophagus, as in molluscivores like Pumpkinseed (*Lepomis gibbosus*) or the molariform morph of the Cuatro Ciénegas Cichlid (*Herichthys minckleyi*) (Lauder 1983c; Hulsey et al. 2005; Helfman et al. 2009).

In teleosts, the lower pharyngeal jaw elements are formed from the hypertrophied fifth gill arches—the first four arches support the gill filaments and gill rakers (Evans and Deubler 1955). The upper pharyngeal jaw elements in teleosts consist of 1–5 paired tooth plates that may or may not be fused to the gill-arch elements; the upper tooth plates are often reduced to 1–2 pairs in more derived groups (Lauder 1982). In addition to elements associated with gill arches, the upper biting surface can be formed from processes originating on neurocranial bones, such as the masticatory process of the basioccipital bone in cyprinids (Harrington 1955). Ventrally, lower teleosts have paired tooth plates that are aligned but not necessarily fused with the ceratobranchial bones of the fifth gill arch; in more derived teleosts, these elements become fused to the fifth cerato-branchials (Lauder 1982; Helfman et al. 2009). Pharyngeal jaws are particularly well developed in the cyprinids, catostomids, certain sunfishes, and cichlids.

Cypriniform Pharyngeal Jaws

Fishes in the order Cypriniformes, a group of ostariophysan fishes containing the families Cyprinidae and Catostomidae, are the first group of the teleosts (Chapter 7; Figure 7.5) to have well-developed pharyngeal dentition (Nelson 2006; Helfman et al. 2009). The pharyngeal teeth, which make up the ventral part of the pharyngeal jaws, are in one to three rows in cyprinids and in a single row with more than 16 teeth in catostomids (Berra 2001). In native North American minnows there are no more than two rows of teeth, and the major row (innermost row) has no more than six teeth and usually four to five teeth (Evans and Deubler 1955; Page and Burr 1991). Shapes of pharyngeal teeth vary from sharp, hooked teeth or conical teeth in piscivores and insectivores, to flat crushing tooth plates in detritivores and omnivores, to molariform teeth in certain herbivores and molluscivores, although some species such as carp have different tooth shapes in the same arch (i.e., a heterodont condition) (Evans and Deubler 1955; Sibbing 1988, 1991a). Pharyngeal teeth are replaced continually throughout life (Evans and Deubler 1955). The upper dentition consists of the enlarged masticatory process, which is covered with a horny pad (i.e.,

the chewing pad) in minnows and a soft pad in suckers (Sibbing 1982; Nelson 2006).

Pharyngeal teeth work against the dorsal chewing pad and are capable of complex grinding, crushing, and shredding movements, depending on the species (Sibbing 1982). In Common Carp, the pharyngeal arch is moved by modified gill-arch muscles; by epaxial muscles, which indirectly act to raise the pharyngeal bones; and by hypaxial muscles inserting on the pectoral girdle, which act to pull the teeth downward and backward, resulting in a sliding motion of the teeth against the chewing pad. Even though cypriniform fishes lack a retractor dorsalis muscle that acts to retract the upper pharyngeal jaws, as in more derived teleosts, the chewing pad is also mobile relative to the lower pharyngeal teeth, moving with the base of the skull and being powered by the epaxial body musculature (Sibbing 1982, 1988; Lauder and Wainwright 1992). However, the lower pharyngeal jaw is most important in prey transport and swallowing in minnows as well as in other lower teleosts such as esocids and salmonids (Lauder 1983b). (As an aside, the powerful shearing and tearing motions of the pharyngeal jaws explain why food habit studies of minnows are made challenging by the general lack of intact prey items.)

Sunfish Pharyngeal Jaws

In contrast to minnows, both the lower and upper elements of the fifth branchial arches support pharyngeal jaws in centrarchids and other neoteleostean fishes (Chapter 7; Figure 7.5), and there is a large muscle, the retractor dorsalis, that originates on the anterior vertebra and inserts on the upper pharyngeal jaws—in fact, this muscle is one of the defining characters of the Neoteleostei (Lauder 1983b; Lauder and Wainwright 1992). The retractor dorsalis acts to move the upper pharyngeal jaws backward and is in part responsible for the upper pharyngeal jaws taking over the major role in food transport from the lower pharyngeal jaws in the neoteleosts (Lauder 1983b). Among the neoteleosts, the degree of development of the pharyngeal jaws, teeth size and shape, and muscle development differs widely depending on the species and the kind of food eaten.

Among sunfish species, in addition to backward movement of the upper pharyngeal jaws by the retractor dorsalis, the jaws are moved ventrally against the lower pharyngeal jaws by rotation of upper gill-arch bones (the epibranchials). A general pattern of pharyngeal jaw muscle activity during prey mastication and movement into the esophagus, identified in insectivorous and piscivorous centrarchids and Yellow Perch (Percidae, *Perca flavescens*), is for a rhythmic back-and-forth movement of the lower pharyngeal jaws and strong retraction of the upper pharyngeal jaws that starts halfway through the backward movement of the lower jaws and continues into the forward lower pharyngeal jaw movement—this is referred to as the transport pattern (Lauder 1983a, c; Wainwright and Lauder 1992). Two sister species of Sunfish, Pumpkinseed and Redear Sunfish, have hypertrophied pharyngeal jaws, dentition, and musculature allowing them to crush snail shells and consume snails, a diet that is uncommon among other species of *Lepomis* (Lauder 1983a, c). The two species have a generally nonoverlapping natural range, with Pumpkinseed restricted to the northern half of the eastern United States and southern Canada, and Redear Sunfish restricted to the southeastern United States (Lee 1980a, b). The Pumpkinseed retains the generalized rhythmic pattern when feeding on nonmollusk prey; however, when fed snails it is able to alter the pattern of muscle activity so that the upper and lower pharyngeal jaw muscles contract synchronously—referred to as the crushing pattern. Redear Sunfish have lost the ancestral rhythmic transport pattern of muscles firing and only show the synchronous crushing pattern—to a Redear Sunfish, all prey are processed as though they were snails (Lauder 1983a, c). Although both Pumpkinseed and Redear Sunfish are specifically adapted for feeding on mollusks, the more specialized Redear Sunfish is capable of generating 50–100% more crushing force and switches at a smaller

size, compared to Pumpkinseed, to a diet composed largely of snails (Wainwright 1996; Huckins 1997). Pumpkinseed also show polymorphism in the size and strength of the pharyngeal muscles and bones that seem related to the presence of snail prey as the fish develop. In a lake with few snails, Pumpkinseed had smaller pharyngeal muscles and less robust jaws compared to fish in a lake with abundant snail prey, although tooth morphology, other than increased wear in the snail-eating morph, did not differ (Wainwright et al. 1991).

Cichlid Pharyngeal Jaws

In contrast to minnows, both the lower and upper elements of the fifth branchial arches hypertrophy in cichlids and, in contrast to both minnows and sunfishes, the lower pharyngeal jaws are fused, resulting in greater versatility of movements and lesser individual muscle loadings (Liem 1973; Sibbing 1991a). The presence of a complex pharyngeal jaw structure has freed the oral jaws from the role of food manipulation and processing, leaving oral jaws with a single role of prey capture. This has led to a high diversity of form and function in the oral jaws (Liem 1973).

The Cuatro Ciénegas Cichlid, endemic to the clear spring pools of the Cuatro Ciénegas basin in Coahuila, Mexico, displays at least three morphs: a piscivorous morph involving body shape and cranial structure, and two morphs involving the size and shape of the pharyngeal teeth (Swanson et al. 2003). Of the latter two morphs, the papilliform morph has a more delicate lower pharyngeal jaw; reduced size of pharyngeal jaw musculature; and narrow, needle-like teeth in contrast to the molariform morph, which is characterized by a more robust lower pharyngeal jaw; hypertrophied pharyngeal jaw musculature; and broad, crushing molariform teeth. The papilliform morph feeds primarily on detritus, algae, and soft-bodied invertebrates; the molariform morph also consumes soft-bodied invertebrates, algae, and detritus when food is abundant but polarizes on snails when food becomes limiting and uses the molariform teeth and hypertrophied musculature to crush the shells (Liem and Kaufman 1984; Swanson et al. 2003; Hulsey et al. 2006). In fact, the molariform morph of Cuatro Ciénegas Cichlids produces the highest size-specific crushing force known among freshwater or marine fishes that use their pharyngeal jaws to crush mollusks (Hulsey et al. 2005). In some pools the papilliform and molariform morphs also differ in feeding behaviors and locations, with the former primarily scraping or scooping prey from the surface of travertine formations and the latter primarily diving into the soft substratum to engulf both hard and soft-bodied prey, although differences in behaviors are modulated by availability of hard and soft substrata. The occurrence of the two pharyngeal tooth morphs is related both to a genetic component and to phenotypic plasticity (Swanson et al. 2003, 2008).

SUMMARY

Fishes employ a high diversity of feeding modes, and jaw structures show a general trend toward increased complexity and degrees of freedom of movement over evolutionary time. One of the early changes in jaw morphology was the freeing of the maxillary bones from the skull so that the posterior ends could rotate forward during mouth opening and thus aid in the development of suction during feeding. Suction feeding represents the most basal, and most widespread, feeding mode among fishes. Other trends include the development of a posterior process on the premaxillary bones (toothed in more derived fishes) and the exclusion of the maxillary bone from the gape (and loss of teeth on the maxilla). The development of protrusile jaws, accompanied by the development of an ascending process on the premaxilla, has occurred in about half of all bony fishes as well as in some living chondrosteans. Jaw protrusion can aid in capture success by increasing the rate of approach to a prey item, creating a rounder mouth opening to aid in suction feeding, thus increasing the volume of the mouth and pharyngeal cavities and allowing

faster jaw closing with less chance of prey expulsion. It also allows for a more horizontal body orientation in benthic feeding fishes. Fishes that use suction also employ various degrees of ram feeding.

Suspension feeding is less common among freshwater fishes than in marine fishes, but there are freshwater examples among North American species. Suspension feeders can use gill rakers as a dead-end sieve where water flow is largely perpendicular to the gill rakers and particles are collected in the spaces between the gill rakers, or by using the gill rakers and other structures as crossflow filters where water flow is largely parallel to the gill-raker surface and prey are concentrated toward the back of the throat while water exits through the gill rakers.

Several groups of more derived fishes have jaws that are specialized for picking and scraping—among North American freshwater fishes picking and scraping morphologies and behaviors are best developed among the Cyprinodontiformes. Pickers have the force application of the jaws shifted anteriorly and there is increased muscular control and flexibility of the jaws.

Prey processing is largely accomplished by a second set of jaws—the pharyngeal jaws. In some groups of fishes, such as those in the order Cypriniformes, pharyngeal jaws are highly developed and function to shred or macerate prey, in addition to transporting food into the esophagus. In the Cypriniformes, the lower pharyngeal jaws take the major role in prey transport. In more derived fishes such as centrarchids and cichlids, the upper pharyngeal jaws are primarily responsible for prey transport. In these groups, the jaws of some species are greatly hypertrophied and have enlarged molariform teeth, which allow them to feed on hard-bodied prey such as snails.

Although the literature on functional morphology of fishes is large and rapidly growing, information on the two largest families, Cyprinidae and Percidae, is surprisingly limited. Within the percids, only one genus (*Perca*) seems to have been studied. Within the cyprinids, most of the information is based on European taxa, and even then on only a few species. The one exception includes the landmark studies of suspension feeding in a North American cyprinid, the Sacramento Blackfish.

SUPPLEMENTAL READING

Higham, T. E., S. W. Day, and P. C. Wainwright. 2005. Sucking while swimming: Evaluating the effects of ram speed on suction generation in Bluegill sunfish *Lepomis macrochirus* using digital particle image velocimetry. *The Journal of Experimental Biology* 208:2653–60. An example of using Digital Particle Image Velocimetry and high-speed video to understand fluid movements during suction feeding.

Ferry-Graham, L. A., and G. V. Lauder. 2001. Aquatic prey capture in ray-finned fishes: A century of progress and new directions. *Journal of Morphology* 248:99–119. Highlights changes in the understanding of prey capture coincident with technological innovations.

Lauder, G. V. 1982. Patterns of evolution in the feeding mechanism of Actinopterygian fishes. *American Zoologist* 22:275–85. A classic paper on feeding mechanisms in ray-finned fishes.

Sanderson, S. L., A. Y. Cheer, J. S. Goodrich, J. D. Graziano, and W. T. Callan. 2001. Crossflow filtration in suspension-feeding fishes. *Nature* 412:439–41. Details the complexity of mechanisms in suspension-feeding fishes.

NINE

Life History and Reproductive Ecology

CONTENTS

THE PREVIOUS TWO CHAPTERS in this unit dealt with body form and function and included information on evolutionary trends in locomotion, effects of natural selection on body shape, basic jaw structure and function, and morphological specializations associated with prey capture. This chapter continues the study of form and function in relation to reproduction and the role of natural selection in shaping life-history patterns. In fact, evolution of body size and shape of fishes can often be best understood in terms of the selective pressures of life histories.

Fishes demonstrate a fascinating array of reproductive modes and life-history patterns and perhaps show the greatest range of variation among all vertebrates (Winemiller 1995). Also, in contrast to birds and mammals, most fishes continue to increase in body length and mass after reaching sexual maturity (termed indeterminate growth) so that there is a trade-off between funneling production into somatic growth or into reproduction (Charnov et al. 2001). Although most fish species are gonochoristic (meaning that sexes are separate and sex is genetically fixed), there are various examples of unisexual species and species that change sex as a function of age, environment, or social conditions. Reproductive modes vary from fishes with external fertilization, where the eggs are fertilized and develop totally outside the body of the female (termed oviparity), to the opposite extreme of internal fertilization, with varying degrees of maternal nutrient contributions to the eggs (termed viviparity). Parental care of eggs and young spans the range from none to the extreme of viviparity, with numerous intermediate examples of nest construction, nest guarding, and oral brooding. Among these categories there is a huge range of variation among different taxa (reviewed by Balon 1975, 1981).

LIFE-HISTORY PATTERNS AND THE ALLOCATION OF ENERGY RESOURCES

Life-history studies of fishes are concerned with such variables as lengths of spawning seasons, single and annual clutch sizes, sizes of embryos and larvae, degree of parental care, age at first reproduction, reproductive lifetime, adult body size, ova diameter and energy storage, maximum longevity, the presence or absence of spawning migrations, and the relationship of adult versus larval and juvenile survival (Stearns 1976, 1977; Gotelli and Pyron 1991; Winemiller and Rose 1992). The actual life history shown by an organism is the distillation of competing demands or trade-offs on growth, survival, and reproduction. An ideal organism—a "Darwinian Demon"—would, in terms of reproduction, maximize reproductive performance throughout all life-history stages irrespective of any constraints (Partridge and Harvey 1988). However, in nature an organism must deal with the realities of historical (phylogenetic) constraints, physiological constraints, demographic constraints, and ecological constraints of obtaining sufficient food for growth and reproduction while minimizing exposure to predation (including parasitism), all accomplished within the context of long- and short-term environmental variation (Gotelli and Pyron 1991; Winemiller and Rose 1992). The outcome of these interactions is the development of a specific life history for an organism that includes traits shaped by natural selection in response to local ecological factors; however, as also pointed out in previous chapters, the actions of natural selection need not have occurred in the terminal taxon but could just as well have occurred earlier in the evolution of a lineage and represent synapomorphies (shared, derived characters) (Stearns 1976, 1977; Gotelli

and Pyron 1991). Consequently, the life histories of many species can be expected to be a mosaic of contemporary responses to natural selection and historical traits that have been retained. For instance, among 21 species of North American minnows, one trait (length of spawning season) was significantly related to current ecological conditions (measured by latitude), but maximum body size had a strong historical component (Gotelli and Pyron 1991).

To elucidate this point further, some aspects of a species' life history have evolved through physical or biotic selective forces originating outside of their gene pool, such as the effects of environmental factors shown previously. In contrast, other life-history aspects have evolved in response to the presence of other individuals of the same species or sex, such as frequency-dependent intraspecific competition. As shown later, for alternative reproductive strategies, the life histories of male fishes are often influenced by what other males are doing, whereas female life histories can be less influenced by frequency dependent selection and more so by factors originating outside of the gene pool (E. L. Charnov, pers. comm. 2011).

Living and Dying

Understanding trade-offs in life-history evolution as well as responses of fish populations to fishing pressure, or efforts at stock enhancement, requires knowing some basic population demography—the parameters dealing with age structure, mortality, and reproduction. A convenient way to organize age-specific survivorship and mortality data is with a life table. Depending on how the data are collected, life tables can be of two types, dynamic and static. Dynamic life tables, also referred to as horizontal or cohort life tables, are built on data obtained by following a single cohort (age-class) through time until all members have died. This is the preferable approach in terms of having few required assumptions, other than random sampling, but often not possible given the longevity of many species. In contrast, static life tables, also referred to as vertical, current, or time-specific, are usually based on a census taken over a short time period, such as a day or a season, and provide a snapshot of the age structure of a population. Static life tables, although the most common, are less desirable because of the number of necessary assumptions. The method assumes that samples are obtained randomly and that survivorship and birth rates are constant between cohorts, which would only occur if a population has a stable age distribution (Deevey 1947; Krebs 1985).

Construction of a life table requires knowing the number of animals of each age surviving or dying within a given time interval. Depending on the species and how short generation times might be, the time interval could be determined in days, months, or years. For most fishes, the time intervals used would be months or years. The type of input data depends on the particular system being studied. In an exploited population where fishing mortality is high relative to natural mortality, the number dying might be more readily obtained; in nonexploited populations the number of survivors in a given age class (i.e., the age structure of a population) is likely to be the starting point. Because columns in a life table are interconvertible, whether the initial data are based on numbers surviving or numbers dying does not matter.

Somewhat surprisingly, well-documented life tables, covering all life stages from fertilized eggs or newly hatched larvae to adults, are rare for North American freshwater fishes and for fishes in general—a fact pointed out by Deevey (1947), echoed by Weatherly (1972), and seemingly true today. This is not to say that extensive data on survivorship and mortality within certain segments of the life span are lacking. Examples include the detailed study of larval and adult (but not juvenile) survivorship of Pumpkinseed (*Lepomis gibbosus*) (Bertschy and Fox 1999), and numerous studies of fishes exploited in recreational or commercial fisheries. The difficulties in constructing complete life tables for fishes are due to the enormous ranges in body size and habitats occupied over a life span, contributing to greatly differing capture efficiencies over different life stages (i.e., random samples of the entire age structure are nearly impossible). In addition, most fish populations are open and subject to emigration and immigration, and there are often large swings in year-class strengths (Deevey 1947; Weatherly 1972). Even closed populations, such as farm ponds and small lakes, still suffer from the problem of unequal sampling efficiencies among life stages, leading Deevey (1947) to suggest that "a life table constructed from the available observations would certainly be lacking both head and tail."

To provide an example of a life table, I combined data from several studies and time periods for the Fantail Darter (*Etheostoma flabellare*) (Table 9.1). A life table constructed in this manner is, of course, only an approximation of an actual life table that would present data on a single population. Survivorship from ripe, unfertilized eggs to age-1, information that is lacking in most studies of fishes, was 3.5% (Halyk and Balon 1983; Paine and Balon 1986). Age-1 to adult survival of 18.4% was based on population age structure determined from scale annuli by analyzing changes in the number of individuals in successive age groups (Box 9.1). This approach, termed a catch-curve, is useful over a wide range of age groups, from eggs and larvae to adults. Age-specific survival can also be obtained by direct observation and by mark-recapture studies, where different age groups are marked with a unique code or tag, released, and then recaptured on one or more later dates; however, this approach is better suited for larger fish where it is possible to apply external or internal tags.

The life table for Fantail Darters shows a rapid decline with increasing age in both the numbers of individuals (n_x) and proportions (l_x) of individuals remaining. There is also a concomitant decline in the absolute number of individuals dying (d_x) during successive year classes, simply because there are fewer individuals left to die, but a relatively constant finite mortality rate (q_x) and an increase in age specific survivorship (s_x) from age-0 to ages-1 and 2,

BOX 9.1 • Catch Curves, Age Structure, and Survivorship

Early fisheries biologists devised various ways to estimate survivorship and mortality rates of populations from the catches of fish across uniform size groups or age classes. Such approaches are collectively referred to as catch-curve techniques. Various algorithms that incorporate changes in age-group size have been proposed, including those based on proportions of the relative abundance of consecutive age groups and those based on linear regression analysis (Ricker 1975; Miranda and Bettoli 2007). Both approaches require the assumption of constant recruitment and mortality rates over time and that catchability of age groups remains the same once they are vulnerable to the sampling gear.

To estimate the survival (S) of Fantail Darters, I used an approach modified from Stearns (1976) by Bertschy and Fox (1999) that weights within-group survival estimates by the number of fish in that interval, where

$$S = \frac{\sum_{x=v}^{x=\omega} n_x s_x}{\sum_{x=v}^{x=\omega} n_x}$$ (Equation 9.1)

and n_x = number of individuals in age group x, s_x = the proportion of individuals surviving from age x to x+1, v = the first age group vulnerable to the sampling gear (in this case age 2), and ω = the oldest age group. Using the number of fish in successive age groups (based on Karr 1964),

S is estimated as 26%, as shown in the following example.

Age	n_x	s_x
0	14	
1	21	
2	59	0.322
3	19	0.105
4	2	0.000

$$S = \frac{(0.322 \times 59) + (0.105 \times 19) + (0 \times 2)}{59 + 19 + 2} = \frac{21}{80} = 0.26$$

(Equation 9.2)

For fishes with greater longevity, one of the most useful techniques for estimating survival is to plot the log of fish number (catch) versus age class and then determine the slope of the descending part of the curve. Only the descending arm is used in the calculations because the initial part of the catch curve shows size classes or ages that are not fully vulnerable to the sampling gear. Given that the previously mentioned assumptions are met, the slope is equal to the instantaneous mortality rate (Z), which is related to the finite mortality rate (A) as $(1 - A) = S = e^{-z}$, where e is the base of natural logarithms (Ricker 1975). Age structure of Walleye (*Sander vitreus*), based on scale annuli, provides an example of such a catch curve (see the following figure). The slope of the regression line fitted to the natural logarithm of the number of individuals in the fully recruited age groups is –0.366, which is equivalent to 69% survival over the period from age 3 to age 12.

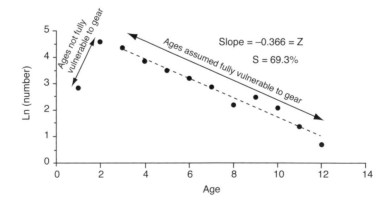

A catch curve based on the age structure of a population of Walleye (*Sander vitreus*) collected by experimental gill nets (i.e., variable mesh sizes) from Lake Sakakawea, North Dakota, and aged using scale annuli. Data from Isermann et al. (2003).

TABLE 9.1

Approximate Life Table for Fantail Darter (Etheostoma flabellare) *Based on Adult Survival Determined for an Iowa Population and Survival from Ripe Eggs to Age-1 for an Ontario, Canada, Population*

Age class (x)	n_x	l_x	d_x	q_x	s_x
0	1000	1.000	965	0.96	0.035
1	35	0.035	26	0.74	0.263
2	9	0.009	7	0.74	0.263
3	2	0.002	2	1.00	0.000
4	0				

SOURCE: Iowa data provided by Karr (1964); Ontario, Canada data provided by Paine and Balon (1986, table 2).

NOTE: Numbers are arbitrarily adjusted to a starting population size of 1,000 and numbers of individuals have been rounded to the nearest whole number.

n_x = number of individuals at the beginning of age x

l_x = the proportion of the population surviving to the beginning of age x ($l_x = n_x/n_0$)

d_x = number dying during the age interval x to x+1 ($d_x = n_x - n_{x+1}$)

q_x = proportion of population dying during age interval x to x+1 ($q_x = d_x/n_x$)

s_x = age-specific survival rate = (n_{x+1}/n_x)

before dropping to 0. Because values of q_x are relative to population size, they are comparable among different populations, in contrast to d_x values. Note that the terminology used in life tables and in treatments of mortality and survivorship varies within and among disciplines.

In general, survival of eggs, embryos, and larvae, relative to the survival of juveniles and adults, is low for most freshwater fishes, especially for those having external fertilization and lacking parental care (Dahlberg 1979). Comparisons of survivorship and mortality among populations and species are difficult because of the great variation in the methods used among studies and the way in which the data are reported. For instance, some studies present survivorship data over short life-history intervals, such as yolk-sac larvae, whereas others might estimate survival from the time of spawning to young-of-year or age-1, and others present data based on a specific number of days or weeks. Because fishes are exposed to forces of mortality on a daily basis, the time over which survivorship is determined strongly influences the value. Consequently, depending on the question being asked, survivorship values can be compared across specific durations or life-history intervals or as instantaneous rates. The instantaneous mortality rate (Z; the mortality in an instant of time; see also Box 9.1) approaches the percent annual or seasonal mortality (A) as the time interval over which A is determined approaches zero. Instantaneous rates are advantageous because they can be added, multiplied, or divided and hence, if the time interval is known, allow conversion of annual or other time-specific rates to daily rates. For example, given a survivorship of yolk-sac larvae of 55% over a 10-day period, the daily survival rate of 94.2% is obtained by determining the instantaneous rate (Z) for the 10-day time interval:

$Z = \ln(0.55) = -0.598$; dividing by 10,

and finding the antilog: (Equation 9.3)

$e^{(-0.0598)} = 0.942$.

As an example of changing survivorship values with life-history stage, daily percent survivorship of Walleye (*Sander vitreus*) gradually increases from 80% for eggs and yolk-sac larvae to 96.5% for post-yolk-sac larvae, and 99.9% for late juveniles and adults (Figure 9.1A). Although a daily survival rate of 80% might seem high, when carried over 24 days from spawning to

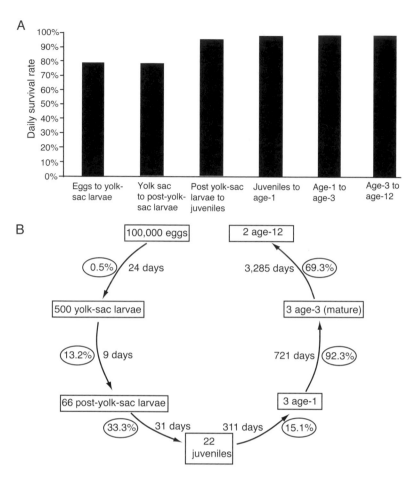

FIGURE 9.1. An example of how survival varies among life-history stages and the impact that this has on cohorts of eggs.

A. Daily percent survival values for six life-history stages of Walleye (*Sander vitreus*).

B. The impact of stage duration and mortality forces on the successive numbers of fish in each of the six life-history intervals (rectangles) of Walleye from fertilized eggs to age-12. The numbers are determined from interval survival probabilities (ovals) based on survival probabilities and stage duration, and assuming that spawning occurs in April and that first reproduction occurs at age-3. The data are from multiple northeastern localities (Dahlberg 1979; Henderson et al. 1996; He and Stewart 2001; Pratt and Fox 2002; and Isermann et al. 2003).

larval emergence it results in an interval survivorship of only 0.5% (Figure 9.1B). By combining daily survival rates and the duration of each stage (assuming that survival from age-1 to age-3 is the same as survival from age-3 to age-12) and starting with an arbitrary initial number of 100,000 spawned eggs, three adults can be expected to reach sexual maturity and two adults can be expected to survive to age-12.

Survival probabilities from egg production to larval emergence tend to be lowest for fishes without any form of parental care and higher for those burying eggs or depositing them in nests, or building a nest and guarding eggs and larvae, although the available data that are directly comparable among these three life-history modes are extremely limited (Figure 9.2A). In contrast, when survival probabilities are compared among life-history categories, but over a longer time interval that extends well beyond any time of protection from a nest or parental guarding, there is no clear pattern and survival values are grouped much more closely together (Figure 9.2B). Fantail Darters, in which males locate nests beneath rocks and guard the eggs until they hatch (Page 1983; Page and Swofford 1984), do show the highest percent survival among the few species listed in Figure 9.2B, but other nest guarders have values less than species that only bury eggs or broadcast the eggs without any parental care. Even though the number of examples is extremely limited, the take-home message is that egg burying or nest guarding can improve survival probabilities, but mortality during later life-history stages can fully or partially offset early survival advantages (Bain and Helfrich 1983; Paine and Balon 1986).

Survivorship curves, constructed by plotting the log of the number of survivors (n_x) against age, provide an efficient way to look at mortality and survivorship across year-classes and among species because a straight line indicates constant age-specific survival and mortality rates (Deevey 1947). The challenge, of course, is in the details of getting appropriate data. Three forms of survivorship schedules are usually recognized: type 1 has low mortality until when organisms approach their maximum life span, type 2 has a constant mortality rate over the entire life span, and type 3 has high mortality early in life followed by low mortality over the majority of life until senescence (Deevey 1947; Gotelli 2001). Type 1 is purported to represent most mammals, including humans, and some lizards. Although few organisms, including fishes, have constant survivorship throughout most of the life span, the type 2 pattern is approached by most birds and some lizards (Pianka 1988). Most fishes, as well as invertebrates and plants, fit a type 3 pattern and, as argued in the following section, most other organisms typically assigned to types 1 and 2 do as well. A survivorship curve for Fantail Darters shows an initially lower survivorship followed by higher survivorship of fish in ages-1–3 and thus falls between types 2 and 3. A major problem associated with the three types

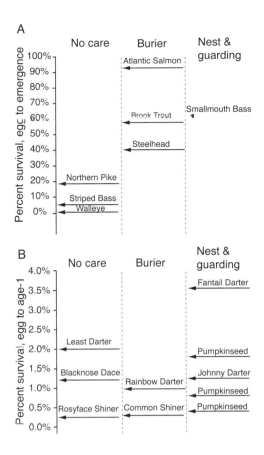

FIGURE 9.2. Benefits of hiding or guarding eggs on survival from (A) egg stage to larval emergence of seven species of fishes, and (B) egg stage to age-1 of eight species of fishes. Pumpkinseed data are from three populations. Based on data from Dahlberg (1979), Paine and Balon (1986), and Bertschy and Fox (1999).

of survivorship curves is that almost all published survivorship curves for vertebrates (as well as other organisms) usually have the initial life-history stages missing, and in virtually all organisms, including humans, mortality is highest among early life-history stages (Levitis 2011). Consequently, type 1 and 2 survivorship curves are most likely curves that omit the life-history stages with the greatest mortality. In fact, when survivorship curves of fishes are based on incomplete life tables, for example, only juvenile and adult stages, they often approximate a type 2 schedule. Assuming that mortality of prereproductive individuals is not maladaptive, the general pattern of high mortality rates early in life is likely due to the early

TABLE 9.2
Approximate Life Table for Fantail Darter (Etheostoma flabellare)

Age class (x)	n_x	l_x	m_x	$l_x m_x$	$x l_x m_x$
0	500	1.000	0	0	0
1	18	0.035	20	0.704	0.704
2	5	0.009	100	0.924	1.848
3	1	0.002	100	0.243	0.728
4	0				
			$\sum_{x=0}^{x=4}$	1.871	3.280

NOTE: Females mature at age-1, have an average clutch size of 40, and large females may produce five clutches per breeding season (Lake 1936; Halyk and Balon 1983). For illustration purposes, I assumed a 1:1 sex ratio, and that age-1 females produce one clutch and age classes 2 and 3 produce five clutches per breeding season (half of which are assumed females). In contrast to Table 9.1, this life table only contains data on females. Other terms remain the same as in Table 9.1.

n_x = number of individuals at the beginning of age x

l_x = the proportion of the population surviving to the beginning of age x ($l_x = n_x/n_0$)

m_x = age-specific natality

elimination of the most frail individuals, the progressive acquisition of greater robustness by surviving individuals, and the concentration of dangerous transitions early in life such as the change from endogenous or yolk feeding to exogenous feeding (Balon 1986; Levitis 2011).

Age-Specific Survivorship and Reproduction

Thus far, the sample life table (Table 9.1) only includes age-specific survivorship and mortality data, and includes information on males and females. To be able to use life tables to provide the information necessary to further understand the evolution of life-history patterns, we must add age-specific natality (m_x), the average number of age-0 female offspring produced by a female of age (x). This provides a way to estimate the per capita rate of increase (r) of a population and, more important, leads to an understanding of how gene substitutions affecting survivorship (l_x), age-specific natality (m_x), and generation time (T) interact in the evolution of life-history patterns. Age specific natality is used because age groups of fishes generally differ in their reproductive output. Fishes are not immediately sexually mature, but depending on the species,

require weeks, months, or even years before maturity. In addition, once females become sexually mature, egg or embryo production is likely to vary with age and especially body size. The modified life table only represents females, assuming a 1:1 sex ratio (Table 9.2).

Fantail Darters mature at age-1 (Lake 1936) and females produce an average clutch size of 40 (Halyk and Balon 1983), of which 20 are assumed to be females. Large females may produce five clutches in a breeding season (Lake 1936); for purposes of illustration, I assumed that age-1 females produced one clutch and ages-2–3 produced five clutches, half of which are assumed to be females (Table 9.2). From the life table, the net reproductive rate, R_0, is estimated as follows:

$$R_0 = \sum_{x=0}^{x=4} l_x m_x = 1.871 \text{ offspring.} \quad \text{(Equation 9.4)}$$

R_0 is interpreted as the number of female offspring produced by a female during her lifetime or as the average number of female offspring produced by the population per year, taking into account the probability of a female surviving from year to year (Wilson and Bossert 1971; Gotelli 2001). In a stable population, R_0 would equal 1; values > 0 and < 1 indicate a declining population and those > 1 an increasing population. The net reproductive rate provides a measure of population increase *per generation* and is also a measure of fitness for nongrowing populations, albeit one that does not consider the importance of generation time (Gotelli 2001; Charnov and Zuo in press). Another measure of fitness includes the effect of generation time on reproductive output and is also scaled to absolute time rather than generation time. To do this, R_o is divided by an estimate of cohort generation time, T_c, where

$$T_c = \sum_{x=0}^{x=4} xl_x m_x / R_o = 1.753 \text{ years.} \quad \text{(Equation 9.5)}$$

Knowing both the net reproductive rate and the cohort generation time, the intrinsic rate of increase (r) is estimated as

$$r \simeq \frac{\ln(R_o)}{T_c} = 0.357. \quad \text{(Equation 9.6)}$$

For a population having a stable age distribution and exponential growth, the intrinsic rate of increase allows determination of population sizes over time, and can range from a negative value for a declining population, to zero for a stable population, to a positive value for an increasing population. For example, a Fantail Darter population with r = 0.36 and an initial size of 100, would reach 3,660 individuals in 10 years based on the following relationships:

$$N_t = N_o e^{rt}; N_{10} = 100 \times e^{(0.36 \times 10)} = 3,660.$$

$$\text{(Equation 9.7)}$$

There are more exact methods for determining r (see Wilson and Bossert 1971; and Gotelli 2001); however, the main interest here is to use the general formula for r to understand forces shaping life-history evolution.

Life-History Theory

Life-history theory attempts to predict how an organism might answer the challenges of survival and reproduction in a specific environment, or how fitness can be maximized for that environment. The answers to these challenges, such as how many offspring to produce annually and in a lifetime, is it better to produce a few large eggs or many small eggs, what should be the age at first reproduction, and how much energy should be allocated to parental care, represent trade-offs in response to environmental forcing functions such as habitat stability, annual temperature and rainfall patterns, resource availability, and predation levels (Stearns 1976; Charnov et al. 2001; Vila-Gispert et al. 2002).

Over the last 30–40 years, numerous hypotheses relative to life-history tactics have been proposed, and many have been eliminated or incorporated into other hypotheses as more information has become available (Stearns 1976; Winemiller 2005). For instance, the habitat template model (Chapter 4; Figure 4.4) is an example that includes life-history predictions of body size and longevity relative to habitat heterogeneity and the frequency of disturbance. One of the most influential models in the development of life-history theory was that of r- and K-selection, formally proposed by Robert MacArthur and Edward O. Wilson (1967) following earlier work by MacArthur and colleagues (Box 9.2) (Reznick et al. 2002). The model was originally developed to understand the colonization and persistence of organisms on islands, but was soon applied widely to other habitats (Pianka 1970). Although this model has been largely supplanted in life-history studies, it had a major role in stimulating extensive interest in the evolution of life histories (Reznick et al. 2002). The r-K model, which is a deterministic model because it does not consider fluctuations in variables such as adult and juvenile survival, has been replaced by stochastic models of life-history evolution that incorporate age-specific survival probabilities and consider multiple, and interacting, axes of selection (Stearns 1976, 1977; Reznick et al. 2002).

A recent life-history model uses the three components of the estimator (Equation 9.6) for the per capita rate of increase, l_x, m_x, and T_c, to define its three endpoints (Figure 9.3A). The model is empirically based, using ordination

BOX 9.2 • Life-History Evolution and the Saga of r- and K-Selection

The understanding of how life-history traits are influenced by natural selection received a major impetus with the publication of the r- and K-selection model by MacArthur and Wilson (1967). The model was influenced by earlier work of Dobzhansky (1950) contrasting how mortality affected populations in temperate versus tropical zones. As proposed by MacArthur and Wilson, r- and K-selection represented the changing selection pressures of habitats as organisms first colonized islands (selection for rapid growth, high r) compared to organisms in saturated island habitats (selection for high competitive ability; high carrying capacity, K). Pianka (1970) greatly expanded the application of the r- and K-selection model to mainland systems and contributed greatly to its widespread acceptance by showing how differing selective regimes affected the evolution of different life-history strategies. As presented by Pianka, the r- and K-selection model can be viewed as two endpoints of life-history evolution that would follow from contrasting environmental conditions. As environments change (e.g., from the

original concept of resource-rich, largely unoccupied habitats to resource-limited saturated habitats) then selection pressures would change, resulting in a change in the life-history strategy of an organism along the r-K continuum (Reznick et al. 2002). The r- and K-selection model became a paradigm, albeit relatively short lived, that provided major stimulation to the developing subdiscipline of life-history evolution (as shown in the table below) (Reznick et al. 2002).

Models of life-history evolution can be used as a way to classify life histories, as a way to predict life-history traits given certain selective pressures, and as causal explanations of life histories (Stearns 1977). As such, the r- and K-selection model remains useful as a shorthand way of describing a suite of life-history tactics. The r- and K-selection model has suffered from criticisms of the theory and from criticisms of application, often in the form of unsubstantiated assumptions. Criticisms of the theory were related to the expanded r- and

(continued)

Contrasts in Life-History Tactics as Predicted for r- and K-Selected Organisms

Life-history tactics	r-selected[a]	K-selected[b]
Generation time	Short	Long
r_{max}	High	Low
Semelparity versus iteroparity	Semelparity	Iteroparity
Fecundity and clutch size	High	Low
Reproductive effort (gametes)	High	Low
Reproductive effort (parental care)	Low	High
Population size	High variability	Low variability
Body size	Small	Large
Longevity	Short	Long

SOURCE: Based on Pianka (1970).

a. r-selected organisms would be characterized by unstable environments, density-independent mortality, with the greatest mortality early in life and often catastrophic.

b. K-selected organisms would be characterized by stable environments, density-dependent mortality, mortality relatively constant over the life span or relatively low until senescence, and rarely catastrophic.

BOX 9.2 *(continued)*

K-selection model based on Pianka and others, as well as the original, more limited model by Mac-Arthur and Wilson (1967).

APPLES VERSUS ORANGES

Criticisms of the basic r-K continuum argue that r and K are not directly comparable. The intrinsic rate of increase (r) can be determined from basic life-history parameters: age-specific survivorship, age-specific natality, and generation time. In contrast, K is not determinable from basic life-history parameters but is a complex interaction of a population and its resource base (Stearns 1977). However, R_0, the number of female offspring produced by a female during her lifetime, can be substituted as a measure of fitness for K-selection (E. L. Charnov, pers. comm. 2011).

SINGLE VERSUS MULTIPLE CAUSATION

Criticisms of the expanded theory focused on the attempt to explain life-history evolution on the basis of one axis of selective pressure (density independent versus density dependent selection) while ignoring other selective pressures such as predation and age-specific mortality. As such, empirical tests of the theory were not necessarily supportive of the predictions (Stearns 1977). Finally, the r-K model was not derived from an age-structured population model (Gotelli 2001). More recent models of life-history evolution acknowledge multiple causation of life-history evolution by incorporating a broader range of selective agents, including age-specific mortality and fecundity, resource limitation, density dependence, density independence, environmental variation, and extrinsic mortality (Stearns 1977; Reznick et al. 2002).

and cluster analyses of life-history variables to identify patterns in North American freshwater and marine fishes and tropical freshwater fishes (Winemiller 1989, 1992, 2005; Winemiller and Rose 1992). The three life-history endpoints proposed by the model are different adaptive suites associated with maximizing the intrinsic rate of increase, r (Equation 9.6), and are opportunistic, periodic, and equilibrium.

Opportunistic fishes occur in areas with greater environmental disturbance and lower resource predictability and can rapidly recolonize disturbed habitats. In reference to Equation 9.6, higher fitness has resulted from short generation times (T_c) rather than large clutch sizes, although they do tend toward prolonged breeding seasons (another way of increasing m_x). They also have small body sizes, short life spans, and high reproductive effort relative to body size, but have lower investment per offspring compared to equilibrium species. Opportunistic species generally have population levels well below the carrying capacity.

Periodic fishes occur in habitats with decreased environmental variation and moderate resource stability. In periodic species,

selection has been in the direction of increasing fecundity (m_x) and higher adult survival (l_x). They have low parental investment per offspring, but increased fecundity is associated with larger body sizes and greater longevity and generation times. Such fishes tend to have short breeding seasons and one or a few reproductive events per season and also tend to undergo long spawning migrations. Age structure of periodic fish populations is strongly influenced by early life-history stages and the occurrence of occasional strong year classes. In addition, periodic fishes are disproportionately represented in recreational and commercial fisheries.

Equilibrium species are typical of habitats with low disturbance levels, high resource stability, and often high levels of competition and predation. In reference to Equation 9.6, equilibrium species have increased juvenile survival (l_x), moderate to long generation times (T_c), and smaller clutch sizes (m_x), but with greater parental investment per offspring, such as parental care and/or large eggs. This endpoint is largely consistent with K-selection of the r-K model, but differs in that equilibrium species often have small body sizes and short to moderate

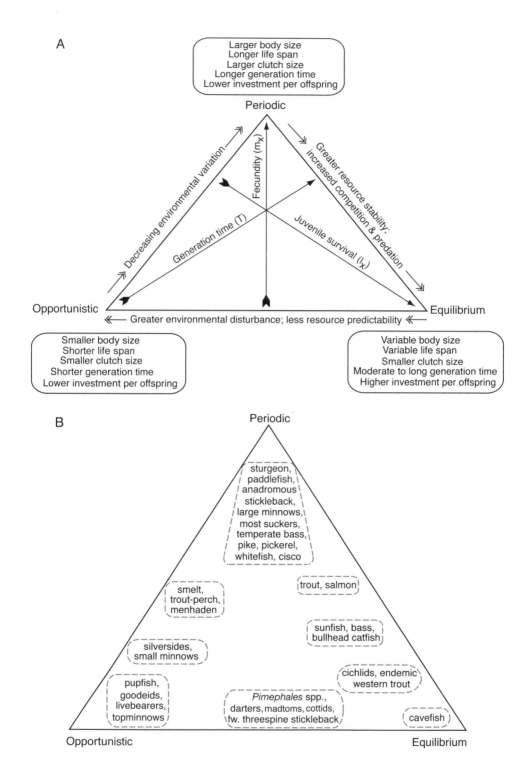

FIGURE 9.3. A. The three-endpoint model of life-history evolution, as defined by juvenile survivorship (l_x), age specific fecundity (m_x), and generation time (T) (inner arrows). Outer double arrows show abiotic or biotic forces that would select for different life-history patterns. B. Approximate positions of some North American freshwater fishes in the three-dimensional life-history space. Based on Winemiller (1989, 1992, 1995, 2005), Winemiller and Rose (1992), and Mims et al. (2010).

life spans. In fact, there is generally a trade-off between reproductive effort and the probability of future reproduction (Sargent and Gross 1993; Poizat et al. 1999; Charnov et al. 2001).

More recent support for the three-endpoint model of life-history variation includes an analysis of North American freshwater fishes (excluding Mexico); a study of freshwater fishes of the continental United States; and a comparative study of freshwater fishes in Europe, South America, and North America, and marine fishes in North America (Vila-Gispert et al. 2002; Mims et al. 2010; Olden and Kennard 2010). The data also show a strong phylogenetic signal, emphasizing the importance of recognizing lineage divergences in understanding life-history patterns.

Because the three-endpoint pattern is based on the extremes of life-history differences, most fishes fall somewhere in between the endpoints. Large, long-lived fishes with no parental care and limited investment per offspring are grouped at or near the periodic endpoint (Figure 9.3B). Many of these fish groups, such as sturgeons, Threespine Sticklebacks (*Gasterosteus aculeatus*), large minnows (e.g., Pikeminnows), temperate bass, whitefish, cisco, and suckers (e.g., Razorback Sucker, *Xyrauchen texanus*; Redhorses) are migratory, moving between salt and fresh water, or showing extensive movements within freshwater (Lucas and Baras 2001). Some, such as sturgeon and Paddlefish (*Polyodon spathula*), are not considered extreme periodic strategists because of increased parental investment shown by large egg sizes. Trout and salmon tend to be intermediate between periodic and equilibrium endpoints because of the increased investment in offspring shown by large egg sizes, nest construction, and brood hiding, and the often great distances traveled to reach suitable spawning habitats. Fishes intermediate between opportunistic and equilibrium endpoints include most darters, madtom catfishes, freshwater populations of Threespine Sticklebacks, and sculpin. These fishes tend to have short life spans, short generation times, and smaller clutch sizes, but also show increased parental investment through nest construction, egg burying, or nest guarding. Fishes intermediate between opportunistic and periodic endpoints have small body sizes and short generation times but larger clutch sizes (Winemiller and Rose 1992).

Not surprising given the major geological and climatic events that have been a constant feature of the world's continents and aquatic habitats (Part 2), there are differences in life-history patterns among geographical regions. South American freshwater fishes appear skewed toward opportunistic patterns whereas European and North American freshwater fishes are intermediate between opportunistic and periodic patterns, and North American freshwater fishes occupy reduced life-history space compared to North American marine fishes. A broad-based study of North American freshwater fishes (excluding Mexico and Alaska) also shows geographic differences in the occurrence of the three endpoints (Mims et al. 2010). Opportunistic species make up 69% of southeastern U.S. fish species, followed by periodic (19%) and equilibrium (12%) species (Olden and Kennard 2010). The southeastern region is dominated by minnows and darters and includes numerous topminnows and livebearers (Ross 2001; Boschung and Mayden 2004). Although the southeastern region was not glaciated during the Pleistocene, it was subjected to major changes in flow patterns, rainfall, river discharge, and temperature; coastal streams were also impacted by major changes in sea level (Chapter 3; Ross 2001). Especially for fishes in small streams, selection for rapid population growth and rapid colonization of habitats made available through increased flow and access to floodplains, or from the resumption of flow following droughts, would have favored opportunistic species (O'Connell 2003; Adams and Warren 2005; Olden and Kennard 2010).

Periodic species tend to occur more in northwestern and northern North America, and in the Rocky Mountain area—regions that were exposed to the full force of Pleistocene glaciation (Chapter 3). Because present-day fishes are generally those that migrated long distances from refugia, selection generally favored larger body size, migration ability, and a long reproductive life span to survive unfavorable reproductive conditions—all traits of periodic species (Mims et al. 2010). Equilibrium species are common along the Pacific Coast and along rivers draining northward from Canada, both faunas dominated by salmonids, and along drainages of the Texas Gulf coast, a fauna with strong representation of centrarchids and ictalurids. Inland from the Pacific Coast, especially along the Rocky Mountains, periodic species are common. Although there is considerable variation and numerous exceptions, opportunistic species tend to occur more in lower latitudes and periodic and equilibrium species at higher latitudes (Mims et al. 2010). At least some of the apparent differences in the patterns are likely caused by strong biases in the habitats and species studied and by hard-classifying families in one of the three endpoints (Winemiller and Rose 1992; Vila-Gispert et al. 2002; Mims et al. 2010).

Approximately 30% of the North American freshwater fish species are near the opportunistic end of the model, 54% are near the periodic endpoint, and 17% are near the equilibrium endpoint. However, opportunistic and equilibrium taxa compose more of the threatened and endangered species listed by the U.S. Fish and Wildlife Service than would be expected by chance. Opportunistic and equilibrium species make up 47% of the continental fauna but 64% of the U.S. threatened and endangered species (Winemiller 2005).

Models of life history, as with the Winemiller-Rose model, can start with a pattern and then infer process, or begin with a process, such as basic metabolism, and then predict a pattern (Mangel 1996). The Winemiller-Rose model is useful for visualizing different adaptive suites of life-history variation. A second approach is to derive a general growth model starting with first principles based on how metabolic energy is allocated between tissue maintenance and the growth of new tissue (i.e., increasing body size) (West et al. 2001). The model can then be

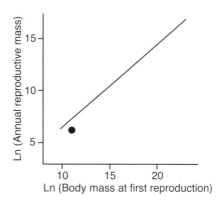

FIGURE 9.4. The allometric relationship between the average annual body mass allocated to reproduction and the average body mass at first reproduction. The solid line is a regression (R^2 = 0.74) based on 139 species of North American marine and freshwater fishes; the closed circle shows data from 48 species of darters not included in the regression calculation. Based on Charnov et al. (2001).

used for developing various allometric relationships linking growth rates and the timing of life-history events, and the predictions of the model can then be tested against real data. By adding reproduction to the basic metabolic model, Charnov et al. (2001) developed a series of allometric scaling predictions relating age at maturity, average adult life span, and the relative body mass allocated annually to reproduction. One prediction that follows is that the logarithm of the average body mass allocated to annual reproduction is linearly related with a slope of approximately 0.75 to the logarithm of body mass at first reproduction (essentially a dimensionless invariant). The prediction has thus far been tested using the Winemiller and Rose (1992) data set of 139 North American marine and freshwater fishes, showing a very close fit (R^2 = 0.74) (Figure 9.4). Basically, fishes that delay initial reproduction until very large body sizes devote a high level of annual mass to reproduction in contrast to fishes that mature at small body sizes. Although likely, the predictive relationship has not been tested for strictly North American freshwater fishes; however data from darters (Percidae) that were not included in the regression calculation also fall near the regression line.

Offspring Number, Size, and
Spawning Frequency

ANNUAL AND LIFETIME FECUNDITY AND REPRO-
DUCTIVE EFFORT Annual fecundity, the number of mature ova produced within a single
reproductive season, is a function of clutch size
and the number of times a female spawns. The
number of eggs is one part of a female's reproductive output—the other is egg size (mass).
For both marine and freshwater fishes, for any
body size there usually is a strong negative correlation between clutch size and egg mass so
that there is a trade-off between the two (Elgar
1990; Fleming and Gross 1990). The relationship of clutch size to egg mass is not always
clear and can vary among individuals of the
same species, as well as among species and
genera (Marsh 1986).

The same annual reproductive output
(yearly egg number × egg mass) can be achieved
by numerous small eggs or by fewer large eggs.
Demersal spawning fishes produce fewer large
eggs compared to pelagic spawning fishes, and
(probably because demersal spawners are more
common in freshwater compared to marine
habitats) freshwater fishes produce fewer large
eggs relative to body size compared to marine
fishes, which produce more small eggs (Duarte
and Alcaraz 1989; Elgar 1990). Egg number is
positively correlated with body size whereas egg
mass is not. Annual reproductive output also
has a strong positive correlation with body size,
much stronger than for egg number, for both
freshwater and marine fishes; however, when
annual reproductive output is adjusted for body
size, there is no difference between marine and
freshwater fishes—the difference is in the way
reproductive energy is allocated (Wootton 1984;
Elgar 1990). The same pattern is shown within
benthic and limnetic morphotypes of Threespine Sticklebacks in Benka Lake, Alaska, where
limnetic females produce a larger relative
clutch mass comprising more, smaller eggs
in contrast to benthic females, even though
the total allocation to reproduction is the same
for both (Baker et al. 2005). Based on a modeling exercise using pelagic marine eggs and
larvae, food-rich environments favor the production of more, small eggs whereas food-poor
environments favor the production of few,
large eggs (Winemiller and Rose 1993). The
same pattern may occur for freshwater fishes,
although the production of larger eggs in benthic Threespine Sticklebacks is likely caused by
selection for larger larvae. Other things being
equal, sizes of newly hatched larvae decrease
as developmental temperatures increase.
Because benthic fish nest in warmer water
than limnetic fish in Benka Lake, and thus
would have smaller larvae, there apparently is
selection for larger egg sizes, which contribute
to larger larval sizes, to counteract the temperature effects during development (Baker et
al. 2005).

Because of the positive relationship between
clutch mass and the size of the female, age specific natality (m_x) tends to increase with age.
Fishes near the periodic endpoint (Figure 9.3)
are characterized by larger body sizes as well
as larger clutch sizes, although clutch sizes in
some large migratory fishes are reduced from a
theoretical maximum because of large ova.

Lifetime fecundity, the number of eggs or
young produced over a female's lifetime, is a
function of age specific natality and the probability of survival to the next spawning season (l_x).

CLUTCH PRODUCTION A single clutch comprises a cohort of propagules whose development
is triggered by a single hormonal signal, that
quickly increases in size through yolk deposition
(vitellogenesis), and that is completely oviposited
in a short time interval (Heins and Rabito 1986;
Heins et al. 1992; Baker et al. 2008). Patterns of
ovarian clutch production fall into three basic
groups: synchronous, group synchronous, and
asynchronous (Wallace and Selman 1981). In
synchronous clutch production, all oocytes
grow and ovulate from the ovary in unison,
and there is no replenishment of one ovarian
stage by an earlier stage. Synchronous clutch
production is typified by fishes with single lifetime breeding periods (i.e., semelparous fishes)
such as eels or salmon. In group-synchronous

clutch production, there are at least two groups of oocytes developing in the ovary. Typically, there is one group of larger oocytes that will be released as a single clutch and a second group of smaller, more heterogeneous oocytes from which the large ovarian class is recruited. This is the most common pattern in fishes and is characteristic of those with multiple clutches and multiple breeding seasons (i.e., iteroparous fishes). Finally, in asynchronous clutch production, oocytes of all stages are present in the ovary without dominant size groups. This pattern is common in topminnows (*Fundulus*) (Selman and Wallace 1986).

SPAWNING WITHIN A SINGLE SEASON
Species and individual populations within a species can vary in the number of times that spawning occurs in a single season or in an individual's lifetime. Clearly, knowing whether fishes produce single or multiple clutches within a spawning season and whether egg mass varies within and among spawning seasons is crucial to an understanding of reproductive effort and age-specific natality (Heins and Rabito 1986; Heins et al. 2004).

Various North American fish groups are known to occasionally spawn a second time during the breeding season, such as catfishes (Least Madtom, *Noturus hildebrandi*) and pickerel (Chain Pickerel, *Esox niger*) (J. G. Miller 1962; Baker and Heins 1994). This occasional repeat spawning is not the same physiologically, or in terms of its life-history implications, as the normal production of multiple clutches produced at short intervals throughout the breeding season. Although the production of multiple clutches in freshwater fishes has been known since the 1950s through work by Clark Hubbs and his students on darter (*Etheostoma*) species, the full appreciation of this on estimates of annual fecundity has only occurred in the last several decades. In aquaria, both Greenthroat Darters (*Etheostoma lepidum*) and Rio Grande Darters (*E. grahami*) produced clutches of mature eggs every 4–10 days (Strawn and Hubbs 1956; Hubbs and Strawn 1957). More

recent studies have shown that the ovaries of many species are extremely dynamic and that a sexually mature female can cycle eggs through the various stages quite rapidly (Box 9.3). Although the list is no doubt incomplete, multiple spawning by an individual female in a single spawning season is documented (or strongly suspected) in 8 families, 13 genera, and 23 species of North American freshwater fishes (Table 9.3). Multiple clutches may occur in North American populations of the tetras (Characidae), but this remains to be documented.

The production of multiple clutches is generally viewed as an adaptation for producing more clutches per spawning season, and thus increasing age-specific natality, without increasing body or ovary sizes (Burt et al. 1988). Fishes with multiple spawning typically are small-bodied, are generally not of recreational or commercial importance, and have parental care only by the male or lack parental care altogether so that multiple clutch production is not constrained by time spent with parental care. Pumpkinseed are an apparent exception to this general pattern because they have a larger body size and, at least in an Ontario, Canada, study population, the quantity of eggs produced by females over an average of 2.1 spawning events per season was small enough to be produced in a single clutch. Instead, multiple spawning in this population of Pumpkinseed is perhaps due to bet-hedging selection in a variable environment (Fox and Crivelli 1998). Fox and Crivelli also suggested that parental care might constrain repeated spawning by females, but in Pumpkinseed, as in other centrarchid species (with the exception of a southeastern population of Largemouth Bass, *Micropterus salmoides*), parental care is only provided by the male and, at least in some species, a male may attract and mate with more than one female during a single nesting interval (Cooke et al. 2006; Warren 2009).

Various approaches have been used to determine the presence of multiple spawning in fishes. For instance, based on laboratory studies, female Bannerfin Shiner (*Cyprinella leedsi*), which are batch synchronous, can produce egg

BOX 9.3 • Gonadal Stages in Teleosts

Gonadal development can be described by using external features or by histological procedures, and both approaches are useful depending on the questions being asked and the resources available. Examples of ovarian and oocyte staging, using histological techniques, were shown by Selman and Wallace (1986); Grier (1981) provided an example of histological staging of teleost testes. An improved terminology for histological staging in fishes, applicable to both sexes of elasmobranchs and teleosts, includes the following six categories (Brown-Peterson et al. 2007).

IMMATURE Never spawned

DEVELOPING Gonads starting to develop

SPAWNING CAPABLE Spawning possible within current breeding season

ACTIVELY SPAWNING Spawning currently happening, just completed, or will happen soon

REGRESSING Spawning activity completed for season

REGENERATING Sexually mature but not currently reproductively active

External staging, especially of ovaries, can be useful because it can be rapidly accomplished in the field as well as in the laboratory (Ricker 1968). A useful system of external staging of ovaries is based on six categories and was developed for minnows and darters but is applicable to other North American fish families as well (Heins and Rabito 1986; Heins and Baker 1993).

EXTERNAL STAGING OF OVARIES

LATENT (LA) The transparent to slightly translucent ovaries are very small and thin. The maturing oocytes may be without visible yolk or with some yolk present (vitellogenic) but with the nucleus still visible.

EARLY MATURING (EM) The translucent to opaque ovaries are small to moderate in size. The maturing oocytes are small or of moderate size, translucent to opaque, and with the nucleus obscured by yolk.

LATE MATURING (LM) The white to cream ovaries are small to greatly enlarged and may occupy much of the space in the body cavity. The maturing oocytes are of moderate to large sizes and white to cream or yellow.

MATURE (MA) The cream to yellow ovaries are moderate in size to greatly enlarged and may occupy a large portion of the body cavity. There are two separate groups of follicular oocytes, a group of smaller maturing oocytes that are translucent to opaque and a group of larger mature oocytes that are usually opaque and cream to yellow but without the vitelline membrane (the membrane surrounding the yolk) separated from the yolk.

RIPENING (MR) The cream to yellow ovaries are moderate sized to greatly enlarged. There are two distinct groups of follicular oocytes, a group of small oocytes and a group of larger ripening oocytes that are usually translucent (sometimes transparent) with the vitelline membrane obviously separated from the yolk.

RIPE (RE) The cream to yellow ovaries are moderate sized to greatly enlarged and may occupy much of the body cavity. There are two groups of relatively large oocytes, one group includes white- to cream-colored maturing follicular oocytes of moderate to large size, and a second group of translucent to transparent ripe ova concentrated in the lumen of the ovary with the vitelline membrane separated from the yolk. These ovulated oocytes are ready to be oviposited and fertilized (in species with external fertilization).

clutches every 3–10 days (mean = 4.6) (Figure 9.5). The spawned ova, represented in the top panel, were already oviposited and were recovered from crevices of the spawning substrata in the aquaria. The intermediate size group of ripe, ovulated oocytes were found in the lumen of the ovary, and the smallest size group were all intrafollicular oocytes (Heins and Rabito 1986). In reference to ovarian categories of Box 9.3, females sacrificed before spawning were MA,

TABLE 9.3

Examples of North American Freshwater Fishes Showing Multiple
Clutch Production by a Female within a Single Reproductive Season

Family/Species	Climate (location)	Source
Atherinopsidae		
Menidia audens, Mississippi Silverside	Temperate (OK)	Hubbs 1976
Menidia menidia, Atlantic Silverside	Maritime (MA)	Conover 1985
Centrarchidae		
Lepomis gibbosus, Pumpkinseed	Cool temperate (Ontario)	Fox and Crivelli 1998
Characidae		
Hyphessobrycon pulchripinnis[a], Lemon Tetra	Warm temperate/tropical	Burt et al. 1988
Cyprinidae		
Cyprinella analostana, Satinfin Shiner	Cool temperate (PA)	Gale and Buynak 1978
Cyprinella leedsi, Bannerfin Shiner	Warm temperate (FL)	Heins and Rabito 1986
Cyprinella nivea, Whitefin Shiner	Warm temperate (SC)	Cloutman and Harrell 1987
Cyprinella lutrensis, Red Shiner	Cool temperate (PA)	Gale 1986
Notropis longirostris, Longnose Shiner	Warm temperate (AL and MS)	Heins 1990, 1991
Pimephales notatus, Bluntnose Minnow	Cool temperate (PA)	Gale 1983
Pimephales promelas, Fathead Minnow	Cool temperate (PA)	Gale and Buynak 1982
Elassomatidae		
Elassoma zonatum, Pygmy Sunfish	Warm temperate (KY)	Walsh and Burr 1984
Fundulidae		
Fundulus notatus, Blackstripe Topminnow	Cool temperate (MI)	Carranza and Winn 1954
Gasterosteidae		
Gasterosteus aculeatus, Threespine Stickleback	Circumpolar (AK, MA)	Wallace and Selman 1979, 1981; Baker et al. 2008
Apeltes quadracus, Fourspine Stickleback	Cool temperate (MA)	Wallace and Selman 1979
Percidae		
Etheostoma grahami, Rio Grande Darter	Warm temperate (TX)	Strawn and Hubbs 1956
Etheostoma lepidum, Greenthroat Darter	Warm temperate (TX)	Hubbs and Strawn 1957
Etheostoma lynceum, Brighteye Darter	Warm temperate (MS)	Heins and Baker 1993
Etheostoma olmstedi, Tessellated Darter	Cool temperate (PA)	Gale and Deutsch 1985
Etheostoma rafinesquei, Kentucky Snubnose Darter	Cool temperate (KY)	Weddle and Burr 1991
Etheostoma spectabile, Orangethroat Darter	Warm temperate (TX)	Hubbs 1985
Nothonotus rubrum, Bayou Darter	Warm temperate (MS)	Knight and Ross 1992
Percina vigil, Saddleback Darter	Warm temperate (MS)	Heins and Baker 1989
Poeciliidae		
Poecilia latipinna, Sailfin Molly	Warm temperate (FL)	Snelson et al. 1986

a. The genus *Hyphessobrycon* occurs in Mexico, although *H. pulchripinnis* is native to South America.

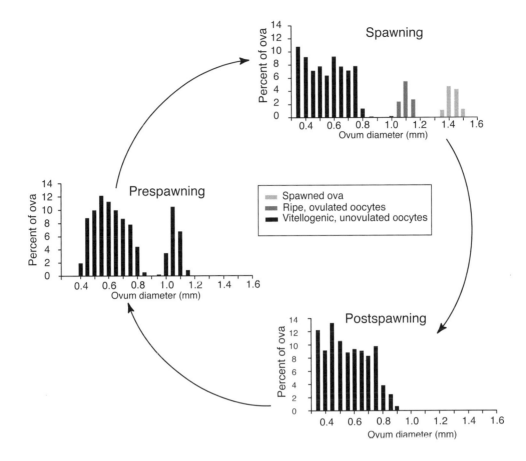

FIGURE 9.5. Ovarian cycling, illustrated by the Bannerfin Shiner (*Cyprinella leedsi*). The graphs show the percent of ova in 0.05 mm size classes in prespawning, spawning, and postspawning females. The average time of the cycle, from spawning to spawning, is 4.6 days. The data are an average of three fish in each category. Based on Heins and Rabito (1986).

those during spawning were RE, and those after spawning were LM. In Bannerfin Shiner, 26–228 mature ova are deposited per clutch. To maintain the cycle, there is continual recruitment from the oogonia, so that the eggs spawned during a season would be much more than the complement of oogonia and oocytes present at the start of the season.

Another approach to determining multiple clutch production is to use outdoor enclosures, where a male-female pair is placed in a cage that contains spawning substrata. By daily examination of the substrata, the occurrence of multiple spawning under natural temperature and light regimes can be determined. Using plastic pools (1.1 or 1.5 m diameter) suspended in a larger outside pool,

11 pairs of Bluntnose Minnows (*Pimephales notatus*) produced from 7 to 19 clutches over a three-month period, with an average spawning interval of 5.3 days and an average clutch size of 93 to 239 (Gale 1983). Over the three months, each pair produced from 1,112 to 4,195 eggs (average = 2,396). Impressively, the total volume of water-hardened eggs over three months was 1.6 times the volume of a single female. Fathead Minnows (*Pimephales promelas*) produce an even greater volume of eggs relative to the female's body size during a spawning season, with egg volumes 3.8–6.8 times greater than the female's volume (Gale and Buynak 1982).

Based on the behavior of nine females held in instream cages that were provided with spawning substrata, Kentucky Snubnose Darter (*Etheostoma rafinesquei*) can produce a clutch on average every 3.7 days and can spawn seven or more times in 26 days (Weddle and Burr 1991). Females also can spawn partial clutches over a period of several days.

In addition to laboratory studies, multiple clutch production can be determined from histological studies of wild-caught specimens. For example, based on the presence of postovulatory follicles, two Alaskan populations of Threespine Stickleback had spawning intervals of 2.2–7.8 days. The spawning interval increased as the rate of clutch production slowed with the advance of the breeding season and fish became more energy limited as body condition declined (Bagamian et al. 2004; Brown-Peterson and Heins 2009). In fact, a major controller of the interval between clutches in Threespine Stickleback is food availability (Ali and Wootton 1999). Female Sticklebacks can also retain a clutch of ovulated eggs up to 48 h while they search for a suitable mate (Brown-Peterson and Heins 2009).

The appreciation of multiple production of clutches during a single spawning season in the two most speciose families of North American freshwater fishes, in addition to other families (Table 9.3), has necessitated a reevaluation of annual fecundity estimates for many species (Conover 1985). Because of this, estimates of annual fecundity based on preseason counts of vitellogenic follicles are likely to greatly underestimate age-specific natality. Estimates of annual fecundity and reproductive effort are further complicated because of the potential for changes in clutch size and/or ovum size over the spawning season, with a general negative relationship between ovum size and clutch size (Heins et al. 2004). Ovum size is often inversely related to temperature, so that larger ova tend to be produced earlier in the breeding season (Marsh 1984, 1986; Heins et al. 2004). This is perhaps an adaptation for providing greater energy stores for developing embryos and larvae at a time when food availability might be lower (parental

investment hypothesis), or providing fewer, high-quality eggs at a time when environmental uncertainty might be greater (bet hedging hypothesis) (Heins et al. 2004). Changes in clutch size, adjusted for female body size, over a breeding season of North American freshwater fishes support three of four basic patterns, and patterns may differ even within the same genus. In one pattern, clutch size peaks midway through the breeding season. This is shown by *Etheostoma rafinesquei* (Weddle and Burr 1991), *E. caeruleum* (Heins et al. 1996), and *Menidia menidia* (Conover 1985). In a second pattern, clutch size declines throughout the breeding season as shown by *Fundulus heteroclitus* (Kneib and Stiven 1978). In a third pattern, clutch size increases throughout the breeding season. This is shown by *Etheostoma lynceum* (Heins and Baker 1993; Heins et al. 2004) and *Percina vigil* (for one of two years; Heins and Baker 1989). A fourth pattern, in which clutch size is relatively invariant over the breeding season, has not been linked with North American freshwater fishes, but is shown by marine species and may be most likely under relatively constant food availability and environmental conditions (Conover 1985). Overall, changes in clutch size seem to be proximally driven by variation in temperature or nutrient availability, or from selection resulting in the greatest offspring production when the probability of survival is greatest (Conover 1985).

SPAWNING WITHIN A LIFETIME: SEMELPARITY VERSUS ITEROPARITY A key to understanding how reproductive effort is allocated over time is the relationship between adult and larval/juvenile survival, which can be affected by environmental variation such as floods or droughts, size-specific predation that might impact one life-history stage more than another, and the cost to the adults, including survival costs, of producing offspring (Partridge and Harvey 1988). As the probability of adult survival increases relative to survival of larvae/juveniles, iteroparity (reproducing over multiple seasons) should evolve along with the reduced cost of annual reproductive output (Periodic Strategy of Figure 9.3) (Partridge and Harvey 1988). Iteroparity can represent bet hedging and be a form of resilience in response to environmental variation (Chapter 6), and is predicted to be favored as environmental uncertainty increases (Orzack and Tuljakarpur 1989). When there is low environmental variation, so that the survival probability of young is likely to be high, coupled with extensive parental investment such as numerous large eggs, nest construction, and risky, long-distance migration (intermediate between Periodic and Equilibrium of Figure 9.3), selection should favor semelparity—"big-bang" reproduction. This is a single lifetime breeding event with maximal egg production at the expense of future survival (Winemiller 1992; Crespi and Teo 2002). Semelparous North American freshwater fishes include some populations of American Shad (*Alosa sapidissima*), most populations and species of Pacific Salmon (*Oncorhynchus* spp., excluding nonmigratory Cutthroat and Rainbow trout), American Eels (*Anguilla rostrata*), and all Lampreys (Hardisty and Potter 1971; Leggett and Carscadden 1978; Haro et al. 2000).

American Shad populations along the Atlantic coast of North America vary in the extent of repeat breeding, with populations in lower latitudes (< 32 N), where moderate temperatures make riverine spawning habitats more predictable, being totally semelparous. American Shad have a narrow temperature range for spawning and egg development of 13–19° C, and optimal growth of larvae and juveniles occurs at 20–25° C (Leach and Houde 1999; Bilkovic et al. 2002). At higher latitudes, the increased temperature variation reduces the probability of juvenile survival, relative to that of adults, and the proportion of repeat breeding increases, reaching 60–80% in New Brunswick populations. The southern semelparous populations put more energy into reproduction, in contrast to more northern populations, and have greater relative and absolute fecundities. This results in an inverse relationship between the expected lifetime fecundity,

which is greatest in low latitudes, and the frequency of repeat breeding, which is greatest in high latitudes (Leggett and Carscadden 1978).

Females of all five North American Pacific Salmon species (Coho, *Oncorhynchus kisutch*; Chinook, *O. tshawytscha*; Sockeye, *O. nerka*; Pink, *O. gorbuscha*; and Chum, *O. keta*) are semelparous, as are most males of these species. There are several instances, however, of freshwater-resident males of Chinook, Coho, and Sockeye salmon that sometimes will breed more than once (Crespi and Teo 2002). Most other salmonid species, including Steelhead, are iteroparous, although there are exceptions.

In salmonids, the semelparous life cycle evolved from an iteroparous life cycle in association with a larger body size, greater ovum mass, anadromy, and strong development of secondary sex characteristics. In addition to being able to produce more eggs, larger fish are also more efficient swimmers (Chapter 7), having a lower mass-specific cost of locomotion, and thus are capable of more extensive migrations. Larger eggs result in larger fry and juveniles, with concomitantly greater survivorship, and are thus a form of increased parental care (Crespi and Teo 2002). Semelparous salmonids, the five species of Pacific Salmon, are also unique among the subfamily Salmoninae in showing nest guarding once the nest is completed and in continuing nest guarding until they die (Fleming 1998). Finally, semelparous salmonids take parental care to yet another level because the carcasses of adults increase the productivity of streams and thus help to increase the survivorship and growth of young salmonids (Crespi and Teo 2002).

Semelparous salmonids generally show greater reproductive costs (as percent energy loss) for males and females compared to iteroparous salmonids. Energy losses in anadromous populations are determined from the time of entry into coastal areas or into freshwater to the time of postspawning or death (in the case of semelparous species), or between pre- and postspawning in freshwater resident populations. Thus determined, the costs of reproduction do not include those of migration, which would be partially compensated by feeding in the ocean (until feeding ceases in fresh water), but do include those of territory defense, courtship, secondary sex characteristics (primarily in males), nest construction (females), nest guarding (females), and gamete production (Hendry and Berg 1999; Fleming and Reynolds 2004). Although quite variable, average energy losses of females exceed those of males by 13% (for semelparous populations) to 24% (for iteroparous populations). Atlantic Salmon (*Salmo salar*), which have relatively higher energy losses during reproduction than other iteroparous salmonids, actually have a low level of iteroparity, with repeat breeding on the order of only 10% in most populations (Fleming and Reynolds 2004).

In spite of the overall greater energy loss at reproduction for semelparous versus iteroparous populations, the gonadal-somatic index (GSI) does not differ between the two life-history patterns (Figure 9.6A). Consequently, semelparous salmonids invest more in other aspects of reproduction, including migration, nest preparation, nest guarding, courtship, and sexual dimorphism (Fleming 1998). Although there is again considerable overlap, greater differences in female GSI are shown between resident and anadromous populations of salmonids, reflecting the greater investment in current reproduction given the reduced chance of breeding again in anadromous populations (which is even further reduced by the current increased reproductive investment of migration) (Figure 9.6B; Fleming and Reynolds 2004). At least in marine fish populations, and likely so for other fish populations, GSI is strongly and positively correlated with the adult instantaneous mortality rate (Gunderson 1997).

To Travel or Not: Spawning Migrations

Fishes near the periodic endpoint of Figure 9.3, or those intermediate between periodic and equilibrium endpoints, often undertake long-distance spawning migrations. Spawning migrations, as discussed previously, can represent a

major parental investment in offspring, where adults move from feeding or resting areas to areas that are thought to be more suitable for egg, larval, and juvenile survival (see Chapter 5; Figure 5.5). The majority of migratory fishes of inland habitats restrict their movements to fresh water, termed potamodromous, whereas relatively few species show some form of periodic, physiologically mediated movement between the sea and fresh water as part of their life cycle (termed diadromous). Two subgroups of diadromy shown by North American fishes have different biomes for late-juvenile and adult feeding and for reproduction (Myers 1949; McDowall 1997). Anadromous fishes have the major feeding and growth areas at sea and move from the sea to fresh water for purposes of spawning, with generally little or no feeding once adult fish enter fresh water. Catadromous fishes use fresh water as the major feeding and growth biome and move from fresh water to the sea as adults for spawning, with little or no feeding in the sea (Figure 9.7) (McDowall 1988, 1992, 1997).

Anadromous fishes include representatives within the Salmonidae, Osmeridae, Clupeidae, Acipenseridae, and Moronidae (McDowall 1988, 1997). Catadromous fishes are less common in North America, with only three primary examples, Striped Mullet (*Mugil cephalus*, Mugilidae), American Eel (Anguillidae), and Hogchoker (*Trinectes maculatus*, Achiridae) (Dovel et al. 1969; De Silva 1980; T. L. Peterson 1996; McDowall 1988, 2007). Worldwide, only about 0.9% of fishes are diadromous, of which 48% are anadromous and 25% are catadromous (McDowall 1997; Nelson 2006). The remainder are amphidromous, a third type of diadromy first described by Myers (1949) and then refined by McDowall (1987, 2007), in which the reproductive and major growth biomes are the same. Amphidromous fishes can be either fresh water or marine, although the former appears much more common. Larvae of fresh water, amphidromous fishes migrate to the sea soon after hatching, where early feeding and growth occur for a few weeks to several months,

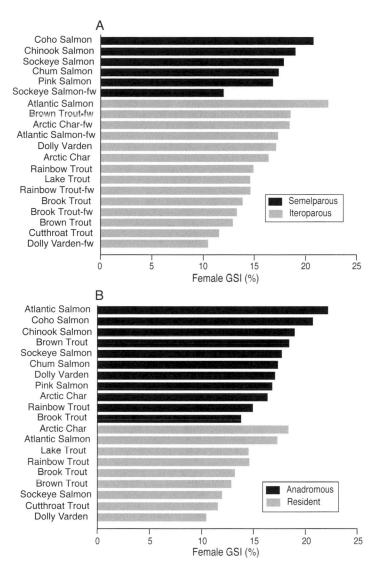

FIGURE 9.6. Female GSI [(gonad weight/somatic weight) × 100] compared between (A) semelparous or iteroparous and (B) anadromous or resident populations of salmonids. FW = freshwater populations. Based on Fleming (1998).

followed by a return migration to fresh water while still small juveniles. Most feeding and growth occur within fresh water, as do matura- tion and spawning. In contrast to anadromy and catadromy, the return migration to fresh water has a trophic base in contrast to a gametic base (McDowall 1997, 2007). Application of the term amphidromy to North American fishes has been complicated by various usages; the term does not refer to aperiodic movement between the sea and fresh water but to an actual migration

(see Chapter 5). Although amphidromy is more common in the tropics and subtropics, there are North American examples including some populations of three cottid species, Prickly Sculpin (*Cottus asper*), Coastrange Sculpin (*C. aleuticus*), and Pacific Staghorn Sculpin (*Leptocottus armatus*) on the Pacific Coast (McDowall 1988; Brown and Moyle 1997; Moyle 2002). Some populations of the Mountain Mul- let (*Agonostomus monticola*), which occur along the southern Atlantic and Gulf coasts likely also

Fresh water

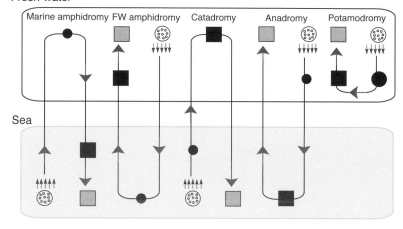

Sea

FIGURE 9.7. Migratory patterns of fishes. In potamodromy, migration takes place entirely within fresh water, such as moving from a lake or large river where adult feeding and growth occur to a smaller stream for spawning, hatching, and early growth of larvae and juveniles. In anadromy, movement is from the adult feeding and growth area at sea into fresh water for spawning, hatching, and early growth of larvae and juveniles, followed by the return migration to the sea. The situation is reversed in catadromy, where most growth occurs in fresh water, and spawning, hatching, and early development of young occur at sea. Amphidromous fishes grow, mature, and spawn in the same biome, either fresh water or marine. The size of the closed boxes and circles indicates the relative amount of feeding and growth in a particular habitat. Based in part on Gross (1987) and McDowall (1988, 1992, 2007).

are amphidromous (Matamoros et al. 2009). Other Gulf and Atlantic species that may have amphidromous populations include the Bigmouth Sleeper (*Gobiomorus dormitor*, Eleotridae) and the River Goby (*Awaous banana*, Gobiidae) (Musick et al. 2000).

The evolution of migratory behavior in fishes, including those restricted to fresh water and those that show diadromy, can be understood by R_o, the simple model of fitness introduced earlier in this chapter, where

$$R_o = \sum l_x m_x. \qquad \text{(Equation 9.8)}$$

Age-specific natality (m_x) is influenced by such things as differences in productivity and temperature of a habitat, and survivorship (l_x) is influenced by such things as predation and disease levels and environmental stress. Migratory fishes, moving from one habitat to another at different stages in their life cycle, incur costs

and benefits that change values of l_x and m_x and thus alter fitness (Figure 9.7). For instance, in a potamodromous fish, migration from the adult feeding ground, such as a lake, into a tributary stream for spawning, incurs some costs to the adult fish that could lower adult survivorship. This can be balanced by increased hatching success and better larval and juvenile growth and survival in the small stream as a consequence of greater food availability, lowered predation or disease, or less environmental stress. In addition, assuming the adult feeding ground allows faster growth and thus larger body size, the number of eggs spawned would be increased, thus raising age-specific natality. Selection for migration would occur if fitness (R_o) increases as a result of such behavior.

Among diadromous fishes, anadromy is most common in temperate to polar regions of both the Northern and Southern Hemispheres, whereas catadromy is greatest in the tropics. The pattern seems to be driven by the relationship of freshwater to marine productivity, referred to as the food availability hypothesis (FAH), which emphasizes selection for adult growth (Gross et al. 1988; Maekawa and Nakano 2002). When ocean productivity exceeds that of fresh water, as occurs in temperate to polar regions, anadromy is more common; when freshwater productivity exceeds marine productivity, as occurs in the tropics, catadromy is more common. Although the FAH has been criticized for being overly simplistic by ignoring selection on survival and growth of young, and because of questions concerning the classification of certain fish groups as catadromous or anadromous and concerns about the relationship of productivity measures to actual food availability (Dodson 1997), the hypothesis is supported by work on growth and body size in resident freshwater and anadromous populations of Dolly Varden (*Salvelinus malma*) (Maekawa and Nakano 2002).

Even though diadromy or potamodromy benefited fitness at some point in the evolutionary history of a group, changing environmental conditions, including those caused by humans, might reduce or eliminate the benefit, resulting in the decline or extirpation of populations/species showing a particular migratory behavior (Gross 1987). Consequently, a life history dependent on multiple habitats and access to them can be more fragile than a simple life history (Gross 1987). For example, consider the Gulf Sturgeon (*Acipenser oxyrinchus desotoi*), an anadromous fish that has three principal habitats required in the life history, in addition to requiring access to the habitats (i.e., no barriers along rivers): a specific spawning ground that can be 100–200 km upstream, a summer-fall holding area located 40–60 km upstream, and a marine feeding area located primarily in barrier island passes of coastal waters (Heise et al. 2004, 2005; Ross et al. 2009). For purpose of example, assume that there is a 90% probability that each habitat remains suitable and a 10% chance that each will not. This scenario corresponds to a 73% chance (0.90³) that all three habitats remain suitable and a 27% chance that

all three will not remain suitable. Thus, all else being equal, compared to a single habitat, diadromy, in this case, lowers the chance of survival by 17% (27%–10%). Not surprisingly, many North American diadromous fishes are considered endangered, threatened, or of special concern, a pattern also true worldwide (Williams et al. 1989; McDowall 1999; Warren et al. 2000).

SEX, MATING, AND INVESTMENT IN OFFSPRING

As with other aspects of their life history, fishes show amazing variation in male behavior and mating tactics. In species where male reproductive success is dependent on male-male competition and aggression, and where there is extensive male parental care, there may be several types of males (Dominey 1981; Gross 1982). Male reproductive strategies are context dependent and are solutions of how to maximize fitness with given amounts of energy and probabilities of future survival (Dominey 1984). Often the trade-off is between energy spent on somatic growth for larger body size, secondary sex characteristics, and female attraction behaviors, versus energy spent on gonadal growth.

Mating Tactics

Fishes in the family Centrarchidae are known for nest construction and defense, and in the genus *Lepomis*, only the male is involved in these behaviors (Warren 2009). After nest construction, the male attracts females to his nest for spawning through visual displays in conjunction with well-developed secondary sex characteristics. These parental males (also referred to as bourgeois males) are highly aggressive, have a relatively large body size, and a light body color with a dark yellow to orange breast (Gross and Charnov 1980; Dominey 1981; Gross 1982; Avise et al. 2002). After spawning, the male guards the nest for 1–2 weeks against potential predators on the developing embryos and early stage larvae, and such male care is critical for survival of young (Gross and MacMillan 1981).

Parental males spend energy on somatic growth associated with female attraction and securing mates, and relegate relatively less energy to gonadal development. In Bluegill (*Lepomis macrochirus*), parental male gonad weights average 1.1–1.9% of body weight (Dominey 1980; Gross 1982). Parental males occasionally enter the nests of other parental males and fertilize eggs of females that they are courting, perhaps made easier by habituation of males to other males in nearby nests of breeding colonies (Avise et al. 2002; Jennings and Philipp 2002; Mackiewicz et al. 2002).

Male mating tactics and parentage of young have been studied thus far in six sunfish species. In some populations of at least four species, Bluegill, Pumpkinseed, Longear (*L. megalotis*), and Spotted Sunfish (*L. punctatus*), there are two mutually exclusive mating strategies: parentals and peripherals (Box 9.4), which are defined by male size and behavior (Gross 1982; Warren 2009). Other sunfish species, especially those occurring in nesting colonies such as Redear Sunfish (*L. microlophus*), are likely to have similar alternative reproductive strategies.

Peripheral males can sometimes achieve similar male fitness as parentals through cuckoldry (or kleptogamy; Box 9.4) (Neff and Clare 2008). By stealing fertilizations from the parental male, the cuckolder is acting as a reproductive parasite. Two forms of cuckolders shown within sunfishes are sneakers and satellites (Gross and Charnov 1980; Gross 1982). Sneaker males, which generally are smaller than parental males, dart in between a spawning pair just at the moment of egg deposition and attempt to fertilize the female's eggs. The sneakers are chased by the larger, nest-defending male and often show missing scales, wounds, and damaged fins at the end of the breeding season. Compared to parental males, sneakers pay nothing for nest construction, female attraction, and defense of young, instead putting much more energy into sperm production, with the gonad weight in Bluegill sneakers averaging 4.6% of body weight (Gross 1982). In addition to Bluegill, sneaker males are documented in

BOX 9.4 • Cuckoos, Stolen Fertilizations, and Deception: Peripheral Males, Cuckoldry, and Kleptogamy

From the fourteenth century, cuckoldry has been used to describe the situation where a husband (the cuckold) unknowingly supported offspring resulting from adulterous behavior by his wife (Power et al. 1981). The original term derives from the habit of Cuckoos depositing their eggs in the nests of other birds and then leaving eggs and young to be supported by the unsuspecting parents, even though in the case of the cuckoo, both the male and female of the host nest provide care to young that are not their own. In biology, cuckoldry has generally been used to describe instances where a male provides parental care to offspring that are not his own, usually through stolen fertilizations by another male (the cuckolder) (Power et al. 1981). Although the term cuckold, as historically used, follows the concept of a male raising offspring that are not his own, it suffers from the human connotation of the female's unfaithfulness. To correct for this, the term kleptogamy has been proposed as a more neutral, as well as more specific, term for stolen fertilizations (Gowaty 1984). Peripheral male is another, even more general, descriptor that has been used to describe the same breeding behavior in fishes (Blanchfield and Ridgway 1999). For example, sneaker and satellite males in sunfishes are both peripheral males engaging in cuckoldry through kleptogamy. Peripheral males among salmonids, where females build and defend nests, engage in kleptogamy but not cuckoldry.

some populations of Pumpkinseed, Spotted, and Longear sunfishes (Jennings and Philipp 1992; Warren 2009).

Another approach to mating is shown by satellite males that hover in the water column above a parental male's nest (hence satellite). Satellite males have only been observed for Bluegill (Warren 2009). Satellite males reduce, but do not eliminate, attacks by the defending parental male by mimicking the dark vertical bars, dark background, and dark eyes of females and by descending slowly into a nest—normally while a female is paired with the male. Satellite males attempt to position themselves between parental male and female and thus fertilize the eggs. Although less common, Bluegill satellites will also engage in apparent homosexual behavior by pairing with a parental male in the absence of a female (Dominey 1980, 1981). Again, because they are not investing in somatic growth associated with nest construction and defense and mate attraction, more reproductive energy is allocated to gonadal development, with gonad weight averaging 3.3–4.2% of body weight (Dominey 1980; Gross 1982). The evolution of this system may be asymmetric, with satellites gaining matings by deceiving parental males as to the sex of female mimics, or symmetrical, where the parental male benefits from the nest site being more attractive to a female because a "female" is already on the site, and the satellite male benefits by having access to females. Disadvantages to the parental male of sharing incoming females with satellite males may be balanced with the advantages of producing a spawning center (Dominey 1980).

In Bluegill, but not necessarily other *Lepomis* species, whether an alternative male is a parental or a cuckolder seems to be the result of a genetically controlled polymorphism—the same male does not practice both behaviors, and parentals and cuckolders have different growth trajectories (Figure 9.8). Parental males are larger, mature later, and have lower mortality rates compared to sneakers and satellites. In contrast to the discrete reproductive pathways of parentals and cuckolders, sneakers and satellites are an ontogenetic progression. Although body size generally forms a progression of increasing size from sneaker to satellite to parental, in Cazenovia Lake, New York, parentals and satellites had similar age distributions (Dominey 1980; Gross 1982). In Bluegill populations, parental males are generally less common than cuckolders, although the frequency of cuckoldry decreases with decreases in overall population density (Gross 1982). In Lake Opinicon, Ontario, offspring in nests of Bluegill

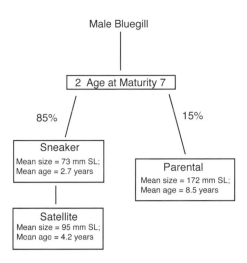

FIGURE 9.8. Male mating systems in Bluegill showing two discrete pathways of cuckolders and parental males, and the ontogenetic pathway of peripheral males from sneaker to satellite. Percents of peripheral and parental males are for a Lake Opinicon, Ontario, population. Based on Gross (1982).

paternal males were sired by the paternal male an average of 76.5% (range 14–100%), with the level of paternity highest early and late in the breeding season, and lower during midseason when reproductive activity in the breeding colony was greatest and paternal male condition was lowest because of increased nest defense (Neff and Clare 2008). Even within Lake Opinicon, different breeding colonies varied in the level of cuckoldry (0–59%) based on the percentage of young sired by nonparental males (Philipp and Gross 1994). In general, a single parental male Bluegill will spawn with several females over the course of the reproductive season (Mackiewicz et al. 2002).

Only sneaker males have been documented in Pumpkinseed and Spotted Sunfish, the two other sunfish species known for having peripheral and parental males (DeWoody et al. 2000a; DeWoody and Avise 2001; Neff and Clare 2008). In Pumpkinseed, young in nests were sired by the parental male an average of 62% (range 0–100%). In Spotted Sunfish, young sired by the parental male made up 98% of the young in a nest, but almost half of the nests contained at least a few young sired by peripheral males. Whether or not the two male morphs are behaviorally mediated or the result of a genetic polymorphism, as is thought to be the case in Bluegill, is not known (DeWoody et al. 2000a).

Redbreast Sunfish (*L. auritus*) and Dollar Sunfish (*L. marginatus*) apparently lack peripheral males, although field observations suggest that sneaker males might be present in a Virginia population of Redbreast Sunfish (Lukas and Orth 1993). Stolen fertilizations appear to be rare, with the parental male generally siring 95% of offspring in the nests of both parental species (DeWoody et al. 1998; Mackiewicz et al. 2002). Stolen fertilizations that do occur are caused by parental males spawning with females in the nests of other nearby parental males.

Several females normally spawn in the nests of a parental sunfish male over the course of a single reproductive season. For instance, young in Dollar Sunfish nests were from an average of 2.5 females, young in nests of male Redbreast Sunfish were from an average of at least 3.7 females, and young in nests of male Spotted Sunfish were from at least 4.4 females (DeWoody et al. 1998; DeWoody et al. 2000a; Mackiewicz et al. 2002).

In addition to sunfish, peripheral males have been genetically documented in Percidae in the Tessellated Darter (*Etheostoma olmstedi*), although parental males sired an average of 86% of young in their nests, and observed in Orangefin Darter (*E. bellum*) (Fisher 1990; DeWoody et al. 2000b). Other North American freshwater fishes showing types of alternative mating systems in at least some populations are found in the families Cyprinidae (Mayden and Simons 2002), Cyprinodontidae (Kodric-Brown 1986), Gasterosteidae (Blais et al. 2004), Poeciliidae (Snelson 1985; Plath et al. 2007), and Salmonidae (Blanchfield and Ridgway 1999).

Sex and Mating Systems

The great majority of fishes (roughly 88%) are gonochoristic, having separate sexes with sex determined genetically and/or environmentally, and the sex of an individual does not change once maturation has occurred (Avise et al. 2002; Patzner 2008). Although rare, environmental sex determination (ESD) has been documented in Atlantic Silversides (*Menidia menidia*) and in laboratory populations of the Clearfin Livebearer (*Poeciliopsis lucida*) (Sullivan and Schultz 1986; Conover and Heins 1987). In these fishes, more females are produced under cooler temperatures and more males under warmer conditions. This is presumably adaptive in allowing females more time for development of larger body sizes and consequently greater fecundity, although other factors may also be involved. In both species, the degree of ESD is apparently controlled by a genetic polymorphism.

Although gonochorism is the most common, and is the ancestral condition among fishes, some fishes are hermaphroditic and a few are even unisexual (Schultz 1977; Avise et al. 2002; Avise and Mank 2009). Unisexuality is rare among animal species, and among vertebrates there are only 22 known genera. Over a third of these (eight) are fishes and most occur in freshwater habitats of North America, where at least 17 unisexual forms have been discovered (Table 9.4) (Schultz 1977; Vrijenhoek et al. 1989; Kraus 1995). In fact, the first known unisexual vertebrate was the Amazon Molly (*Poecilia formosa*), documented as an all-female taxon by Carl and Laura Hubbs (Hubbs and Hubbs 1932, 1946). All unisexual fishes are of hybrid origin between two gonochoristic species (Vrijenhoek 1994). However, this is not to say that these clones are short lived, at least in an ecological sense. *Poeciliopsis monacha-occidentalis*, for instance, has apparently survived for some 60,000 years (Avise et al. 1992), although the gonochoristic progenitors have existed for much longer, perhaps 4–12 million years (Vrijenhoek 1994).

In diploid unisexuals of *Poeciliopsis*, reproduction occurs by hybridogenesis, where a female mates with a male congeneric and only female offspring develop. The male genome is expressed in the offspring but is eliminated during meiosis, whereas the female genome passes

TABLE 9.4

Genera of Freshwater Unisexual Fishes Occurring in North America

Family	Genus	Number of forms
Cyprinidae	*Chrosomus*	4
Atherinopsidae	*Menidia*	1
	Fundulus	1
Poeciliidae	*Poeciliopsis*	7
	Poecilia	4

SOURCE: Based on Schultz (1977), Vrijenhoek et al. (1989), and Kraus (1995).

intact to the haploid egg, creating a hemiclonal lineage. Hybridogenesis does not form true clones because paternal genomes are added afresh each generation (Vrijenhoek 1979, 1994; Moore 1984). The remaining unisexual fishes, including some *Poeciliopsis*, are typically triploid and reproduce through gynogenesis, where sperm from a congeneric is needed for egg activation, but the male genome is not incorporated and the eggs develop matroclinously.

Gynogenetic and hybridogenetic populations form ecologically fascinating systems because the unisexual clones, although likely interacting competitively with gonochoristic, congeneric species, are also dependent on heterosexual males for reproduction, even though the males' genomes are not incorporated in the offspring's genome (gynogenesis) or are incorporated but not transmitted between generations (hybridogenesis) (Vrijenhoek 1979). Assuming equal fecundity, an all-female species can achieve twice the population growth rate of a gonochoristic species, at the expense of donor males from conspecific populations (Moore 1984). Also, in environments that are intermediate and relatively unchanging, sex is costly, as only 50% of one's genome is transferred to offspring. Consequently, unisexuality and clonal populations might be expected to be successful in such habitats. A critical question, therefore, is how do clones keep from out-competing and eliminating their gonochoristic host species? Although

numerous hypotheses have been proposed, three nonmutually exclusive hypotheses relating to this are the behavioral regulation/sperm limitation hypothesis, the resource partitioning hypothesis, and the Red Queen hypothesis (Van Valen 1973; Moore 1984; Vrijenhoek 1994). The behavioral regulation hypothesis involves mate recognition and the tendency of heterosexual males to prefer female conspecifics over clones. At low population densities of sexually reproducing fish, clones would be sperm limited. As population sizes of the sexual fish increase, breeding males form dominance hierarchies and subordinate males tend to mate more often with the clones. In support of this hypothesis, the occurrence of male preference for conspecifics has been widely documented (Moore 1984).

There is also some support for versions of the resource partitioning hypothesis. One form of this hypothesis is the "frozen niche-variation model," where each clone has a fixed adaptive complex that includes a small portion of the niche space occupied, but underutilized, by nonclonal gonochoristic species (Vrijenhoek 1979). Predictions of this model have been supported by coastal populations of the Texas Silverside (*Menidia clarkhubbsi*), a clonal, all-female species inhabiting coastal areas from Alabama to Texas, and by all-female clones of *Poeciliopsis* occurring in streams of northern Mexico (Vrijenhoek 1979; Echelle and Echelle 1997). Coexistence of these clones with each other and with the gonochoristic species is perhaps facilitated by trophic resource partitioning. For instance, during the dry season, one member of a pair of *P. monacha-lucida* clones specialized as a scraper feeding on organic detritus derived from leaf litter in shaded, rocky pools. The lower jaw dentition consisted of numerous, small tricuspid teeth. The other clone had fewer, larger teeth in the lower jaw and browsed on floating mats of organic detritus in nutrient-rich, unshaded, shallow pools (Vrijenhoek 1978, 1979). However, the extent of resource partitioning between the clones has been questioned because stomach contents showed high overlap (Moore 1984).

Why sexual reproduction should be favored over asexual reproduction is addressed by Van Valen's Red Queen hypothesis (Van Valen 1973). The term comes from Lewis Carroll's Alice in Wonderland, where the Red Queen shows Alice that she must run fast to stay in the same place. Biologically, the Red Queen hypothesis predicts that the increased genetic diversity, and potentially more rapid evolutionary change, allowed by sexual reproduction compensates for any genetic or ecological disadvantage of sex (Salathé et al. 2008; Morran et al. 2011). Consequently, unisexual organisms should be at a disadvantage in terms of responding over the long-term to parasites, predators, or competitors. The prediction that clonal forms should have greater parasite loads in contrast to sexual forms has been tested with two clones of *Poeciliopsis monacha-lucida* and the syntopic, sexually reproducing Headwater Livebearer, *Poeciliopsis monacha*. The most common clone in a stream pool showed significantly greater parasite burdens compared to the less common clone or the sexual population, except where a sexual population had low genetic diversity because of a recent founder effect caused by complete drying and then recolonization of a pool (Lively et al. 1990). When a few Headwater Livebearer females were added to the pool after 15 generations, the genetic diversity again increased with a concomitant drop in parasite load (Vrijenhoek 1994).

At least 12 families of teleostean fishes, in nine orders, are hermaphroditic (C. L. Smith 1967; Avise and Mank 2009). Three forms of hermaphroditism are protogynous hermaphrodites, in which an individual is first a female and then becomes a male; protandrous hermaphrodites, in which an individual is first a male and then becomes a female; and synchronous hermaphrodites, in which functional male and female cells are present in the gonads at the same time. Hermaphroditic fishes are almost all marine species and most are not self-fertilizing (C. L. Smith 1967). An exception is the Mangrove Rivulus (*Kryptolebias marmoratus*, Aplocheilidae), a small cyprinodontiform occurring in fresh to brackish water in tropical and subtropical areas of the Gulf, Atlantic, and Caribbean, including southern Florida (Kallman and Harrington 1964; Miller 2005). The Mangrove Rivulus is a synchronous hermaphrodite and most populations are obligate, self-fertilizing clones (Kallman and Harrington 1964). However, some populations do include functional males and thus gain increased genetic variation through outcrossing. The presence of males is apparently triggered by some unknown ecological factor (Weibel et al. 1999). In theory, synchronous hermaphroditism would allow for the most rapid rate of population increase compared to other forms of hermaphroditism and to gonochoristic species—if longevities were equivalent, which is not often the case (C. L. Smith 1967). Synchronous hermaphroditism would also be favored where populations are small and isolated so that finding mates is difficult. In this case, synchronous hermaphroditism provides "reproductive assurance" because reproduction can occur without another individual (Charnov et al. 1976; Weeks et al. 2006; Avise and Mank 2009).

Parental Care

Parental care of young can include prefertilization activities, such as substratum cleaning and nest construction, and postfertilization activities, such as egg hiding or egg burying; parental defense of nest and offspring; caring for eggs by fanning, cleaning, and removing dead or diseased embryos; mouth brooding; pouch brooding; and internal gestation (Baylis 1981; Blumer 1982; Sargent et al. 1987). Overall, approximately 22% of actinopterygian families show parental care, which is represented in basal fishes such as Bowfin (*Amia calva*) as well as in derived lineages such as darters, sunfishes, and livebearers (Blumer 1982; Gross and Sargent 1985; Mank et al. 2005). The evolution of parental care is a function of the relative maturity and survival rates of young and the survival rates of adults. Parental care is favored when the survival rate of eggs and juveniles is low in the absence of care, the adult death rate is relatively high, the egg maturity rate is relatively low, and the duration of the juvenile stage is relatively short (Klug and Bonsall 2009). In general, fishes with parental care are near the equilibrium end of the three-endpoint life-history model, characterized by smaller clutch sizes and increased parental investment in offspring (Figure 9.3). Egg size is often positively correlated with the quality (i.e., intensity and duration) of parental care in fishes, and a life-history model predicts that increased egg size is favored when parental care reduces the instantaneous mortality rate of eggs (Sargent et al. 1987).

Among families of bony fishes showing parental care, male-only care is most common (~50%), followed by female-only care (~32%) and biparental care (~18%) (Figure 9.9A). This is quite different from the majority of vertebrate taxa where there is internal fertilization and the care of young is provided by the female. However, internal fertilization has evolved in at least 21 families of bony fishes and 14 of these families have internal gestation of embryos—an extreme form of parental care (Mank et al. 2005). Parental care is more prevalent in freshwater compared to marine fishes, occurring in 60% of freshwater fish families versus only 16% of marine families (Baylis 1981).

The increased frequency of parental care in freshwater habitats compared to marine habitats, and the preponderance of male care over female care, are both explained by a model based on the differential survival of zygotes and the rates of gametogenesis males and females (Baylis 1981). Freshwater environments tend to show greater heterogeneity in physical and chemical factors per unit distance compared to marine environments. Demersal and/or adhesive eggs are likely favored in fresh water because pelagic eggs would run the risk of being swept out of favorable conditions—in fact, the only North American freshwater fish groups with pelagic eggs are those in large lakes or those that migrate extensive distances upstream to spawn (Balon 1975). In contrast, pelagic eggs in a marine environment have a much greater chance of remaining within the same water mass during their developmental period, and even in coastal species tidal influences would put demersal/adhesive eggs at risk because of periodic changes in the water mass (Baylis 1981). Male residency is favored when suitable spawning sites are limited because a male is able to produce fertile gametes more rapidly than a female and can leave more offspring than a female through the potential of multiple matings over a short time period (Baylis 1981).

The kind of parental care is strongly influenced by whether fertilization is internal or external. In fishes with external fertilization, nest construction is a preadaptation for male-only parental care; close to 80% of such families have care of young provided solely by the male (Gross and Sargent 1985; Mank et al. 2005). Internal fertilization is a preadaptation for internal gestation, and 90% of families with internal fertilization show female-only parental care, most through internal gestation but several with external guarding (Mank et al. 2005). Although various scenarios have been proposed for the evolutionary sequence of parental care (e.g., C. Smith and Wootton 1995), a recent phylogenetically based approach suggests that the three forms of parental care (male-only, female-only, and biparental) have arisen independently from an externally fertilizing ancestor without parental care (Figure 9.9A) (Mank et al. 2005; Mank and Avise 2006).

Parental care has been documented in 54% of North American freshwater fish families (27 of 50), with male-only care shown in 63% of the families, female-only care in 52%, and biparental care in 33% (Table 9.5). In some cases, multiple types of care are shown by different taxa in the same families. Within the Centrarchidae, male-only parental care occurs in most species and both sexes are usually polygamous (Cooke et al. 2008; Warren 2009). Attendant males remain on the nest for only 1–7 days (Sacramento Perch, *Archoplites interruptus*; Pumpkinseed; and Bluegill) to nearly a month (Smallmouth, *Micropterus dolomieu*; and Largemouth bass), depending on the species (Gross

TABLE 9.5

Types of Parental Care Shown by North American Freshwater Fish Families
Families are listed in decreasing order of species richness

Family	Provider	Egg bury	Nest	Egg guard	Larval guard	Fan
Cyprinidae	m	x	x	x		x
Percidae	m, f	x		x		x
Catostomidae	bp	x				
Poeciliidae	f					
Ictaluridae	m, f, bp		x	x	x	x
Goodeidae	f					
Atherinopsidae	f					
Salmonidae	f, bp	x	x	x		
Cyprinodontidae	m	x		x		x
Centrarchidae	m, bp		x	x	x	x
Cottidae	m		x	x		x
Cichlidae	m, f, bp		x	x	x	x
Eleotridae	bp			x		
Osmeridae	f	x				
Amblyopsidae	f					
Elassomatidae	m			x		
Gobiidae	m, f, bp		x	x		x
Gasterosteidae	m		x	x	x	x
Characidae	m, f, bp		x	x	x	x
Umbridae	m, f		x	x		
Syngnathidae	m					
Gobiesocidae	m[c]			x		
Amiidae	m		x	x	x	
Aphredoderidae	bp		x	x[d]		
Embiotocidae	f					
Synbranchidae	m		x	x		
Ariidae	m, f					

SOURCE: Unless otherwise listed, the source is Blumer (1982) as are the descriptions of care giving.

bp = both parents at the same time

Brood pouch = incubation of eggs in an external brood pouch

Clean = removing debris from eggs, usually by manipulating eggs in the mouth

Cull = removing dead or diseased eggs from the nest

Egg bury = egg-burying behavior

Egg guard = egg guarding from potential predators

f = female only

Fan = moving water over the developing eggs by fin movements or from the mouth and opercular chamber

IG = internal gestation of embryos in the ovary or oviduct

Cull	Clean	MRE	MRL	Brood pouch	Oral brood	IG	Source
x							
x							
							Page and Johnston 1990
						x	
	x						
						x	
						x[a]	Grier et al. 1990
							Balon 1975
x							
							DeWoody et al. 2000b
x		x					Fiumera et al. 2002
x	x	x	x		x[b]		McKaye and McKaye 1977
							McKaye et al. 1979
					x		
							Walsh and Burr 1984
x							
x	x	x	x				
x		x					
				x			
							Ridley 1978
							Fletcher et al. 2004
						x	
					x		Miller 2005

Larval guard = guarding the larvae from potential predators

m = male only

MRE = moving or retrieving eggs that have moved from the nest

MRL = moving or retrieving larvae that have moved from the nest

Nest = nest construction

Oral brood = keeping eggs and or larvae in oral or opercular chambers

a. Fertilized eggs are only held for a short time before being deposited

b. Oral brooding is not documented in North American cichlids

c. Documented in marine species of the genus *Gobiesox*; no information on freshwater species

d. Male Pirate Perch only defend the nest area for ≤ 2 h

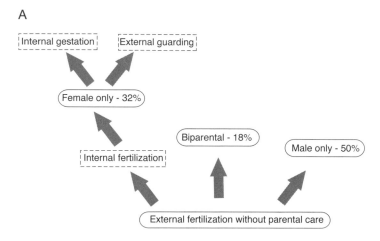

A

Internal gestation ← Female only - 32% → External guarding

Female only - 32% ← Internal fertilization

Biparental - 18%

Male only - 50%

Internal fertilization ← External fertilization without parental care

Biparental - 18% ← External fertilization without parental care

Male only - 50% ← External fertilization without parental care

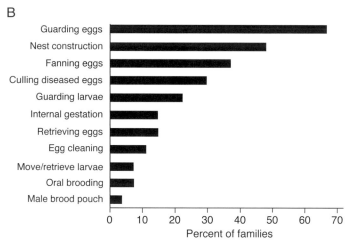

B

Type of care	Percent of families
Guarding eggs	~67
Nest construction	~48
Fanning eggs	~37
Culling diseased eggs	~30
Guarding larvae	~22
Internal gestation	~15
Retrieving eggs	~15
Egg cleaning	~11
Move/retrieve larvae	~7
Oral brooding	~7
Male brood pouch	~4

Percent of families

FIGURE 9.9. Parental care in fishes. A. Evolutionary pathways of parental care in bony fishes, showing the three independent origins of the three forms of parental care (solid ovals) from an external fertilizing ancestor lacking parental care. Percents show the families exhibiting each form of care. Based on Blumer (1979), Sargent and Gross (1993), Mank et al. (2005), and Mank and Avise (2006). B. Types of parental care shown by families of North American freshwater fishes. Data from Table 9.5.

and MacMillan 1981; Sargent et al. 1987; Cook et al. 2006). However, in at least one population of Largemouth Bass, located in the Savannah River drainage of South Carolina, males and females are mostly monogamous and females also participate in providing care (DeWoody et al. 2000c).

The most common form of parental care is egg guarding, followed by nest construction, fanning eggs with the fins or by expelling water from the mouth or opercular chambers, removing dead or diseased eggs from the nest, guarding larvae against predators, and internal gestation of the embryos by the female (Figure 9.9B). Pipefishes, family Syngnathidae, provide the only North American example of parental care being provided by placing embryos in a brood pouch.

SUMMARY

Life-history patterns of North American freshwater fishes are the distillation of numerous competing demands of growth, survival, and reproduction. Although life histories are shaped by natural selection, any life history is a mosaic of traits shaped by contemporary selection and traits shaped by selection at various stages in the evolution of the lineage. Life tables are a convenient way to organize the basic parameters of life histories that include age-specific survivorship, age-specific natality, and cohort generation time. Survivorship usually increases with age and is lowest in the egg and early larval stages. As a consequence, survivorship curves for fishes are generally closest to a type 3 shape.

A three-endpoint model of life histories effectively captures the variation in life histories among freshwater and marine fishes. The model is based on how fitness can be maximized by increased juvenile survival and generation times and small clutch sizes (equilibrium species), shorter generation times and extended reproductive seasons (opportunistic species), or increasing age-specific natality and adult survival (periodic species). In general, opportunistic species are more common at lower latitudes whereas periodic and equilibrium species are more common at higher latitudes.

A model derived from basic metabolic principles predicts that the annual mass devoted to reproduction scales allometrically with body mass at first reproduction, with strong empirical support from a data set including North American freshwater and marine fishes. The body mass devoted to reproduction can be packaged in single large clutches, multiple clutches produced within the same reproductive season, or single clutches per reproductive season produced over several or more years. An increasing number of fishes are now known to produce multiple clutches within a single spawning season so that single measurements of fecundity underestimate annual egg production. Iteroparity is more common in fishes that occupy varying environments and that have high adult (relative to young) survivorship. In contrast, semelparity occurs in fishes that experience lower environmental variation and that have high reproductive investment, including long-distance migrations. Spawning migrations are common in fishes intermediate between periodic and equilibrium life histories. Among fishes that show regular migrations, movement within fresh water (potamodromy) is most common among North American freshwater fishes, followed by anadromy, amphidromy, and catadromy. Although complex migratory behaviors evolved because of the increased fitness realized by migration, changes in habitats, including those caused by humans, often put such lifestyles at greater contemporary risk.

Male fishes of numerous species often show a trade-off in how reproductive energy is allocated. Parental males expend high amounts of energy in such things as territory defense, nest construction, and mate attraction, and relatively less on actual gonadal size. In contrast, peripheral males shunt most of their reproductive energy into gonadal growth and attempt to steal fertilizations from the parental males. Although less common, some fishes are hermaphrodites and others are unisexual. Unisexuality is shown within several groups of North American freshwater fishes, especially the Poeciliidae and the Atherinopsidae. Hermaphrodites are more common in marine fishes and only one species, the Mangrove Rivulus, occurs in North American fresh water.

Parental care is documented in over half of North American freshwater fish families and the most common type of care is egg guarding. Care is most often provided by the male; biparental care is least common. The kind of parental care is strongly influenced by whether fertilization is external or internal. In the latter case, care is generally provided only by the female.

SUPPLEMENTAL READING

Avise, J. C., A. G. Jones, D. Walker, J. A. DeWoody, et seq. 2002. Genetic mating systems and reproductive natural histories of fishes: Lessons for ecology and evolution. *Annual Review of Genetics* 36:19–45. A thorough review of mating systems in fishes.

Balon, E. K. 1975. Reproductive guilds of fishes: A proposal and definition. *Journal of the Fisheries Research Board of Canada* 32:821–64. A detailed classification of reproductive guilds of fishes.

Balon, E. K. 1981. Additions and amendments to the classification of reproductive styles in fishes. *Environmental Biology of Fishes* 6:377–89. Further clarification of reproductive guilds.

Gotelli, N. J. 2001. Age-structured population growth, 50–80. In *A primer of ecology*. Sinauer Associates, Sunderland, Massachusetts. A lucid treatment of growth in natural populations of organisms.

Miranda, L. E., and P. W. Bettoli. 2007. Mortality, 229–77. In *Analysis and interpretation of freshwater fisheries data*. C. S. Guy and M. L. Brown (eds.). American Fisheries Society, Bethesda, Maryland. A thorough coverage of how mortality is determined in fish populations.

Winemiller, K. O., and K. A. Rose. 1992. Patterns of life-history diversification in North American fishes: implications for population regulation. *Canadian Journal of Fisheries and Aquatic Sciences* 49:2196–218. One of the foundation papers on the three-endpoint model of fish life history patterns.

WEB SOURCES

Brown-Peterson, N., S. Lowerre-Barbieri, B. Macewicz, F. Saborido-Rey, J. Tomkiewicz, and D. Wyanski. 2007. An improved and simplified terminology for reproductive classification in fishes. http://digital.csic.es/handle/10261/11844.

Frimpong, E. A., and P. L. Angermeier 2009. FishTraits database. http://www.cnr.vt.edu/fisheries/fishtraits.

PART FOUR

Interactions among Individuals and Species

NDIVIDUALS AND POPULATIONS interact in various ways, including courtship and reproduction, competition for resources (nesting sites, mates, cover, food), predator-prey relationships, facilitation, and mutualism. Such interactions among individuals of the same or different species are often based on some form of communication. The chapters in this unit explore the mechanisms and roles of communication (Chapter 10); trophic position and interactions during resource acquisition, such as competition among individuals of the same or different species (Chapter 11); predator-prey interactions during resource acquisition, including avoidance of predation (Chapter 12); and facilitation and mutualism as part of resource use (Chapter 13). These chapters cover most of the major ways in which individuals and populations might interact (Part Table 4.1). Some interactions are symmetrical, either benefitting both parties (+, +) or harming both (–, –), whereas some are asymmetrical, where one member benefits at the expense of the other. However, representing interactions by zeros, pluses, or minuses is an oversimplification. Directions of interactions, although symmetrical in terms of sign, may differ in the strength of the interaction, such as when competition affects one member of the interaction

PART TABLE 4.1

Some Possible Types of Interactions among Individuals and Populations

Type of interaction	Effect on sender/giver	Effect on receiver
Amensalism	–	0
Communication	+	+
Commensalism	+	0
Competition	–	–
Mutualism	+	+ (obligatory)
Neutralism	0	0
Parasitism	+	–
Predation	+	–
Protocooperation	+	+ (non-obligatory)
Spying	0	+

SOURCE: Based in part on Pianka (1988).

+ = positive effect

– = negative effect

more strongly than the other. Also, a negative interaction, such as resource competition, may in the long term positively affect fitness if it contributes to maintaining a stable population size (R. Ryel, pers. comm. 2011).

TEN

Communication among Individuals

CONTENTS

AN INTEGRAL PART of the biology of fishes is their ability to communicate, both among individuals of the same population or species, as well as among different species. In fact, communication is a critical aspect, both in terms of helping to maintain reproductive isolation between species, and also to facilitate interactions among and between species. Communication, by definition, refers to the interaction between two or more individuals. Technically, communication is a phenomenon of one organism producing a signal that, when responded to by another organism or organisms, confers some advantage to the signaler or to its group. The signal can be visual, chemical, auditory, electrical, or a combination of modalities. Although fascinating, electrical communication is not known for North American freshwater fishes, being documented in relatively few fish groups in Africa, South America, and the marine environment (Moller 2006).

As with other life functions, communication evolves as a trade-off among conflicting demands. For instance, coloration in fishes is a balance between factors maximizing signal value for communication and those that make an individual inconspicuous and at less risk for predation. Communication also requires the integration of the signaling, receiving, and behavioral systems, all in the context of the physical and biotic components of the environment. All else being equal then, the signal mode with the highest rate of information transmission should prevail over modes transmitting less information. For example, visual signals contain temporal and spatial information compared to sound, which only has temporally varying information, but visual signals only work where there is sufficient water clarity or sufficient open water not obstructed by plants, debris, or other materials (Endler 1992).

The term sensory drive refers to all processes that might cause biases in the direction of evolution of sensory systems. For example, signal characteristics of the sender should evolve to best exploit the signal-reception characteristics of the receiver. However, the signal is also affected by predation risk, so the direction of evolution should favor a balance between selection for communication and selection against discovery by a predator. In addition, the transmission and quality of the signal are strongly affected by the environment, such as water chemistry and turbidity, ambient light penetration, and ambient sound. Consequently, natural selection should favor systems that have high information content, maximize the signal relative to background noise, minimize the degradation of the signal, and have signals that stimulate the sensory system of the receiver most effectively, while also minimizing threats of predation (Endler 1992). Because of strong environmental constraints on all forms of communication, signaling systems often vary greatly, even within the same species in different habitats. In addition, anthropogenic changes to aquatic systems can have major impacts on the expression, transmission, and receipt of signals, a potentially very important conservation concern that is often overlooked (Sluijs et al. 2011).

Compared to the terrestrial environment, the aquatic environment offers challenges and advantages among the various forms of communication. Because water is quite similar to living tissue in terms of optical and acoustic density, signal detection can be more difficult. Signals can also be distorted or attenuated by the physical properties of water and the nature of aquatic habitats. Although much less an issue in most freshwater habitats compared to marine systems, water acts as a color filter, alters light quality, and causes greater light attenuation per unit distance compared to air. Turbidity also can reduce or eliminate visual

signals. Water also selectively filters sound waves—higher frequencies, with lower wave amplitudes, are attenuated faster than lower frequencies, which have higher wave amplitudes. In addition, signal disturbances caused by sound reflection and reverberation are problems in shallow water (Kasumyan 2008).

On the positive side, low-frequency sound waves are transmitted almost five times faster in water than air and equivalent acoustic oscillations require less energy to produce in water compared to air (Kasumyan 2008; Garrison 2009). Water is also a better medium for the transfer of chemicals that are characteristically associated with living organisms, although depending on current flow, the transmission can be unidirectional and chemical signals may persist in the environment after the stimulus for their release has long since passed.

CHEMORECEPTION

Compared to terrestrial organisms, fishes rely heavily on chemical signals, something that is not surprising considering that water is an excellent solvent and they are living in a milieu of waterborne compounds. Among North American freshwater fishes, chemoreception is used in a number of ways, including food location, sexual recognition, identification of individuals or species, predator recognition and avoidance, recognition of young, orientation to habitat, and cues in migration. In some cases, the use of chemosensory cues does not fit within the definition of communication in the strict sense, although the responses of the receivers can be relevant and adaptive.

Morphologically, chemoreception can be divided into three general categories of olfaction, taste, and a general chemosense, with the latter residing in free epidermal nerve endings of trigeminal (fifth cranial nerve) or spinal origin (Kotrschal 2000). Olfactory receptors are innervated by the paired first cranial nerves. These nerve bundles also include branches going to the retina, which allow olfactory stimuli to alter visual responsiveness of retinal cells (Stell et al.

1984; Kotrschal 1991). Taste receptors are innervated by cranial nerves 7, 9, and 10, as well as spinal nerves, and include taste buds, which are compound structures that also include tactile receptors, and solitary chemosensory cells (SCC) (Kotrschal 1991, 2000). In contrast to terrestrial vertebrates, taste receptors are not confined to the inside of the mouth and are also on the fins and the body, especially around the mouth and head but also laterally (Kotrschal 1991; Gomahr et al. 1992; Sorensen and Caprio 1998). Densities of SCCs, which can account for 60–90% of all epidermal sensory cells, can be as high as 4,000 per mm^2 in cyprinids and 1,000–2,000 per mm^2 in Bullhead Catfish (*Ameiurus*) (Kotrschal 1991). Taste buds are also widespread on the surface of the body and fins. Based on a study of European cyprinids, taste bud density is greatest around the head and decreases posteriorly. In addition, benthic fishes and fishes living in turbid water generally have higher taste bud densities than pelagic fishes or fishes living in clear water. The highest densities of taste buds, 200–300 per mm^2, are along the throat area; in fins, the high densities are 100–150 per mm^2. Especially for benthic species, taste bud density is greater along the ventral surface of the body and fins compared to the dorsal body surface and fins (Gomahr et al. 1992). The Sicklefin Chub (*Macrhybopsis meeki*), which inhabits large, turbid rivers of the Mississippi River drainage, also has high densities of taste buds on the body and fins (Davis and Miller 1967). As yet, little is known about the kinds of olfactory signals perceived by free nerve endings. Although chemosensory receptors can be distinguished morphologically, they are highly synergistic in function and functions often overlap among the three types (Kotrschal 2000).

Chemosensory Communication

Pheromones are external chemical stimuli (odors) that are secreted to the outside by an individual and received by a second individual or individuals of the same species in which they elicit a specific adaptive response that is not dependent on prior learning or experience (Liley 1982; Sorensen and Stacey 2004; Burnard et al. 2008). Allomones are chemical signals that carry information between different species. Communication via pheromones or allomones is advantageous because they can transmit through darkness or turbid water and around obstacles, metabolically they require little energy to produce, they have a range from close contact to several km, and they can be relatively persistent in the water column. On the negative side, the transmission of pheromones or allomones is dependent on diffusion or currents, so communication can be unidirectional or relatively slow. The slow fade-out limits possibilities for switching messages, and the message may remain long past the time of response and have unintended consequences, such as attracting predators (Liley 1982; Stacey and Sorensen 2005). Environments with different flow characteristics consequently have varying qualities of chemosensory signals (Sherman and Moore 2001). Although pheromones are involved in various behaviors, most work has been based on those involved in reproduction (Burnard et al. 2008).

The evolution of pheromones is thought to have involved three discrete steps, starting with the ancestral stage—the release of chemical substances from a donor fish without eliciting a response in another conspecific (the putative receiver; Figure 10.1) (Stacey and Sorensen 2005). In the second, or spying stage, the selection favors a response to the olfactory cue by the receiver, along with the possibility of increased sensitivity to the cue. At this point the donor remains unchanged, although the spying stage is a prerequisite for the third stage. Communication can evolve if there is positive selection on the donor for the production of a chemical cue responded to by conspecifics. In the final stage, both the donor and receiver are respectively specialized for the production and detection of the chemical signal. Historically, fish alarm substances have been considered pheromones. However, more recent studies suggest that alarm substances are primarily involved in immune functions. As such, alarm substances

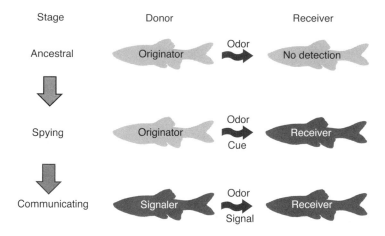

Stage   Donor   Receiver

Ancestral   Originator   Odor   No detection

Spying   Originator   Odor / Cue   Receiver

Communicating   Signaler   Odor / Signal   Receiver

FIGURE 10.1. Three stages in the evolution of pheromonal communication in fishes. In the ancestral stage, the donor releases various odors into the water, but both the donor and receiver are unspecialized (indicated by light gray shading) and the receiver does not respond to the released odors. In the spying stage, the donor remains unspecialized, but the receiver has become specialized (indicated by dark gray shading) for detection and response to the chemical cue. In the communication stage, the donor is specialized for the production of the chemical cue and the receiver is specialized for the detection and response to the cue. Based on Stacey and Sorensen (2005).

are treated in the "Predator Avoidance" section of Chapter 12.

Mate Attraction, Courtship, and Spawning

Pheromones are involved, or at least implicated, in mate attraction, courtship, and spawning among various fish groups. Olfactory responses to pheromones likely evolved relatively few times and have diversified slowly (Stacey and Cardwell 1995). Among North American freshwater fishes, examples of males responding to pheromones released by females include species in the families Petromyzontidae, Clupeidae, Cyprinidae, Characidae, Ictaluridae, Salmonidae, and Poeciliidae (Liley 1982; Burnard et al. 2008). In at least some cases, the response is due to gonadal hormones (i.e., prostaglandins and steroids) or their metabolites that are released into the environment and are thus active as hormonal pheromones (Stacey and Cardwell 1995; Stacey and Sorensen 2005; Burnard et al. 2008). This discovery of hormonal pheromones shows that the importance of hormones and their metabolites is not limited to intraindividual "communication"

but is extended as well to nearby conspecifics. In fact, although hormones might act slowly within an individual in the development of secondary sex characteristics and sexual behavior, hormonal pheromones can very rapidly communicate hormonal status of an individual to its conspecifics.

Mature male Goldfish (*Carassius auratus*) display increased reproductive behavior when exposed to water occupied by female Goldfish that have ovulated eggs in their ovaries but do not respond in such a way to mature females that have not ovulated (Partridge et al. 1976; Liley 1982). Later work has shown that males are responding to prostaglandins (F prostaglandins), indicative of ovulation, and to various hormonal steroids (Stacey and Cardwell 1995). In addition to stimulating or affecting ovulation in fishes, F prostaglandins are involved in eliciting female sexual behavior and as a pheromone serving to attract males (Sorensen and Goetz 1993). Male Fathead Minnows (*Pimephales promelas*) show increased courtship behavior when exposed to water occupied by females injected with F prostaglandins, but do not respond directly to water with F prostaglandins but lacking in female Fathead Minnows, suggesting that it is not the presence of the prostaglandin alone, but that the prostaglandin causes females to release some other chemical substance that is attractive to males (Cole and Smith 1987). Female Sailfin and Dwarf mollies (*Poecilia latipinna* and *P. chica*), as well as females of other molly species, release a pheromone from the ovary during or subsequent to ovulation. The pheromone increases male swimming activity and social interactions among males, as well as increasing the frequency of mating attempts (Thiessen and Sturdivant 1977; Brett and Grosse 1982; Liley 1982). Similar responses have been shown in Sea Lampreys (*Petromyzon marinus*), Atlantic Salmon (*Salmo salar*), and Rainbow Trout (*Oncorhynchus mykiss*) (Liley 1982; Stacey and Cardwell 1995).

Female attraction to male chemical cues also occurs in a number of North American freshwater fish families, including Petromyzontidae, Salmonidae, Cyprinidae, Ictaluridae, and Cyprinodontidae (Stacey and Cardwell 1995; Kodric-Brown and Strecker 2001; Burnard et al. 2008). Male Sea Lampreys build nests in streams and then use a pheromone based on a specific bile acid to communicate their reproductive status and their location to ovulated female lampreys down current from them. This seems to be a case of active signaling by the males rather than the females responding to a metabolic by-product (i.e., spying). Male Sea Lampreys do not feed as mature adults, and thus the bile acids, produced in the liver, are not needed for digestion; in fact, the adult lampreys lack bile ducts and gall bladders. The bile acids are released by diffusion across the gill membranes and can attract females from distances of at least 65 meters in a stream with a discharge of 2.3 m³s⁻¹ (Li et al. 2002). The mature female olfactory system is highly sensitive to the specific bile acid released by the males and can discriminate bile acid from mature males from other bile acids produced by conspecifics (Siefkes and Li 2004).

Male Fathead Minnows also produce a chemical substance (or substances) that is attractive to females during the reproductive season as well as to females with regressed gonads. Females show greater responsiveness to the male chemical stimulus in the morning, the typical spawning time for Fathead Minnows, compared to the afternoon when they become more responsive to chemical signals from other females. The male pheromone likely serves to guide reproductively ready females to nest sites of defending males (Cole and Smith 1992). There is also some evidence that male Channel Catfish (*Ictalurus punctatus*) use chemicals to identify the nest site and to attract females, as do Rainbow Trout and Lake Trout (*Salvelinus namaycush*) (Liley 1982; Zhang and Hara 2009). Although not demonstrated by field experiments, female Lake Trout are highly sensitive to bile acids that are produced by mature male conspecifics, strongly suggesting a sexual attraction system analogous to that of lampreys (Zhang et al. 2001; Zhang and Hara 2009). In addition to communication between sexes, pheromones also are used for same-sex communication. For instance, male Goldfish respond to the gonadal condition of other males so that they are synchronized in terms of milt production (Stacey and Sorensen 2005).

On a cautionary note, however, with the exception of Sea Lamprey, all the results mentioned previously were based on experiments done in static tank systems with consequently high concentrations of the pheromones. Evidence from the tropical Guppy (*Poecilia reticulata*) is cautionary in how olfactory signals might operate in nature. In both a natural stream and in a flowing-water laboratory stream, female Guppies did not show any evidence of attraction to chemical cues released by male or female conspecifics, although they did respond when tested in a static system. These results emphasize the importance of concentration and perhaps the importance of short-range effects in chemical signals (Archard et al. 2008). They also emphasize the need for testing the role of pheromones under more natural conditions.

Social Status and Individual Recognition

In addition to being involved in reproductive behavior, pheromones are also used to communicate individual identity (including kin recognition), social status, or other social signals. In a pioneering study, blinded individual Yellow Bullhead (*Ameiurus natalis*) were positively and negatively conditioned to respond to water from two different conspecifics held in separate aquaria. For the positive conditioning, fish were fed when water from one conspecific's tank was introduced. For negative conditioning, fish were given a mild shock when water from the other conspecific's tank was introduced. Once conditioned, the test fish responded to the positive stimulus by moving to the surface and showing feeding behavior; test fish responded to the negative stimulus by moving to the bottom of the aquarium and hiding. Overall, the test fish discriminated between the two conspecifics in 95.5% of the trials but were unable to discriminate between the two individuals when their sense of smell was blocked (Todd et al. 1967). In addition, Yellow Bullhead seemed to be able to recognize changes in social status through olfactory cues. In a similar study, Brown Bullhead (*A. nebulosus*) were able to correctly identify a conspecific 90% of the time as shown by positive responses to water taken from conspecifics' aquaria (Carr and Carr 1985).

An aspect of social behavior, the level of aggression between individuals, is controlled by a protein-based pheromone in juvenile Brown Bullhead. When held at high population densities, juvenile Brown Bullhead show reduced aggressive behavior; however, juveniles held at low population densities are highly aggressive toward one another. When aggressive fish from low-density populations received water from high-density populations, they showed a significant decline in aggressive behavior (Carr and Carr 1986).

Chemosensory cues are used by fishes to identify kin, with North American examples from Cyprinidae, Ictaluridae, Salmonidae, and Poeciliidae (Brown and Brown 1996; Griffiths 2003; Ward and Hart 2003). Because the releaser apparently does not modulate the signal relative to the receiver, these cases better fit the case of spying rather than communication (Figure 10.1).

Work on European Threespine Sticklebacks (*Gasterosteus aculeatus*) indicates that odor is involved in kin recognition but that it is insufficient for kin recognition in the absence of visual cues. Kin recognition may be important in mate selection and the avoidance of inbreeding, in the avoidance of cannibalism, in school or shoal formation, and in the reduction of stress (Ward and Hart 2003). Olfactory determination of kin is particularly well demonstrated in salmonids, including the genera *Oncorhynchus*, *Salmo*, and *Salvelinus* (Griffiths 2003; Ward and Hart 2003).

Kin recognition appears to be based to a large extent on waterborne chemosensory cues associated with major histocompatability complex (MHC) genes (Olsén et al. 1998; Griffiths 2003). MHC genes are important in the immune system of vertebrates, including fishes, and are highly polymorphic such that individuals often differ in their MHC genotypes. The odor of individuals is related to their MHC genes, making kin recognition and recognition of individuals possible. Various hypotheses linking the MHC complex to specific odors have been proposed, although the exact linkage remains uncertain (Rajakaruna et al. 2006). High relatedness increases similarity among MHC genes and allows kin recognition. Among Atlantic Salmon, Arctic Char (*Salvelinus alpinus*), and Brook Trout (*S. fontinalis*), siblings show a preference for those that carry more similar MHC genes compared to siblings with more dissimilar MHC genes (Olsén et al. 2002; Rajakaruna et al. 2006). Overall, there is conflicting evidence on whether kin recognition is an innate or learned behavior, but at least in Arctic Char, some early association with kin is required for kin recognition using MHC to occur (Brown and Brown 1996; Olsén et al. 1998, 2002). The period of association likely occurs after hatching but prior to emergence from the redd (Brown and Brown 1996).

Given the known survivorship value of schooling, a fish could increase its inclusive fitness by joining a school made up of its siblings (Box 10.1) (Quinn and Busack 1985). A pioneering study on Coho Salmon (*Oncorhynchus kisutch*) showed that juvenile fish preferred water from siblings over water from nonsiblings or from other sources. In addition, there was only a slight decrease in preference for water from unfamiliar siblings compared to familiar siblings, indicating that the waterborne cue was not based solely on familiarity (Quinn and Busack 1985). Other studies, including work on Atlantic Salmon parr, have shown a reduction in aggression toward kin compared to nonrelated conspecifics and that the reduced aggression corresponded with better growth under laboratory conditions (Brown and Brown 1996; Griffiths 2003). In addition, kin recognition could play a role in the avoidance of inbreeding (Ward and Hart 2003). Although fishes have the ability to recognize kin and tend to associate with them in laboratory situations where the concentration of chemical cues is relatively great, initial evidence from natural systems generally did not indicate that fish tend to associate more with kin than nonkin. Evidence for the lack of relatedness within fish associations or schools comes from studies on European minnows (*Phoxinus*), a widespread North American minnow, the Common Shiner (*Luxilus cornutus*), and European populations of Threespine Sticklebacks (Dowling and Moore 1986; Griffiths 2003). Because Threespine Sticklebacks remain in the nest for several days after hatching, the potential for schools to consist of closely related fish seems likely. However, in nature, individuals in schools show low or no relatedness (Peuhkuri and Seppä 1998). These earlier studies of kin associations were necessarily limited in power by the techniques available, such as using protein (allozyme) electrophoresis. A more recent study on Brook Trout using microsatellite DNA suggests that fish schools often do show a greater degree of relatedness than that based on chance. At least in Brook Trout, schools seem to be hierarchically

The fitness of an individual is measured by the proportion of the individual's genes present in the subsequent generation and is the basis for the operation of Darwinian natural selection. However, the genetic contribution of an individual is not fully dependent on its own survival to reproductive age and successful breeding and survival of offspring to adulthood. The reason for this is that individuals share some portion of their genotype with siblings, cousins, second cousins, and so on, with the proportion of shared genes decreasing in more distantly related relatives (Hamilton 1964a). This led early population geneticists such as Fisher and Haldane to realize the importance of relatedness in the understanding of the evolution of social behaviors, especially altruism. Hamilton (1964a) proposed a quantity, called inclusive fitness, that "allows for interactions between relatives on one another's fitness." Inclusive fitness provides a genetic answer for the evolution of limited sacrifices for kin by an individual based on the degree of relatedness shown by the kin. In the genetically modeled world of Hamilton (1964a), the benefit to a sibling must minimally average twice the loss to the altruistic individual, the benefit to a half-sib must average at least four times the loss, and the benefit to a cousin must minimally average eight times the loss to the individual. Put even more succinctly, from the standpoint of a genetically modeled world where behaviors are solely controlled genetically, no individual will sacrifice its life for any single individual but will sacrifice it when it can save more than two brothers, four half-brothers, or eight first cousins. Thus, given the ability to recognize kin, an individual fish should modulate its behavior toward an individual relative to the degree of relatedness. For instance, an individual could raise its inclusive fitness by being less aggressive to kin than non-kin or by performing behaviors such as alerting kin to the presence of a predator (Hamilton 1964b; Greenberg et al. 2002).

grouped, including familiar but unrelated fish, fish from the same population, and groups of kin, although more school members were unrelated than were kin. In addition, based on age-size relationships, groups of kin or population members remain together for up to four years (Fraser et al. 2005).

Given the few populations and species that have been studied using modern approaches, the degree that fish schools include groups of kin is still largely unknown and provides an exciting prospect for further work. However, in terms of expectations for the occurrence of highly related fish within a single school, there are trade-offs between the advantages of kin selection and the advantages of associating with less related individuals or, in reproductively active fishes, the avoidance of inbreeding (Griffiths and Armstrong 2001; Ward and Hart 2003). In general, the intensity of resource competition is thought to increase as relatedness among individuals increases, so that negative selection for kin-structured schools might also be occurring and that schools comprised of mixed genotypes would be favored (Griffiths and Armstrong 2001).

Migration

Migratory fishes employ chemical cues to locate streams, spawning sites, or feeding areas. Chemical cues can emanate from physical or biological aspects of the environment, including odors released by conspecifics. Although the use of these chemicals in migration generally does not appear to constitute communication, there are instances where true communication and/or spying could be occurring.

Extensive studies of Pacific and Atlantic salmon species clearly demonstrate that these fishes rely primarily on olfaction to identify their natal rivers and to return to natal spawning sites. In fact Buckland (1880), in reference to Atlantic Salmon homing to streams of the British Isles, said, "When the salmon is coming in from the sea he smells about till he scents the water of his own river." Seventy-one years later, Arthur Hasler and his graduate student

Warren Wisby provided the first experimental evidence for what became known as the olfactory hypothesis when they showed that fishes (in this case Bluntnose Minnow, *Pimephales notatus*) could differentiate between odors of two different streams (Hasler and Wisby 1951; Magnuson and Quinn 2005). This was followed by similar conditioning on Coho Salmon, showing that they could also discriminate between water from the same two streams used in the Bluntnose Minnow study (Hasler 1956). The olfactory hypothesis proposed that (1) streams differ in their chemical characteristics and that these characteristics remain consistent over time, (2) fishes are able to detect these differences, (3) the home-stream odors are learned by young salmon before they migrate to sea and young fishes do not respond to odors of nonnatal streams, and (4) the memory of the odors elicits upstream migration in adults. In a follow-up paper, Wisby and Hasler captured sexually mature Coho Salmon during their upstream migration from two connected streams. They temporarily blocked the nares of approximately one-half of the fish to stop the flow of water over the sensory tissue, and transported both the treated and untreated fish back downstream where they were released below the juncture of the two streams (Figure 10.2). Fish with blocked nares were unable to consistently enter the tributary where they were initially captured, whereas those with functioning nares generally made the correct choice (Wisby and Hasler 1954).

The olfactory hypothesis comprises two principal, but not mutually exclusive, hypotheses regarding the origin of the chemical signal: the imprinting hypothesis and the pheromone hypothesis (Yamamoto and Ueda 2009). The imprinting hypothesis states that fishes learn the chemical odors of their natal stream during a critical period (or periods) in their life cycle. The period when fishes imprint on the home-stream odor seems to be associated with elevated levels of thyroxine. Early studies of hatchery-reared Coho Salmon showed that imprinting occurred during the parr-to-smolt

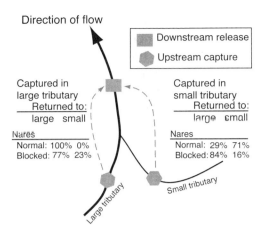

Direction of flow

Downstream release
Upstream capture

Captured in large tributary
Returned to:
large   small

Captured in small tributary
Returned to:
large   small

Nares
Normal: 100% 0%
Blocked: 77% 23%

Nares
Normal: 29% 71%
Blocked: 84% 16%

Large tributary
Small tributary

FIGURE 10.2. The role of olfaction in salmon homing. Coho Salmon with blocked nares (olfactory pits) were less successful in repeating their original choice of streams. Fish were initially captured during their upstream migrations at weirs in the large and small tributaries and then taken back downstream below the confluence and released. Based on Wisby and Hasler (1954).

transformation (PST). However, in wild fish, which experience natural environmental cues and which, depending on individual populations and species, may leave the redd much earlier than the time of PST, imprinting generally occurs much earlier, often at the end of the embryonic development period (Dittman and Quinn 1996). In addition, because peaks in thyroxine occur several other times during the freshwater stage, such as during periods of rapid growth and during the PST, young salmon likely use multiple imprinting on stream odors to recognize sequential waypoints on their return migrations. Once olfactory cues have guided adult salmon back to their natal stream and the approximate section of the stream where they emerged from the redd, they use ecological and behavioral cues such as substratum size, current speed, intrasexual competition, and mate availability to determine the exact spot for the initiation of redd construction (Dittman and Quinn 1996). The chemical cue is contained in the organic fraction and likely produced from a mix of vegetation and soil types in the stream's drainage basin.

Recent studies indicate that Pacific salmon species strongly respond to naturally occurring amino acids in streams (Yamamoto and Ueda 2009). Mature Chum Salmon (*O. keta*), in fact, are attracted to test water that has received an amino acid complement that matches their home-stream water. Most likely, the amino acid complement of streams is influenced by complex biological interactions occurring within the watershed such as microbial biofilms, soil and vegetation types, and litter (Yamamoto and Ueda 2009). Studies have also shown that salmon can be imprinted on artificial organic substances such as morpholine (a heterocyclic amine) so that they can be induced to return to whatever stream is receiving the artificial imprinting substance (Hasler and Scholz 1980).

The pheromone hypothesis proposes that adult migratory fishes are attracted to their home stream by pheromones released by juvenile conspecifics from the same population (Nordeng 1977). In addition to releasing pheromones from the natal stream by juvenile fish, the downstream seaward migration of smolt would, in most species of salmonids, overlap with part of the upstream migration of adults, providing an olfactory trail from coastal waters to the spawning grounds. If true, the pheromone hypothesis would constitute either actual communication or spying (Figure 10.1). Because the odorant produced by juvenile fish could contribute to part of the overall odors of the natal stream, this hypothesis is not mutually exclusive with the imprinting hypothesis. However, it does require that juvenile fish be in the stream system when the mature adults are migrating—something that is not true for some species of salmonids, such as Chum Salmon or Pink Salmon (*O. gorbuscha*) (Yamamoto and Ueda 2009).

Although Pacific and Atlantic salmon species have the ability to recognize kin and conspecifics (see preceding paragraphs), support for the pheromone hypothesis is mixed. In experiments on adult Coho Salmon, fish selected water conditioned by juveniles of their own population when tested against city water, but they did not select water conditioned by juveniles of their own population over water conditioned by juvenile Coho Salmon from another population (Quinn et al. 1983). A follow-up study did show that Coho Salmon can distinguish between populations based on chemosensory cues, so that pheromones could be involved, but not required, for successful homing to natal streams (Quinn and Tolson 1986).

In contrast, there is evidence that Norwegian Arctic Char seem to use pheromones produced by juveniles of their own population to locate their natal stream. Parental fish were removed from the Salangen River and transported to a hatchery in southern Norway with a different water source where eggs were fertilized and the young fish were raised. When fish reached the smolt stage, 96 were released in the marine environment. Subsequently over 1–3 years, 27 fish were recaptured with 63% of the recaptures in the upper Salangen River, including the natal area. The putative olfactory cues were pheromones produced by conspecifics in the natal area and by the natural downstream migration of smolts that coincided with the upstream movement of adults (Nordeng 2009).

Counter to most of the studies on salmonids, the pheromone hypothesis is strongly supported by work on lampreys (Petromyzontidae). Lamprey genera typically include both parasitic and nonparasitic species, with parasitic species generally showing longer migrations than nonparasitic taxa (Hardisty and Potter 1971). In addition, migratory lampreys do not necessarily return to their natal stream. Based initially on studies of Sea Lampreys, returning migrants select spawning rivers and locate spawning sites based primarily on their innate recognition of bile acid–based pheromones released by stream-resident larvae (Bjerselius et al. 2000; Vrieze and Sorensen 2001; Sorensen and Vrieze 2003). In addition, pheromones can act synergistically with natural stream odors. Adults are able to discern the odors at picomolar concentrations representing realistic concentrations in natural stream water and respond by swimming against current flow (rheotaxis). Once Sea Lampreys change from the migratory stage to the mature stage, they no longer respond to the larval pheromone, but do respond to odors released by mature members of the opposite sex.

The pheromone system in lampreys seems to be evolutionarily conserved, with high similarities among tested species in bile acid production and in detection responses by adults. The absence of species-specific differences in the production of bile acid pheromones is not surprising given the similar spawning and nesting requirements among lamprey species (Fine et al. 2004). The lamprey pheromone system has conservation and management applications. In areas such as the Great Lakes where Sea Lampreys are not native, the pheromone could be employed to attract and remove returning adult fish. In conservation efforts, the generalized lamprey pheromones could be used to attract rare native species of lampreys to appropriate spawning sites (Sorensen and Vrieze 2003; Fine et al. 2004).

VISUAL COMMUNICATION

Communication by visual means involves both the evolution of the receptors and the signals. It can include responses to color and color patterns, body shape, and body and fin movements. Given the wide range of colors exhibited by fishes, it is no surprise that fishes are able to see colors. In addition, many fishes are responsive to wavelengths in the ultraviolet (UV) range. Although most of the research on UV reception has involved tropical marine or freshwater fishes, UV reception occurs in juvenile salmonids (with the sensitivity of receptors for UV lost or reduced in adult fishes), in some minnows, and in cyprinodontids and goodeids (Siebeck et al. 2006).

Both scotopic (rod based, dim light) and photopic (cone based, bright light) vision arose early in vertebrate evolution, perhaps more than 540 mya (Collin and Trezise 2006). Fishes differ in their visual pigment systems, ranging from benthic fishes, such as ictalurids, that essentially lack sensitivity to low and midrange wave lengths, and which have high rod retinas that are sensitive in low light conditions (i.e., primarily scotopic vision), to fishes with 4–5 different visual pigments in the cones, resulting in well-developed color sensitivity (photopic vision). Examples of the latter include surface-inhabiting genera such as *Cyprinodon*, *Fundulus*, *Gambusia*, *Poecilia*, and *Xiphophorus* (Levine et al. 1980).

Colors in fishes can be caused by specialized pigments that are generally contained in cells, called chromatophores, but occasionally by free pigments in the body tissues. Common pigments are carotenoids (producing yellows, oranges, and reds) and melanins (producing browns, grays, and blacks). However, because fishes cannot synthesize the pigments or their precursors, pigment-based colors are diet dependent (Price et al. 2008). Structural colors, schematochromes, are colors produced by reflective structures such as layers of guanine, and cells containing such structures are iridophores (Bond 1996). Schematochromes are responsible for short-wavelength colors (blues and violets) and silvery colors (Price et al. 2008). In contrast to other vertebrates, pigment cells of bony fishes are under neuroendocrine control, so that the color signals can be changed rapidly (Moyle and Cech 2004).

The use of visual communication is lessened in visually complex environments, or in deep or turbid water. However, with the exception of Cavefishes (Amblyopsidae), lamprey ammocoetes (which burrow into the substratum), and blind catfishes, all North American fishes have well-developed eyes. Both larval and adult lampreys have lateral photoreceptors on the body that are responsive to light levels but are not likely involved in communication (Deliagina et al. 1995).

As a light beam passes through water, it is altered by scattering and absorption caused by water molecules and by dissolved and suspended matter (Loew and Zhang 2006). The quality of downwelling light in pure water or fresh water without significant amounts of dissolved organic compounds or particulate matter has a transmission maximum centered in the blue range (wavelengths of 450–490 nm). Water with small amounts of chlorophyll would appear green; whereas water with large amounts of tannins, chlorophylls, and lignins, as would be typical of marshes, swamps, and lowland rivers, has transmission maxima in the brownish-yellow, orange to red range. Compared to most marine habitats and clear, tropical lakes, many freshwater habitats are so strongly colored that spectral quality of light can change in only a few meters, resulting in microhabitat differences in color reception and the use of visual signals. Depending on the viewing direction, and thus the path distance of light through water, the quality of light can vary dramatically (Levine et al. 1980; Loew and Zhang 2006; McLennan 2007).

Defense of Territory

Visual signals, including frontal and lateral displays that make an individual look larger than it really is (i.e., inflationary displays), are part of agonistic behavior in many species. Although details of inflationary displays vary within and among taxa, there are some general characteristics, as shown by examples from salmonids, cyprinids, and elassomids.

Juvenile Atlantic Salmon actively defend feeding territories in streams during the day but settle to the bottom at night. Visual aggressive displays include frontal and lateral displays, both of which are accompanied by elevated fins (Keenleyside and Yamamoto 1962). During frontal displays, the more visual paired fins are all fully open, although the less visible dorsal fin is only partially raised. In lateral displays, all fins are fully extended. In addition to frontal and lateral displays, young Atlantic Salmon also charge or flee, nip, and chase. Besides the fin and body posturing, complex color changes occur during aggressive encounters among juvenile Atlantic Salmon. Active, aggressive fish tend to be pale; whereas a nonaggressive, fleeing fish turns pale yellow above the lateral line and gray below, and the parr marks are blurred and there are dark bands on the back.

Aggressive fish have a pronounced vertical band through the eye in contrast to the overall dark eyes of submissive fish. Similar visual displays occur in other salmonid genera such as *Salvelinus* and *Oncorhynchus* (Puckett and Dill 1985; Grant et al. 1989).

Among cavity-nesting minnows, such as Fathead Minnow, a nesting male will confront an intruding male by performing a lateral display in which the fins are all elevated, similar to that described for salmonids. Rival males tend not to try to usurp the nest of obviously robust, dark-colored males with strong lateral banding, but show a greater propensity to attack males that appear weakened. As with salmonids, if the display fails to dissuade the intruder, the defending fish rams or nips it (Unger 1983). Surprisingly, male Fathead Minnows that have nests in close proximity to other males do not show a loss of body weight or apparent robustness over the breeding season in contrast to males that have solitary nests—even though solitary and group males do not leave the nest to feed. Successful male Fathead Minnows that are in close association with other males use a form of deceit in that they maintain a stable body weight and apparent bulk throughout the nesting period by using water to replace weight lost by depletion of energy reserves and muscle mass. Males nesting in isolation do not do this and thus, without the sham bulking with water, look much less robust, even though both group- and solitary-nesting males have equivalent dry weights and lose dry weight at similar rates. Group nesting males also maintain more intense color patterns. Even deceit has its limits, however, and solitary males generally have longer average nesting periods (19–20 days) than group-nesting males (11 days) (Unger 1983).

Among Everglades Pygmy Sunfish (*Elassoma evergladei*), dominant males sometimes defend feeding territories when prey are economically defendable (Rubenstein 1981). In addition to actual nipping, males often use lateral displays that are similar to those described previously. As two males approach each other, the caudal fins are spread, the dorsal fins are erect, pelvic fins are fully spread and held perpendicular to the long axis of the body, and the pectoral fins are used to position the fish by actively vibrating. In addition, the dominant fish may raise and lower the dorsal fin while at the same time alternatively moving the pelvic fins up and down. The lateral displays help to communicate the individual's fighting ability in the absence of actual physical encounters.

Schooling and Aggregation

Visual signals often play an important role in the formation and maintenance of fish schools and aggregations. Fish associations can be formed on the basis of many factors, including familiarity, kinship, species, body size, parasite load, habitat, and diet features, and the choice may be based on chemosensory as well as visual cues (Hoar et al. 2000; Frommen and Bakker 2004; Ward et al. 2007).

Among centrarchids, laboratory experiments show that juvenile Rock Bass (*Ambloplites rupestris*) apparently use vision to recognize conspecifics (Brown and Colgan 1986). As a consequence, vision is likely important in the formation and maintenance of conspecific associations. Vision is also important in shoaling decisions in Threespine Sticklebacks, in addition to chemosensory cues. Based on European studies, nonreproductive Threespine Sticklebacks use their ability to see in the UV range to choose shoals. Threespine Sticklebacks and European Minnows (*Phoxinus phoxinus*) also choose shoals with body sizes most similar to themselves (Griffiths 2003; Modarressie et al. 2006).

Courtship and Reproduction

Where conditions permit, visual communication often has important roles in reproduction. Sexual status and quality of mates can be conveyed by colors or body shapes, and the types of signals include movements involved with initiation of courtship and recognition of the partner's social status (Guthrie and Munz

1993). Visual communication using sexually dimorphic color patterns has been studied intensively in African cichlids and species of *Poecilia*, especially Guppies in the New World tropics (Endler 1992; Dalton et al. 2010). Among species occurring in North American freshwater, studies of visual communication have emphasized Sticklebacks; Bluefin Killifish (*Lucania goodei*); Pupfishes (*Cyprinodon*); and various minnow, sunfish, and darter species (Guthrie and Munz 1993; Foster 1995; Kodric-Brown and Strecker 2001; Fuller and Noa 2010).

The behavior and coloration of Threespine Sticklebacks have been intensively studied since the pioneering work of the German ethologist Niko Tinbergen (1952a, b). Male Threespine Sticklebacks defend a breeding territory, build a tubular nest, and use a combination of coloration and behavioral displays to entice females into the nest for spawning, followed by male parental care of the eggs and yolk-sac larvae and defense of the nest (Tinbergen 1952b; Foster 1994). What has only been understood more recently is the tremendous amount of adaptive variation in the components of the Stickleback courtship display system, as well as other aspects of their biology, in sharp contrast to the earlier view that behavioral components were basically invariant across populations of the same species (Foster 1995). Such variation can occur across small spatial scales, likely in response to changing selective pressures, and is an important reminder of the need to be cautious in making broad generalizations of sensory communication among populations of the same species, and among species within genera and families (Fitzgerald 1993).

Coloration in male anadromous Threespine Sticklebacks typically progresses through four distinct stages during the interval from nest construction, to courting, to spawning, to care of young. During nest construction the male is dull colored, with pale blue eyes, dull red throat, and a medium-gray body, and is similar in appearance to a nonterritorial male (McLennan and McPhail 1989; McLennan

2007). A courting male is bright overall; develops an intense red color across the ventral and lateral surfaces of the head and body; and has bright, cerulean blue eyes. In the latter stages of courtship the male develops a white flush over the body, perhaps as a signal of spawning readiness. Spawning males show a snowy white flush over the whole body so that the red colors are masked. Finally, the nest-tending and nest-guarding male has an increase in melanism, which masks the bright colors and gives the body an overwash of medium to dark gray, except for the blue eyes and the intense red on the throat. Behavioral displays include the zigzag dance, in which the male makes exaggerated movements from side to side during his approach to a female, and dorsal pricking, in which the male presses his dorsal spines into a female's abdomen during backward-swimming thrusts or a female is positioned with only her snout above the male's dorsum so that his spines do not actually prick her (Tinbergen 1952a, b; Foster 1995). Dorsal pricking can be initiated by the male or female.

The intensity of the red throat coloration is, at least in some instances, a visual signal communicating male quality to a female. For example, in laboratory experiments of European Threespine Sticklebacks, parasitized males have reduced color intensity and are not selected by females when given a choice between a healthy male and a parasitized male (Milinski and Bakker 1990). In addition, female preference for a male changes when the formerly healthy male is experimentally infected with parasites. The basis of the female's choice is perhaps based on an indication of a male's ability to care for young in his nest but could also indicate selection for genes carrying parasite resistance. In contrast, an anadromous population of Threespine Sticklebacks from the St. Lawrence Estuary, Quebec, did not show any relationship between male color intensity and the level of parasite infestation, nor between color intensity and aggressive dominance over other males. Females did prefer to mate with the most brightly colored males, but parasites have apparently not played a major role in the evolution of male nuptial coloration in this population (Fitzgerald et al. 1994).

The nature of the visual courtship signals in Threespine Sticklebacks is also influenced by local predation pressure. Recall that some populations of Threespine Sticklebacks include benthic and limnetic forms that differ in body shapes and foraging modes (Chapter 7). Benthic forms often are cannibalistic, forming foraging groups of several or more individuals (Foster 1994). In these populations, male courtship is less conspicuous, the zig-zag dance is absent or minor, and dorsal pricking is more common (Foster 1995). In addition, dorsal pricking is initiated more often by the female in cannibalistic populations, compared to noncannibalistic populations. Thus differences in foraging mode and cannibalistic tendencies affect the nature of visual signals used in courtship behavior.

Water characteristics have a strong effect on the transmission quality of visual signals. Because high levels of tannins filter out short-wavelength light (400 nm; blues and violets) more strongly than long-wavelength light (reds), the outcome is red-shifted horizontal light in tannin-stained habitats. In such an environment, the red nuptial coloration of male Threespine Sticklebacks would be rendered inconspicuous as seen against a reddish background. As predicted by the sensory drive hypothesis, Threespine Stickleback populations in high-tannin environments generally show an intense black nuptial coloration instead of the normal red color, and the color differences are heritable (Reimchen 1989; Scott 2001). The loss of red nuptial coloration and the development of melanistic males in the Chehalis River, Washington, was first attributed to convergence in threat displays between Threespine Sticklebacks and the Olympic Mudminnow (*Novumbra hubbsi*), endemic to the Chehalis River, that were thought to be competing for limited space (Scott and Foster 2000). However, studies in both the Chehalis River drainage, and in numerous lakes on the Queen Charlotte Islands, British Columbia,

now show that the probability of populations with red nuptial coloration decreases as the level of tannins increases (lower transmission of blue light), while the probability of melanistic populations increases (Reimchen 1989; Scott 2001). Although the results seem to support the sensory drive hypothesis, the pattern could also reflect variation in dietary carotenoid pigments, which are often low in tannin-stained water, given that fishes cannot directly synthesize carotenoid pigments but are dependent on their dietary source. If carotenoid pigments are limiting, this could favor the loss of red nuptial coloration (Scott and Foster 2000). The occurrence of melanistic males in water with high transmittance at 400 nm might be due to recent habitat changes, such as clear-cutting resulting in the loss of tannin input into the stream (Scott 2001).

Bluefin Killifish, found in both clear springs and tannin-stained swamps in the southeastern United States, show a similar pattern to Threespine Sticklebacks of how sensory drive can impact visual communication. Males have distinctive red, yellow, blue, red-blue, or yellow-blue anal fins that are part of a visual courtship signal to females. Males with blue, or predominantly blue, anal fins are more abundant in tannin-stained swamps where the transmission of UV-blue light is low, thus causing the horizontal light quality to be red shifted. Under these conditions, red or yellow signals seen against a reddish horizontal background would be less visible. Males with predominantly red or yellow anal fins are more abundant in clear springs, where UV-blue penetration is high. Under these conditions, blue visual signals would be less distinct when seen against a bluish background in contrast to red or yellow signals (Fuller 2001, 2002). Female mating preferences also differ between clear and tannin-stained habitats. Females in clear springs show a general preference for males with red anal fins in contrast to females in tannin-stained swamps, which show a preference for males with blue anal fins (Fuller and Noa 2010). Female preference for males with red or yellow anal fins was much

FIGURE 10.3. Male nuptial coloration in the Redspot Darter (*Etheostoma artesiae*). See also corresponding figure in color insert.

more subtle, although in female choice experiments, red males showed a slightly higher spawning success than yellow males, perhaps indicating an innate preference of females for red color. However, there was no effect of rarity on a female's choice of red or yellow male color morphs (Fuller and Johnson 2009).

The two examples of sensory drive emphasize how visual communication systems, in these cases courtship displays, can vary over space and time. For instance, in habitats occupied by Bluefin Killifish, wet years can increase the influx of tannins into streams, resulting in poorer communication of red visual signals and selecting for males with blue anal fins. In dry years, water would be clearer because of a decrease in incoming organic material, resulting in poorer communication of blue signals and selecting for red or yellow anal fins (Fuller and Noa 2010).

Darter species, especially within the large genera *Etheostoma* and *Nothonotus*, have an amazing array of color patterns—patterns that typically are shown by males and displayed during the breeding season (Figure 10.3) (Page 1983). This strongly suggests that visual signals based on male color patterns are important in reproduction and are a consequence of strong sexual selection. Somewhat surprising given the large number of darter species, there are relatively few studies that have investigated the role of male color patterns in darters and the importance of sexual selection, and even fewer that have tested the hypothesis that male breeding colors are part of behavioral isolating

mechanisms in the evolution of darter taxa (Williams and Mendelson 2010).

Female Rainbow Darters (*Etheostoma caeruleum*) use visual cues to select males, with the most likely cue being male coloration (Fuller 2003). Sexual selection is also suggested in Splendid Darters (*E. barrenense*) and Banded Darters (*E. zonale*), where females use visual cues to select conspecific males when given a choice between conspecific and heterospecific males. Female choice based on male colors also indicates that visual signals are an important aspect of behavioral isolation in these two darter species and perhaps have been a major contributor to the extensive radiation of darters overall (Williams and Mendelson 2010).

Even though sexual selection seems to be responsible for male breeding colors in some darter species, this is not always the case. During the breeding season, male Orangethroat Darters (*E. spectabile*) have orange throats and develop blue and red stripes on body and dorsal fins. However, females do not show any preference for brightly colored males over dull males, nor do they select larger over smaller males. Instead, the male breeding coloration is perhaps related to male-male interactions (Pyron 1995).

The role of visual signals in reproductive behavior is, of course, not limited to color patterns or body size. Females of various species of fishes choose to spawn in nests that already contain eggs, with freshwater examples including darters, minnows, sculpins, sticklebacks, and sunfish (Porter et al. 2002). One way to suggest

FIGURE 10.4. Egg mimics on the dorsal fin (arrow) of the Fantail Darter (*Etheostoma flabellare*). See also corresponding figure in color insert.

the presence of eggs is to have egg-mimicking structures or pigments (Knapp and Sargent 1989). Among darters, structures mimicking eggs develop during the breeding season on the first or second dorsal fins of males in most species within the subgenus *Catonotus*, and on pelvic and pectoral fins in the subgenus *Boleosoma*. Egg-mimicking pigments, but not structures, develop during the spawning season on the pectoral fins of the Striped Darter (*E. virgatum*) (and in the subgenus *Catonotus*) (Page and Bart 1989; Page and Knouft 2000; Porter et al. 2002). Although putative egg-mimicking structures can be present in both sexes, they are most developed in males (Knapp and Sargent 1989). Female Fantail Darters (*E. flabellare*) prefer to spawn with males that have egg mimics (Figure 10.4) over males that had their egg mimics experimentally removed, and female preferences for males with egg mimics have been shown in other species as well (Knapp and Sargent 1989; Porter et al. 2002). Although knobs on fins have been suggested to have other roles, current evidence indicates that the egg mimics are primarily a visual signal by the male suggesting to the female that he has successfully mated and has eggs in his nest (Bart and Page 1991).

ACOUSTIC COMMUNICATION

Fishes produce sounds to locate and choose mates (potentially important in reproductive isolation), and to show submission or aggressive responses and willingness to fight to competitors for territory, mates, or food (Ladich 1997; Amorim 2006). Although the mechanism of purposeful sound production is different in fishes compared to air-breathing vertebrates, both are similar in that all produce various unintentional sounds as part of their daily lives. For fishes, these include hydrodynamic sounds caused by swimming, pneumatic sounds caused by movement of air along the pneumatic duct in physostomous fishes, stridulatory sounds created by friction of mobile bony elements including jaw and pharyngeal teeth, cavitational sounds produced by the negative pressure gradient inside the mouth during suction feeding, stringed sounds caused by vibrations of stretched tendons in the fins, respiratory sounds caused by regular movements of the opercular bones during breathing, and percussive sounds as a consequence of striking the substratum with the body or fins. Although a basic part of daily life, unintentional sounds can also be preempted for use in signaling and communication (Fine et al. 1977; Kasumyan 2008). Fishes can perceive underwater sounds through the lateral line and through vibrations picked up by the air bladder, in those fishes having one. In addition, fishes can receive vibrations through the substratum (Fine et al. 1977; Whang and Janssen 1994).

Worldwide, specialized sound production is known in some 800 species of marine and freshwater fishes from 109 families (Kasumyan 2008). In addition to the possible use of normally unintentional sounds, purposeful sounds are produced by the contraction of specialized sonic muscles that are attached to the wall of the swimbladder in some species and by specialized bony elements used in stridulatory sounds. Fish sounds typically consist of low-frequency pulses that can vary in duration, number, and repetition rate, and higher states of arousal sometimes are accompanied by more rapid pulse rates. High-frequency sounds, more common in courtship than aggression, are typical of otophysans (North American orders: Cypriniformes, Siluriformes, and Characiformes) because of their specialized hearing mechanism, the Weberian apparatus. Sound repertoires are usually limited to one or two distinct sound types, but there are a number of examples of more complex sounds as well (Ladich 1997; Amorim 2006; Phillips and Johnston 2008a).

Although low-frequency sounds can propagate long distances in deep water, such as the open ocean, low-frequency communication is greatly constrained in shallow water. In addition, shallow water often has a high background noise caused by wind and wave action. The highest background noise in streams is where the water's surface is broken; smooth runs, even when swift, are likely quieter than riffles. As a consequence, sensory drive is as important in acoustic signals as in the visual signals described previously. Water depth and ambient noise level are strong selective forces on the evolution of acoustic signaling and reception in freshwater fishes (Lugli and Fine 2003). For example, two freshwater goby species living in swift, rocky Italian streams show their maximum hearing sensitivity in a narrow band around 100 Hz and produce low-frequency sounds with most energy from 70 to 150 Hz, in the range of their maximum hearing sensitivity. A window in the ambient noise spectrum of the streams occupied by the gobies occurs at around 100 Hz at most noisy locations. This "quiet window" in the sound frequency spectrum falls between low-frequency turbulent noise and higher-frequency noise from bubbles associated with breaking water. Consequently, there is a match between the quiet window in ambient noise, the frequencies used in sound production, and the greatest hearing sensitivity (Lugli et al. 2003). A similar match between the frequency of the signal and the optimal hearing frequency occurs in Coosa Bass (*Micropterus coosae*) (Holt and Johnston 2011). Because of the generally high ambient noise and the shallow depth, fish sounds in streams or in the littoral zone of lakes are primarily limited to short-distance transmission on the order of 40–50 cm or less (Lugli and Fine 2003; Phillips and Johnston 2008b).

By far, the most information on sound production and audio communication is for marine fishes. Well-studied freshwater examples include South American and African cichlids, Asian gouramis, and several families of South American catfishes (Fine et al. 1977; Amorim 2006). In North American fresh waters, sound production and sonic communication have been studied in at least 32 species and eight families, including sturgeons, minnows, catfishes, cottids, pupfishes, sunfishes, darters, and drum (Table 10.1). However, this number likely greatly underestimates the actual number of fishes using acoustic communication. For instance, although nine species within the minnow genus *Cyprinella* are shown in Table 10.1—and a total of 19 examined so far, including unpublished data, have been shown to produce purposeful sounds—it is thought that all the species of *Cyprinella* (27–30 species) are sound producers (Phillips and Johnston 2009).

With few exceptions, sounds produced by North American freshwater fishes are made in the context of male-male aggression associated with reproduction—especially during male-male encounters over territory and during courtship and spawning (Table 10.1). Sunfishes (genus *Lepomis*) use sound as part of complex courting and spawning displays. Bluegill (*L. macrochirus*) and Pumpkinseed (*L. gibbosus*) both use a grunting or rasping noise, produced by the grinding of pharyngeal teeth, to mark transitions between aggressive and nonaggressive behaviors. The sounds apparently convey to the female the male's high level of aggression and sexual arousal

TABLE 10.1
Examples of Acoustic Communication in North American Freshwater Fishes

Family/species	Sound	Mechanism	Sex	Purpose	Source
Acipenseridae					
Scaphirhynchus albus	Squeaks, chirps, knocks, moans	Swimbladder?	Both	Courtship and reproduction?	Johnston and Phillips 2003
S. platorynchus	Squeaks, chirps, knocks, moans	Swimbladder?	Both	Courtship and reproduction?	Johnston and Phillips 2003
Cyprinidae					
Clinostomus funduloides	Knocks	Unknown	Male	Male-male aggression	Winn and Stout 1960
Codoma ornata	Short bursts, low-frequency pulses	Unknown	Male	Male-male aggression associated with courtship and reproduction	Johnston and Vives 2003
Cyprinella analostana	Knocks, knock trains, purrs	Unknown	Male	Male-male aggression associated with courtship and spawning	Winn and Stout 1960; Stout 1975
C. callisema	Short pulses, knocks	Unknown	Male	Male-male aggression associated with courtship and spawning	Phillips and Johnston 2009
C. galactura	Knocks, short knocks, pulses	Unknown	Male	Male-male aggression associated with courtship and spawning	Phillips and Johnston 2008a, 2009
C. gibbsi	Short pulses, chirps, rattles	Unknown	Male	Male-male aggression associated with courtship and spawning	Phillips and Johnston 2009
C. lepida	Low-frequency bursts	Unknown	Male	Male-male aggression associated with courtship and spawning	Phillips et al. 2010
C. lutrensis	Series of knocks	Unknown	Female	Courtship	Delco 1960; Stout 1975
C. spiloptera	Knocks	Unknown	Male	Male-male aggression	Winn and Stout 1960
C. trichroistia	Short pulses, chirps, rattles	Unknown	Male	Male-male aggression associated with courtship and spawning	Phillips and Johnston 2009
C. venusta	Single knocks	Unknown	Female	Courtship	Delco 1960; Stout 1975

(continued)

TABLE 10.1 (*continued*)

Family/species	Sound	Mechanism	Sex	Purpose	Source
Margariscus margarita	Knocks	Unknown	Male	Male-male aggression	Winn and Stout 1960
Pimephales notatus	Aggressive knocks, purrs, single knocks	Unknown	Male	Male-male aggression associated with courtship and spawning	Johnston and Johnson 2000
Ictaluridae					
Ameiurus nebulosus	Ratchet	Pectoral stridulation	Unknown	Agonistic behavior	Rigley and Muir 1979
Ictalurus punctatus	Rapid pulses	Pectoral stridulation	Unknown	Courtship, reproduction, and agonistic behavior	Fine et al. 1996, 1997
Cyprinodontidae					
Cyprinodon bifasciatus	Not described	Unknown	Male	Aggression associated with courtship and spawning	Johnson 2000
Cottidae					
Cottus bairdi	Knocks, head slap	Hitting substratum	Male	Courtship and reproduction	Whang and Janssen 1994
C. paulus	Knocks	Unknown	Male	Courtship, reproduction, and agonistic behavior	Kierl and Johnston 2010
Centrarchidae					
Lepomis cyanellus	Grunts	Unknown	Male	Courtship and spawning	Gerald 1971
L. gibbosus	Rasps	Pharyngeal teeth	Male	Courtship and spawning	Ballantyne and Colgan 1978
L. humilis	Grunts	Unknown	Male	Courtship and spawning	Gerald 1971
L. macrochirus	Grunts, rasps	Pharyngeal teeth	Male	Courtship and spawning	Gerald 1971; Ballantyne and Colgan 1978
L. megalotis	Grunts	Unknown	Male	Courtship and spawning	Gerald 1971
L. macrochirus x gibbosus	Rasps	Pharyngeal teeth	Male	Courtship and spawning	Ballantyne and Colgan 1978
L. microlophus	Pops	Jaw snapping	Male	Courtship and spawning	Gerald 1971
L. punctatus	Grunts	Unknown	Male	Courtship and spawning	Gerald 1971

(*continued*)

TABLE 10.1 *(continued)*

Family/species	Sound	Mechanism	Sex	Purpose	Source
Micropterus coosae	Low-frequency pulses	Unknown	Both	Aggressive encounters	Johnston et al. 2008; Holt and Johnston 2011
Percidae					
Etheostoma flabellare	Drums and knocks	Unknown	Male	Male-male aggression associated with reproduction	Speares et al. 2010
E. nigripinne	Drums and knocks	Unknown	Male	Male-male aggression associated with courtship, reproduction, and spawning	Johnston and Johnson 2000
E. crossopterum	Drums and knocks	Unknown	Male	Male-male aggression associated with courtship, reproduction, and spawning	Johnston and Johnson 2000; Speares et al. 2010
E. nigripinne x E. crossopterum	Drums and knocks	Unknown	Male	Male-male aggression associated with courtship, reproduction, and spawning	Johnston and Johnson 2000
Sciaenidae					
Aplodinotus grunniens	Drums	Swimbladder	Male	Courtship and spawning	Schneider and Hasler 1960; Schneider 1962; Fine 1977

(Gerald 1971; Ballantyne and Colgan 1978). Calls of *Lepomis* species vary in duration, pulse rate, duration of sound, and number of sounds produced per unit time (Gerald 1971).

Sound production associated with reproduction, including male-male agonistic displays, is common in the shiner genus *Cyprinella*, as well as related taxa. Male Whitetail Shiner (*C. galactura*) make complex sounds that are associated with aspects of reproduction, including male-male or male-female aggression and male-female courting. Males vocalize during agonistic behaviors, such as dominance establishment, territory defense, and male-male assessment, but generally do not make sounds during highly aggressive actions such as lip locking and circle swims, perhaps because of competing energy demands. Sound production is also important in recruitment of females to a nesting site (*Cyprinella* are crevice spawners) and mate attraction displays, and to a lesser extent prior to and during spawning. Sounds made during courtship had higher dominant frequencies than those made during aggressive behaviors. Because higher-frequency sounds should attenuate more rapidly, calls used in

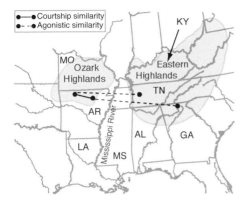

FIGURE 10.5. Geographic variation in courtship and agonistic calls in the Whitetail Shiner (*Cyprinella galactura*). Circles indicate approximate locations of the four populations. The Ozark Highland populations were more similar in courtship signals compared to those in the Eastern Highlands; adjacent populations grouped with more distant populations in agonistic signals Based on Phillips and Johnston (2008b).

courting might be more short range than aggressive calls (Phillips and Johnston 2008a).

Whitetail Shiners have a disjunct distribution occurring west of the Mississippi River in streams of the Ozark Plateau and Ouachita Mountains of Arkansas and Missouri, and in streams east of the Mississippi River including the Cumberland and Tennessee river drainages (Figure 10.5) (Gilbert and Burgess 1980; Mayden 1989). Although all populations share the same acoustic repertoire, calls varied among populations, primarily in pulse rate, pulse duration, and pulse interval. Courtship signals were more stereotyped than agonistic signals, and the Ozark populations showed greater call similarity compared to the more distant Eastern Highlands populations in the Tennessee drainages. Because they are used in mate recognition, courtship signals are more likely under stabilizing selection compared to agonistic signals, which may be more affected by genetic drift (Phillips and Johnston 2008b).

Microhabitats occupied by stream-dwelling darters can vary greatly in the level of background noise. Fringe Darters (*Etheostoma crossopterum*) occupy quiet pools with little or no flow, and the males make drumming and knocking

sounds associated with male-male aggressive encounters during the reproductive season. The dominant frequencies of aggressive sounds vary rather widely from 78 to 496 Hz (mean = 151). Because the ambient background noise shows a quiet area in the range of 90–390 Hz, Fringe Darters are reducing the signal-to-noise ratio of their calls by generally fitting them within the quiet area. Sounds produced by Fringe Darters also include higher-frequency harmonics. Fantail Darters occupy fast-moving riffles, among the noisiest of aquatic habitats, and, as with Fringe Darters, produce drums and knocks during male-male aggression. In the riffles, the quiet window of ambient noise ranges from 150 to 330 Hz and the dominant frequencies of the aggressive sounds range from 172 to 374 Hz, again closely matching the quiet window. However, in contrast to Fringe Darters, Fantail Darter sounds generally lack harmonics, the higher frequencies of which would be masked in their noisier environment. Consequently, differences in signal characteristics of the two darter species, both in the subgenus *Catonotus*, seem to be shaped by the ambient noise spectra of their environments (Speares et al. 2010).

SUMMARY

Communication is an integral part in the daily lives of North American freshwater fishes. Among other roles, communication is involved in establishment and maintenance of dominance relationships and territories, mate attraction and selection, spawning, maintenance of social groupings, migration, and avoidance of predation. As with other organisms, including humans, true communication requires both a signaler and a receiver. The evolution of communication systems is the outcome of competing demands of maximizing signal value versus excessive risk of predation to an individual or to its young. Sensory drive refers to all processes causing biases in the evolution of communication systems, such as the constraints on the evolution of signal production by the nature of the receptor system, the potential for "eavesdropping" on signals by predators or competitors, and the impact of the environment on transmission and signal quality.

North American freshwater fishes use chemical, visual, and acoustic means of communicating, with each system having advantages and disadvantages. Chemical signals received via olfaction or taste are widely used, in part because water is an excellent solvent for chemicals associated with living organisms. However, the transmission and persistence of chemical signals are influenced by the amount and direction of water flow. A specific class of chemical signals, pheromones, are secreted to the environment by the signaler and received by a conspecific in which they elicit a specific adaptive response. Pheromone-signaling systems are particularly important in reproduction, including mate attraction, courtship, and spawning, in a variety of fish groups. Pheromones are also involved in communicating social status and individual identity, the latter being linked to the evolution of MHC genes. Chemical communication, or sometimes spying, is also important in migration.

Visual communication is also widely used in fishes in territory defense, aggressive displays, courtship, spawning, and maintenance of schools and aggregations, and is advantageous because of the quantity of information that can be conveyed. However, visual communication is also strongly influenced by environmental conditions. As light passes through water, it is affected by the distance it travels and by what is in the water. In the highly tannin-stained waters of marshes, swamps, and certain lowland rivers, the shorter wavelengths, including blues and violets, are rapidly attenuated in contrast to the longer wavelengths of browns, yellows, oranges, and reds so that horizontal light is strongly redshifted. One effect of this is that yellows and reds that are used in visual signals in clear water become less visible in the reddish background of tannin-stained water. Because of the strong linkage of visual systems with water quality, the nature of signals and their efficacy can vary both temporally and spatially within species.

Acoustic communication is perhaps the least known communication modality of North American freshwater fishes. Recently, the importance of acoustic communication, especially in behaviors associated with reproduction, has been demonstrated in a diversity of groups but especially in minnows, sunfishes, and darters. Purposeful sounds can be produced by the scraping of bony elements against each other, called stridulatory sounds, or by sounds associated with the swimbladder, either by passage of air through the pneumatic duct or by muscles attached to the swimbladder. In the majority of species, the actual mechanisms of sound production remain unknown. As predicted by the sensory drive hypothesis, the frequency of acoustic signals can be matched to optimum receptor frequency and to quiet areas in the background sound spectrum of the environment.

SUPPLEMENTAL READING

Burnard, D., R. E. Gozlan, and S. W. Griffiths. 2008. The role of pheromones in freshwater fishes. *Journal of Fish Biology* 73:1–16. A current review of pheromone research.

Fine, M. L., H. E. Winn, B. L. Olla. 1977. *Communication in fishes*, 472–518. In How animals communicate. T. A. Sebeok (ed.). Indiana University Press, Bloomington. An excellent background paper for understanding fish communication.

Ladich, F., S. P. Collin, P. Moller, and B. G. Kapoor. 2006. Communication in fishes. Vol. 1 and 2. Science Publishers, Enfield, New Hampshire. A current review of fish communication.

Fuller, R. C., and L. A. Noa. 2010. Female mating preferences, lighting environment, and a test of the sensory bias hypothesis in the Bluefin Killifish. *Animal Behaviour* 80:23–35. A fascinating study of the interrelationship of water quality, nuptial coloration, and the role of genetics and phenotypic plasticity in female mate choice.

Speares, P., D. Holt, and C. Johnston. 2010. The relationship between ambient noise and dominant frequency of vocalizations in two species of darters (Percidae: *Etheostoma*). *Environmental Biology of Fishes* 90:103–10. An intriguing study of how quiet windows in ambient sound levels shape the frequencies of fish vocalizations.

WEB SOURCES

Fish sounds, Dr. Carol Johnston, Auburn University. http://www.ag.auburn.edu/fish/about/facilities/ichthyology/sound-production.

ELEVEN

Interactions in Resource Acquisition I

NICHES, COMPETITION, AND TROPHIC POSITION

CONTENTS

THE OCCURRENCE AND PERSISTENCE of individuals and populations depend on acquiring the spatial or trophic resources, all related to energy acquisition, required by each life-history stage. Potential impacts of changes in resource availability were covered in Part 2,

and the functional morphological adaptations involved in obtaining food or using particular habitats were covered in Part 3. This chapter focuses on possible interactions involved in obtaining trophic or spatial resources, resource linkages in fish assemblages (i.e., food chains and food webs), and trophic positions. The rate of evolution of morphological structures and behaviors associated with feeding is in part influenced by competitive interactions among species (Liow et al. 2011), hence the close association between material in this chapter with chapters in Part 3. For instance, among darters in the genus *Percina*, the rate of evolutionary change in morphological structures associated with feeding is correlated with the number of co-occurring congeners, albeit the evidence does not identify whether greater co-occurrence of *Percina* species is a cause or consequence of the greater rate of morphological change (Carlson et al. 2009).

CHEMICAL ECOLOGY

Historically, the principal means of understanding energy sources, food web relationships, and trophic positions of fishes has been through direct analyses of stomach or intestinal contents (i.e., gut contents), and through visual observations of feeding. Such analyses provide resolution of taxa contributing to the diet and, depending on the study, can be used to determine spatial and/or temporal variation in resource use (Hynes 1950; Windell 1971; Bowen 1996). Over the last several decades, and especially since 1995, a new approach, stable isotope analysis (SIA), has been added to studies of trophic position, linkages, and trophic resource use (Box 11.1) (Peterson and Fry 1987; Dalerum and Angerbjörn 2005; del Rio et al. 2009). SIA and direct approaches to diet, such as gut analysis or visual observations of feeding, are best considered as complementary tools for understanding food habits, trophic position, and food webs. Direct approaches provide information on recently consumed food items, and, because ingested prey are generally killed whether they are eaten and assimilated or not, these direct approaches are perhaps better at showing the impact of predation on ecosystems (Franssen and Gido 2006). In contrast, SIA potentially provides time-integrated dietary information on prey items that are assimilated, if the food is allocated to the tissue being analyzed, and does not require identification of gut contents (Peterson and Fry 1987; Perga and Gerdeaux 2005).

Although direct approaches to dietary analysis show recently consumed items, and SIA provides a long-term view, there is much variability in how long a time period is actually integrated by SIA. The time required for isotopes in fish tissues to reflect changes in isotopes of food sources is related to rates of metabolic replacement and growth. Whereas new growth reflects the isotopic values of current food items, turnover in existing tissues through metabolic replacement can be much slower. The turnover of isotopes in fish tissues is related to the rate of protein turnover, which in turn is related to the specific metabolic rate of each tissue and allometrically related to the body size of the animal. In general, structural elements such as collagen, striated muscle, and red blood cells should have lower rates of isotopic incorporation compared to visceral organs, such as the liver, the digestive tract, and plasma proteins. Also, protein intake generally has a positive effect on protein synthesis and thus the isotopic incorporation rate. The degree that different tissues reflect dietary isotopic ratios is also variable and can be related to lipid content and amino acid composition. Consequently, each tissue potentially differs in the time lag before it reflects dietary isotope values, and each tissue can provide a different measure of dietary isotopes (Tieszen et al. 1983; Peterson and Fry 1987; Dalerum and Angerbjörn 2005; del Rio et al. 2009).

In general, changes in isotopic ratios of tissues require 3–12 months or longer before they reflect changes in diet, depending on species and the tissue being analyzed (Jardine et al. 2003). In Swedish lakes, Roach (*Rutilus rutilus*), Perch (*Perca fluviatilis*), and Bream (*Abramis brama*) all showed an approximate three-month lag before the $\delta^{15}N$ value of consumer tissue was highly correlated with the diet as determined by gut analysis (Persson and Hansson 1999). A one-year feeding experiment using a Canadian population of Broad Whitefish (*Coregonus nasus*) determined how fast sulfur, carbon, and nitrogen isotope ratios responded to a change in isotopic ratios of the diet (Hesslein et al. 1993). Most of the change in SIA following the change in isotope ratios of the diet was due to growth rather than metabolic replacement, and, in contrast to the Swedish study, the half-life (the median residence time in the tissue) for metabolic replacement of sulfur, carbon, and nitrogen isotopes in muscle tissue was greater than one year.

The importance of growth periodicity on tissue changes in stable isotopes is emphasized by a study of European Whitefish (*Coregonus lavaretus*) in Lake Geneva (Perga and Gerdeaux 2005). Growth occurs from March to September and the study tested predictions that skeletal muscle isotopic composition would more closely reflect isotopic values of prey, primarily zooplankton and benthic chironomids, during the growth period, compared to the liver isotopic value, which would reflect current prey and thus show

BOX 11.1 • Stable Isotope Analysis (SIA)

By the 1980s, stable isotope ratios, first employed by geochemists, were starting to be appreciated by ecologists as another important tool in understanding complex ecological processes. Stable isotope ratios can provide clues about sources and transformations (fractionation) of organic matter, which in turn leads to inferences about habitat use, nutrient sources, food webs, and trophic position (Fry and Sherr 1984; Peterson and Fry 1987). Most simply, isotopes are mass unit variations of elements, such as ^{13}C and ^{12}C, and stable isotopes are those that do not decay over time (Jardine et al. 2003). Ecological studies primarily use isotopes of carbon, nitrogen, and sulfur. The common isotope of carbon is ^{12}C (98.89%) and the rarer isotope is ^{13}C (1.11%). The most abundant nitrogen isotope is ^{14}N (99.64%) and the rarer isotope is ^{15}N (0.36%). Sulfur exists in four stable isotopic forms, the most common of which are ^{32}S (95.02%) and ^{34}S (4.21%). Isotope ratios (amounts of heavier/lighter isotopes) are determined from mass spectrometry and are reported relative to established standards in the general format of

$$\delta X = [((^{heavier}X_{sample} / {}^{lighter}X_{sample}) / (^{heavier}X_{standard} / {}^{lighter}X_{standard})) - 1] \times 1000,$$

(Equation 11.1)

where δ refers to $\pm$ differences from known standards and X refers to a particular isotope (Cabana and Rasmussen 1996; Jardine et al. 2003).

Fractionation means a change in the ratio of heavy to light isotopes and usually arises when similar molecules of slightly different mass react at different rates. The resulting isotopic distributions reflect reaction conditions (process information) such as links in a food web. If fractionation is low or minimal, stable isotope distributions provide information on the origin of samples (source information),

such as the origin of terrestrial or aquatic primary production that might be supporting a particular population. No change with increasing trophic level occurs with sulfur, so sulfur isotopes are good indicators of plant or bacterial sources of food and are used to track anthropogenic sulfur in sediments and food webs. Enrichment of the heavier carbon isotope (^{13}C) in trophic level transfer, such as from primary producer to primary consumer tissues, is low (usually in range of one part per thousand [‰] and typically 0.2‰ for freshwater consumers). Consequently, $\delta^{13}C$ also is used for source information (Peterson and Fry 1987; Jardine et al. 2003).

In contrast, the heavier nitrogen isotope (^{15}N) is enriched in consumer tissues from 3 to 5‰ relative to the diet because the lighter isotope is preferentially excreted (Peterson and Fry 1987; del Rio et al. 2005). As a result of the fractionation difference between carbon and nitrogen isotopes, $\delta^{15}N$ is generally used as a time-integrated measure of trophic level among populations of the same consumer species (i.e., process information). There is the potential for local variation in the ^{15}N value, so that it is sometimes important to adjust for between habitat variation of primary producers or other organisms that are in low trophic levels. Overall, $\delta^{15}N$ provides a measure of food-chain length related to bottom-up mass transfer and, in contrast to food chain lengths based on stomach content analyses, does not require detailed taxonomic information on all species present in the food web. In addition, some trophic links based on food habits can be of negligible importance in terms of mass transfer because, although the items are eaten, they may not be assimilated or may be only slightly assimilated (Peterson and Fry 1987; Cabana and Rasmussen 1996; Jardine et al. 2003).

---

greater annual variability. The isotopic composition of prey varied continuously throughout the year and isotopic turnover in European Whitefish tissues, based on the $\delta^{15}N$ value, occurred in liver tissue after one month, but took 4–5 months for muscle tissue. The results supported the prediction that the isotopic ratio of liver tissue would be more variable than for muscle tissue, with carbon and nitrogen isotopes in the

liver having three times the variation of those in muscle tissue. Overall, the results suggested that the $\delta^{13}C$ portion of the diet was not being incorporated into muscle tissue in autumn and winter. European Whitefish muscle $\delta^{15}N$ varied during the winter, but some variation was likely caused by the fish catabolizing tissue proteins and not fully related to changes in the isotopic ratio of the prey. Consequently, dorsal muscle tissue provides a time-integrated image of the isotopic composition of food consumed during the period of active growth (March to September in Lake Geneva). Feeding does occur in the winter but is allocated to basal metabolism and reproduction, emphasizing how isotopic ratios of different consumer tissues are affected by how nutrients are routed in the body once assimilated. The take-home message from this study is that "isotope composition of muscle cannot provide reliable information about the food consumed during seasons when there is no somatic growth" (Perga and Gerdeaux 2005).

Stable isotope analysis can integrate a rather long (well more than one year) feeding period, especially in slow-growing fishes or populations, and the response time can also vary greatly among consumer tissues. It is also important to recognize that isotopic variation can be caused by various factors other than a change in the diet of the focal species. It can be due to a dietary change of the prey species, without any change in the diet of the focal species, or even by changes occurring several trophic levels removed from the focal species (Dalerum and Angerbjörn 2005).

Thus depending on the tissue analyzed, SIA integrates long-term assimilation of nutrients and does not necessarily reflect short-term feeding patterns. The latter would only occur if organisms tended to specialize on particular food types rather than feeding opportunistically. For instance, the general correspondence of SIA and gut content analysis in Northern Pike (*Esox lucius*) inhabiting various boreal lakes indicates that short-term feeding may not be simply opportunistic but that it reflects long-term individual differences so that some individuals are long-term invertebrate specialists whereas others are long-term fish specialists (Beaudoin et al. 1999).

RESOURCE LIMITATION AND COMPETITION

In many instances, access to resources, or their availability, can be strongly influenced by individuals of the same or different species.

Competition occurs when the negative interactions depress individual fitness, population growth rates, or overall population sizes of the two groups (individuals, populations, and species). In freshwater fishes, this can occur primarily through the use, and consequent depletion, of shared resources, or by the aggressive responses of individuals that might keep others from having access to resources. In the first case, resources are limiting by their common use by the same (intraspecific) or different (interspecific) species, without active interactions between the individuals. This is referred to as exploitation competition, or sometimes as consumptive or passive competition, such as the common use of drifting insects by several species of fishes. In contrast, when resources are limited because of access to them, such as when the defense of a feeding or nesting territory by the territory holder keeps other individuals from using the resources, interference competition takes place; this is also known as active competition (Schoener 1974; Pianka 1988; Begon et al. 1996). Interference competition can be subdivided into territorial competition, in which an individual aggressively defends, or shows the potential to defend, some spatial resource against other individuals, and encounter competition, in which interactions among mobile individuals results in some type of harm, such as lost time or energy, theft of food, or injury or death (Schoener 1983). A third type, preemptive competition, is recognized as separate from exploitation and interference, although it shares elements of each.

In preemptive competition, some type of space, such as a nesting site or foraging site, is passively occupied by an individual so that other individuals do not occupy the space unless the occupant leaves. This is similar to exploitation because the shared use of a resource pool (such as all the potential nesting holes), limits overall resource availability, but unlike units of food, units of space can be reused. Because preemptive competition also includes avoidance of use, it fits the category of interference as well (Schoener 1983; Gotelli 2001).

Importantly, although in at least some cases the potential for competition increases with the overlap in resource use, high overlap by itself does not indicate that competition is occurring. The key aspect is whether the resources in question are in limited supply or the organisms show evidence of negative interactions, such as reduced growth, survival, or reproductive success (Wiens 1977; Schoener 1983; Begon et al. 1996). Competition is usually viewed as being symmetrical, with both sides affected negatively; however, as shown later in this chapter, the impact of competition, although still a negative response, is not necessarily equal on both sides; often one group of competitors is affected much more strongly than the other.

The presence and importance of competition in fish assemblages have been assessed through four principal means. Most of the early studies of resource use and the potential for competition occurred via observational field studies of resource partitioning (which cannot prove the existence of competitive interactions, but can show the potential for competition if large overlaps occur in resource use and if resources are limiting), and through so-called "natural experiments," including character displacement, in which different combinations of species occur in relatively close proximity. Manipulative field studies and/or laboratory experiments have also been important in assessing the role of competition in fish assemblages. Although all approaches have their strong and weak points, perhaps the strongest approach is a combination of well-designed observational studies that suggest important hypotheses that can then be tested by properly designed field or laboratory experiments.

Ecological Niches

The niche concept is central to ecology, and niche terminology continues to pervade the ecological literature. For instance, one-third of the papers in the 2010 issues of *Ecology*, one of the leading ecological journals, mentioned the term niche. The principal issue with the niche "concept" is that there is not one concept—in fact many different niche concepts have been described, some appropriate to population studies and others to community-level studies, so that an indication of what concept is being applied is needed (Box 11.2) (Hurlbert 1981). In addition, attempts to synthesize niche concepts have tended to oversimplify usages, further muddling the understanding of niches (James et al. 1984). Finally, most niche concepts theoretically require determination of a species' position (use or fitness) along a large number of resource axes, something that is logistically difficult or impossible (Godsoe 2010).

Recently, the multitude of niche concepts have been distilled down to three: the recess/role niche, the population-persistence niche, and the resource-utilization niche (Schoener 2009). The recess (i.e., cubbyhole)/role niche has its major focus on the environment and corresponds with Grinnell's autecological use of the ecological niche in that a species has a set of behavioral, physiological, and morphological adaptations for exploiting particular food and habitat resources in a community. If such resources are not used, then the recess niche would be empty. This concept also leads to the consideration of ecological equivalents—different species in disparate localities that have the same or similar niche characteristics.

The population-persistence niche has its focus on species' populations rather than the environment. It emphasizes interactions among and within species, especially competition, and corresponds closely with the multidimensional Hutchinsonian niche. With this concept, the niche is a property of the species so that there can be no empty niches. The synecological population-persistence niche offers a quantitative way to characterize niches, with a focus on macrohabitats, and lends itself well to multivariate statistical approaches (Green 1971; Hutchinson 1978).

The resource-utilization niche has its origin in an influential paper by Robert MacArthur

BOX 11.2 • Historical Development of Niche Concepts

Identifying the first appearance of a scientific term is often fraught with challenges and the ecological niche is no exception. The first use of niche in an ecological sense was in 1910 by an experimental geneticist, Roswell H. Johnson, writing about limits to the distribution of ladybugs (Gaffney 1975; Hutchinson 1978). However, Johnson did not use the term again and his first mention of niche apparently escaped notice by his contemporaries (Cox 1980). The strongest candidate for being the father of the ecological niche concept is Joseph Grinnell, a towering figure in vertebrate biology at the University of California in the late nineteenth and early twentieth centuries. In a paper on the distribution of a small bird, the Chestnut-backed Chickadee, Grinnell (1904) essentially described the role of niche overlap and resource competition in limiting species' distributions, although he did not use the term niche. His first use of the term "ecological niche" occurred in a 1913 paper published with H. S. Swarth on distributions of birds and mammals at the limits of their ranges in the San Jacinto mountains of Southern California (Cox 1980). Grinnell further refined his use of ecological niche in two papers published in 1917, stating for instance, "It is of course, axiomatic that no two species regularly established in a single fauna have precisely the same niche relationships" (Grinnell 1917a). It is clear from this paper, and also from Grinnell (1917b), that he was not just equating niche with habitat or place as some recent authors have proposed, but that his view included food, climate, habitat, and so on. Thus, in these and later papers, he used niche to designate the place in an association of biotic and physical factors occupied by a single species. In essence, the Grinnellian niche comprises "the range of values of environmental factors that are necessary and sufficient to allow a species to carry out its life history" and determination of these values requires studies of variation in resource use among populations over the geographical range of the species (James et al. 1984).

Charles Elton, another leader in ecological thinking in the early and mid-twentieth century, first used the term ecological niche in a 1924 paper where he referred to lemmings as occupying the same ecological niche as mice and rabbits of lower latitudes (Cox 1980). Elton viewed the niche as an organism's "place in the biotic environment, its relations to food and enemies" (Elton 1927).

Consequently, his was more of a functional concept; in particular, he tended to emphasize the position of an organism in a food chain, and the niche was not necessarily concerned with competitive exclusion. Other contributors to the niche concept include the Russian ecologist Georgii Gause who formally developed the concept of the competitive exclusion principle, although naturalists, such as Grinnell and Darwin, had already expressed similar thoughts that two species with identical niches could not coexist (Kingsland 1985). Thus the niche became "a unit structure over which species fought for possession" (Gause, in Kingsland 1985). Gause's work stimulated G. E. Hutchinson and later his student Robert MacArthur to look more closely at ecological relationships of organisms to see how resource separation might occur.

Hutchinson (1957b) proposed a formal definition of the niche as being an n-dimensional, geometric space or hypervolume. Within the hypervolume an individual or population could survive indefinitely. In spite of challenges offered by the Hutchinson's hypervolume concept, it greatly stimulated a quantitative approach to studies of the niche (Hurlbert 1981). Hutchinson viewed the preinteractive, or fundamental, niche as the entire set of conditions under which a given organism can live and replace itself, and the interactive, or realized, niche as the actual set of conditions under which an organism exists, as influenced by interactions of other species. As envisioned by Hutchinson, the realized niche volume is always less than the fundamental niche. However, the concept of facilitation (see Chapter 13) alters this view because it shows that neighboring species do not necessarily have a negative impact. When neighboring species have a positive impact, this can lead to the paradox that realized niches are larger than fundamental niches (Bruno et al. 2003).

Actual measurement of niche volume is problematic. The Hutchinsonian niche is basically a synecological concept, determined among species within a single community or several communities, and reflecting the interactions of other species. As such, it would be a realized niche. The Grinnellian niche is an autecological concept, being determined among many different populations of a species with many different realized niche states; it comes closest to approximating the fundamental niche (James et al. 1984).

and Richard Levins (1967). It is also quantitative and multidimensional, but rather than focusing on fitness per se, this concept emphasizes how organisms are actually using resources and the limiting similarity among niches to allow for coexistence of species. The relative use of a particular resource can be plotted along an axis—similar to the Hutchinsonian niche (Box 11.2), but with the response variable being relative use instead of fitness (the latter being much more difficult to measure). This concept is also similar to the Grinnellian niche in terms of what is being measured; the difference is that Grinnell was focusing more on what the environment offered, rather than resources used by an organism (Schoener 2009). Recent approaches to niche modeling using GARP or other algorithms (see Chapter 4) are based primarily on Grinnellian and resource-utilization niche concepts because they quantitatively describe resources used by a species, allow for empty niches, and can be used to examine limiting similarity. They include aspects of the population-persistence niche in that both usually (but certainly not always) focus on the macrohabitat scale.

Compared to vertebrate organisms with extended parental care so that young are able to reach near-adult size before actively competing on their own, most fishes are actively foraging over several orders of magnitude in body size, with a concomitant wide range of food size and kind, and often progress through a range of habitats as they increase in size. Thus niche width, or the breadth of resources used by an organism throughout its life (the ontogenetic niche), would be much greater in fishes compared to most other vertebrates. This would seem to be especially true along the food-size axis, so that resource separation among species along trophic axes should be less common (Werner 1977; Werner and Gilliam 1984). However, the prediction of reduced importance of trophic separation in fishes is not well supported, perhaps because studies of resource partitioning have not addressed the full age/size range of species (Ross 1986).

Resource Partitioning and Other Observational Studies

Resource partitioning refers to how organisms within an ecological community differ in their use of resources and derives from the basic questions of how species coexist and how many species can occur together (Ross 1986; Schoener 1974, 1986). During their initial popularity, the basis for studies of resource partitioning was to determine the limiting bounds placed by interspecific competition on the number of stably coexisting species (Schoener 1974). More recently, the importance of competition as the force behind observed differences in resource use is not accepted *a priori* (Schoener 1986).

Observational studies of resource partitioning in North American freshwater fishes have been based about equally on habitat, food, and temporal (seasonal and diel) variables (Ross 1986). An early quantitative study of spatial resource separation examined habitat use by fishes greater than about 20 mm TL in Lawrence Lake and Three Lakes, two small Michigan lakes. The common fish species showed horizontal (distance from shore) differences in occurrence as well as vertical (water column position) differences. Green Sunfish (*Lepomis cyanellus*) occurred closest to shore in shallow water, followed by Longear Sunfish (*L. megalotis*) and Pumpkinseed (*L. gibbosus*) in slightly deeper water and normally within a half-meter or less from the bottom. Minnow species and Yellow Perch occurred more in the water column but still in water ≤ 1 m deep, and Largemouth Bass (*Micropterus salmoides*) and Bluegill (*L. macrochirus*) occurred farthest from shore in deeper, more open water (Werner et al. 1977). Based on other studies, these species also show separation by food kind and size. For instance, among the centrarchids, Bluegill feed on small invertebrates primarily using suction feeding, Largemouth Bass feed on larger prey using both ram and suction feeding, and Green Sunfish use prey intermediate in size between Largemouth Bass and Bluegill (see Chapter 8). Both Pumpkinseed and Longear Sunfish are benthic specialists, but only Pumpkinseed has enlarged molariform teeth allowing it to feed on heavy-bodied mollusks. There are also temporal differences in feeding, with Black Crappie (*Pomoxis nigromaculatus*) primarily being nocturnal predators on a broad range of prey occurring in the open water column (Werner et al. 1977; Ross 2001).

Observational data also show evidence of strong habitat separation among stream fishes. For instance, eight minnow species in a section of a southeastern, blackwater stream separate along a habitat dimension that primarily reflects their association with aquatic vegetation. Bluenose (*Pteronotropis welaka*) and Flagfin (*P. signipinnis*) shiners have greater association with submerged aquatic vegetation, and Cherryfin (*Lythrurus roseipinnis*) and Blacktail (*Cyprinella venusta*) shiners are less associated with vegetation. Species most similar in terms of association with vegetation show separation along the second resource axis of water column position (Figure 11.1). The only species pair that was essentially the same along the two resource dimensions differed in time of feeding activity, with the Longnose Shiner (*Notropis longirostris*) feeding during the day and the Longjaw Minnow (*N. amplamala*) feeding primarily at night (Baker and Ross 1981). Temporal niche separation could occur if resources, primarily aquatic insect drift in this case, were replenished between day and night, although whether the complementarity in foraging time is actually related to resource limitation is unknown (Ross 1986).

Vertical habitat segregation of fishes seems to be common and, as the previous studies have suggested, potentially could be a result of interactive segregation. However, such differences could also be the consequence of phylogenetic history or other noninteractive factors, such as individual-based optimal foraging. For instance, nine species (all in different genera) in a 37-m section of an Appalachian stream, Coweeta Creek, North Carolina, segregated into a water column guild (five species) and a benthic guild (four species) (Grossman and Freeman 1987).

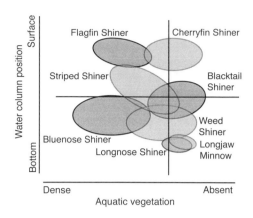

FIGURE 11.1. An estimation of realized niche shapes for eight species of southeastern cyprinids using multiple discriminant analyses. Based on Baker and Ross (1981); used with permission from Ross and Matthews (in press).

Guilds were stable seasonally and annually, and guild members apparently did not differ in habitat use. The differences in microhabitat use between guilds were attributed to common selective pressures or to phylogenetic constraints on the species and not to ongoing biotic interactions, with the possible exception of Warpaint Shiner (*Luxilus coccogenis*) and Rosyside Dace (*Clinostomus funduloides*). However, in a later paper, individual-based optimal foraging models better explained habitat differences between these two species than did interspecific interactions (Grossman et al. 2002).

Water column fishes in Coweeta Creek show high overlap in microhabitat use, suggesting two possibilities—resources might not be limiting or that species might be competing for food (Grossman and Freeman 1987). The importance of competition in structuring microhabitat use of fishes comprising the water column guild of Coweeta Creek were further studied by underwater observations, with the major focus on the introduced Rainbow Trout (*Oncorhynchus mykiss*) and the native Rosyside Dace, the most abundant water column species in the assemblage (Freeman and Grossman 1992). Rosyside Dace spent less time in feeding areas and had lower feeding rates when Rainbow Trout were present compared with when they were absent.

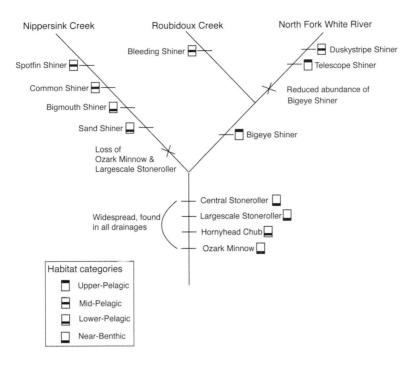

Nippersink Creek
Roubidoux Creek
North Fork White River

Bleeding Shiner
Duskystripe Shiner
Telescope Shiner

Spotfin Shiner
Common Shiner
Bigmouth Shiner
Sand Shiner

Reduced abundance of
Bigeye Shiner

Bigeye Shiner

Loss of
Ozark Minnow &
Largescale Stoneroller

Central Stoneroller
Largescale Stoneroller
Hornyhead Chub
Ozark Minnow

Widespread, found
in all drainages

Habitat categories

Upper-Pelagic
Mid-Pelagic
Lower-Pelagic
Near-Benthic

FIGURE 11.2. The importance of evolutionary history in the interpretation of ecological characters. Illustrated by an area cladogram of three streams and showing predominant minnow species overlain with their water column positions. Modified with permission from Gorman 1992; copyright Leland Stanford University.

The Rainbow Trout were too small to act as predators and, furthermore, rarely were overtly aggressive toward the dace. The presence of Rainbow Trout apparently lessened the value of a particular feeding site for Rosyside Dace, perhaps by depleting local food resources. In contrast, Rosyside Dace entered and left foraging sites independent of other cyprinids and aggression was rare (Freeman and Grossman 1992). The impact of Rosyside Dace on Rainbow Trout was analyzed using a laboratory stream system. In this study, Rosyside Dace did not impact habitat selection by Rainbow Trout (Grossman and Boulé 1991). A priori, the overall results suggest asymmetrical, exploitative competition of Rainbow Trout on Rosyside Dace; however, the presence of exploitative competition would depend on the availability and the foraging value of alternative sites—that is, are Rosyside Dace simply moving away from Rainbow Trout without suffering a loss in energy intake. The fact that both species seem to be foraging at water velocities close to their energetic optima supports this, although whether or not Rosyside Dace suffer a fitness penalty in the presence of Rainbow Trout (thus indicating competition) has not been tested (Freeman and Grossman 1992).

Vertical habitat segregation is also apparent in a suite of minnow species in two streams of the Ozark Highlands, Roubidoux Creek and North Fork White River (see Chapter 3; Figure 3.5) and one in northern Illinois, Nippersink Creek (Gorman 1992). However, in addition to field observations, Gorman also included laboratory studies of habitat preferences and interactive habitat, and examined historical influences on ecological patterns using phylogenetic relationships of the minnows and an area cladogram showing relationships of the streams the minnows inhabited (Figure 11.2). Four species (Central Stoneroller, *Campostoma anomalum*; Largescale Stoneroller, *C. oligolepis*; Hornyhead Chub, *Nocomis biguttatus*; and Ozark Minnow, *Notropis nubilus*) occur in all three streams and, based on the area cladogram, composed part of the ancestral fauna. All of these species are bottom or near-bottom inhabitants that shift little in vertical water column position in the presence of other species. In contrast, species occupying the middle and upper regions of the water column are more recent additions to the fauna and show greater evidence of interspecific interactions. This suggested to Gorman (1992) that "niche partitioning involving mid-pelagic species has not yet reached an evolutionary equilibrium." The near absence of Bigeye Shiner (*Notropis boops*) in the North Fork is perhaps due to its replacement by the Telescope Shiner (*N. telescopus*), another large-eyed pelagic minnow. However, resolution of this question would require knowing if the Bigeye Shiner occurred in the drainage before the arrival of the Telescope Shiner. In contrast to purely observational studies, studies such as Gorman (1992) allow a fuller understanding of the differences in resource use among co-existing species, in this case showing that ecological differences among the benthic guild are not due to contemporary interactions, but that ecological differences among the more recent water column guild could be due to interactive segregation.

Most studies of resource overlap among freshwater fishes have focused on juvenile and adult life-history stages, yet intra- and interspecific interactions among early life-history stages (larvae and postlarvae) are potentially critical (Ross 1986). In arid, western rivers, such as the Rio Grande, habitats for larval fishes can be ephemeral, and reproduction is often tied closely to the hydrograph. Larval and juvenile stages of six fish species in a section of the Rio Grande in New Mexico showed high habitat overlap, occurring primarily in slow or no velocity habitats (Pease et al. 2006). These habitats occurred during high discharge, when backwater areas were inundated, or during low discharge, when flow was greatly reduced in the main channel. Even though they used similar, low-velocity habitats, species partitioned habitats temporally based primarily on differences in peak spawning periods related to discharge patterns. For instance,

White Sucker (*Catostomus commersonii*) initiated spawning in early April at generally higher river discharge compared to Red Shiner (*Cyprinella lutrensis*), which began spawning near mid-June at low river discharge. Temporal separation of larvae of native and nonindigenous fishes is also common in other southwestern streams (Gido and Propst 1999). In terms of trophic separation of fishes in the Rio Grande, there were no obvious differences in carbon and nitrogen isotope ratios among larval and juvenile stages of the six studied species (four cyprinids and two catostomids), and carbon isotopes for larvae were nearly identical to those obtained from adult fishes. Carbon isotope ratios indicated that 80% of dietary carbon for larval and juvenile fishes was from benthic algae and 20% from emergent macrophytes. The six species also overlapped in nitrogen ratios, indicating that they were all feeding at approximately the same trophic level and thus supporting the idea of general overlap in trophic position. Larvae and early juveniles of all six species feed on microinvertebrates, including species of rotifers, copepods, and cladocerans; however, without supporting studies of actual food items, overlap in trophic position and in carbon sources does not necessarily imply overlap in how the carbon is packaged (i.e., the actual food taxa) (Pease et al. 2006).

Observational field studies addressing interactive segregation among fishes continued to be prevalent in the ecological literature through the 1980s and mid-1990s. Of 37 papers reviewed by Matthews (1998) that dealt with habitat and/or food segregation of North American freshwater fishes, 32 (86%) found evidence of segregation and 14 (44%) of these studies inferred that competition was involved. However, there has been a trend to support field studies with historical information and/or manipulative field or laboratory studies, as some of the examples in this section have already shown.

A problem with observational studies, relative to the question of competition, is their usual lack of a clear null hypothesis from which to judge observed differences among cohabiting taxa in such things as resource use. In non-experimental research, data are collected and evaluated for their consistency with specific hypotheses or with the operation of a specific causal process, but without the knowledge of what values might be attributed to similar data in the absence of these causal processes. This often results in a nonnull hypothesis being the tested hypothesis. Preferably, null models (models generating the distribution of values for the variable of interest in the absence of a putative causal process) should be used as a first step in evaluating nonexperimental evidence (Connor and Simberloff 1986). However, null models are not without their own problems, especially because the way a null model is constructed can control the outcome, and a profuse literature has formed around this issue (see review by Gotelli and Graves 1996).

Opportunistic Distributions and "Natural Experiments"

"Natural experiments"—that is, the comparison of nonmanipulated areas—are offered in situations where a species occurs in different combinations with other species or resources. Although such opportunistic, observational studies usually lack suitable controls, as well as having other problems with experimental design (Hairston 1989; Underwood 1998), they often can suggest the presence of competition and lead to more tightly focused, experimental tests. Interpretations of such opportunistic distributions are strengthened if multiple sympatric and allopatric sites are available, a situation that is typically rare.

In a broad sense, "natural experiments" include niche shifts and character displacement, where niche shifts allow organisms to rapidly alter their trophic resource use or reproductive behavior in response to the presence or absence of heterospecifics, and character displacement refers to the reduction of overlap in resource use or in reproductive phenotypes between species through the process of natural selection (Brown and Wilson 1956; Werner and Hall 1976;

Pfennig and Pfennig 2009). Phenotypically variable traits that remain under persistent, strong selection can lose their variability and become genetically fixed, so that phenotypic shifts can lead to evolutionary divergence resulting in character displacement (Pfennig and Pfennig 2009).

Niche Shifts

The spread of Red Shiner into watersheds in Arizona and New Mexico where it is not native has coincided with a dramatic decline of Spikedace (*Meda fulgida*), a small minnow endemic to the Gila River basin of Arizona and New Mexico (Colorado River drainage). Two hypotheses have been proposed for the rapid decline of Spikedace—displacement by competition with Red Shiner or replacement of Spikedace by Red Shiner following major environmental alteration (Douglas et al. 1994). By studying habitat use in allopatric and sympatric populations, Douglas et al. demonstrated that Spikedace shifts its habitat use where it co-occurs with Red Shiner. Importantly, the initial part of the study was a rigorous comparison of habitat availability in the sympatric and allopatric sites, thus reducing the chance that any observed differences in habitat use were simply caused by differences in habitat availability. In allopatry, both species seem to be selecting less common but similar habitats that are significantly shallower and have slower current flow and finer substrata than the average habitats available. When the populations of the two species are in sympatry, Red Shiners do not change their microhabitat but Spikedace are displaced, apparently through interference competition, into deeper microhabitats with swifter currents and larger particle sizes.

The hypothesis of interference competition is also supported by interactions of the Tessellated Darter (*Etheostoma olmstedi*), native to Atlantic drainages, including the Susquehanna River of Pennsylvania, and the Banded Darter (*E. zonale*), which has been introduced from the Lake Michigan and upper Mississippi watersheds. The Banded Darter is more of a habitat generalist compared to the Tessellated Darter, and excludes the Tessellated Darter from riffle and run habitats, restricting it to shallow pools and stream margins (Van Snik Gray et al. 2005). By comparing three sites where the Tessellated Darter did not co-occur with the Banded Darter, and one site where both occurred, and after controlling for habitat availability, Van Snik Gray et al. showed that in allopatry the Tessellated Darter occurred in significantly deeper microhabitats, with higher current speeds and larger particle sizes than when in sympatry. Tessellated Darters also showed niche compression in sympatry with the Banded Darter. Results of the field study corroborated an earlier laboratory study of habitat selection. In addition, the laboratory study supported the hypothesis of interference competition, showing that Banded Darters performed agonistic behaviors to Tessellated Darters but that the latter did not show aggressive responses toward Banded Darters (Van Snik Gray and Stauffer 2001).

Character Displacement

Studies of character displacement and release take a similar approach to natural experiments but with an evolutionary rather than ecological time scale. Ecological character displacement occurs in sympatric species through genetically controlled, phenotypic divergence (Schluter and McPhail 1992; Schluter 2000; Pfennig and Pfennig 2009). Ecological character release is a special case of character displacement involving a shift in characters in allopatric compared with sympatric populations caused by the absence of other restricting species (Robinson and Wilson 1994; Robinson et al. 2000). Both displacement and release generally invoke competition as the driving force so that the distinction between the two is simply the historical pattern of events. Although it is also true that most studies of character displacement and release are unable to provide independent evidence that competition is responsible for the shifts in phenotype, the current consensus is that competition is most often the driving factor (Robinson and Wilson 1994; Schluter 2000; Pfennig and Pfennig 2009). In addition, character release and the associated trophic polymorphism are overwhelmingly

more common in species-poor, relatively recent, high-diversity (i.e., large, deep lakes) environments, with 94% of the examples found above 39° N latitude, not coincidentally the maximum southern extent of the Wisconsinan ice sheet (Griffiths 1994; Robinson and Schluter 2000).

Examples of ecological character displacement or release involving benthic and limnetic forms are particularly well demonstrated among populations of Threespine Sticklebacks (*Gasterosteus aculeatus* complex, Gasterosteidae) and Sunfishes (*Lepomis* spp., Centrarchidae) (Schluter and McPhail 1992; Robinson et al. 1993). Apparent character release occurs in Arctic Char (*Salvelinus alpinus*), with some morphotypes differing on the basis of body depth and size (Griffiths 1994).

At least six lakes in southwestern British Columbia, along the Strait of Georgia, have species pairs of the Threespine Stickleback complex, with one member of a pair primarily benthic and the other limnetic (see also Chapter 7) (Schluter and McPhail 1992; Rundle et al. 2000; Baker et al. 2005). In these, as in other Stickleback populations in North America, colonization of freshwater habitats was by the marine anadromous form over the last 12,000–20,000 years as the Cordilleran ice sheet began to retreat from northwestern North America (Bell and Foster 1994; Clague and James 2002; Bell et al. 2004). Mitochondrial DNA analysis indicates that neither limnetic nor benthic types form a single lineage among lakes. Instead, relationships of species pairs are closer within lakes, indicating that species pairs apparently evolved independently on a lake-by-lake basis from the ancestral anadromous stock (Taylor and McPhail 1999; Wund et al. 2008). Limnetic sticklebacks obtain most of their prey from the plankton, except during the breeding season when males move into the littoral zone of lakes to establish nests on the bottom and attract females. The larger and deeper-bodied benthic forms, which also have fewer and shorter gill rakers, occur in structurally complex habitats and obtain most of their food from benthic invertebrates (Bentzen and McPhail 1984; Wootton 1994). In lakes with only a single species or ecotype, resource use tends to be more generalized over benthic and planktonic habitats. Overall, the results strongly suggest that the divergence of benthic and limnetic forms of Sticklebacks occurred via resource competition, a finding supported experimentally by Schluter (1994) and Bolnick (2004) (see the section that follows).

In the Adirondack region of the northeastern United States, numerous lakes were formed as Pleistocene glaciers withdrew some 17,000 years ago (Dyke et al. 2002). Some lakes were colonized by both Bluegill and Pumpkinseed, but others were colonized only by Pumpkinseed. Where they co-occur, Bluegill occupy the water column and feed primarily on planktonic prey, and Pumpkinseed occupy the benthic littoral zone and feed mostly on benthic invertebrates (see also Chapter 7) (Robinson et al. 1993, 2000). In lakes where Bluegill are absent, Pumpkinseed show character release, retaining the benthic form but also having a limnetic form that acts like a Bluegill—occurring in the water column and feeding primarily on plankton. The role of competition-driven, disruptive selection is suggested by better condition and faster growth rates of the more extreme limnetic forms of Pumpkinseed compared to fish with intermediate body forms (Robinson et al. 1996).

Arctic Char, found at high latitudes in northern Europe and in northern North America, show various polymorphisms in body shape and size, growth rate, age and size at sexual maturity, body coloration, and food and habitat use. At least some of these polymorphisms are thought to be caused by reduced interspecific competition (i.e., character release) in concert with intense intraspecific competition (Griffiths 1994; Power et al. 2009). Different phenotypes generally are found in relatively young systems that offer discontinuous foraging niches and low potential for interspecific competition. Widely documented in Europe, only five instances of sympatric morphs of Arctic Char have been studied in North America, although there are likely numerous systems where polymorphisms exist. Lake Aigneau, a moderately deep lake in northern Quebec, supports two phenotypes of Arctic Char, but is somewhat unusual for lakes with multiple Arctic Char phenotypes in having a relatively diverse fish fauna (10 species, including three salmonids) (Power et al. 2009). There are two size groups of Arctic Char: a small size group with a modal fork length of 210–220 mm having terete bodies and residual parr marks, and a large size group with a modal fork length of 600–640 mm, having deeper, more robust bodies; large heads; and dorsal humps (Figure 11.3A). Fish in the two size groups show almost no overlap in occupied depths during the summer, with fish in the smaller size group at depths ≤ 5m in the littoral zone of Lake Aigneau, and those in the larger size group typically at depths > 15 m in the profundal zone. The two groups also differ in age at maturity, with 50% maturity occurring by age 4 in the littoral zone fish and age 11 in the profundal fish, and in longevity, with littoral zone fish reaching six years and profundal fish to 21 years. Minimal interbreeding and genetic divergence, as determined by mtDNA haplotypes and microsatellites, characterize the two size groups. Not surprisingly given the different habitats, trophic niches also differ. Littoral-zone fish consumed a summer diet dominated by chironomid larvae and pupae whereas profundal fish had empty, shrunken stomachs; no evidence of recent feeding activity; and depleted tissue-nitrogen levels—all of which suggest summer fasting. Stable isotope analysis shows different isotopic ratios for the two groups, with profundal fish averaging one trophic level higher than littoral fish (Figure 11.3B). Further, the isotope ratios indicate heavy reliance by the profundal Arctic Char on the littoral Arctic Char. Overall, the large, profundal fish seem to be fasting during the summer and then feeding on the small morph of Arctic Char in the winter when ice drives the small fish into winter refugia in the limited area of deep water, providing the large fish with an abundant food supply. The pattern is likely driven by intense exploitative competition for available summer resources, favoring the adoption of winter feeding by the large fish. The genetic differences support the possibility

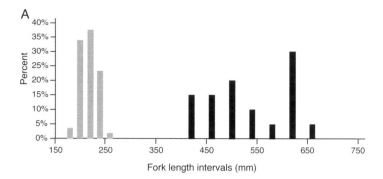

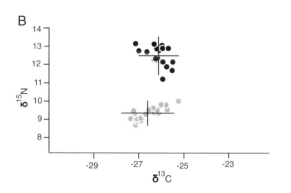

FIGURE 11.3. Character release and intraspecific competition. A. Length frequency distributions (fork length) of littoral (light gray) and profundal (black) Arctic Char (*Salvelinus alpinus*). Littoral fish are grouped in 30 mm size intervals; profundal fish in 50 mm size intervals. B. Trophic separation based on stable isotopes of carbon and nitrogen of littoral (light gray) and profundal (black) Arctic Char. Error bars are ± 2 SE. Based on Power et al. (2009).

of different allopatric origins for littoral and pro-fundal fish, with more recent colonization by the latter, but the evidence is not complete enough to eliminate the competing hypothesis of ecologically driven sympatric divergence. The Lake Aigneau Arctic Char are distinct from most polymorphic forms of Arctic Char—normally the small form occurs in the energetically limited habitat of deep water and the larger form occurs in littoral or pelagic habitats (Power et al. 2009).

Field and Mesocosm Experiments

Well-designed field or mesocosm experiments are not without their own problems, but they generally are considered to provide the best evidence for competition (Schoener 1983; Connell 1983; Hurlbert 1984; Eberhardt and Thomas 1991). In addition to the need for experimental arenas to closely match major attributes of natural habitats, other key points for experimental studies of competition include the need for proper controls so that only the effects of an added competitor are being measured and not potential impacts from capture and handling.

Experiments also need to have proper and sufficient replication so that any effects from a competitor are not masked or attributed to differences among the sites used for controls and treatments—in fact, "the need for replication in any field experiment is paramount" (Underwood 1986). The same need for replication applies to mesocosm experiments, laboratory streams, and so on. A still too common problem of ecological experiments is pseudoreplication—the result of confusing experimental units and evaluation units, or the case of not replicating treatments (the experimental units), although samples (evaluation units) might be replicated, and then testing for experimental effects using inferential statistics. Pseudoreplication is a problem of the experimental design in combination with using an inappropriate statistical analysis for testing the hypothesis of interest. Finally, the assignment of replicates to treatments and controls must be randomized so that potential biases are avoided (Hurlbert 1984, 2009; Fausch 1998). Basically, experimental designs used in competition experiments fall into two groups, additive and substitutive (Box 11.3).

BOX 11.3 • Design of Competition Experiments

Competition experiments, as with all experiments in ecology, are only as meaningful as their design, analysis, and interpretation, which are, of course, based on the hypothesis to be tested. In addition, successful field, mesocosm, or laboratory experiments require a strong measure of good judgment, sound biological insight of the subject organisms and their habitats, and a solid dose of artistry as well (Hurlbert 1984, 2009; Underwood 1997). Other useful references on ecological experiments include Hairston (1989), Underwood (1990), and Fausch (1998).

The two basic questions concerning interspecific competition—does competition occur when two species are placed together, and is the level of this competition greater than just adding the same number of a single species—are most simply tested by two different types of experiments (see part A of the following figure). Additive designs are useful in testing for the presence of competition, especially in the situation of competition between native and nonnative species, or between potential competitors that differ greatly in size. In this design, the only change is the addition of equal numbers of the two species to the test habitat (stream, pond, enclosure, etc.); intraspecific competition is held constant. However, because the number of individuals is greater in the treatment than in the control groups, this design cannot differentiate between intra- and interspecific competition. Substitutive designs (B) maintain the same number of individuals across the treatment and control groups, thus allowing the level of interspecific competition to be adjusted relative to the level of intraspecific competition. For instance, this design would be used where competition is known to occur between native species having similar sizes, habitats, and food use. Ideally, the two designs would be combined so that the experiment would test both the existence and the relative strength of interspecific competition as well as intraspecific competition (see part C below) (Underwood 1978; Fausch 1998).

A. Test for presence of competition

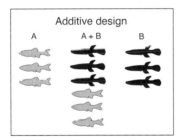

B. Test level of interspecific vs. intraspecific competition

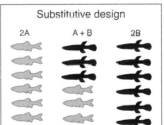

C. Test for presence of competition and strength of interspecific relative to intraspecific competition

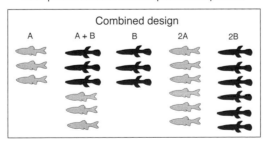

Experimental designs in the study of competition showing the proportions of individuals of each of two species that might be used in additive, substitutive, and combined designs. The number of individuals are, of course, chosen by the experimenter. Based on Underwood (1978, 1986), Connell (1983), and Fausch (1998).

Interspecific Competition

An early additive experiment testing the occurrence of competition used juveniles of three sunfish species (*Lepomis*) to test for niche shifts in the presence and absence of interspecific competition, while keeping intraspecific competition relatively constant (Werner and Hall 1976). The experiment was conducted in four small ponds that were as identical as possible. One pond received 900 individuals each of Bluegill, Green Sunfish, and Pumpkinseed for a total of 2700 fishes. In the three other ponds, 900 fish of each species were stocked alone. The principal response variables were growth (changes in body weight over the experiment), prey size, and habitat use. An increase in growth rate and/or breadth of habitat use in allopatry versus syntopy would indicate a release from interspecific competition. All species showed increased weight gain and use of larger prey in allopatry (measures that relate to overall fitness), but the level of competitive release was greatest for Bluegill, which showed more than a 200% increase in the size of prey it consumed (Figure 11.4A). In allopatry, the three species used similar foraging habitats, as determined from the nature of prey in stomach content samples, with all three species getting $\geq$ 40% (dry weight) of their diets from vegetated areas. However, Bluegill and Pumpkinseed shifted their habitat use when all three species were together (Figure 11.4B). When together, Green Sunfish continued to feed on invertebrates characteristic of vegetation, but Bluegill shifted primarily to open water prey, and Pumpkinseed shifted to benthic prey. The principal conclusions were that sunfishes show considerable niche flexibility, that such flexibility would be adaptive in allowing fish to respond to seasonal changes in resource availability, and that such niche shifts suggest the importance of competition in structuring natural assemblages that include these species. Although aspects of the experimental design were criticized, but surprisingly not for the total absence of replication, the overall conclusions have been supported by subsequent studies (Maiorana 1977; Werner and Hall 1977a).

The apparent asymmetric competition between Bluegill and Green Sunfish was further examined with a substitutive, but still unreplicated, design by establishing three approximately 50 m³ enclosures in the vegetated zone of a small pond. Overall densities were kept the same, with one enclosure receiving 250 individuals of each species, one receiving 500 Bluegill, and one receiving 500 Green Sunfish (Werner and Hall 1977b). Although both Green Sunfish and Bluegill favored vegetation-dwelling prey, the more territorial and aggressive Green Sunfish was able to force Bluegill to shift to alternative habitats offering different prey spectra. There was little effect of Bluegill on the growth in length of Green Sunfish, but Green Sunfish had a strong impact on growth of Bluegill, thus confirming the strong asymmetry in competition (most likely interference competition) favoring Green Sunfish in vegetation. Bluegill showed greater niche flexibility and were able to forage efficiently on smaller prey typical of the open water column.

A later study tested the hypothesis that movement of juvenile Bluegill and Pumpkinseed from open water to vegetation, which occurs as a consequence of predator avoidance, resulted in competition among small fishes in the vegetation (Mittelbach 1988). Twelve cages arrayed along the vegetated littoral zone of a lake were stocked with two juvenile Pumpkinseed and varying numbers of Bluegill, allowing for two replicates. Growth of both species declined linearly in response to increased fish density in the cages, indicating that the concentration of small fishes in the vegetated littoral zone can lead to competition for food. In addition, increases in fish density resulted in decreases in invertebrate body length and decreases in the density of large invertebrate prey in the cages. Further, the density of large invertebrates was related positively to growth rates of both sunfish species. The mode of competition among these small sunfish was most likely exploitative (passive) (Mittelbach 1988).

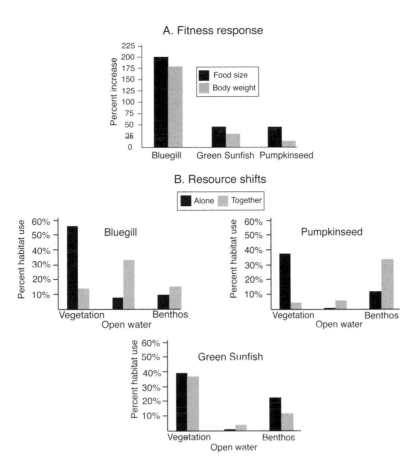

FIGURE 11.4. Experimental tests of competition in juvenile sunfish species.
A. Measures of competitive release, shown by percent weight change and percent change in prey size, of Bluegill (*Lepomis macrochirus*), Green Sunfish (*L. cyanellus*), and Pumpkinseed (*L. gibbosus*) when raised together or in isolation.
B. Directions of habitat shifts of the three species when grown together or separately. Habitat use was determined from the nature of prey consumed; percentages for each species do not sum to 100% because prey not typical of one of the three foraging habitats are not included. Based on Werner and Hall (1976).

Even though competition is considered a negatively symmetrical response, the impact can vary greatly between pairs of competing species, as shown by the interaction between Green Sunfish and Bluegill. The strength of competition can also be unequal between size groups or individuals of the same or different species, and the direction of the impact can change with life-history stage. Asymmetry in both inter- and intraspecific competition has been demonstrated in studies of adult-sized Mottled Sculpin (*Cottus bairdi*), Kanawha Sculpin (*C. kanawhae*), and Fantail Darter (*Etheostoma flabellare*), species that are common riffle inhabitants of many southeastern streams and co-occur in some. Using 16 replicated, outdoor experimental streams, Mottled Sculpin showed the effects of intraspecific competition with decreases in growth, condition, and survivorship when their density was increased from 5 to 10 fish per stream. Interspecific competition showed the same negative outcomes when five Mottled Sculpin were placed together with four Kanawha Sculpin, a somewhat larger species. Surprisingly, when five Mottled Sculpin were placed with five Fantail Darter, the Mottled Sculpin showed negative responses in survival, body mass, and total length that were stronger than their response to Kanawha Sculpin, although the actual mechanism of competition (active versus passive) is not known. This finding is counter to the generally held view that competition is strongest between closely related species and the conclusions of the study are particularly strong given the robust experimental design (Resetarits 1995).

The level of competition and even the type of interaction between Mottled Sculpin and Fantail Darter change with life-history stage. Based on findings from 20 replicated, outdoor experimental streams, juvenile Fantail Darter and Mottled Sculpin compete at densities that approximate natural conditions, although the nature of the response differs between the species. Juvenile Fantail Darter had a negative effect on growth of juvenile Mottled Sculpin, but did not affect sculpin survival or relative condition (determined by the final individual body mass compared to predicted body mass of a wild fish of the same length). In contrast, juvenile and adult Mottled Sculpin did not affect survival or growth of juvenile Fantail Darter, and juvenile sculpin only had a slight negative effect on the relative condition of darters. Adult Fantail Darter facilitate survival of juvenile Mottled Sculpin, although they had a negative effect on final total length and body mass. The reason for the increase in survival is not known. These results emphasize the natural complexity of species interactions. Overall, the impact of competition between Fantail Darter and Mottled Sculpin is asymmetrical—Fantail Darter impact Mottled Sculpin much more than the reverse, even though the darters are smaller than the sculpin. In addition, the intensity of competition and even the nature of the response (i.e., competition versus facilitation) change depending on the response variables (Resetarits 1997).

Competitive interactions within an assemblage composed of native and nonindigenous salmonids of Brook Trout (*Salvelinus fontinalis*), Brown Trout (*Salmo trutta*), and Coho Salmon (*Oncorhynchus kisutch*) were studied in natural streams flowing into Lake Michigan, and in a laboratory stream (Fausch and White 1986). In syntopy, Coho Salmon (introduced from western North America) could defend profitable foraging sites against Brook Trout (native to Michigan streams) and Brown Trout (introduced from Europe). Coho Salmon emerged earlier than Brown Trout and were 7–28% larger than Brown Trout throughout the first growing season. Coho Salmon also emerged earlier than Brook Trout and again were larger, although sympatric populations of Coho Salmon and Brook Trout were uncommon in the tributaries, likely because Brook Trout were being displaced by Brown Trout (Fausch and White 1981, 1986). In the experimental stream, Brook Trout, in contrast to observations in natural streams, could displace Brown Trout from the most profitable foraging sites. Under allotopic conditions, both trout species increased their breadth of foraging sites, including those sites with the highest profitability, but Coho Salmon, the dominant competitor, showed little change in position. Shifts in habitat use under syntopy were due to aggressive behavior of the dominant fish, thus making this an example of asymmetrical interference competition (Fausch and White 1986). In fact, asymmetry is the general rule among competitive interactions (Schoener 1983).

The impact of introduced species of Pacific salmon, and especially Coho Salmon, as well as native fishes on native populations of Atlantic Salmon (*Salmo salar*) along the Atlantic Coast have been addressed experimentally in a number of studies. However, many studies have lacked sufficient replication, making the results difficult to evaluate. Of studies suitably replicated, there seems to be sparse evidence of competitive effects at any spatial or temporal scale. The most likely competitor of juvenile Atlantic Salmon would be juvenile Coho Salmon, given their size advantage (earlier emergence and greater size at hatching) and their innate aggressiveness (Fausch 1998).

The role of interspecific competition in driving vertical habitat use of Cutthroat Trout (*Oncorhynchus clarkii*) and Dolly Varden

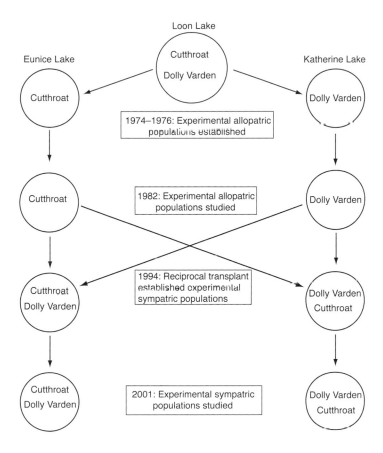

Loon Lake

Eunice Lake

Katherine Lake

1974–1976: Experimental allopatric populations established

1982: Experimental allopatric populations studied

1994: Reciprocal transplant established experimental sympatric populations

2001: Experimental sympatric populations studied

FIGURE 11.5. Design of a 28-year reciprocal transplant experiment to study interactive segregation in Cutthroat Trout (*Oncorhynchus clarkii*) and Dolly Varden (*Salvelinus malma*). The study involved Loon Lake, where the two species are naturally sympatric and sequential experimental allopatric, and then sympatric populations in two formerly fishless lakes, Eunice Lake and Katherine Lake. Based on Andrew et al. (1992), Northcote (1995), and Jonsson et al. (2008).

(*Salvelinus malma*) has been assessed over a 28-year period from 1974 to 2001 using whole-lake transplants (Figure 11.5). From 1974 to 1976, sympatric populations of the two species in Loon Lake, British Columbia, were used to create two allopatric populations in two similarly sized nearby Lakes—Eunice Lake received Cutthroat Trout and Katherine Lake received Dolly Varden (Andrew et al. 1992; Northcote 1995). In sympatry, Cutthroat Trout foraged in the shallow littoral and epipelagic (< 5 m) zones whereas Dolly Varden foraged in the deeper pelagic and epibenthic zones. In laboratory studies, Cutthroat Trout are more effective at capturing surface prey and littoral benthos compared to Dolly Varden, but the latter have better visual acuity under low light conditions and thus are more effective at finding prey in deep water (Northcote 1995). After eight years, the experimental allopatric population of Dolly Varden had expanded their habitat use in Katherine Lake to

include shallow littoral habitats. In contrast, the vertical distribution of Cutthroat Trout in the allopatric population remained essentially the same as that of the naturally sympatric population. The results suggested that the change in habitat use of Dolly Varden could be the result of release from interspecific competition, but the observed difference in habitat could also be due to differences between the three lakes. In addition, the evolved differences in foraging and visual acuity suggest why Dolly Varden were using pelagic and epibenthic habitats, but not why they were apparently restricted in resource use by Cutthroat Trout (Andrew et al. 1992; Northcote 1995; Jonsson et al. 2008). To control for lake effects, in 1994, reciprocal transplants were done between trout and char in the two allopatric populations and the habitat use of the trout and char was assessed in 2001. The results showed that Dolly Varden were displaced from the littoral and epipelagic zones in the presence of Cutthroat Trout but that the vertical distribution of Cutthroat Trout remained essentially the same between the two experimental sympatric populations. The asymmetric competition presumably results from reduced foraging success in shallow water caused by interference from Cutthroat Trout so that deeper foraging habitats become relatively more profitable to Dolly Varden (Jonsson et al. 2008).

Although effects of interspecific competition on fishes most often focus on interactions with other fish species, interspecific competition between fishes and organisms in other taxa can also be important. Crayfish of various species have been introduced into the Colorado River basin and may feed at the same trophic level as fishes. The Virile Crayfish (*Orconectes virilis*), an opportunistic omnivore that has one of the broadest natural ranges of North American crayfish east of the continental divide, has been introduced into western drainages, including the Colorado River basin (Taylor et al. 1996; Lodge et al. 2000; Larson and Olden 2011). Competitive interactions among the Virile Crayfish, Gila Chub (*Gila intermedia*), and Flannelmouth Sucker (*Catostomus latipinnis*) were studied in 200 aquaculture tanks using a replicated, combined substitutive and additive design (Box 11.3) (Carpenter 2005). All tanks received the same daily food ration of chironomid larvae, and the response variable was the change in body mass of the three species over the course of the experiment. Although Virile Crayfish reduced the growth of both fish species by competition for food, the degree of competition varied between the two species. Gila Chub were more affected by intraspecific competition than interspecific competition, whereas Flannelmouth Sucker were more affected by interspecific competition with Virile Crayfish than by intraspecific competition. Virile Crayfish growth was not altered by the presence of either fish species. The increased level of interspecific competition on Flannelmouth Sucker compared to Gila Chub likely is a consequence of the former species being more benthic. Because of the inherent aggressiveness of

Virile Crayfish, competition most likely occurs via interference (Carpenter 2005).

Intraspecific Competition

Intraspecific interactions also can impact the structure of fish assemblages (Fausch and White 1986; Freeman and Stouder 1989; Matthews 1998). The effect of intraspecific competition was tested using Red Shiners, an extremely widespread minnow in the midwestern United States extending into northern Mexico (Matthews 1985b). Red Shiners were stocked at densities from about 3 to 30 individuals/m² in large experimental streams (with two replicates) (Matthews et al. 2001). Fish showed moderately reduced overwinter survival that was negatively related to initial stocking density (Figure 11.6). Although fish showed little growth between initial fall stocking through the end of the year at any density level, fish density did have a strong, negative linear effect on survival and spring growth (Figure 11.6). Also, at high population densities, growth was skewed with fewer individuals growing large enough to reach sexual maturity. In a subsequent experiment (W. J. Matthews, unpublished data), Red Shiners with free access to a series of pools and riffles in predator-free experimental streams typically formed groups with densities lower than those that had resulted in low modal growth in the groups in the previous experiment (Matthews et al. 2001). This finding suggests that fish may avoid excessively dense aggregations (unless factors like predators or differences in habitat quality provide stronger stimuli). The actual mechanism or mechanisms responsible for the growth depression relative to fish density is unclear. Fish density did not have any effect on algal standing crop or density of benthic invertebrates and, in fact, had a positive effect on benthic primary productivity. The latter effect disappeared after 5–7 months and its occurrence was likely caused by the transfer of nutrients from surface prey (Red Shiners feed on surface and water-column prey) to the benthic compartment, with the disappearance of the effect occurring after other nutrient sources

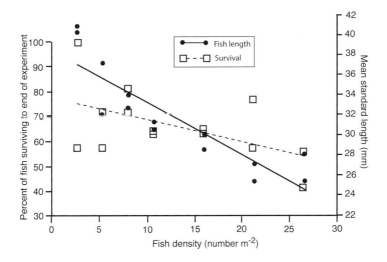

FIGURE 11.6. Effects of intraspecific competition in Red Shiner (*Cyprinella lutrensis*) on growth and survival of fish in a replicated experimental stream. Based on Matthews et al. (2001).

became more available (Gido and Matthews 2001). The actual mechanisms for growth suppression could have been due to interference or exploitation competition for surface prey (which were not measured), or to interference competition for benthic prey, but the suppression was not due to exploitation competition for benthic prey (Matthews et al. 2001).

Morphological and behavioral divergence of Threespine Sticklebacks in northwestern lakes also provide opportunities to assess the presence of competition and its role in promoting such divergences. In various lakes, Threespine Sticklebacks have diverged into benthic and limnetic morphologies, whereas in other lakes, populations are polymorphic with intermediate as well as extreme morphologies (See Chapter 7). The role of competition as a driver in morphological divergence was first shown experimentally by a pond experiment using fish from a polymorphic (both benthic and limnetic forms) population of Threespine Sticklebacks in Cranby Lake, British Columbia, using a basic additive design (Schluter 1994). The study tested the prediction (based on a competition hypothesis) that individuals at the limnetic extreme would suffer disproportionately when limnetic fish from Paxton Lake (having distinct benthic

and limnetic forms) were added. Because fish at the two extremes are uncommon in Cranby Lake, additional fish at the limnetic and benthic extremes were produced by hybridizing Cranby Lake fish with benthic and limnetic fish from Paxton Lake. The study used two ponds, each divided into equal treatment and control sections (Figure 11.7A). In both of the ponds, the added numbers of the limnetic form resulted in lowered survival and decreased growth in the more extreme limnetic fish but the treatment effect diminished gradually in fish that were morphologically more distant from the limnetic body form (Figure 11.7B). Although the study only had two replications, the results are strong support that morphological diversification in Threespine Sticklebacks could be driven by resource competition (Schluter 1994, 2010).

Another pair of studies with Threespine Sticklebacks further tested the role that competition might play in disruptive selection, testing the prediction that populations with elevated intraspecific competition should show faster niche expansion or stronger diversifying selection. In the first study, pairs of large enclosures were established in one lake in 2001 and in five different lakes in 2002—all on Vancouver Island, British Columbia

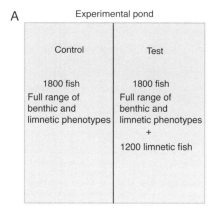

A

Experimental pond

Control	Test
1800 fish Full range of benthic and limnetic phenotypes	1800 fish Full range of benthic and limnetic phenotypes + 1200 limnetic fish

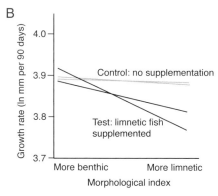

B

FIGURE 11.7. Competition within limnetic morphologies of Threespine Stickleback (*Gasterosteus aculeatus*). A. Control and treatment in the two halves of an experimental pond (two ponds were used in the experiment). B. Growth responses of different morphologies of fish in response to supplementing the numbers of limnetic fish in two different ponds. Based on Schluter (1994, 2010).

(Bolnick 2004). The enclosures went from shore to open water and thus included a range of habitats. Some fish were removed from one randomly chosen enclosure out of a pair of enclosures to create a low-density treatment; fish were added to the other enclosure of the pair to form a high-density treatment. Fish in both enclosures of a pair had natural phenotypic distributions. At the end of the 2001 experiment, fish in the high-density experiments showed evidence of disruptive selection in gill-raker length (limnetic forms have longer gill rakers than benthic forms), were smaller, and had lower relative gonad mass than those from the low-density experiments. The high-density enclosure in 2001 also

had lower zooplankton density than the low-density enclosure. Results were more variable in 2002, but generally provided support for the hypothesis of competition-driven disruptive selection (Bolnick 2004).

The second study used five pairs of enclosures placed in the shallow water of a single lake on Vancouver Island; one enclosure of a pair received 30 fish (low density) and the other 90 (high density) (Svanbäck and Bolnick 2007). After two weeks, the high-density treatments had lower densities of benthic invertebrates and zooplankton and fish in this treatment had reduced gut contents and lower growth rates. The fish in the low-density treatments did not differ in growth rate or gut content from fish taken outside of the enclosures; the only artifact of caging was a lower zooplankton density than outside the low-density enclosures. Importantly, fish in the high-density treatment had greater variation in diet due to increased individual specialization leading to an overall increase in population niche breadth. In contrast, the diet variation and niche breadth in wild-caught fish were similar to fish in the low-density enclosures. Fish survival did not differ among the treatments so that differential survival was not the reason for changes in niche characteristics. The study confirms that competition for limited resources can lead to increased diet variation among members of a single population and that such changes can occur in a matter of weeks, essentially as soon as fish can detect a change in prey availability (Svanbäck and Bolnick 2007).

Intraspecific interactions between larval and adult fishes are also important, and can have both positive and negative effects, but are as yet poorly studied. Adult male Bluegill build circular nests in the littoral zone of lakes and ponds, or quiet areas of streams, and the male defends the developing embryos and early larvae against predators (Avila 1976). However, once the larvae leave the nest, their interactions with adult Bluegill can be positive or negative. Adults can have a positive impact on conspecific larvae by consuming predators of larvae, such as predatory insects, or by increasing the abundance of small zooplankton through removal of large zooplankton. Potential negative effects include competition for large zooplankton and cannibalism. In pond experiments conducted on adult and larval Bluegill (during the experiment larvae transitioned to postlarval YOY) at the Kellogg Biological Station, Michigan, the presence of adult Bluegill had a strong negative impact on the final mass of larvae through competition for large zooplankton prey, such as *Chaoborus* and *Daphnia*, which were the dominant larval prey in absence of adults (Rettig and Mittelbach 2002). As adult density increased, the prey biomass in stomachs of larval Bluegills decreased, resulting in a decline in average mass of larvae. However, larval survival was greatest at intermediate adult densities, with the low survival in the absence of adults being related to the high density of predatory insects. Because adult Bluegill consume the large-bodied, predatory insects, the density of predatory insects declined as adult density increased; however, because of the competitive interaction, the overall production of larvae peaked at intermediate adult densities (Rettig and Mittelbach 2002).

FOOD WEBS, TROPHIC POSITION, AND COMPETITIVE EXCLUSION

Food webs and food chains describe the complexity of trophic interactions occurring within a community and thus indicate how matter and energy move through an ecosystem (Krebs 1985; Post 2002). The understanding of these food webs has been greatly influenced by the pioneering works of Elton (1927), on trophic pyramids, and Lindeman (1942), on the dynamics of energy flow in communities (Cook 1977). Food webs are dynamic, changing spatially as well as temporally, and resulting in the formation and breakage of linkages among trophic groups. Temporal changes include seasonal and annual variation related to variation in the physical environment (temperature, water chemistry, precipitation amounts, runoff, water level, stream discharge, and floodplain access in rivers) or biotic environment (spawning and recruitment of young, migratory patterns, energy demands, and impacts of fisheries) (Power et al. 1995; Winemiller 1996). Also, the introduction of nonindigenous fishes and invertebrates can have widespread, and often unexpected, impacts on food webs, providing the potential for elimination of native species (Vander Zanden et al. 2003).

Webs and Chains

Food webs and food chains comprise trophic groups and the linkages between them. There are several types of food chains and food webs, based on what is emphasized in the data. A connectance web essentially shows all known linkages and trophic groups from primary production and detritus up to the top-level consumers (essentially who eats whom) and, because it is based on presence/absence data, is qualitative (Figure 11.8A). Two other webs are quantitative and can represent the size of the trophic groups and the degree of energy flow or interaction between them. An energy web emphasizes the amount and direction of energy flow between trophic groups, again starting from primary producers and detritus and moving upward to top-level consumers, essentially following the work of Lindeman (1942) (Figure 11.8B). Often, only surrogates of energy flow are available, such as the number, volume, or biomass of consumed organisms. The importance of a linkage or the size of a trophic group can change based on the currency used to evaluate them. For instance, trophic group size based on the number of individuals will be biased toward small organisms, whereas trophic group size based on biomass will be biased toward large organisms (Winemiller 1996; Post 2002). Finally, an interaction web (or functional food web) is based on the strength of the interaction between levels and is generally a top-down web that is determined by experimental manipulation. Interaction webs offer an estimate of the functional food-chain length as shown by the impact of a species or a group of species on some

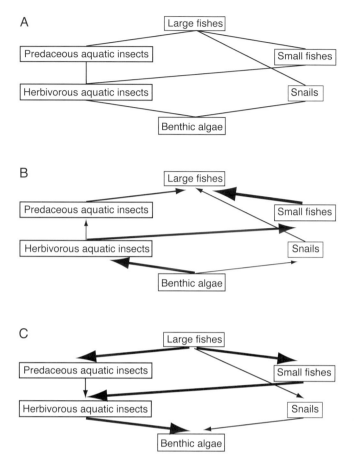

FIGURE 11.8. Hypothetical aquatic food webs.
A. Connectance web.
B. Energy web, with the amount of energy transfer between levels shown by the width of the arrows.
C. Interaction web, with the strength of the interaction between trophic levels shown by the width of the arrows. Based on Post (2002) and Woodward and Hildrew (2002).

community attribute, such as the population size of herbivores (Paine 1980) (Figure 11.8C).

A food chain refers to a single pathway showing the linkages and trophic groups between the bottom and top of a food web. As such, a single food web can contain multiple food chains. Both connectance and energy webs provide estimates of realized food-chain length, which is based on all the different trophic pathways that lead from the base of a food web to a consumer. For instance, the three possible realized food chains from Figure 11.8B are link-1 (benthic algae to herbivorous aquatic insects to predaceous aquatic insects to large fishes—four levels), link-2 (benthic algae to herbivorous aquatic insects to small fishes to large fishes—four levels), or link-3 (benthic algae to snails to large fishes—three levels). Stable isotope analysis provides an improved way to estimate energy

webs and food-chain length by providing a summary of energy flow and integrating the outcome of complex interactions. A functional food chain is based on an interaction web (Post 2002). Because interaction webs and functional food chains emphasize the role that a predator has on community properties, they are treated in more detail in Chapter 12.

The control over the length of a food chain has been a topic of interest at least from the pioneering works of Elton and Lindeman. Factors thought to be important include the history of community organization (what organisms have had access to a community), the amount of available resources, habitat stability, and the size of the ecosystem (Post 2002; Power and Dietrich 2002). The effect of resource availability seems to only prevail at extremely low resource levels, such as might occur in deep caves, extremely high altitudes or latitudes, or extreme deserts, and is not likely a factor in most North American freshwater ecosystems. Because most aquatic food webs are size-structured (i.e., body size increases with trophic position), food chain length is influenced by the range in body size between organisms at the bottom and top of the food web. Although intuitively appealing, shorter food chains have not generally been linked with increasing levels of disturbance, although disturbance has been associated with the loss of top predators in small ecosystems (Post et al. 2000; Post 2002). Undoubtedly, part of the problem is in identifying what constitutes a disturbance (see also Chapter 6). For instance, in rivers of northern California, and likely in other Pacific coast streams, the lack of flooding, rather than flooding, is the disturbance. The lack of flooding shortens the dominant food chain to two levels—primary producers and a large, predator-resistant herbivorous caddisfly, in contrast to a 3–4 level food web that occurs during years with winter flooding, moving from primary producers, to small herbivorous aquatic insects, to larger aquatic insects, to juvenile salmonids (Power 1992a; Wootton et al. 1996; see Chapter 12). There is also an effect of ecosystem size on food-chain length, although where changes/additions occur is not well explained (Post et al. 2000; Post 2002).

In contrast to a clearer separation of trophic position just shown, a study comparing trophic positions of small-bodied fishes from four streams in Kansas, Oklahoma, and New Mexico showed much more variability and overlap. Trophic position was based on $\delta^{15}N$ from muscle tissue, a synthesis of published studies on food habits of the species, and gut content analyses from a single collection of fishes from each stream. For purposes of comparison, data were analyzed using the broad trophic categories of algivore/detritivores, omnivores, and insectivores. All three approaches to determining trophic position generally agreed, but the relationship between the percentage of invertebrates consumed and $\delta^{15}N$ was only weakly correlated (r = 0.42). Trophic separation was blurred,

primarily because the small-bodied fishes in these four streams occupied only two trophic levels and all streams were dominated by omnivores. Omnivorous fishes can show high variability in feeding depending on time of year, flow patterns, and food availability. Because of the highly variable hydrology of the study streams, there has likely been strong selection for omnivory (Franssen and Gido 2006).

Changes in food webs as a consequence of nonindigenous fish and invertebrate introductions can be extensive. One of the better-studied systems is that of Lake Tahoe, a large lake located in the Sierra Nevada Mountains on the California/Nevada border. The changes in the food web of Lake Tahoe over a 130-year period from the late 1800s and early 1900s to 1998–2000, have been studied using stable isotope analysis of muscle tissue from formalin-preserved fishes and aquatic invertebrates (after correcting for the effects of chemical preservation) and by historical studies on fishes and invertebrates in the system (Vander Zanden et al. 2003). Historically, the top predator was the Lahontan Cutthroat Trout (*Oncorhynchus clarkii henshawi*), which evolved in Pleistocene Lake Lahontan (see Chapter 3 and Figure 3.3). By the late 1930s, this species, which fed on schools of pelagic fishes, was extirpated from Lake Tahoe as a consequence of combined effects of Lake Trout and other nonindigenous fish introductions starting in the late 1800s, overexploitation, dam construction, and loss of spawning habitat. Lake Trout essentially occupied the same place in the food web as Cutthroat Trout. In the 1960s, a nonindigenous, large zooplankter, *Mysis relicta*, was introduced and became well established a decade later, corresponding with the extirpation of two native cladocerans, *Daphnia* and *Bosmina* (Vander Zanden et al. 2003). In spite of the long history of introduction of nonindigenous fishes, the most significant change in the Lake Tahoe food web occurred subsequent to the introduction of *Mysis*, in concert with the onset of eutrophication. Historically, although the food web was driven by both benthic and planktonic primary production, benthic primary production predominated. The contemporary food web has been significantly shifted toward pelagic-based production (Vander Zanden et al. 2003; Schmidt et al. 2007; Turner et al. 2010).

Competitive Exclusion

As exemplified previously, competitive interactions among or within species can affect fish assemblages on ecological and evolutionary time scales. On ecological time scales, competition can result in changes in resource use, changes in growth rates and condition, and presumed changes in fitness. On evolutionary time scales, competitive interactions can provide strong selective forces resulting in genetically fixed differences in morphology and behavior, leading, at least in some cases, to speciation. However, unequivocal evidence of competitive exclusion of one species by another, other than in laboratory or mesocosm experiments, is difficult to obtain and is uncommon for any taxonomic group (Ricklefs 1990; Begon et al. 1996), including North American native fishes. In fact, various studies suggest that physical variation in flow and temperature, recolonization dynamics, and temporal variation in reproduction have greater impacts on North American fish assemblages than competitive exclusion (Schlosser 1982; Grossman and Freeman 1987; Grossman and Boulé 1991).

A seemingly excellent opportunity to document competitive exclusion in nature is the long-term investigations mentioned previously on Dolly Varden and Cutthroat trout (reviewed by Northcote 1995). However, even though the work provides a convincing example of asymmetric competition, competitive exclusion did not occur. The loss of species due to competition may be more prevalent in terms of impacts of nonnative species on native species (reviewed by Fausch 1988; Ross 1991; Fuller et al. 1999). However, even for the relatively well-studied cases of interactions of native and introduced salmonids, clear evidence of the mechanism involved where interactive segregation in habitat use was demonstrated is generally lacking

(Fausch 1988). Where mechanisms of interaction are documented, impacts of direct predation are better supported than competition (Ross 1991).

The strongest support for competitive exclusion comes from a study of the replacement of Rio Grande Silvery Minnow (*Hybognathus amarus*) by the Plains Minnow (*H. placitus*) (see also Chapter 15). Rio Grande Silvery Minnow is native to the Rio Grande and Pecos river systems and the Plains Minnow is widespread in midwestern streams. Both use similar habitats and have equivalent life histories (described in Chapter 5, Ontogeny and Movement). The time of introduction of the nonindigenous Plains Minnow into the Pecos River, as determined from museum records, occurred in the early 1960s. Based on genetic data, the colonizers came from two sources, and the total number of original colonizers was around 32–115 (Moyer et al. 2005). Within only 10 years, the native Rio Grande Silvery Minnow was gone from the Pecos River. Given that the Plains Minnow successfully became established in the Pecos River, there are two competing hypotheses for the loss of the Rio Grande Silvery Minnow—hybridization or competition. Genetic assignment tests and phylogenetic analyses have excluded hybridization as a primary factor in species replacement; in fact, there are very few individuals that are potentially of hybrid origin. Population modeling indicates that species replacement could have occurred within 10 years, given the initial number of colonizers and a fitness difference between the native and nonindigenous *Hybognathus* species. Consequently, given the lack of support for the hybridization hypothesis, the alternative hypothesis of competition, most likely exploitation competition, is probably the cause of the species replacement (Moyer et al. 2005).

SUMMARY

Studies of species' interactions as a consequence of resource acquisition have long remained a major area of ecological research. One line of evidence for the potential occurrence of competition has been the degree of dietary overlap, traditionally determined by studies of gut contents or by direct observations of feeding. Over the last several decades, the analysis of stable isotopes has proven an additional and effective means of determining the trophic bases of fish populations, especially where a time-integrated approach is needed.

Competition can occur within or among species through active interference that limits access to resources by another individual or species. Competition can also occur passively through the common use of a resource. When competition occurs, it often is asymmetrical, impacting one individual or group more than another. Competition is typically evaluated through observational field studies, such as resource partitioning, through "natural experiments," including character displacement and release, and through manipulative field and/or laboratory studies, all of which can be useful approaches. Especially with observational studies or "natural experiments," evidence of competition is often inferred from shifts in niche overlap, as shown by the differentiation of sticklebacks and sunfishes in species-poor lakes into benthic and limnetic forms. Niche concepts are widely used in ecology, although their use is complicated by competing niche definitions. In interpreting any observational study of differential resource use, such as the commonly observed vertical habitat segregation by fishes, it is important to first eliminate the possibility that the pattern is a consequence of the history of the species (e.g., phylogeny), rather than caused by contemporary interactions. Studies of resource use and niche overlap in fishes are further complicated by the great range of body sizes, and perhaps habitats, shown by fishes as they transition from larval, to juvenile, to adult stages.

Trophic interactions in a community are complex, shaped by competitive or predatory interactions. Food webs and food chains (food chains show only a single pathway) can be in the form of connectance (essentially who eats whom) or

the amount of energy transferred at each link or level. Interaction webs or chains comprise a third group and show the strengths of interactions, such as the impact of a predator on lower trophic levels. The number of trophic levels is sensitive to environmental disturbance and to the introduction of nonindigenous species. Competitive interactions have varying effects on fish assemblages, both over ecological and evolutionary time scales. Although a long-established concept in ecology, competitive exclusion of one group by another is rarely documented in nature.

SUPPLEMENTAL READING

Bowen, S. H. 1996. Quantitative description of the diet, 513–29. In *Fisheries Techniques*, 2nd ed. B. R. Murphy and D. W. Willis (eds.). American Fisheries Society, Bethesda, Maryland. An excellent source of information on the methods of dietary analyses in fishes.

Hurlbert, S. H. 1981. A gentle depilation of the niche: Dicean resource sets in resource hyperspace. *Evolutionary Theory* 5:177–84. An interesting attempt to categorize the multitude of different niche concepts.

Hurlbert, S. H. 1984. Pseudoreplication and the design of ecological field experiments. *Ecological Monographs* 54(2):187–211. An influential paper on problems of experimental design in ecology—I consider this a "must read" paper for students in any area of ecology.

Pfennig, K. S., and D. W. Pfennig. 2009. Character displacement: Ecological and reproductive responses to a common evolutionary problem. *The Quarterly Review of Biology* 84:253–76. A thorough current review of the concept of character displacement.

Post, D. M. 2002. The long and short of food-chain length. *Trends in Ecology and Evolution* 17:269–77. A clear discourse on the types of food webs and food chains and on factors influencing food-chain length.

Schoener, T. W. 2009. Ecological niche, 3–13. In *The Princeton Guide to Ecology*. S. A. Levin (ed.). Princeton University Press, Princeton, New Jersey. A concise overview on three current niche concepts.

Wiens, J. A. 1977. On competition and variable environments. *American Scientist* 65:590–97. A timeless paper emphasizing the aperiodicity of competitive interactions in naturally variable environments.

TWELVE

Interactions in Resource Acquisition II

PREDATION, AVOIDING PREDATION, AND
PREDATOR EFFECTS ON ECOSYSTEMS

CONTENTS

HOW ENERGY IS ALLOCATED among individuals and species, and thus how assemblages and communities are shaped, is affected by competitive interactions (Chapter 11) and by predator-prey interactions (this chapter). Predator-prey interactions impact communities by altering competitive balances, affecting foraging behaviors and habitat use, and changing community structure through top-down effects (Kitching 1986). Unlike the negatively symmetric relationship of competition, predation is an asymmetric relationship (+, −), where one member of the interacting pair benefits and the other is negatively affected (see Part Table. 4.1). In the broad sense, predation includes herbivory, carnivory, parasitism, and cannibalism—essentially all are situations where one organism totally or partially consumes another organism that is alive when first attacked. Consequently, for heterotrophs, predation includes all ways of obtaining food other than consuming dead or dying organic matter (scavengers and detritivores) (Pianka 1988; Begon et al. 1996). Scavengers and detritivores are often treated along with predation but differ in the nature of the interaction (+/o), given that survival of dead or dying organisms is/was not related to their consumption.

Compared to those that are less successful, more efficient predators should show higher fitness owing to more resources for maintenance, growth, predator avoidance, and reproduction. Natural selection thus favors any component of predation that contributes to greater feeding success (Pianka 1988). Concurrently, selection should favor prey species that are better able to interrupt the predation sequence. The essentially reciprocal selection for increased predation success and increased success at avoiding predation leads to a classic "Red Queen" situation (see also Chapter 9), where the continued evolution of both parties is required to maintain the status quo (Van Valen 1973; Liow et al. 2011).

Predator-prey relationships typify interception and avoidance situations, referred to as differential games, and are made even more intriguing because each member of the pair is likely to be both potential predator and potential prey (either concurrently or at some time during ontogeny), and this is particularly true for fishes (Gans 1986; Webb 1986). This limits the degree of development each organism may invest in improving its capacity for success as a predator or in predator avoidance, given that individual traits, such as those enhancing success of a predator or those enhancing predator avoidance, are often linked genetically and selection acts on an organism as a functional unit and not on individual traits.

The nature of the predator-prey relationship is affected by the relative sizes of prey and their predators, with two basic approaches—to be a relatively larger predator than the prey, or to be a relatively smaller predator than the prey. As fish predator sizes become smaller than their fish prey, tactics such as scale eating or fin nipping can occur. At the other extreme, at least for terrestrial predators, as prey and predator sizes become similar, predator tactics may switch to multiple predators attacking a single large prey. Importantly, social hunting in fishes seems not to be associated with the taking of large prey but is rather an adaptation for capturing schooling prey (Griffiths 1980). Social hunting is well documented in marine and tropical freshwater fishes (Major 1978; Potts 1983; Sazima and Machado 1990; Magurran and Queiroz 2003), but it is apparently undocumented and perhaps rare in North American freshwater fishes.

Although most successful predation events result in the death of the prey, partial predation, where only a portion of the prey is consumed and the prey generally is not killed, occurs in instances of herbivory on multicellular plants, carnivory on colonial organisms, parasitism (including exoparasitism such as fin nipping, scale eating, and eye picking), and bivalve siphon cropping (Vermeij 1982). The first two categories of partial predation (herbivory on multicellular plants and carnivory on colonial organisms) are commonly represented among North American freshwater fishes. Siphon cropping occurs primarily in estuarine and coastal waters, although at times by species such as Threespine Stickleback (*Gasterosteus aculeatus*) that are also common in freshwater habitats; it is apparently undocumented in freshwater habitats (Peterson and Quammen 1982; Delbeek and Williams 1987). Forms of exoparasitism, such as scale eating and fin nipping, are also better documented and occur primarily in fishes of marine and tropical freshwater habitats (Sazima 1983; Grutter 2002). Scale eating is a derived feeding mode usually characterized by modified jaw teeth, with freshwater examples including Neotropical characiform fishes, African cichlids, and several genera of Neotropical, Asian, and African catfishes (Sazima 1983). In addition, one species of pupfish (*Cyprinodon*) from the Bahamas is thought to be the only known scale eater among some 1,500 atheriniform species (Martin and Wainwright 2011). Eye-picking attacks by the Cutlip Minnow (*Exoglossum maxillingua*), native to northeastern North America, are documented on other fish species as well as conspecifics. The Cutlip Minnow is characterized by an unusual lower jaw that has three lobes, with the inner lobe raised, bony, and tongue-like, and the outer two lobes soft and fleshy (Scott and Crossman 1973; C. L. Smith 1985). Attacks on other fishes by Cutlip Minnow increase as the density of the minnows rise (such as in a bait bucket), and attacks in captivity have been documented on Cutlip Minnows as well as various other minnows, suckers, salmonids, and Yellow Perch (*Perca flavescens*) (Johnson and Johnson 1982; C. L. Smith 1985; Dale and Pappantoniou 1986).

The apparent rarity of the derived modes of feeding, such as forms of exoparasitism, in North American freshwater fishes is due in part to the historical development of the fish fauna (see Chapter 2), especially the lack of characiform fishes, which have jaw teeth, and the numerical dominance of cypriniform fishes, which lack jaw teeth (see Chapter 8). There is also a latitudinal gradient in the importance of most biological interactions, including cleaning behavior (treated in Chapter 13) (Schemske et al. 2009); however, it is still not possible to ascertain if the various modes of derived feeding are truly as rare among North American freshwater fishes as their limited documentation would suggest.

In instances where prey are not normally killed, natural selection for predator avoidance may be less intense than in situations where prey are frequently killed outright. However, even though selection for predator avoidance might be weaker when prey are not killed, antipredation traits should still evolve. Plants, for instance, show a multitude of adaptations for predator avoidance (Ames 1983; Begon et al. 1996).

TROPHIC GUILDS

A useful approach for understanding the trophic organization of fishes is to recognize functionally defined trophic guilds. The term guild was formally proposed as "a group of species that exploit the same class of environmental resources in a similar way" (Root 1967). Defined in this way, the guild concept encompasses both the food resource and the way it is obtained (e.g., piscivory could be achieved by actively chasing prey or by ambush; see also Chapter 8). Importantly, although groups of species with highly similar functional roles within a community would belong to the same guild, say planktivore, in a different community the same guild might comprise different species. The placement of fishes into feeding guilds is necessarily approximate. Given that fishes are often opportunistic feeders, their diets may change spatially and temporally, and most fishes show strong ontogenetic changes in food habitats with growth (Poff and Allan 1995). Feeding guilds, as well as other guild categories, are widely used in the determination of aquatic system health via the Index of Biotic Integrity (Karr 1981; Bramblett and Fausch 1991). Feeding guilds are also important for understanding functional responses of different communities to hydrological or geomorphic changes, as well as temporal changes within a single community (Grossman et al. 1982; Poff and Allan 1995). To enhance the utility of feeding and other guilds (e.g., habitat), it is sometimes desirable to recognize two kinds of guilds related to food habits—a feeding guild as defined by the foraging habitat (i.e., benthic, water column, or surface) and a trophic guild as defined by the nature of the prey (i.e., detritivore, herbivore, or insectivore), as an alternative to combining prey and feeding habitat (Noble et al. 2007). However, most trophic guilds involving fishes combine food and feeding location characteristics, often because the information required for designating two separate guilds is simply inadequate. Species that form a specific guild would have similar trophic positions (see also Chapter 11), but depending on how narrowly a guild is defined, fishes that feed at the same trophic level may still differ in both the kinds of food and their feeding locations, and thereby be placed into different guilds (following Root 1967). For instance, the fishes that consume herbivorous aquatic insects could potentially be separated into two feeding guilds, one characterized by feeding over a soft substratum versus the other feeding over a rocky substratum (see Chapter 11; Figure 11.8).

Although assignments of fishes to feeding guilds are usually based on the dominant prey items from gut contents (usually restricted to large juveniles and adults), other approaches also include multivariate analysis of gut contents and stable isotope studies (Vander Zanden et al. 1997; Olden et al. 2006; Gido and Franssen 2007; Noble et al. 2007).

Based on data from nine faunal studies that are representative of different regions of North America, I have assigned 224 species in 28 families to nine commonly recognized feeding guilds (Table 12.1). If more data were available, most of these categories could be further broken down into subcategories (Goldstein and Simon 1999). For example, the herbivore guild could be subdivided into grazers, which crop food very close to the substratum and often ingest other bottom materials, and browsers, which feed on epiphytic algae and macrophytes by biting off pieces of plants above the substratum. Overall, the data from Table 12.1 show that omnivores, benthic invertebrate feeders, and macrocarnivores/piscivores dominate the North American freshwater fish fauna, along with generalized invertebrate feeders and surface/water-column feeders (Table 12.2). Parasites (including only adult parasitic lampreys), planktivores, and herbivores are relatively rare. Not surprisingly, cyprinids, the most species-rich group, show the greatest range of guild membership, but most are omnivores followed by generalized invertebrate feeders and surface/water-column feeders (Figure 12.1). Catostomids are about equally divided among omnivores, benthic invertebrate feeders, and detritivores (including the consumption of biofilm organisms such as bacteria, algal cells, and protozoans). The second-largest family of North American fishes, the percids, are primarily benthic invertebrate feeders (the darters), with macrocarnivores/piscivores limited to Yellow Perch, Sauger (*Sander canadensis*), and Walleye (*S. vitreus*). Three families, ictalurids, centrarchids, and salmonids, are dominated by macrocarnivores/piscivores. In fact, the main line

of evolution of ray-finned fishes since the early Mesozoic has progressed through a series of generalized carnivores characterized by having relatively large mouths and adapted to directly approach and seize prey that are fully exposed (see also Chapter 8) (Hobson 1974). Often such predators are nocturnal or crepuscular in feeding and from this main evolutionary line there have been numerous specialized offshoots.

Ontogenetic diet changes occur in most freshwater as well as marine fishes. In fact, almost all bony fishes start out as selective predators on zooplankton, or other small invertebrates such as chironomid larvae, before switching to other food sources as they grow (Browman and O'Brien 1992; Childs et al. 1998; Wootton 1998). For instance, Bluegill (*Lepomis macrochirus*) show strong ontogenetic dietary shifts in prey size and kind in addition to habitat. Exogenous feeding begins at about 5.0–5.9 mm TL as the yolk sac is absorbed. At this stage Bluegill move to open water and feed on very small food items such as rotifers and copepod nauplii. By 7–10 mm TL they rely less on rotifers and more on the larger cyclopoid copepods and cladocerans. However, by 14–17 mm TL, to avoid open-water predators, Bluegill move back into the more protected vegetated habitats where they feed on aquatic insects such as midge larvae, as well as copepods and cladocerans. Fish larger than 50 mm TL include terrestrial and aquatic insects in their diet and only the larger forms of copepods and cladocerans are consumed (Ross 2001).

One of the more extreme examples of changes in trophic ontogeny occurs with Striped Mullet (*Mugil cephalus*), a primarily marine-estuarine species that also has been categorized as catadromous because of feeding movements of hundreds of kilometers into fresh water (De Silva 1980). As with other bony fishes, larval Striped Mullet, especially those of 5–15 mm SL, feed primarily on animal prey, including microcrustaceans such as copepods, and on small aquatic insects such

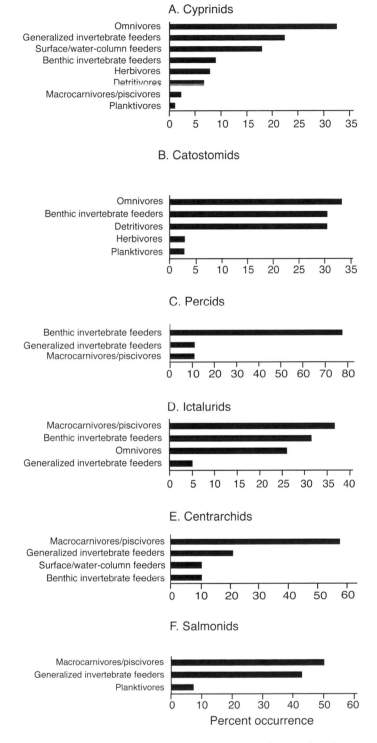

FIGURE 12.1. Feeding guilds occupied by six species-rich families of North American freshwater fishes: A. Minnows, family Cyprinidae; B. Suckers, family Catostomidae; C. Perches, family Percidae; D. Catfishes, family Ictaluridae; E. Sunfishes, family Centrarchidae; F. Trouts and salmons, family Salmonidae. Based on data from Table 12.1.

TABLE 12.1

Examples of Feeding Guilds for 224 Species of Freshwater Fishes from Nine North American River Basins
Guild placement is for adult or late-juvenile fishes

Guild	Description	Family	Species
Herbivore	Algae, vascular plants, and seeds	Cyprinidae	Blackside Dace, Central Stoneroller, Chisel-mouth, Grass Carp, Largescale Stoneroller, Ozark Minnow, Tennessee Dace
		Catostomidae	Bridgelip Sucker
Detritivore	Organic bottom material from various sources: rotting vegeta-tion, animal remains, feces, and pseudofeces; live materials: bacteria, attached algae, benthic diatoms, and resting cysts of plankton	Petromyzontidae	American Brook Lamprey
		Clupeidae	Gizzard Shad
		Cyprinidae	Bluntnose Minnow, Common Carp, Fathead Minnow, Mississippi Silvery Min-now, Suckermouth Minnow, Weed Shiner
		Catostomidae	Bigmouth Buffalo, Black Buffalo, Bluehead Sucker, Desert Sucker, Flan-nelmouth Sucker, Highfin Carpsucker, Hogsucker, Mountain Sucker, River Carpsucker, Smallmouth Buffalo, Sonora Sucker
Omnivore	Wide range of animal and plant food	Cyprinidae	Blacknose Shiner, Bluehead Chub, Blunt-nose Minnow, Bonytail Chub, Brassy Minnow, Bullhead Minnow, Common Carp, Common Shiner, Creek Chub, Eastern Blacknose Dace, Fathead Min-now, Gila Chub, Golden Shiner, Goldfish, Headwater Chub, Hornyhead Chub, Little Colorado Spinedace, Longfin Dace, Mimic Shiner, Northern Redbelly Dace, Redfin Shiner, Roundtail Chub, Sand Shiner, Spottail Shiner, Southern Red-belly Dace, Striped Shiner, Utah Chub, Virgin Spinedace, Woundfin
		Catostomidae	Alabama Hogsucker, Black Redhorse, Blacktail Redhorse, Golden Redhorse, Longnose Sucker, Mountain Sucker, Quillback Carpsucker, Razorback Sucker, Rio Grande Sucker, Spotted Sucker, Utah Sucker, White Sucker
		Ictaluridae	Black Bullhead, Brown Bullhead, Snail Bullhead, Yellow Bullhead, Channel Catfish
		Cyprinodontidae	Desert Pupfish

(continued)

TABLE 12.1 *(continued)*

Guild	Description	Family	Species
		Fundulidae	Southern Studfish
		Poeciliidae	Gila Topminnow, Guppy, Sailfin Molly, Shortfin Molly
		Cichlidae	Blue Tilapia, Convict Cichlid, Mozambique Tilapia, Redbelly Tilapia, Rio Grande Cichlid
Planktivore	Catch and/or filter zooplankton and phytoplankton	Polyodontidae	Paddlefish
		Clupeidae	Threadfin Shad
		Cyprinidae	Golden Shiner
		Catostomidae	Bigmouth Buffalo
		Salmonidae	Bloater
Generalized invertebrate feeder	Variety of invertebrate food from bottom, surface, and water column, including terrestrial and aquatic insects, zooplankton, and benthic microcrustacea	Cyprinidae	Bigmouth Shiner, Bonytail Chub, Creek Chub, Finescale Dace, Hornyhead Chub, Lake Chub, Lined Chub, Loach Minnow, Moapa Dace, Pearl Dace, Red Shiner, Redside Shiner, River Shiner, Rosyface Shiner, Roundtail Chub, Silver Chub, Speckled Chub, Spinedace, Spottail Shiner, Striped Shiner
		Ictaluridae	Stonecat
		Salmonidae	Brook Trout, Brown Trout, Cutthroat Trout, Gila Trout, Mountain Whitefish, Rainbow Trout
		Umbridae	Central Mudminnow
		Aphredoderidae	Pirate Perch
		Percopsidae	Trout-Perch
		Gadidae	Burbot
		Gasterosteidae	Brook Stickleback
		Centrarchidae	Bluegill, Green Sunfish, Longear Sunfish, Orangespotted Sunfish
		Percidae	Blackbanded Darter, Blackside Darter, Iowa Darter
Surface/ water-column feeder	Terrestrial insects, emerging insects, and zooplankton	Hiodontidae	Goldeye, Mooneye
		Cyprinidae	Alabama Shiner, Bandfin Shiner, Bigeye Shiner, Blacktail Shiner, Coosa Shiner, Emerald Shiner, Lake Chub, Rainbow Shiner, Redfin Shiner, Redside Dace, Ribbon Shiner, Silverstripe Shiner, Spotfin Shiner, Steelcolor Shiner, Tricolor Shiner, Yellowfin Shiner

(continued)

TABLE 12.1 *(continued)*

Guild	Description	Family	Species
		Atherinopsidae	Brook Silverside, Mississippi Silverside
		Fundulidae	Blackspotted Topminnow, Blackstripe Topminnow, Plains Killifish, Plains Topminnow
		Poeciliidae	Eastern Mosquitofish, Western Mosquitofish
		Centrarchidae	Bluegill, Longear Sunfish
Benthic invertebrate feeder	Aquatic insect larvae and other small invertebrates associated with hard and soft substrata; snails and clams	Acipenseridae	Lake Sturgeon, Shovelnose Sturgeon
		Cyprinidae	Flathead Chub, Longjaw Minnow, Longnose Dace, Longnose Shiner, Riffle Minnow, Silverjaw Minnow, Speckled Dace, Suckermouth Minnow
		Catostomidae	Creek Chubsucker, Greater Redhorse, Hogsucker, Golden Redhorse, Lake Chubsucker, Quillback, River Redhorse, White Sucker, Shorthead Redhorse, Silver Redhorse, Smallmouth Buffalo
		Ictaluridae	Brindled Madtom, Freckled Madtom, Speckled Madtom, Stonecat, Tadpole Madtom, Yellow Bullhead
		Umbridae	Central Mudminnow
		Percopsidae	Trout Perch
		Cottidae	Banded Sculpin, Mottled Sculpin, Paiute Sculpin
		Centrarchidae	Pumpkinseed, Redear Sunfish
		Percidae	Banded Darter, Bronze Darter, Channel Darter, Cherokee Darter, Etowah Darter, Fantail Darter, Gilt Darter, Greenbreast Darter, Greenside Darter, Johnny Darter, Leopard Darter, Logperch, Mobile Logperch, Mud Darter, Orangebelly Darter, Orangethroat Darter, Rainbow Darter, River Darter, Slenderhead Darter, Speckled Darter, Western Sand Darter
Macro-carnivores/ piscivores	Fishes; large active invertebrates such as crayfish and odonate larvae	Amiidae	Bowfin
		Lepisosteidae	Alligator Gar, Longnose Gar, Shortnose Gar, Spotted Gar

(continued)

TABLE 12.1 *(continued)*

Guild	Description	Family	Species
		Anguillidae	American Eel
		Clupeidae	Alabama Shad, American Shad
		Cyprinidae	Creek Chub, Colorado Pikeminnow
		Ictaluridae	Black Bullhead, Blue Catfish, Brown Bullhead, Channel Catfish, Flathead Catfish, White Catfish, Yellow Bullhead
		Esocidae	Chain Pickerel, Grass Pickerel, Northern Pike, Muskellunge
		Salmonidae	Arctic Grayling, Brook Trout, Brown Trout, Cutthroat Trout, Lake Trout, Rainbow Trout, Sockeye Salmon
		Gadidae	Burbot
		Moronidae	Striped Bass, White Bass, Yellow Bass
		Centrarchidae	Black Crappie, Coosa Bass, Green Sunfish, Largemouth Bass, Redbreast Sunfish, Rock Bass, Shadow Bass, Smallmouth Bass, Spotted Bass, Warmouth, White Crappie
		Percidae	Sauger, Walleye, Yellow Perch
		Sciaenidae	Freshwater Drum
Parasite	Attach to host fishes and feed on host tissue	Petromyzontidae	Chestnut Lamprey, Sea Lamprey, Silver Lamprey

SOURCE. Guilds as defined by Benke and Cushing (2005). Colorado River (Olden et al. 2006); Gulf Coast Southwest (Zeug and Winemiller 2008a); Gulf Coast Southeast (Walters et al. 2003); Missouri (Horwitz 1978; Bergstedt and Bergersen 1997); Nelson/Churchill (Poff and Allan 1995); Ohio (Horwitz 1978; Grossman et al. 1982); Southern Plains (Orth and Maughan 1984); Upper Mississippi (Horwitz 1978; Schlosser 1982; Poff and Allan 1995); St. Lawrence (Horwitz 1978); General (Goldstein and Simon 1999; Westneat 2001)

NOTE: Some species are assigned to more than one trophic guild primarily because of size, seasonal, or regional differences, resulting in a total of 261 entries in the table.

as mosquito larvae (Harrington and Harrington 1961; De Silva 1980). The amount of animal prey decreases with size to the point that young Striped Mullet (20–30 mm SL) feed on large amounts of organic matter, bacteria, algae, and diatoms. As growth continues, the feeding mode shifts progressively from browsing on exposed organisms to grazing on surface or subsurface materials. Fish larger than 40 mm SL begin to dig into the substratum, such that sand grains make up 50–60% of the bulk of the gut contents. Larger Striped Mullet (110 mm SL) scrape down an average of 5–7 mm into the sediment (Odum 1970; Eggold and Motta 1992). Finally, Striped Mullet ($\geq$ 200 mm SL) can filter over 450 kg of bottom sediment per year, retaining bacteria, protozoans, algae, vascular plant fragments, and sand grains, the latter being rich in adsorbed organic matter (Odum 1970; Moriarity 1976; Eggold and Motta 1992).

PROBLEMS FOR PREDATION

During the process of predation, a successful predator progresses through a series of steps (the actual number can vary depending on the situation), each with its own issues and

TABLE 12.2

Percent Occurrence of Feeding Guilds Occupied by a Sample of North American Freshwater Fish Species

Trophic guild	Percent occurrence
Omnivores	22
Benthic invertebrate feeders	21
Macrocarnivores/piscivores	18
Generalized invertebrate feeders	15
Surface/water-column feeders	11
Detritivores	7
Herbivores	3
Planktivores	2
Parasites	1

SOURCE: Based on data from Table 12.1.

challenges, collectively known as the predation cycle (Eggers 1977) (Figure 12.2). Following consumption, the predator might pause for digestion or other systemic needs as well as other behaviors, so the cycle begins again at the level of search, where the predator is faced with questions of what resources are available and where they are located. A hungry fish performs searching activity in a way that increases the probability of finding food. The initiation of the search behavior is dependent on the hunger state as controlled by the balance between the amount of food in the stomach and the systemic need (Curio 1976; Gill 2003). Detection and identification involve locating an item and then recognizing it as a potential food source, followed by a decision to pursue. For instance, with large prey items or highly aggressive prey organisms, the benefit of energy gain must be weighed against the cost (in terms of energy, time, or the risk of injury) of obtaining the item. If the decision is not to pursue, then the search process resumes. If the decision is to pursue, then the cycle moves to the next step of approaching the prey and aiming the jaws so that a strike can be made. Again, depending on the prey type and size, the process of subduing the prey may be a negligible (e.g., planktivores

and herbivores) or a major part of the predation cycle (for large, aggressive prey).

Selective Predation or Not?

A basic question is whether fishes feed on food items in proportion to their encounter frequencies, or do they actively choose certain prey over others. If the latter is true, what is the basis for the decision and how does this impact the aquatic community? In addition to rates of encounter, how and whether or not predators are selective can depend on the density of the prey and predator populations, how prey react to the predator, the density and quality of alternative foods, and characteristics of the predator including the level of satiation, attack efficiency, and morphological or behavioral specializations (Leopold 1933; Ivlev 1961; Curio 1976).

In 1980, the eminent ecologist G. E. Hutchinson, in the forward to a book titled *Predation and Freshwater Communities* (Zaret 1980), wrote, "Fifteen years ago, no data in limnology made less sense than those relating to the specific composition of the zooplankton faunas of different lakes. In 1965, a classic paper by Brooks and Dodson showed the way to resolve much of the difficulty of the subject. Since then a flood of light has been thrown on what had been exceedingly obscure." As an aside, the impressive study published by Brooks and Dodson, which has now been cited over 1800 times, was based on Dodson's undergraduate honors thesis at Yale University under the direction of Brooks (Havel 2009; Peckarsky 2010).

Zooplankton faunas of lakes can vary greatly in species composition and/or size composition, and initially as Hutchinson alluded to, the patterns of variation among lakes made little sense. Classic studies by Brooks and Dodson (1965), Galbraith (1967), and Brooks (1968) indicated that prey-size selection by planktivorous fishes was important in controlling both the kinds of plankton in lakes as well as the co-occurrence of fishes. For instance, Brooks and Dodson showed that the mean size of zooplankton in a Connecticut lake shifted downward from

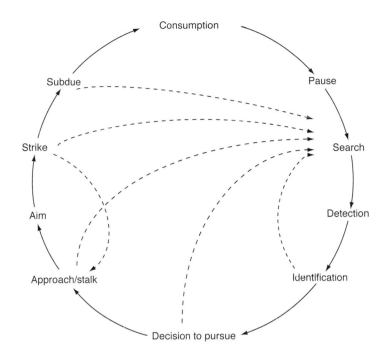

FIGURE 12.2. Elements of predation. Solid lines indicate successful transition from step to step; dashed lines indicate the return to earlier elements because of unsuitable prey or failure to complete a step. Based on Curio (1976), Endler (1986), O'Brien et al. (1986), and Webb (1986).

approximately 0.8 mm to 0.3–0.4 mm after the introduction of Alewife (*Alosa pseudoharengus*). Galbraith looked at the size of prey items consumed by Rainbow Trout (*Oncorhynchus mykiss*) and Yellow Perch and found that both primarily consumed *Daphnia*, but mainly those over 1.3 mm, and ignored the often more numerous smaller individuals. However, gill-raker spacing indicated that smaller plankton would have been retained if simple filtration were the only mechanism involved, suggesting that Rainbow Trout and Yellow Perch were selective predators on certain kinds and sizes of zooplankton.

Numerous other studies have provided additional evidence of selective predation by fishes, through both experimental studies and observations of prey eaten relative to available prey sizes (Ivlev 1961; Hall et al. 1970; Rakocinski 1991; Gutowski and Stauffer 1993). For instance, in pond experiments with Bluegill, consumption of the largest prey-size class increased and consumption of the smallest prey-size class decreased as a function of increasing fish size (Hall et al. 1970). Although larger Bluegill generally consumed larger prey, prey density also affected prey-size selection,

so that large predators would switch to highly abundant small prey when available.

Selection by prey size seems to be more common than selection based on prey kind independent of size or individual prey traits (Gill 2003). However, the selective basis of prey choice can appear subtle. For instance, Northern Pikeminnow (*Ptychocheilus oregonensis*) and Cutthroat Trout (*Oncorhynchus clarkii*) were selective for the number of lateral plates in equivalent-sized Threespine Sticklebacks, generally resulting in better survival of sticklebacks with 7 plates compared to fish with 5, 6, 8, or 9 plates (likely related to a balance between protection and flexibility). However, the direction of selection varied with season, and fish with different lateral-plate counts also showed different behavioral responses to the predators (Moodie et al. 1973). In another experiment with Threespine Sticklebacks, Pumpkinseed predators were selective for vertebral number and the ratio of precaudal to caudal vertebrate of larvae. The actual target of selection appeared to be the ratio of precaudal to caudal vertebrae and the optimum ratio decreased as larval size increased. Functionally, the ratio is related to burst-swimming performance, and thus escape ability, of larvae (Swain 1992).

In most studies of fish food habitats in freshwater and marine environments, levels of prey identification are usually at the level of family or higher, which can be an appropriate level of identification depending on the questions being asked. For instance, in a sample of 30 published studies of food habits of North American freshwater fishes (average number of prey categories = 19), the majority of studies did not identify any prey to the generic or specific levels, and 50% or more of prey taxa were identified to genus in only three studies and to species in only two studies (Table 12.3). However, in studies of resource separation (see also Chapter 11) or in categorizing a predator as a generalist or specialist, the level of prey identification can greatly affect the outcome. For instance, the aquatic larval stages of the dipteran family Chironomidae (midges) are a major prey category for many small freshwater fishes, including darters (Strange 1993; Knight and Ross 1994; Ross 2001). With few exceptions, studies do not identify chironomids beyond the family level (identification below the level of family is highly labor-intensive), and any studies of resource separation among fishes feeding on chironomids would infer high dietary overlap.

Speckled Darter (*Etheostoma stigmaeum*), a common species in many southeastern streams of the United States, feeds heavily on larval chironomids (Ross 2001). In a second-order reach of Beaverdam Creek, Mississippi, chironomids usually composed over half of the aquatic invertebrate resource base and included 35 taxa (Alford and Beckett 2006). If chironomids were identified only to the level of family, the data indicate opportunistic feeding by Speckled Darter on various invertebrate taxa, including chironomids. However, when chironomid identifications were resolved below the level of family, the Specked Darter turns out to be a chironomid specialist, selecting only 1–3 chironomid taxa that tend to be more mobile or otherwise conspicuous. Similar results occurred in a companion study that included Brighteye Darter (*E. lynceum*), Gulf Darter

TABLE 12.3

Levels of Identification in 30 Food Habit Studies with an Average of 19 Prey per Study

Percent of prey	Identified to genus	Identified to species
0	11	18
10	5	4
20	5	2
30	2	4
40	4	0
50	0	1
60	1	0
70	0	0
80	1	1
90	1	0
100	0	0

NOTE: Data show the number of studies that identified different percentages of prey to genus or species (e.g., 11 studies did not identify any prey to genus and 1 study identified 90% of the prey to genus). Based on the first 30 fish taxa appearing in an electronic database search of fish food habits. Only papers with tabular data on fish stomach contents were used and the total number of prey taxa was restricted to animal categories.

(*E. stigmaeum*), and Blackbanded Darter (*Percina nigrofasciata*), in addition to Speckled Darter (Alford and Beckett 2007). With chironomid identifications at the level of family, all darter species could be categorized as opportunistic predators feeding on a small number of invertebrate prey taxa. When chironomids were identified to the genus/species levels, all four fish species were clearly specialists, primarily on chironomid taxa. The take-home message from these studies is that the level of taxonomic identification can have a major bearing on the outcome of studies of resource overlap and prey selection.

Basis of Prey Selection

Although it is clear that fishes are often selective predators, the actual basis of how prey selection occurs is less clear. Among planktivorous fishes, selection of prey is determined by the visual conspicuousness and apparent size of the prey (affected by body size, the degree of pigmentation and opaqueness of the prey body, and movement), prey abundance, and encounter rates of the predator and prey (Brooks 1968). In addition, at least some predators seem able to assess the nutritional value among prey in the same size category (i.e., choosing *Daphnia* with eggs in brood cavity over those without eggs) (Werner and Hall 1974; O'Brien 1979; Gliwicz et al. 2010). Detection of prey in piscivores is also related to size and conspicuousness of prey. For example, the reaction distances of Largemouth Bass (*Micropterus salmoides*) to Redfin Shiner (*Lythrurus umbratilis*) and Bluegill prey increase linearly with prey size. In addition, movements of prey decrease the reaction distance, especially for small prey. Light intensity also affects prey detection, but at least in one study of Largemouth Bass preying on Bluegill, the reaction distances remained relatively constant in decreasing light until very low light levels (< 5 lux) (Howick and O'Brien 1983).

The success of the approach, aim, and strike sequence varies widely among fishes and is influenced by fish size, experience, water conditions, prey movement, and so on. Relatively few studies have actually addressed the predation success of young fishes under natural conditions. In a study of recently emerged Brook Trout (*Salvelinus fontinalis*) in three streams in Ontario, Canada, fewer than 42% of the attacks ended successfully with ingestion of the food item (McLaughlin et al. 2000). Importantly, the attack rate was only a poor predictor of ingestion rate because of problems of distinguishing appropriate prey or because of avoidance behavior by the prey. (Anyone who has tossed small sticks or other inedible floating objects into a stream or pond and elicited strikes by young fishes can appreciate their challenges of distinguishing appropriate prey). The capture success of young Brook Trout increased with body size but was less for attacks directed at surface prey than for midwater or bottom-inhabiting prey. Although capture success was greater for moving compared to hovering fish, hovering fish were more likely to ingest a captured item compared to moving fish—perhaps because of better prey identification.

For predators that swallow prey whole, the maximum prey size is constrained by gape size or other size limitations in the mouth and esophagus, and once the size of prey exceeds a critical measurement of mouth size (see also Chapter 8), the chances of being eaten drop to zero (Zaret 1980). Gape-limited predation is particularly important for the early life-history stages of most fish species as well as for macrocarnivores/piscivores, but generally is not important for planktivores once the predator is past the early juvenile stage, nor for many benthic predators (Mittelbach 1981; Christensen 1996; Gill 2003). For macrocarnivores/piscivores, the size range of available prey increases as a function of gape size, and even though predators are generally size selective, the total spectra of available prey sizes is much greater for large compared to small predators (Figure 12.3). The amount of energy per prey item also increases with prey size, but because costs associated with handling also increase, predators are more likely to feed somewhere between the minimum and maximum prey sizes rather than selecting the largest possible prey (Stein 1977; Nilsson and Brönmark 2000; Gill 2003).

Deep-bodied prey species are generally less vulnerable to gape-limited predators, compared to more terete body forms, as are those with well-developed spines (Nilsson and Brönmark 2000). In experimental tanks, Largemouth Bass selected similar sized (in SL) Fathead Minnows (*Pimephales promelas*) over the deep-bodied Pumpkinseed (*Lepomis gibbosus*) when given a choice. The maximum size (both in body depth and SL) of Pumpkinseeds consumed by bass increased for three progressively larger size groups (Hambright 1991). In a classic European study of the effects of armor on prey selection by piscivores (Northern Pike, *Esox lucius*; and Perch, *Perca fluviatilis*), when given equal numbers of minnows (a mixture of four species), Ninespine Stickleback (*Pungitius pungitius*), and Threespine Sticklebacks, both predators preferentially consumed minnows, followed by the Ninespine Stickleback, which

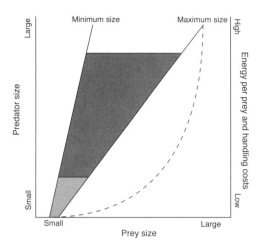

FIGURE 12.3. Prey-size availability, energy gain per prey item, and handling costs in gape-limited predators. Small predators have a reduced size range of prey available to them as well as an overall smaller prey-size spectrum (light shading) compared to the prey-size range and prey-size spectrum (dark shading) of large predators. Both energy gain per prey item as well as handling costs increase with prey size (dashed line). Based on Gill (2003).

has much smaller spines than Threespine Sticklebacks. Only when the other two prey items were largely depleted did the predators switch to preying on Threespine Sticklebacks (Figure 12.4). As further evidence of the role of spines, protection from predation was generally lost for the Threespine Sticklebacks when the dorsal, pectoral, and pelvic spines were clipped off near the base (Hoogland et al. 1956). In general, the impact of gape-limited predators on their prey is to select for rapid growth, defensive structures, or changes in body shape that quickly move them beyond the prey-size range taken by the predator (Rennie et al. 2010).

Optimal Foraging

Decisions by animals on where, when, and how to forage are often driven by optimality—balancing the net gain in energy per unit time against energetic costs in foraging (including search, capture, and handling) as well as risks of predation or injury (Emlen 1966; Schoener 1971; Helfman and Winkelman 1991). Most

simply, the basic hypothesis is that within a range of possible behaviors, animals will forage in ways that maximize fitness. Generally the studied currency is energy or time and the utility of optimal foraging models depends on appropriate estimates of costs, benefits, and constraints (Pyke et al. 1977). Optimal diets are affected by prey abundance, and both theory and empirical data indicate greater selectivity in food choice when the predator is near satiation or when food is abundant, in contrast to lower selectivity when a predator is starved or when prey are scarce (Ivlev 1961; Emlen 1966; Schoener 1971). For instance, when placed in a pool with four apparent density levels and three size classes of *Daphnia magna*, juvenile Bluegill (mean size 73.5 mm TL) were nonselective at the lowest prey density, consuming the three size classes as they were theoretically encountered (Figure 12.5) (Werner and Hall 1974). At the highest prey density, fish fed primarily on only the largest prey-size class and generally ignored the smaller prey. Because detection probabilities are greater for large versus small prey (as previously mentioned), the actual numbers of *Daphnia* in the experimental arena had to first be adjusted so that the probability of detection was the same for all size classes (relative to the largest size class, numbers of intermediate sized prey were set at 1.2 times, and numbers of the smallest prey at 3.7 times).

Studies on optimal foraging, including those on fishes, first appeared in the mid-1960s and have been common in the literature since the 1970s, but have had variable success in correctly predicting diets. Optimality predictions have been most successful for foragers feeding on immobile or weakly mobile prey (e.g., zooplankton) relative to the predator, but generally less successful in predicting diets for predators attacking more mobile prey (Pyke et al. 1977; Sih and Christensen 2001). Two major predation strategies have been defined, time minimizers and energy maximizers (Schoener 1971; Pyke et al. 1977). Time minimizers are likely to occur in species or life stages that have relatively fixed energy requirements so that fitness

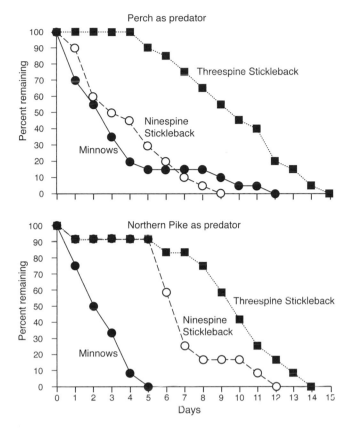

FIGURE 12.4. Predator selection of unarmored or weakly armored prey over armored prey. Twenty individuals of each prey species were used in the Perch predation experiment and 12 individuals of each prey species were used in the Northern Pike experiment. The minnows included a combination of four European species: European Minnow (*Phoxinus phoxinus*), Roach (*Rutilus rutilus*), Rudd (*Scardinius erythrophthalmus*), and Crucian Carp (*Carassius carassius*). The two other prey species were the Ninespine Stickleback (*Pungitius pungitius*) and the Threespine Stickleback (*Gasterosteus aculeatus*). Data from Hoogland et al. (1956).

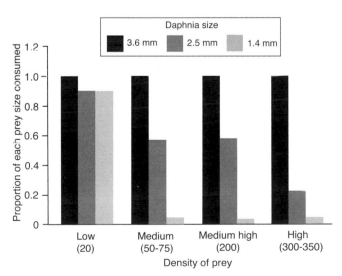

FIGURE 12.5. The role of prey availability in optimal feeding of juvenile Bluegill (*Lepomis macrochirus*). The densities of prey refer to the density of the largest size class; because encounter probabilities decrease with prey size, all else being equal, the number of intermediate prey was 1.2 times the density of the large prey and the number of small prey was 3.7 times the density of the large prey. This resulted in equal encounter probabilities of the three size classes. Based on Werner and Hall (1974).

is best maximized by freeing up time for other activities by minimizing the time spent foraging in the fulfillment of energy requirements. Energy (or number) maximizers are more likely among life stages or species for which fitness increases commensurately with increased food consumption. In this case, the maximum fitness will occur for the animal that achieves the greatest rate of energy gain over an extended time period for feeding.

All darters (Percidae) typically feed on the same general size spectrum of prey, with most prey no larger than 5 mm (Page 1983; Rakocinski 1991). In a comparative study of Dusky Darter (*Percina sciera*), Blackbanded Darter (*P. nigrofasciata*), and the Freckled Darter (*P. lenticula*), three sympatric darter species in the primitive subgenus *Hadropterus* that differ in average and maximum body sizes (Figure 12.6A), prey size typically increased with increasing darter size within and among species (Rakocinski 1991). The maximum prey size was a reflection of gape limitation. Diet optimality in terms of prey-size use was based on handling time/prey mass and diets were compared using an index of divergence of actual prey-size spectra from the optimal prey size (optimal diet index [ODI]) (Figure 12.6B). The ODI represents the average divergence of all prey from an optimum prey size, scaled to the optimum prey size. Higher values of ODI indicate less optimal diets and a value of 0 indicates a fully optimal diet. Thus a value of 15 would indicate that the prey-size composition of the diet was 15 times less valuable than a diet made up solely of optimum prey sizes. Small Blackbanded and Dusky darters were number maximizers and both changed to time minimizers (i.e., prey-size selection) at approximately 1 g wet-weight (medium sized), when the diets became about 10 times lower in value than an optimal diet. The very large Freckled Darter was clearly size selective, although the sample of this species was small and did not include juveniles. In general, small individuals of all three species showed more optimal diets than large individuals. Small prey tend to be much more abundant than large prey, and the geometrically decreasing abundances of larger prey items force large darters to have less optimal diets despite their change to prey-size selection (but large body size can have other advantages; see chapters in Part 3) (Rakocinski 1991). It has been hypothesized that the prey-size spectrum in North American streams has selected for the trend toward smaller body sizes in the more derived groups of darters (Page and Swofford 1984).

In contrast to the three darter species, where small fish had more optimal diets compared to large fish, a classic study of three size classes of Bluegill showed better agreement with optimal diet predictions for the largest size class (101–150 mm SL), compared to two smaller size groups (10–50 mm SL and 51–100 mm SL) (Mittelbach 1981). The optimal diet was based on a model comprising prey encounter rates, prey handling times relative to fish size, prey density, and prey size. The model was parameterized using laboratory data and then tested in a small Michigan Lake. Not only did the actual diet of the largest size class of Bluegill agree closely with model predictions, but the large size group also switched habitats relative to foraging profitabilities as predicted, initially feeding in vegetated habitats and then shifting to the open-water plankton in July. Bluegill < 100 mm SL were apparently constrained from switching to more profitable habitats by the threat of predation from Largemouth Bass (Mittelbach 1981).

PREDATOR FAILURE, OR HOW TO AVOID BEING EATEN

As a potential prey item, an organism faces a different set of problems from a predator. Important information for prey includes the kinds of predators sharing its environment, the encounter frequency with predators, and whether or not predators can be avoided, deterred, or defended against. Efforts to avoid predation can result in direct energy expenditures and also energy lost because the feeding

A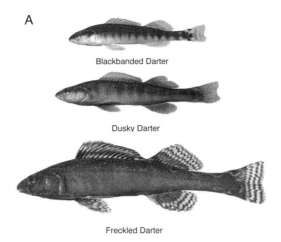

Blackbanded Darter

Dusky Darter

Freckled Darter

B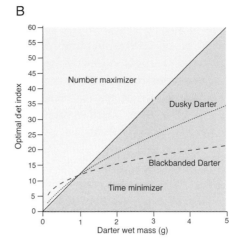

FIGURE 12.6. Optimal feeding in three sympatric darter species. A. Relative average adult sizes of the three species: Blackbanded Darter (*Percina nigrofasciata*), maximum body size 93 mm SL; Dusky Darter (*P. sciera*), maximum body size 110 mm SL; Freckled Darter (*P. lenticula*), maximum body size 168 mm SL. Maximum body sizes from Ross (2001). B. Size changes in feeding tactics of Dusky Darter and Blackbanded Darter based on optimal diet calculations (sample sizes were too low to include the Freckled Darter). The area above the diagonal line indicates a number-maximizing tactic and the area below a time-minimizing tactic. The optimal diet index shows the departure of the actual diet from a diet consisting wholly of optimal prey sizes (high values are increasingly less optimal). Based on Rakocinski (1991).

time or efficiency is reduced by the need for deterrence. Consequently, an organism must balance the risk of predation against the effects that active avoidance of predation will have on its ability to harvest resources, attract mates, reproduce, and so on. Again, because all fishes at some point in their life cycle are both predator and prey, there can be potentially conflicting demands on natural selection.

The predation cycle (Figure 12.2) also provides a framework for predator avoidance mechanisms. Avoidance mechanisms can occur at any (or all) of the stages, although risk to the potential prey generally increases the further along in the cycle (making it better to avoid detection than rely on avoidance methods to be successful at the point of attack).

Avoiding Detection and Identification

Rarity and Predator Behavior

An important aspect of predator avoidance is to be actually or functionally rare. Actual rarity can occur through natural selection for greater dispersion and/or wider habitat tolerances.

Because some predators preferentially attack more common prey (termed apostatic predators), actual rarity can be an effective means of avoiding predation, even though continued apostatic predation will result in making rare prey more common (Endler 1986; Thomas et al. 2010). Apparent rarity can occur through temporal and spatial activity cycles that minimize times of overlap with predators. Apparent rarity can also involve behaviors, morphologies, and color patterns that reduce the signal-to-noise ratio, such as crypsis, or polymorphisms that provide common and rare phenotypes of the same species (Endler 1986).

Most fishes are active during the day or night, but usually not both, in part because of retinal structure and function (Helfman 1986). Retinal structure and function, however, evolve in response to natural selection and such changes can potentially occur rather quickly, as demonstrated in laboratory populations of Guppies (*Poecilia reticulata*) that showed a functional response to selection for vision under blue or red light conditions after only 3–4 generations (Endler et al. 2001). Vision in bright light (photopic vision) is dependent on pigments in cone cells, in contrast to vision in low light (scotopic vision), which is based on pigments in rod cells (Munz and McFarland 1973; Helfman 1986). Although primarily studied in marine tropical-reef systems, there is a general pattern of change between day- and night-active fishes in at least some North American freshwater habitats as well (Helfman 1981; Salas and Snyder 2010; Roach and Winemiller 2011). For instance, in Cazenovia Lake, New York, diurnally active planktivores, surface/water-column feeders, and macrocarnivores, such as Bluegill, Yellow Perch, Mimic Shiner (*Notropis volucellus*), Banded Killifish (*Fundulus diaphanus*), and Smallmouth Bass (*Micropterus dolomieu*), moved out of the water column at night to resting areas near the bottom where they were essentially motionless or became more closely associated with cover until dawn (Helfman 1979, 1981). Nocturnal use of cover was especially pronounced in smaller fishes. Nocturnally active fishes such as Golden Shiner (*Notemigonus crysoleucas*), Rock Bass (*Ambloplites rupestris*), and Brown Bullhead (*Ameiurus nebulosus*) moved from shelter to foraging areas and initiated feeding as light levels dropped.

During evening twilight and dawn, light conditions can rapidly change such that neither scotopic or photopic visual systems are effective, resulting in a time of increased vulnerability to predators by the "visually impaired" prey. As documented with marine fishes, it turns out that many predators have intermediate retinas that are most effective at twilight and dawn (mesopic vision) compared to either light or dark conditions (Munz and McFarland 1973). In both marine and freshwater systems, some predators also have the behavioral advantage of being near the bottom and striking upward into the water column at prey that are backlit. Thus most prey species become inactive during the intermediate light conditions of twilight and dawn, and smaller individuals of diurnal species cease activity and seek shelter earlier in the evening and then initiate activity and leave shelter later in the morning than do larger

individuals of the same species (Helfman 1981; McFarland 1986).

Fish Shoals and Apparent Rarity?

Fishes commonly occur in associations of two or more individuals and, as shown by experiments with Fathead Minnow, prefer the company of other conspecifics over remaining alone (Hager and Helfman 1991). Groups of fishes that stay together for various social reasons are termed shoals. Fish schools make up one category of shoaling fishes defined by having some level of internal organization, such as polarization (all pointing in the same direction) and the constant adjustment of speed and direction to match other members in the school (Breder 1976; Partridge 1982; Pitcher 1986; Pitcher and Parrish 1993). At least half of all known fishes form schools at some part of their life cycle and perhaps a quarter school throughout life (Shaw 1978).

Several predator deterrent roles have been proposed for fish shoals and/or schools, including the avoidance of predation by means of achieving apparent rarity, avoidance of detection and/or attack by the combined vigilance of the shoaling individuals, and confusing predators (Partridge 1982; Magurran and Pitcher 1987). The argument for apparent rarity is that under conditions where the visual range of the predator is low relative to its swimming speed, the envelope of detection surrounding each fish shoal (envelope of detection is the distance away from a prey that a predator can locate it; its radius equals maximum visibility) is not much greater than the envelope of detection for a single fish. Thus the probability of detection by a swimming predator for four fish in a shoal is only slightly greater than one-fourth of the probability of detection for four individual fish (Figure 12.7). In the open sea, or in the pelagic zone of a large lake, the chance of finding a school of 1,000 fish is only slightly greater than finding one fish—apparent rarity results from the overlapping of envelopes of detection (Partridge 1982). The predictions of the encounter effect of aggregation have been tested under

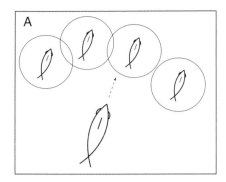

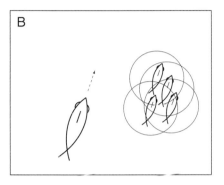

FIGURE 12.7. The creation of apparent rarity through shoaling behavior. Envelopes of detection of individual fish (A) cover approximately four times the area compared to the area covered by overlapping envelopes of detection for shoaling fish (B). The dashed line indicates the direction of movement of the predator.

laboratory conditions using Threespine Stickleback feeding on chironomid larvae. As groups of chironomids became larger, the number of groups decreased, resulting in lower encounter rates between predators and their prey as a function of increasing group size (Ioannou et al. 2011).

Once a shoal is detected, however, the actual fitness of the prey might not be enhanced by shoaling if predators consume multiple prey, and in many cases predacious fishes accompany shoals of prey fish, so that detection by the predator has already occurred even if actual feeding is sporadic. In addition, fish shoals can be more apparent to aerial predators than single fishes, and the detection of shoals by fish predators increases with water clarity (Pitcher 1986; Pitcher and Parrish 1993). Because of this, Pitcher and Parrish argue that it is unlikely

that the creation of apparent rarity has been important in the selection for shoaling behavior in fishes. The experimental support for the encounter rate hypothesis with Threespine Stickleback and chironomids suggests that broad generalities might be inappropriate and that the importance of apparent rarity through shoaling is context dependent.

Crypsis

Apparent rarity is more likely to occur through crypsis compared with shoal formation (Pitcher 1986). Crypsis refers to any trait of an organism that reduces its probability of detection by another organism when it is potentially perceivable. Crypsis can include behavioral (i.e., remaining motionless) as well as morphological and chemical traits and, although it is most often studied with visual predators or prey, it could work in any sensory modality and, of course, applies to all life-history stages, including eggs, larvae, and adults (Ruxton 2009; Stevens and Merilaita 2009). In addition to benefiting prey, various forms of crypsis would also be advantageous to a predator in remaining undetected by its prey until the final approach and strike.

Two forms of crypsis that are usually recognized are general resemblance, where the prey color pattern resembles a random sample of the background over which it is usually seen by a predator, and masquerade or special resemblance where the prey color pattern resembles a specific, normally inedible object (Endler 1978, 1986, 1988). There is not an absolute distinction between general and special resemblance, the differences being instead a matter of degree of how closely the color pattern or morphology matches an object. In the case of special resemblance, the prey may be detected by a predator but not identified as prey (Ruxton 2009). Although masquerade is often considered a form of crypsis, because it also functions at the level of identification rather than detection, it is sometimes considered as a separate category (Skelhorn et al. 2010). Examples of masquerade occur primarily in marine fishes, such as the Leafy Seadragons (Syngnathidae) of the Pacific Ocean off south Australia, and in Neotropical freshwater fishes, such as the Leaffish (Polycentridae) of the Amazon Basin, and have apparently not been documented among North American freshwater fishes (Helfman et al. 2009; Skelhorn et al. 2010).

General resemblance to the stream bottoms is shown by numerous North American bottom-dwelling fishes, including suckers (Catostomidae), madtom catfishes (*Noturus*, Ictaluridae), sculpin (Cottidae), and darters (Percidae), all of which have 3–4 or more large dorsal saddles that extend down the sides (Figure 12.8) (Armbruster and Page 1996). Thus far, at least 49 species of fishes in these four families have been recognized as having the saddle pattern. Species with the dorsal saddle pattern almost always occur in rocky streams and in flowing water, and the lighter spaces between the saddles are thought to resemble rocks while the saddles themselves resemble the shadows and spaces between the rocks.

Glacier Bay, located in southeastern Alaska, provides a natural laboratory to study the response of cryptic fishes to changing environments. The fjord is over 100 km long and was covered by a neoglacial advance that reached its maximum extent around 1700, followed by recession starting between 1735 and 1785. As the glacier has withdrawn up the valley, streams are progressively exposed so that the oldest streams are near the mouth and the newest are farthest up the valley (Milner et al. 2000). The substratum color of the streams varies with stream age; young, less productive streams generally lack algal cover and are dominated by white, gray, and black rocks. In contrast, the greater productivity of older streams results in more algae, submerged macrophytes, and riparian vegetation so that the substratum has more red and green colors (Whiteley et al. 2009). As the streams are exposed, they are gradually invaded by Coastrange Sculpin (*Cottus aleuticus*), a benthic, cryptically colored fish. Coastrange Sculpin show strong phenotypic divergence in coloration among Glacier Bay streams, and fish

color is significantly correlated with the color of the stream bottoms across the chronological gradient.

Color change can occur rapidly, on a scale of minutes, via signals sent from the nervous or endocrine systems and acting on pigment dispersion in chromatophores. As pigments are concentrated at the center of a chromatophore, the color is lighter compared to chromatophores with dispersed pigments (see also Chapter 10). Color change also occurs over longer time intervals (morphological color change) as new pigments are manufactured or broken down (Whiteley et al. 2009). Color changes occur so as to maximize crypsis and avoid predation (most likely at the level of detection). Younger streams, because of greater turbidity and the absence of streamside vegetation generally have reduced predator threats from fishes, birds, and mammals, in contrast to older, clearer streams, so that the selection for crypsis also varies with stream age (Whiteley et al. 2011). In addition to Coastrange Sculpin, various other species have shown rapid (maximum change in 1.5–6 hours) changes in color pattern and color intensity as a function of background changes, including Eastern Mosquitofish (*Gambusia holbrooki*), Sailfin Molly (*Poecilia latipinna*), and Rainwater Killifish (Cox et al. 2009).

Countershading in fishes and other organisms, where the dorsum is dark and the undersides are light, also constitute a form of general resemblance when resulting in background matching (Ruxton et al. 2004). When viewed from below by a predator, the lighter ventral surface blends in with the downwelling light and when viewed from above the darker dorsum matches the darker stream or lake bottom (Helfman et al. 2009; Stevens and Merilaita 2009).

Fishes can also avoid detection through disruptive colorations—markings that form false boundaries or hide existing boundaries, so that detection of overall body shape is made more difficult. Disruptive coloration can function as crypsis, or if detected, can make identification more difficult (Sherratt et al. 2005; Stevens and

FIGURE 12.8. Apparent rarity through general resemblance, a form of crypsis in the Frecklebelly Madtom (*Noturus munitus*). Based on Armbruster and Page (1996). See also corresponding figure in color insert.

Merilaita 2009). Disruptive coloration likely contributes to the crypsis that results from the banding pattern created by dark saddles of certain darters and sculpins (Armbruster and Page 1996; Whiteley et al. 2009).

Reducing Predation Risk at the Approach, Aim, and Strike Phases

Once a prey fish has been detected by a predator, there are still numerous options for reducing the chances of being eaten. Some examples include fleeing; the confusion effect caused by large numbers of prey in a shoal or school; and prey defenses through spines, body shape, and behavior.

A common response of prey to an approaching predator is to use evasive movements, which can be particularly effective if the prey's direction of travel differs from the predator's (Endler 1986). In fact, most all fishes use a rapid acceleration out of the strike path to avoid predation. The escape response is generated by one of two large spinal neurons (called Mauthner cells) that trigger a mainly unilateral contraction of the axial musculature. Because the first phase of the response results in the body forming a "C" shape, it is also referred to as a C-start response. The response, which occurs in less than one second, involves both maximum acceleration and a turn, thus propelling the body in

a trajectory approximately at a right angle to its original direction (Webb 1986; Walker et al. 2005; Tytell and Lauder 2008). Faster C-starts can lead to increased survival probability in an attack, as shown by a study of the Pike Cichlid (*Crenicichla alta*) attacking Guppies (*Poecilia reticulata*) (Walker et al. 2005).

Fathead Minnows, as well as other cyprinids, respond to a predator by fright responses that include dashes and skitters (Mathis and Smith 1993b). Skitters are C-start accelerations that last for 5–10 body lengths, followed by braking and a return to the shoal. Skittering behavior also occurs in synchrony, where most of the shoal members are involved (Pitcher 1986).

An extreme form of aquatic predator avoidance shown by certain small fishes, especially killifishes, is to briefly leave the aquatic habitat. Western Mosquitofish (*G. affinis*) will voluntarily strand themselves on shore to evade predators and are capable of directed movement across damp ground using tail flips (Gibb et al. 2011). Both Eastern Mosquitofish and Rainwater Killifish have been observed avoiding Bluegill predators by leaping onto the surface of lily pads and remaining motionless for 5–10 s. However, in a few instances, Bluegill were still successful in capturing their prey by biting through the lily pad from underneath (Baylis 1982).

When approached by a predator, the response of most shoaling prey fishes is for individuals to move closer together, forming a polarized school, followed by an outward "explosion" of fish (termed flash expansion) if an attack occurs (Nursall 1973; Pitcher 1986; Magurran and Pitcher 1987). The preference of individual fishes for joining shoals of conspecifics, or even heterospecifics, increases under threat of predation. The importance of group size in providing protection from predation is shown by a study of Largemouth Bass predation on Eastern Silvery Minnows (*Hybognathus regius*) conducted in outdoor wading pools. Largemouth Bass rapidly captured single fish, but had poorer success and took progressively longer to capture prey as the group size increased. However, the addition of morphologically distinct fish (dyed with Nile Blue) to a group greatly increased predator success, illustrating the selection against oddity (Landeau and Terborgh 1986). In another study, Fathead Minnow responded more rapidly to joining larger shoals of conspecifics when a predator (Largemouth Bass) was present than in the absence of a predation threat (Hager and Helfman 1991). Similarly, Banded Killifish chose the larger of two conspecific shoals when threatened by an avian predator model, as long as fish were about the same size in the two shoals. In fact, size of individuals in a shoal is often important in shoal choice by an individual. When a Banded Killifish was given a choice between a heterospecific shoal (Golden Shiner) where fish were about the same size as the test fish, and a shoal of larger Banded Killifish, the test fish selected the Golden Shiner shoal instead of the shoal of larger conspecifics (Krause and Godin 1994).

These responses are strong indications that fishes gain some protection from predators by being in a group rather than remaining alone. The selfish herd hypothesis (Hamilton 1971) predicts that the selfish avoidance of a predator, by associating with another potential prey, can lead to aggregation. Individuals that are surrounded by others should reduce their risk of predation and would gain the greatest advantage

by positioning themselves centrally within the group (Morrell and James 2007). The reduced susceptibility to predation offered by fish shoals could also allow fishes in groups to reduce their vigilance toward predators and perhaps gain increased time for various essential activities, such as feeding. One prediction of the group vigilance or "many eyes" hypothesis is that as group size increases there are more eyes on the lookout for predators so that individuals in a large group can spend more time foraging and devote less time to vigilance (Lima 1995). In support of this, European Minnow (*Phoxinus phoxinus*) reacted faster to the stalk of a pike model, but remained on the food patch longer, when in large versus small groups and consequently showed a better balance of effort between watching for predators and foraging (Magurran et al. 1985). In a study of Spottail Shiner (*Notropis hudsonius*) in an Ontario lake, individual fish fled from a sham predator at greater distances compared to fish in groups, suggesting that shoaling fish could gain additional feeding time. Surprisingly, the reaction distance to predators did not change for groups of two or larger (Seghers 1981). Larger shoals may also be smarter—based on studies of Eastern Mosquitofish (*Gambusia holbrooki*), albeit carried out with an introduced population in Australia, larger shoals had greater speed and accuracy of decision making under predation threat than small shoals. The experiments took place in a Y-maze and the response variable was the time it took for a shoal to reach the arm of the maze that did not have a replica of a predator (Ward et al. 2011).

A successful attack by a visually hunting predator generally requires selecting an individual target rather than slashing haphazardly at the shoaling fish (Shaw 1978). One potential reason for increased protection from predators in a group is the confusion effect that shoaling or schooling prey gain because of a neurological information-processing constraint (Shaw 1978; Tosh et al. 2006). As determined primarily from studies on humans, visual information processing has a maximum number of cognitive comparisons that can be performed simultaneously (Fisher 1984). The possible role of multiple prey confusing predators was first mentioned by Miller (1922) in a study on bird flocks, and by Welty (1934), albeit briefly, in a study of Goldfish responding to *Daphnia*. Psychological studies on humans show a limit of 3–7 for the number of items that can be simultaneously recognized (Miller 1956; Fisher 1984; Halford et al. 1998). The constraint results in increasing targeting error as shoal sizes increase beyond 3–7 individuals. In fact, as previously mentioned, predator efficiency is generally lower as shoal sizes increase (Magurran and Pitcher 1987; Godin 1997). In the study of Largemouth Bass predation on groups of Eastern Silvery Minnow, most of the effect of increasing group size on protection from predation occurred at group sizes of 4–8. Consistent with predictions of the confusion model, but with larger group sizes, Threespine Stickleback feeding on *Daphnia* had lower attack success on groups of 500 compared to groups of 20, and the decreased predator success was linked with increased error (Ioannou et al. 2008). In addition to information processing in the central nervous system, even if a predator does decide to attack a particular prey, additional confusion is caused by the continual movement of other prey in the shoal (Pitcher and Parrish 1993).

Reducing Predation Risk at the Strike and Subjugation Phases

Successful predation can be hindered by prey having defensive structures such as spines or teeth, as shown previously for Threespine Stickleback versus minnows in defense against Northern Pike and Perch (*Perca fluviatilis*) (see Figure 12.4). Although most advanced teleosts (Euteleostei, Figure 7.5) as well as ictalurids possess spines that can be used in defense, there is surprisingly little direct evidence of the actual use of spines in defensive behavior, in part because of the interactions of morphology and behavior. However, there is extensive documentation that piscivores select elongate, soft-rayed prey (e.g., minnows and suckers) over deep-bodied prey with spines, such as centrarchids. For

instance, Northern Pike, Muskellunge (*E. masquinongy*), and their hybrid (Tiger Muskellunge) all selected Fathead Minnows and Gizzard Shad (*Dorosoma cepedianum*) over Bluegill in experimental arenas. In addition, handling times for Bluegill were significantly longer compared to the other prey. The presence of dorsal spines apparently acted to direct attacks to the head and tail regions and away from midbody, thus increasing the opportunity for escape (Wahl and Stein 1988). In whole-reservoir experiments, Wahl and Stein also found that all three esocids had reduced growth in a centrarchid-dominated reservoir compared to growth in a reservoir with alternative prey.

Observational studies also provide evidence of selection for morphologies that offer protection against gape-limited predators. For example, in streams of Michigan's Upper Peninsula, Brook Sticklebacks (*Culaea inconstans*) occurred in assemblages with only small, gape-limited predators (11 sites) and in assemblages with a mixture of predators, including gape-limited predators and non-gape-limited predators (15 sites). By one year of growth, Brook Sticklebacks from assemblages with only gape-limited predators had longer bodies, longer spines, and deeper midbody profiles in contrast to those from assemblages with both gape-limited and non-gape-limited predators. In the latter case, defenses against gape-limited predators were minimized and fish maximized their ability to escape predators by having more streamlined bodies. The morphological differences could have arisen through natural selection, through induced phenotypic plasticity, or a combination of both (Zimmerman 2007).

The development and maintenance of morphological defenses against predation can be energetically costly or have detrimental effects on other life functions. In some cases, the defenses are expressed when in contact with predators (inducible defenses), whereas in other instances defenses are always expressed (constitutive defenses). Inducible defenses of prey are phenotypic responses that offer increased protection from predators and arise from exposure to a predator or to chemical cues from the predator. The evolution of inducible defenses would be expected under conditions of variable predation pressure where the defense is not expressed unless needed (Dodson 1989; Tollrian and Harvell 1998; Brönmark and Hansson 2000).

Although long thought to be related to generating negative lift in swift water, the nuchal humps of Razorback Sucker (*Xyrauchen texanus*) (Fig 12.9) and Humpback Chub (*Gila cypha*) appear to be constitutive defenses, most likely evolved in response to the large gape-limited predator, Colorado Pikeminnow (*Ptychocheilus lucius*), native to the Colorado River system. Humpback Chub larger than 210 mm TL and Razorback Sucker larger than 220 mm TL are essentially immune to predation from an average size Colorado Pikeminnow. Not surprisingly, movement of Razorback Sucker and Humpback Chub from nursery areas where they are not syntopic with Colorado Pikeminnow to the adult habitat of deep pools and eddies where they are syntopic with Colorado Pikeminnow corresponds to enlargement of the nuchal hump (Portz and Tyus 2004).

Phenotypic responses to the presence of a predator have been demonstrated in young wild-caught Pumpkinseeds. Benthic and limnetic morphs (see Chapter 7) of Small Pumpkinseed (< 55 mm FL) were raised for six months in plastic tubs either with a Walleye in a screened-off portion of the tub (test) or with a large Pumpkinseed in the screened-off portion (control). Walleye were fed regularly on Pumpkinseed to ensure the presence of a chemical as well as a visual cue. Consistent with expectations, deeper bodies and longer spines were induced for both benthic and limnetic morphs in the presence of predator cues. In addition, the third dorsal spine was longer in the limnetic ecomorph compared to the benthic ecomorph. This was also consistent with expectations, given that limnetic ecomorphs should be more active swimmers and dorsal spines are less energetically costly in terms of drag because they can be folded down during swimming. Contrary to expectations, benthic and limnetic

forms did not differ in body depth, even though deeper bodies can result in greater drag during locomotion. Overall, the predator-induced responses were consistent with expectations—the greater body depth and dorsal spine length afforded a size refuge against gape-limited predators. The predator-induced phenotype resulted in a 70 mm TL Pumpkinseed having a body depth normally associated with an 80 mm TL fish. In addition, the spines have the potential of causing esophageal damage to predators if the Pumpkinseed are attacked. In a companion study, the predator-induced phenotypes had higher survival than controls under simulated benthic but not limnetic conditions (Robinson et al. 2007).

The actual mechanism for the predator-induced changes in body shape are less understood. However, there seems to be gathering evidence that at least some of the morphological responses are caused by a depressed level of activity when in the presence of a predator, resulting in greater amounts of energy available for growth. For instance, Crucian Carp (*Carassius carassius*) grown in treatments with standing and running water and with and without a pike stimulus, showed similar changes in body shape in the enclosure with no pike stimulus and no water current and in the enclosures with a pike stimulus (Johansson and Andersson 2009). A counter argument for fishes in nature is that reduced activity would result in lower food intake and thus reduced growth; however, it might also be that the presence of predators reduces the number of competitors and thus increases the per capita food availability.

Many North American freshwater fishes have spots located posteriorly on the body or fins that range from simple dark spots, as in Blacktail Shiner (*Cyprinella venusta*), to rather complex spots or ocelli, as in the Flier (*Centrarchus macropterus*) (Figure 12.10). As Nursall (1973) suggested, one role of such spots could be to deflect a predator's aim, although there are still scant experimental data for any organismal group that such spots have a predator deflection role (Stevens et al. 2008). The

FIGURE 12.9. Razorback Sucker (*Xyrauchen texanus*). The nuchal hump in this species and in the Humpback Chub (*Gila cypha*) most likely evolved as a defense against Colorado Pikeminnow (*Ptychocheilus lucius*), the large gape-limited predator of the Colorado River system. Illustration by Howard Brandenburg, courtesy of the New Mexico Game and Fish Commission. See also corresponding figure in color insert.

deflection hypothesis proposes that spots on the posterior part of the body can function in drawing predatory attacks to less vital regions of the prey's body, owing to the resemblance of the spots to eyes (Stevens 2005). Most of the research on putative fish deflection marks has been with marine and tropical freshwater fishes. For instance, painting artificial spots on the Panama Tetra (*Hyphessobrycon panamensis*), an otherwise silvery characin lacking distinct spots, served to shift attacks by Gar Characin (*Ctenolucius beani*) posteriorly away from the head and midbody, along with a trend for more missed strikes in the experimentally marked fish (McPhail 1977). Somewhat surprisingly, in spite of Nursall's early suggestion of the role of posterior spots as predator deflection marks in North American freshwater fishes, experimental evidence that they function as deflection marks is still lacking.

Fright and Alarm

Perhaps one of the best-studied, and challenging, examples of predator avoidance in freshwater fishes is that of alarm substances—the chemical group involved in the signaling of fright or alarm. Among North American freshwater fishes, alarm substances in skin extracts and/or alarm responses to skin extracts have been demonstrated in ostariophysans (Cypriniformes, Characiformes, Siluriformes),

esociforms (Umbridae), cyprinodontiforms (Fundulidae, Poeciliidae), gasterosteiforms (Gasterosteidae), scorpaeniforms (Cottidae), and perciforms (Gobiidae, Percidae) (Pfeiffer 1963, 1977; Mathis and Smith 1993a; R. J. F. Smith 1979, 1982, 1992; Wisenden et al. 2007). Conspecific, and sometimes heterospecific, fishes respond to alarm substances by showing fright behavior, such as remaining motionless; showing increased use of shelters; or forming shoals, all of which may lessen chance of predation (R. J. F. Smith 1979, 1992; Mathis and Smith 1993b; Wisenden et al. 2010). Although the majority of studies of fish responses to the alarm substance have been in a laboratory situation, alarm responses have also been detected for several North American freshwater fishes in the field. For instance, after being exposed to a conspecific skin extract, Rainbow Darters (*Etheostoma caeruleum*) in a natural stream showed more than a 60% increase in latency before resuming normal movements compared to controls (Crane et al. 2009). In a study of Northern Redbelly Dace (*Chrosomus eos*), Fathead Minnow, and Brook Stickleback in two small lakes in Minnesota, extract from a 2 cm^2 cyprinid skin was responded to by fishes as an alarm signal for at least a distance of 2 m over a 2-h period (Wisenden 2008).

There is also evidence that prey fishes can also detect the alarm substance from feces of their predators. Northern Pike that were fed

FIGURE 12.10. Examples of posterior fin or body spots in Flier (*Centrarchus macropterus*). See also corresponding figure in color insert.

Fathead Minnows triggered an alarm response when placed near other minnows or when only the pike feces were placed near the minnows (Brown et al. 1995). In addition, the Northern Pike tended to defecate away from their feeding area, suggesting that they were minimizing the chance of alarming their prey.

Studies on alarm substances were pioneered by the German ethologist and Nobel Laureate Karl von Frisch, who named the material Schreckstoff, literally fright substance (von Frisch 1938). Schreckstoff in fishes is stored in club cells, which are large, ductless epidermal cells lacking surface pores. Because they are ductless, the alarm substance is only released when the skin is damaged, a fact that has made the argument for the evolution of alarm substance challenging given that the signaler is generally, although not always, killed (R. J. F. Smith 1992; Williams 1992).

Recently, rather compelling evidence has been presented in support of the hypothesis that epidermal club cells in fishes of the Ostariophysi and Acanthopterygii may be maintained by natural selection owing to protection they confer against various agents that might compromise the integrity of the epidermal layer, such as pathogens, parasites, and UV radiation (Chivers et al. 2007). Because club cells maintain a first line of defense to protect underlying tissues, they are preadapted to be

the first cells that are ruptured in the event of a predatory attack—making the alarm function of the club cells a secondary adaptation. Consequently, rather than viewing the alarm substances as an example of a pheromone involved in communication (see Chapter 10), it now appears they should be viewed as a secondarily evolved predator avoidance response (Magurran et al. 1996; Burnard et al. 2008). Additional research has shown that suppression of the immune system of Fathead Minnows with cortisol injections significantly reduced the number of club cells compared to control fish, as would be predicted if club cells were part of the immune system (Halbgewachs et al. 2009). The role of the alarm substance would seem to fit the definition of exaptations, "features that now enhance fitness but were not built by natural selection for their current role" (Gould and Vrba 1982).

THE BALANCE OF BEING PREDATOR AND PREY

Fishes and other animals must balance the risk of predation against the rewards from foraging, and these trade-offs can affect habitat use and the channeling of energy into growth and reproduction (Gilliam and Fraser 1987; Lima and Dill 1990). For example, studies done in natural lakes and experimental ponds in

Michigan show that Bluegill respond to the risk of Largemouth Bass predation by shifting habitat—a change that has consequences for growth (Werner et al. 1983a, b; Werner and Hall 1988). When they initiate feeding, larval Bluegill leave the nest in the littoral zone and move into the pelagic zone to feed on zooplankton. Once they reach a size that puts them at risk of predation (i.e., 11–14 mm SL), they return to the vegetated littoral zone to feed. The length of time spent in the vegetation depends to some extent on the level of predation risk in the pelagic zone. In ponds with Largemouth Bass, young Bluegill spend several years in vegetated littoral habitats before returning to the more energetically profitable, but riskier, pelagic zone. The switch in habitat occurs between 51 and 83 mm SL, with smaller fish switching when bass density is low and switching at larger sizes when bass density is high. However, the decision to avoid the more profitable but riskier habitats is not without its own costs—Bluegill restricted to the less profitable vegetated habitats had one-third of the foraging return rates and grew 27% less than control fish in the profitable and predator-free control habitat. Usually when Bluegill reach 60–100 mm SL, their risk of predation by Largemouth Bass is substantially reduced. Unlike the smallest size class of Bluegill, two larger size classes (> 50 mm SL) chose habitats to maximize foraging return rates. Because predation threat kept the smallest size class from feeding in open water, there was increased food available for the larger size classes, which showed concomitantly greater growth than in the predator-free control habitat (Werner et al. 1983a, b; Werner and Hall 1988). These studies demonstrate the impact of predator threat on prey behavior, that Bluegill could assess the relative degrees of habitat profitability and risk of predation and balance their choice of habitat accordingly, and the overall complexity of how the threat of predation can differentially affect varying sizes and ages of prey species.

Even though Largemouth Bass become piscivorous while still age-0, they are also at risk from cannibalism by larger Largemouth Bass. In a laboratory study, willingness to forage outside of a refuge and food consumption were studied in a tank divided in half by a clear partition (Parkos and Wahl 2010). The predator, a yearling Largemouth Bass (205 mm TL), was in one half of the tank and visible through a clear partition to three size classes of age-0 Largemouth Bass placed in the other side of the tank. Foraging of the age-0 Largemouth Bass was compared across three kinds of prey: zooplankton, chironomid larvae, and Bluegill, corresponding to a natural ontogenetic progression of prey for the three age-0 size groups. Prey kind did not influence foraging rate. However, the two smaller size classes of Largemouth Bass (20–30 mm TL and 45–55 mm TL) significantly reduced their foraging time and total prey consumption when the predator was visible. The largest size class (70–80 mm TL), although of a size still vulnerable to the predator, did not reduce foraging time nor suffer a loss in prey consumption, perhaps because the increased mobility of this size class reduces the probability of capture (Parkos and Wahl 2010).

Just as risk of predation influences habitat use, foraging areas, and feeding efficiency in lentic systems, fishes in lotic systems also show changes in behavior and habitat use under threat of predation, such as the tendency for bigger fishes to occur in deeper habitats. In small streams in Tennessee, pool depth was a strong predictor of maximum fish size and survivorship. Survival decreased for large individuals of Central Stoneroller (*Campostoma anomalum*), Creek Chub (*Semotilus atromaculatus*), and Striped Shiner (*Luxilus chrysocephalus*) in shallow versus deep habitats, with low survival in shallow habitats attributed to predation by terrestrial predators (e.g., herons and raccoons) (Harvey and Stewart 1991). As documented in various studies, the threat of predation from aquatic predators often causes a shift of small fishes into shallow, vegetated habitats, even though growth rates generally could be optimized by moving to deeper water (Power et al. 1985; Post and McQueen 1988; Werner and Hall 1988; Angermeier 1992).

The interplay between avian and mammalian predation on large fish in shallow water results in large fish seeking out deeper water. Fish predation on small fish in deep water results in small fish reducing predation risk by shifting to shallow water (Power 1987; Gelwick 1990; Harvey and Stewart 1991; Schlosser 1991). The overall result is a concave model of predation risk, with the greatest risk in shallow water for large fish and in deep water for small fish (Power 1987; Schlosser 1991; Harvey and Stewart 1991).

An exception to the rule of small fishes in shallow habitats and large fishes in deep habitats can occur with larval fishes. By eliminating the small fishes that prey on the larvae, large predators can create "predator-free" zones. In northern lakes, the movement of larval Bluegill into the pelagic zone after hatching is likely an effect of Largemouth Bass predation, in that the presence of bass forces juvenile fishes (which would otherwise prey on larval fishes) out of the pelagic and into the littoral zone. Larvae remain in this "predator free" space until about 12–14 mm SL, when their larger size and increased pigmentation raise the risk of predation by bass, resulting in movement back to the littoral zone (Werner and Hall 1988). Similarly, in pools of Brier Creek and Baron Fork of the Illinois River in Oklahoma, survival of larval centrarchids and cyprinids was low in pools that contained juvenile centrarchids and cyprinids, but significantly higher in pools with adult Largemouth or Smallmouth Bass (Harvey 1991a, b). Larvae were generally in deep pools, or deeper sections of pools, where the presence of a predator had reduced or eliminated juvenile fishes. Juvenile fishes were shifted to shallow water habitats where predation from Largemouth Bass would be limited both by access and by risk of predation on the bass by terrestrial predators.

Foraging decisions of fishes under the threat of predation can become more risk tolerant as energy stores are depleted through starvation or through the effects of parasite loads. For example, Threespine Sticklebacks are an intermediate host for the cestode (*Schistocephalus solidus*).

The definitive hosts (i.e., the final host where the parasite reproduces) are fish-eating birds (Barber et al. 2004). As the parasites grow in the body cavity of the Threespine Stickleback, they cause increased swelling of the abdomen and increased demands on energy reserves. The energy demand results in changes in risk tolerance during foraging. Normal Threespine Sticklebacks frightened by a net moving through an aquarium will suppress their feeding motivation for an energy equivalent of 1.3 kilojoules. In contrast, parasitized fish will tolerate only a slight suppression of the feeding response, equivalent to 0.24 kilojoules, or about 20% of the suppression shown by healthy fish (Giles 1987). Similarly, time spent in cover by parasitized Threespine Sticklebacks exposed to a sham avian predator was only a few seconds in fish with well-developed infections, in contrast to the duration of the experiment (ca. 5 min) in healthy fish (Barber et al. 2004). In a study using a fish predator (Cichlidae, *Tilapia*) held in an adjoining, visible aquarium, Threespine Sticklebacks that were infected with *Schistocephalus* foraged at a high rate both away from and near to the predator. In marked contrast, the nonparasitized fish had significantly lower feeding rates and avoided approaching the predator during foraging (Milinski 1985).

PREDATOR IMPACTS ON COMMUNITIES AND ECOSYSTEMS

Predation by fishes or other organisms on aquatic communities can cause changes in species composition, resource use, life-history patterns, competitive interactions, and survivorship of various age classes. These effects can substantially alter both the kinds of species and the sizes of individuals occurring in a local fish assemblage and result in major changes in food web dynamics.

Direct Effects of Predation

Changes in species composition caused by direct effects of predation are perhaps the most obvious impact. The strong effect that size-selective predators have on zooplankton assemblages in lakes has been previously discussed. Phytoplanktivores, such as Gizzard Shad, also can alter phytoplankton density. Both Gizzard Shad and Threadfin Shad (*Dorosoma petenense*) often dominate the standing crop of fishes in impoundments, and Gizzard Shad > 25 mm SL feed on detritus, phytoplankton, and zooplankton using pump filter feeding—suctions that are not visually directed at individual particles (Drenner et al. 1982b). In outdoor pond experiments, Gizzard Shad suppressed populations of *Ceratium*, a large dinoflagellate with relatively slow growth, but had no effect on smaller phytoplankton species and even had an indirect effect of increasing populations of small algae and bacteria (Drenner et al. 1984). Gizzard Shad also suppress some zooplankton species, but, in contrast to the impact of size-selective predators mentioned previously, the pump-feeding of Gizzard Shad selects for species that have little ability to escape. This results in a zooplankton assemblage that is shifted toward more evasive prey (Drenner et al. 1982a).

Changes in species composition are often most evident in species-poor systems. Small, glacier-formed lakes in northern North America are typified by having either large-bodied or small-bodied fish assemblages, with the major differences across regions being the kinds of species in the two groups. In western Canada (Alberta), the large-bodied assemblage comprised Northern Pike and Yellow Perch and the small-bodied assemblage primarily consisted of Brook Stickleback and Fathead Minnow (Robinson and Tonn 1989). In small lakes of eastern Canada (Ontario), essentially the same two types of assemblages occur, except that the large-bodied assemblage is dominated by Smallmouth Bass and other centrarchids such as Rock Bass and Pumpkinseed. Lakes with Smallmouth Bass had an average of 2.3 fewer small-bodied species compared to fish assemblages in lakes where Smallmouth Bass were absent (MacRae and Jackson 2001). Small glaciated lakes in Wisconsin also show the two

distinct assemblages, in this case the large-bodied assemblage is dominated by centrarchids and Northern Pike whereas the small-bodied assemblage is dominated by Central Mudminnow (*Umbra limi*) and cyprinids (Tonn and Magnuson 1982). The centrarchid/Northern Pike assemblage only occurred in lakes with high winter oxygen levels, but the absence of the Central Mudminnow/cyprinid assemblage from such lakes was likely due to predation.

The loss of species because of nonindigenous predator impacts has occurred in species-poor southwestern U.S. streams. For example, widespread declines and extirpations have occurred throughout the native range of the Gila Topminnow (*Poeciliopsis occidentalis*), most likely related to habitat loss and the introduction of nonnative predators including the Western Mosquitofish (*Gambusia affinis*) (Meffe et al. 1983). In field and laboratory experiments on population survival of Gila Topminnow in the presence or absence of Western Mosquitofish, populations of Gila Topminnow without Western Mosquitofish reproduced and grew in both field enclosures and in laboratory aquaria. Field and laboratory populations in the presence of Western Mosquitofish failed to reproduce and steadily declined. Overall, the data provide strong support that the replacement of Gila Topminnow by Western Mosquitofish is primarily due to predation on young and secondarily from stress responses from caudal fin nipping of adult Gila Topminnows by Western Mosquitofish (Meffe 1985). Similarly, in the Colorado River system, adults of the introduced Red Shiner (*Cyprinella lutrensis*) prey on larval Colorado Pikeminnow and may be an important factor in the decline of this large, piscivorous minnow (Bestgen et al. 2006). In the lower Yakima River, Oregon, the introduction of a nonnative predator, Smallmouth Bass, has altered the size-based predation risk on native salmonids. The large native predator, Northern Pikeminnow, which is now rare, did not consume large numbers of salmonids until it exceeded 250 mm FL; in contrast, Smallmouth Bass ≤ 250 mm FL consume more salmonids than do the larger bass.

Because of the size dynamics, predation risk has been reduced for larger prey such as spring Chinook Salmon (*Oncorhynchus tshawytscha*), Coho Salmon (*O. kisutch*), and Steelhead (sea-run Rainbow Trout), but increased for the smaller, ocean type Chinook Salmon (Fritts and Pearsons 2006).

Indirect Effects of Predation

Not surprisingly, impacts of predators on one component (e.g., species or trophic level) of a community can also have effects on other components. In some cases these effects are wide reaching so that a single predator can control the structure of a particular community. Robert Paine (1969) referred to such top carnivores as keystone species—predatory species whose actions greatly modified species composition and physical appearance of communities by holding in check other species that could otherwise dominate the community. Since then, the term keystone species has been applied in numerous different ways, including noncarnivores that for various reasons have strong impacts on ecological communities (see Mills et al. 1993). More recently, Mary Power et al. (1996) expanded on Paine's original use to include species whose impact on a community or ecosystem is disproportionately large relative to its abundance or biomass. In this view, keystone species are not always of high trophic status but can have major effects through consumption at lower trophic levels, or by competition, dispersal, disease, mutualism, and by modifying habitats and biotic factors. Mutualism and the modification of habitats and biotic factors are covered in Chapter 13.

Food webs are ultimately based on primary production by green plants, but also are obviously affected by predators. As such, food webs are influenced by a combination of bottom-up control (i.e., the primary control is by resources) and top-down control (Power 1992b). The top-down view predicts that whether or not organisms are predator or food limited depends on their position in food chains and that plant standing crops are largely regulated by top-down forces. Downward movement through the food web of the effects from a keystone (or other) species is referred to as a trophic cascade. Similar to keystone species, the term "trophic cascade" has been applied to different situations, so that some feel the term has lost meaning (Polis et al. 2000). As used here, trophic cascade refers to impacts of changes in abundance/occurrence of a top predator moving down through the food web, resulting in changes in the biomass or species composition of primary producers. The effects generally alternate between positive and negative as they move downward through different trophic levels (Slobodkin et al. 1967; Carpenter et al. 1985; Power et al. 1985). Lethal interaction is not always required to initiate a trophic cascade because the indirect effects of predation can be transmitted behaviorally via the threat of predation (Lima 1998). Impacts of trophic cascades on the biomass and production of different trophic levels can be dramatic and have been widely demonstrated in marine and freshwater ecosystems (Kitchell and Carpenter 1993a, b).

The impacts of trophic cascades can even extend beyond a single ecosystem, including whether lakes act as net sinks or sources for atmospheric carbon (Schindler et al. 1997). Before experimental manipulation of nutrient levels, four small Indiana lakes received sufficiently high carbon inputs from surrounding terrestrial systems that were all net sources of carbon to the atmosphere. After phosphorus and nitrogen enrichment, the carbon balance changed so that the four lakes were either net sinks for atmospheric carbon or were in approximate equilibrium with the atmosphere. Lakes with piscivorous fishes (*Micropterus* spp.) had lower populations of planktivorous fishes (minnows), which allowed the proliferation of large grazing zooplankton, thus lowering the biomass of primary producers and reducing carbon demand—these lakes were in near equilibrium with atmospheric carbon. Lakes without piscivores had high populations of zooplanktivorous fishes (minnows), resulting in the reduction of large zooplankton grazers (e.g., Cladocerans), triggering an increase in biomass of primary producers thus increasing carbon demand—these lakes functioned as sinks for atmospheric carbon. Consequently, shifts in the trophic structure of lakes caused by changes in fish species can even change the net flux of carbon between air and water.

Interaction Food Webs

An interaction or functional food web (see Chapter 11) is based on the strength of the interaction between trophic levels (usually determined by experimentation) and is generally a top-down web showing the impact of a predator species on some community attribute, such as the population size of herbivores or the standing crop of primary producers (Paine 1980) (Figure 11.8C).

Given the potential complexity of aquatic food webs and their responses to a variety of biotic and abiotic factors, the food web for a single system can change substantially among seasons or years, and the strengths of species interactions can be highly context dependent. Long-term observational and experimental studies are critical for an understanding of any food web in nature. A relatively long-term field study (18 years) combined with five summers of *in situ* habitat manipulations has provided some of the most compelling data on how physical and biotic factors interact in shaping food webs (Power 1990; Wootton et al. 1996; Power et al. 2008).

The South Fork of the Eel River (see also Chapter 1; Figure 1.5) in northern California is subject to a Mediterranean type of climate normally characterized by winter rains followed by a low-flow summer dry period, with flows dropping to < 1 m³s⁻¹. Winter flows sufficient to mobilize the streambed have a recurrence interval > 1.5 years when discharge reaches or exceeds 120 m³s⁻¹. An important impact of periodic winter flood scour is the suppression (by direct mortality and downstream transport) of a large Caddisfly grazer (*Dicosmoecus*), which, because of its armored casing, is generally invulnerable

to fish predation. The absence or near absence of *Dicosmoecus* following winter scouring floods allows the dominant primary producer, a filamentous green algae, *Cladophora glomerata*, to flourish the following spring and summer. The green algae is consumed by various small herbivorous insects, including mayflies and midges, which are consumed by small predators, including dragonfly larvae and young-of-the-year (YOY) California Roach (*Lavinia symmetricus*) and Threespine Stickleback, which in turn are eaten by the young of Steelhead (Figure 12.11A). Thus impacts of fishes on lower trophic levels are generally strong in years with winter floods, although the types of impacts vary depending on predator-specific vulnerabilities of the herbivores. It is also possible that increased growth of filamentous algae following winter floods is related to the release of nutrients, although this has not been specifically tested (Power et al. 2008).

During drought years *Dicosmoecus* is abundant and by its grazing suppresses filamentous green algae. In fact, experimentally removing it rapidly results in a conspicuous algal turf. The trophic cascade generated by Steelhead predation has a minimal effect on the benthos, even though Steelhead consistently suppress odonates and YOY fishes. Instead there are two somewhat decoupled interaction webs in drought years—Steelhead predation on odonates and YOY fishes, and *Dicosmoecus* grazing on filamentous green algae and diatoms (Figure 12.11B) (Power et al. 2008).

Piscivorous Largemouth Bass or Spotted Bass (*Micropterus punctulatus*) can exert strong, top-down trophic effects through at least two levels in small-stream ecosystems, resulting in substantial shifts in assemblage structure. Among pools in Brier Creek, Oklahoma, a strong negative association existed in the distribution of large (>100 mm TL) individuals of the two bass species and the herbivorous Central Stoneroller (Power and Matthews 1983). The impact of grazing by Central Stoneroller was to suppress attached algae (mostly *Rhizoclonium* and *Spirogyra*) so that pools with Central

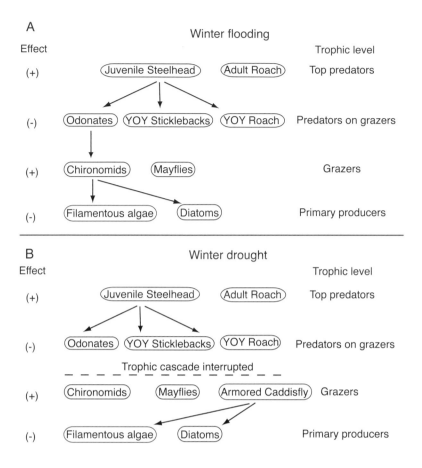

A   Winter flooding

Effect		Trophic level
(+)	Juvenile Steelhead Adult Roach	Top predators
(-)	Odonates YOY Sticklebacks YOY Roach	Predators on grazers
(+)	Chironomids Mayflies	Grazers
(-)	Filamentous algae Diatoms	Primary producers

B   Winter drought

Effect		Trophic level
(+)	Juvenile Steelhead Adult Roach	Top predators
(-)	Odonates YOY Sticklebacks YOY Roach	Predators on grazers
	Trophic cascade interrupted	
(+)	Chironomids Mayflies Armored Caddisfly	Grazers
(-)	Filamentous algae Diatoms	Primary producers

FIGURE 12.11. The effect of floods on interactive food webs in the Eel River, California. Major fish species include California Roach (*Lavinia symmetricus*), Steelhead (*Oncorhynchus mykiss*), and Threespine Stickleback (*Gasterosteus aculeatus*).
A. An example of a summer food web following a scouring winter flood that removed a large armored herbivore (*Dicosmoecus*) that is invulnerable to fish predation.
B. A summer food web that was not preceded by a winter flood, resulting in abundant *Dicosmoecus* and a greatly reduced trophic cascade. The arrows indicate strong impacts of a higher trophic level on a lower trophic level. + and - show the direction of effect at each trophic level. Based on Power et al. (2008).

Stoneroller were generally free of algae, except along the stream margin. In contrast, pools with dense, tall growths of algae contained large bass and the Central Stoneroller was absent or rare. This pattern was tested across time and by bass addition/removal experiments (Power et al. 1985) and persisted throughout a year of study. The addition of Largemouth Bass to a predator-free pool containing Central Stoneroller resulted in a major change in composition and growth form of algae in the pool. Once Largemouth Bass were introduced, Central Stoneroller rapidly emigrated out of the test pools or took shelter and remained in shallow pool edges (Power et al. 1985; Power 1987). As a result, algae grew densely over the next several weeks in the deeper parts of the pools guarded by bass, and algal growth gradually spread into shallow areas as the numbers of Central Stoneroller decreased through emigration. This was interpreted as an example of a three-level trophic cascade with strong effects by the bass controlling ecosystem processes in these pools. The effects of the algivorous Central Stoneroller in stream ecosystems are pervasive. By removal of algae they initiate changes within pools with consequences for invertebrates, processing of particulate organic matter, and bacterial standing crops, causing a total of >20 measurable ecosystem responses (Matthews et al. 1987; Gelwick and Matthews 1992; Matthews 1998).

In Brier Creek, the reciprocal distribution of large bass and Central Stoneroller is temporally persistent. In six of eight surveys (Power et al. 1985; Matthews et al. 1994), the dichotomy in pool-to-pool distribution persisted across 14 pools for >1 year, and 12 additional surveys (1995–2003) indicated the pattern was again persistent (Matthews and Marsh-Matthews 2006b). In addition, these piscivores also result in avoidance of pools by some, but not all, other small-bodied and potential prey species and thus have major effects on local assemblage structure.

Across streams, the impacts of bass species on prey fishes, particularly cyprinids, are variable and depend strongly on the physical setting of a stream and identity of the potential predator. When a search for the existence of the bass-stoneroller-algae trophic cascade was extended to other, larger stream systems with Smallmouth Bass as the dominant predator, the dichotomy between "bass" and stonerollers broke down, and they often were found together in pools (Matthews et al. 1987). Smallmouth Bass apparently were a less controlling predator than the other bass species, and this was confirmed when Smallmouth Bass were moved to an experiment in Brier Creek (Harvey et al. 1988). In seeming contrast, two size classes of another cyprinid, Hornyhead Chub (*Nocomis biguttatus*), reduced their use of deep pools of an experimental stream in the presence of Smallmouth Bass and instead occupied shallow raceways, suggesting the importance of particular species or systems in affecting the strength of the predation response (Schlosser 1988).

Given the nature of trophic cascades, the addition of a new top predator to a system can have major impacts. The native top predator in Yellowstone Lake, Wyoming, was the Yellowstone Cutthroat Trout (*Oncorhynchus clarkii bouvieri*), which relied heavily on zooplankton throughout its life. A large piscivore, Lake Trout

(*Salvelinus namaycush*), was introduced in 1985, with the consequence of creating a four-level trophic cascade. Lake Trout predation reduced Yellowstone Cutthroat Trout resulting in a shift from small zooplankton (i.e., copepods) to large zooplankton (i.e., cladocerans) species. The larger zooplankton species increased grazing pressure on phytoplankton causing a decline in phytoplankton biomass and a concomitant increase in water clarity (Tronstad et al. 2010).

SUMMARY

Predator-prey interactions have widespread impacts on fish assemblages, aquatic ecosystems, and even terrestrial ecosystems. Because fitness of predators is enhanced by successful predation and fitness of prey is enhanced by avoiding predators, both predator and prey species should show continued evolution just to maintain the status quo. In addition, most all fishes are both predators and prey, at least during some phase of their life cycles. Interactions between predators and their prey are strongly influenced by their relative sizes. Most temperate freshwater fish predators are considerably larger than their prey, although there are a few examples of exoparasitism where the opposite is true. Trophic guilds are a convenient way to categorize trophic resources of fishes, even though fishes generally change their diets ontogenetically, seasonally, and spatially. North American freshwater fishes are dominated by omnivores, benthic invertebrate feeders, and macrocarnivores/piscivores.

Successful predation is based on the completion of a series of sequential events (the predation cycle); however, avoidance tactics of prey can disrupt the predation cycle at multiple points. Prey choice by most predators is not random but is based on various selection criteria, of which size, nutrient content, visibility, and absence of defensive structures are most important. At least in many cases, prey selection can be understood by the balance of energy gain versus energy expended or predation risk.

The first level of predator avoidance by prey is to remain undetected. This can be achieved by minimizing activity when predators might be most active, reducing the probability of detection by forming shoals (although this can increase exposure to aerial predators), and by crypsis. Once detected, prey can resort to other defensive measures such as fleeing, confusing the predator by forming tight shoals followed by rapid expansion of the shoal, or defensive structures such as spines. The release of alarm substances from club cells in the skin of a captured conspecific or heterospecific is also important in reducing predation on other fishes in the vicinity of a predator.

Habitat use by fishes is a balance between the rewards from foraging and the risk of predation. Risk exposure to aquatic and terrestrial predators can result in feeding in less than optimal habitats. However, as energy stores are depleted or if a fish is energetically stressed by parasites, it can become more risk tolerant compared to a healthy fish.

The process of predation and prey selection can have direct impacts on species composition and size structure of prey assemblages. Indirect effects are also important and can include the creation of predator free zones for larval fishes. In some cases, the impacts of an apex predator cascade downward through the food web, typically with alternate effects at each successive trophic level, culminating in impacts on primary producers. Nonnative apex predators, consequently, can exert strong changes on aquatic ecosystems. As with other biological interactions, predator impacts, including the presence or strength of trophic cascades, are strongly context dependent and can be altered by floods and droughts as well as other factors.

SUPPLEMENTAL READING

Alford, J. B., and D. C. Beckett. 2007. Selective predation by four darter (Percidae) species on larval chironomids (Diptera) from a Mississippi stream. *Environmental Biology of Fishes* 78:353–64. A important study showing how the level of prey identification affects conclusions of resource specialization.

Chivers, D. P., B. D. Wisenden, C. J. Hindman, T. A. Michalak, R. C. Kusch, S. G. W. Kaminskyj, K. L. Jack, M. C. O. Ferrari, et al. 2007. Epidermal "alarm substance" cells of fishes maintained by non-alarm functions: Possible defence against pathogens, parasites and UVB radiation. *Proceedings of the Royal Society B* 274:2611–19. A landmark paper providing evidence that club cells and their contents evolved first as a defense against pathogens, and only secondarily as alarm signals.

Endler, J. A. 1978. A predator's view of animal color patterns. *Evolutionary Biology* 11:319–64. Thorough treatment of the interaction of cryptic color patterns of prey, competing demands of natural selection, and the visual acuity of predators.

Gibb, A. C., M. A. Ashley-Ross, C. M. Pace, and J. H. Long, Jr. 2011. Fish out of water: Terrestrial jumping by fully aquatic fishes. *Journal of Experimental Biology* 313A:1–5. An intriguing study of terrestrial movement in two distantly related cyprinodontid fishes.

Januszkiewicz, A. J., and B. W. Robinson. 2007. Divergent walleye (*Sander vitreus*)-mediated inducible defenses in the centrarchid Pumpkinseed sunfish (*Lepomis gibbosus*). *Biological Journal of the Linnean Society* 90:25–36. A landmark paper showing how body depth and dorsal spine lengths of Pumpkinseed are induced by exposure to a predator.

Parkos, J. J., III, and D. H. Wahl. 2010. Influence of body size and prey type on the willingness of age-0 fish to forage under predation risk. *Transactions of the American Fisheries Society* 139: 969–75. A recent experimental study showing the balance between willingness to forage and the risk of predation.

Pitcher, T. J., and J. K. Parrish. 1993. Functions of shoaling behaviour in teleosts, 363–439. In *Behaviour of teleost fishes*. 2nd ed. T. J. Pitcher (ed.). Chapman and Hall, New York, New York. Provides thorough coverage on the functions of fish shoals.

Power, M. E., M. S. Parker, and W. E. Dietrich. 2008. Seasonal reassembly of a river food web: Floods, droughts, and impacts of fish. *Ecological Monographs* 78:263–82. A long-term study showing how physical factors (flooding) and species-specific vulnerabilities to predation can affect food webs.

Smith, R. J. F. 1992. Alarm signals in fishes. *Reviews in Fish Biology and Fisheries* 2:33–63. A thorough review on alarm signals in fishes.

Getting Along

CONTENTS

THE PREVIOUS CHAPTERS in Part 4 mainly focused on symmetrical negative interactions like competition or asymmetrical positive/negative interactions such as predation. Indeed, such fully or partially negative interactions have dominated ecological research since Darwin. Much of the earlier literature on competitive interactions, in fact, focused on how different species avoided or reduced spatiotemporal overlap (for instance the concept of resource partitioning). However, interactions that are symmetrically positive or asymmetrically positive and neutral are being increasingly recognized as important in the formation and maintenance of communities (Bruno et al. 2003). This leads to a potential cost-benefit situation of avoiding other species to reduce competition or joining them to reap benefits of group foraging or predator avoidance, and perhaps also to benefit from reduced intraspecific competition (Freeman and Grossman 1992). The occurrence of facilitative interactions among species requires close contact for the various direct interactions treated in the following sections, but less so for indirect interactions.

That different species of North American freshwater fishes frequently occur close together is not at issue. In virtually all aquatic habitats, capturing two or more fish species in a short section of stream or lake shoreline is more common than not. Based on museum records of 291 collections from a variety of aquatic habitats in Minnesota, over 60% of the collections included two or more species, almost a third included three or more, and 9% included five or more species (Mendelson 1975). Similarly, based on 300 museum collections taken over 13 years from a variety of stream and pond habitats in Mississippi, 79% of the collections included two or more species, 66% included three or more, and 54% included five or more (University of Southern Mississippi, Museum of Ichthyology). At the finer scale of individual small seine hauls (ca. 10-m²) in the Canadian River, Oklahoma, 71% included at least two species and 46% included three or more species (Matthews 1977). In a wave-exposed littoral zone of Lake Texoma, a large impoundment on the Texas-Oklahoma border, 86% of seine hauls had at least two minnow species and 60% included three or more minnow species (Matthews 1998). The number of different species likely to be taken in a small area is obviously positively related to regional diversity (see Chapter 4), but these comparisons show that fish species in North American freshwater habitats are commonly in relatively close association with each other.

Statistical tests of fish species' associations also show the potential for interactions if associations are positive or the lack thereof if associations are negative. Based on fishes captured in individual seine hauls, eight minnow species in a Mississippi blackwater stream showed nine significant associations, of which four were positive—Blacktail Shiner (*C. venusta*) and Longnose Shiner (*N. longirostris*), Blacktail Shiner and Longjaw Minnow (*Notropis amplamala*), Longnose Shiner and Striped Shiner (*Luxilus chrysocephalus*), and Longnose Shiner and Longjaw Minnow (Baker and Ross 1981). In a study of habitat associations of nine fish taxa in a small Minnesota Lake, there were three positive associations out of a total of 15 significant species associations—Mimic Shiner (*Notropis volucellus*) and Bluntnose Minnow (*Pimephales notatus*), Bluntnose Minnow and White Sucker (*Catostomus commersonii*), and Common Shiner (*Luxilus cornutus*) and White Sucker (Moyle 1973). Habitat associations were determined by scuba observations of fishes occurring in 1-m² plots in the littoral zone. Although such examples of mixed species occurrences in localized field samples generally do not yield information on actual interactions among species, they do

suggest the potential for interactions among different species of fishes.

The presence of actual mixed species groups provides even stronger support for potential interactions among species. Rosyside Dace (*Clinostomus funduloides*), other cyprinids, and Rainbow Trout (*Oncorhynchus mykiss*) commonly formed mixed-species groups in two small North Carolina streams (Freeman and Grossman 1992), and minnows in a small Ozark stream often formed mixed shoals that were vertically and horizontally intermixed (McNeely 1987). Such interactions among species could include facilitation of feeding, enhanced awareness of potential predators, or compression of species into habitats offering low predation risk (Gorman 1988a, b; Freeman and Grossman 1992). The nature of these interactions and how tightly species are linked in associations is the topic of the rest of this chapter.

FACILITATION

Facilitation, the situation where one organism makes the local environment more favorable for another, is extremely widespread in nature and can occur at numerous points in an ecosystem with widely varying strengths (Bruno et al. 2003; Nummi and Hahtola 2008). Facilitative relationships can be loose associations as well as tightly coevolved and mutually obligate (Box 13.1) (Bruno et al. 2003), and facilitation in fish assemblages can occur between fishes and other taxa, such as riparian vegetation, wading birds, or with other fish species. For example, evidence supporting the model of predation risk of prey fishes in response to terrestrial and aquatic predators (Chapter 12) involves a facilitative interaction between a wading bird and a piscivore. In a study of a wading bird (Heron, *Ardea alba*) and Smallmouth Bass (*Micropterus dolomieu*) predation on Striped Shiner and Central Stoneroller (*Campostoma anomalum*), Heron predation forced the smaller size class of the prey fishes into deeper water where they were more vulnerable to the Smallmouth Bass (Steinmetz et al. 2008). The interaction, which was significant for

With few exceptions, ecologists have focused on asymmetrical positive/negative interactions such as predation and competition—the "tooth and claw" sort of interactions where one organism is killing or hurting another. At the same time, interactions where organisms directly or indirectly help each other have been largely ignored until fairly recently, yet life on Earth as we know it is highly dependent upon such interactions (Roughgarden 1998; Stachowicz 2001; Bruno et al. 2003; Molles 2010). Facilitative behaviors among animals, including mutualisms, have been described since the nineteenth century, so it is no surprise that the terminology of interspecific interactions is complex. The following terminology is largely based on Boucher et al. (1982) and Bruno et al. (2003).

FACILITATION Encounters between organisms that benefit *at least one* of the participants and cause harm to neither (+/o; +/+).

COMMENSALISM Asymmetrical facilitation where one party benefits and the other is not affected (+/o).

MUTUALISM Symmetrical facilitation in which both parties benefit from the interaction (+/+).

DIRECT MUTUALISMS Physical contact between the species.

SYMBIOTIC MUTUALISMS Usually occur between different kingdoms of organisms, such as gut microbes or luminescent bacteria in the skin of some deep-sea fishes.

NONSYMBIOTIC MUTUALISMS There can be direct physical contact, but there are no direct physiological links between partners.

INDIRECT MUTUALISMS Positive effects occur in the absence of direct physical contact.

Facilitative interactions among widely different taxa are prevalent with foundation species and ecosystem engineers—species such as trees, grasses, and emergent and submerged aquatic vegetation, or animals such as Beavers (*Castor canadensis*) that in some way contribute a framework for the entire community by forming and modifying habitats (Jones et al. 1997a; Bruno et al. 2003; Nummi and Hahtola 2008). Although the two terms are sometimes used as synonyms, foundation species (or dominant species) alter the local environment simply by growing, and their effects are roughly in proportion to their abundance (usually expressed as biomass; Figure 13.1) (Power et al. 1996). An example would be the shading of pools in streams by riparian vegetation, resulting in cooler water and a more favorable environment for fishes. Ecosystem engineers (also called keystone modifiers) have impacts on their environment that are much greater than would be predicted by biomass alone (Power et al. 1996). Physical ecosystem engineering involves a process, such as dam building by Beavers, that results in some structural change (an impoundment), which in turn changes the physical environment (impounded versus flowing water), which can result in biotic changes to other species (alteration of invertebrate and fish assemblages) as well as provide feedback to the engineer (reduced risk of predation, etc.) (Jones et al. 1994, 1997a, 1997b, 2010). Although there are concerns about the use of "ecosystem engineer," because of its being burdened by anthropomorphic overtones and implying purpose, over some more neutral term such as "keystone modifier" (Power 1997; Stachowicz 2001), the term continues to be widely used in ecology (Wright and Jones 2006; Jones et al. 2010).

In a situation closely analogous to trophic cascades, foundation species and ecosystem engineers can generate facilitation cascades. The primary difference from a trophic cascade is that the effect is not initiated by a top carnivore, although herbivores are often involved, especially in marine, rocky intertidal systems (Thomsen et al. 2010).

the number of prey consumed, was largely asymmetrical, with Smallmouth Bass being positively affected by the Heron, whereas predation success of Heron remained essentially unchanged in tests with and without the piscivore present.

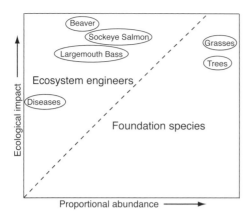

FIGURE 13.1. Foundation species versus ecosystem engineers. Foundation species have ecological impacts roughly proportional to their abundance and occur to the right of the dashed line of 1:1 equivalence. Ecosystem engineers have impacts that are out of proportion to their abundance and occur to the left of the dashed line. Based on data from Power et al. (1996).

Ecosystem Engineers

Species considered to be ecosystem engineers directly or indirectly control the availability of resources to other organisms by their physical modification, maintenance, or creation of habitats; however, the direct provision of resources by living or dead tissue does not constitute engineering (Jones et al. 1994, 1997a). Autogenic engineers function by transforming the environment through endogenous processes, such as tree growth, and remain part of the altered environment. For instance, caddisfly larvae construct armored cases out of gravel or other materials and also build silk nets to trap suspended particulate materials in streams. These biogenic materials can modify water flow in the boundary layer (see Chapter 7) by increasing stream bottom roughness, which translates into enhanced feeding success of the larvae (Nowell and Jumars 1984; Cardinale et al. 2002; Moore 2006).

Allogenic engineers, in contrast, alter the environment by transforming living or nonliving materials from one physical state to another and are not necessarily part of the modified physical structure. The classic example of an allogenic engineer is the Beaver, which harvests

live trees (physical state 1) and converts them into dead trees (physical state 2), which are placed in a dam that creates a pond (Jones et al. 1994). Impacts of allogenic ecosystem engineers are generally of greater importance in streams of low to intermediate energy. In streams with high hydrologic energy, any impacts from ecosystem engineers would likely be overpowered by water energy. Impacts of ecosystem engineers on aquatic habitats are also greater when food webs are heavily subsidized so that the biomass of the engineering species is greater than could be sustained by local productivity, such as occurs with anadromous fishes in streams when their biomass is subsidized by oceanic productivity and, of course, with humans who are highly subsidized energetically and are the only animal to use powerful tools to enhance their impacts on ecosystems (Moore 2006).

Beaver engineering projects act initially by altering stream morphology and hydrology through wood harvesting and dam construction, creating habitat heterogeneity on a large scale by forming patches of lentic habitat in what is otherwise a corridor of lotic habitat. In general, Beaver impoundments are most likely in moderate-sized streams with relatively low gradients (Pollock et al. 2003; Jakes et al. 2007). Their impoundments increase Beaver habitats, expand their food supply, and provide protection from predators. Impoundments also increase retention of inorganic and organic sediments, facilitate groundwater recharge, create and maintain wetlands, alter nutrient cycling, change decomposition dynamics, influence water quality, transform riparian structure and diversity, and affect local plant and animal assemblages. Beaver impoundments flood and kill riparian trees, but following pond abandonment they result in the growth of meadow grasses that are ultimately replaced by shrubs and trees, not necessarily the same trees present before dam construction. A faunal consequence of Beaver engineering is the replacement of running-water invertebrate taxa such as blackflies, Tanytarsini midges, scraping mayflies, and net-spinning caddisflies by pond taxa such as Tanypodinae and Chironomini midges, predaceous dragonflies, tubificid worms, and filtering clams (McDowell and Naiman 1986; Naiman et al. 1988; Snodgrass 1997; Snodgrass and Meffe 1999; Wright 2009).

The sheer magnitude of the historic impact of Beaver on aquatic and riparian ecosystems is difficult to grasp. The North American landscape prior to European contact and colonization was covered by an estimated 25 million Beaver ponds that contained tens to hundreds of billions of cubic meters of sediment. Beaver occurred coast to coast and from the Arctic tundra to deserts of northern Mexico, so that their impacts on stream morphology and hydrology and the aquatic and riparian biota were a common and enduring feature of most North American landscapes (Naiman et al. 1988; Pollock et al. 2003). One consequence of overtrapping from the sixteenth through nineteenth centuries was increased sediment evacuation from abandoned ponds, stream entrenchment, and wetland reduction (Butler and Malanson 2005). In addition, fish habitat availability was markedly reduced in low-order streams and many perennial streams likely became intermittent (Pollock et al. 2003).

Not surprisingly, Beaver impoundments have had substantial impacts on the numbers and kinds of fishes, especially in low-order streams; however, the impacts can change as a function of pond age, location within a drainage, and whether ponds are abandoned or maintained. Forty-eight species of North American freshwater fishes are commonly found in Beaver ponds, and at least 80 species have been documented using Beaver ponds (Pollock et al. 2003).

In a study of low-gradient streams (< 2m/km) on the Savannah River Site on the Upper Coastal Plain of South Carolina, Beaver ponds were initially colonized by small-bodied, insectivorous species such as Eastern Mosquitofish (*Gambusia holbrooki*), Lined Topminnow (*Fundulus lineolatus*), Lake Chubsucker (*Erimyzon sucetta*), Creek Chubsucker (*Erimyzon oblongus*), Sawcheek Darter (*Etheostoma serrifer*),

Savannah Darter (*Etheostoma fricksium*), Ironcolor Shiner (*Notropis chalybaeus*), and Golden Shiner (*Notemigonus crysoleucas*) (Snodgrass and Meffe 1998). Over time, however, these species declined, primarily in response to colonization of the ponds by larger predators such as Redfin Pickerel (*Esox americanus*), Chain Pickerel (*E. niger*), Mud Sunfish (*Acantharchus pomotis*), Dollar Sunfish (*Lepomis marginatus*), Redbreast Sunfish (*L. auritus*), Warmouth (*L. gulosus*), and Flier (*Centrarchus macropterus*). Forty species were collected in the study and 17 species comprised ≥ 2% of the fishes collected. Overall, pond-associated fish species were minor elements of the stream fish assemblage. Of the 17 more common species in the study area, 14 had their greatest abundance in active or abandoned ponds and three had their greatest abundance in streams or recovering streams. Actual patterns of abundance among the four general habitat categories varied, with some species such as Yellowfin Shiner (*Notropis lutipinnis*) and Bluehead Chub (*Nocomis leptocephalus*) found primarily in streams and others such as Eastern Mosquitofish and Lake Chubsucker primarily, or only, in Beaver Ponds (Figure 13.2). In these low-order, blackwater streams, the effect of Beaver impoundments was to increase species richness in headwater streams. The elimination of Beavers from headwater streams would result in reducing species richness by over half. In fact, the presence of Beaver impoundments increased upstream species richness to the point that the positive relationship between fish species richness and stream order was eliminated (Snodgrass and Meffe 1998, 1999). Indeed, Snodgrass and Meffe (1998) suggested that the normally positive and widespread relationship between stream fish species richness and drainage area may be a recent phenomenon resulting from the extirpation of Beavers from much of their historical range.

Beaver dams, especially in later stages when they are larger, can impede the upstream and downstream movement of fishes. In the low-gradient South Carolina streams, passive

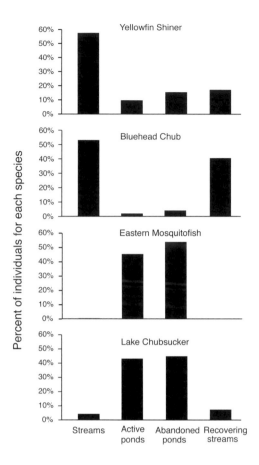

FIGURE 13.2. Examples of frequency distributions of four southeastern fish species in streams and in Beaver impoundments. Overall, more species were most common in impoundments versus streams. Based on data from Snodgrass and Meffe (1998).

downstream movement of age-0 fishes was interrupted by Beaver impoundments, with the fish being retained in the slower current just above the ponds. Further downstream movement was perhaps reduced because of predation risk incurred by traversing the open expanse of water from stream inflow to outflow over the dam (Snodgrass and Meffe 1999). In a small, relatively high gradient (10 m/km) stream in Minnesota, Beaver dams also acted as semipermeable boundaries to fish movement and the degree of permeability likewise varied with direction (upstream versus downstream) and with factors such as pond age and morphology, and stream discharge (Schlosser 1995b). Permeability of the barrier for downstream movement

from the pond to the stream increased as a function of increasing discharge, and declined greatly during low-discharge periods when Beaver seal the dam with mud and vegetation to improve water retention. Upstream boundaries between ponds and streams are more permeable to fish movement but can also decline during times of low discharge, whereas upstream movement from the stream into a pond, although it occurs, is perhaps less likely, especially during low-flow periods (Pollock et al. 2003).

In contrast to the low-gradient South Carolina streams studied by Snodgrass and Meffe, Beaver ponds in Minnesota acted as reproductive source areas for fishes (see also Chapter 4 for source/sink terminology) and the adjacent streams appeared to be reproductive sinks, so that there was high overlap of species between the two major habitats (Schlosser 1995a, b). In fact, only one out of 12 commonly occurring species, Creek Chub (*Semotilus atromaculatus*), actually spawned in the stream. As shown in both the Minnesota and South Carolina studies, newly formed Beaver ponds are organically rich resource patches that are rapidly colonized and exploited by fishes and other organisms, followed by fairly rapid dispersal from the ponds.

Piscine Engineers

Various species of freshwater fishes channel some of their reproductive energy into the building of nest mounds or pits, as shown by various minnows, or circular depressions, as shown by various centrarchids. In addition, species of Pacific salmon (genus *Oncorhynchus*) excavate large nests in the streambed. In the latter case, the principal engineering effect is the alteration of large areas of the streambed. In the case of nests constructed by minnows and sunfishes, the nests are often used as spawning sites by other species—the nest associates.

PACIFIC SALMON EXCAVATIONS The creation of large pebble nesting mounds and bioturbation by nest digging, foraging, and movement constitutes another type of allogenic habitat creation. For example, excavation of the large nests by various species of anadromous Pacific salmon can have a major impact in streams, especially given that the biomass of salmon is highly subsidized by marine productivity. Anadromous salmon accumulate > 99% of their lifetime growth in marine ecosystems (Moore and Schindler 2008). A female salmon excavates egg pockets in the streambed by turning on her side and rapidly undulating her large caudal fin (see also Chapter 9). After the eggs are laid and fertilized, the process is repeated, with the female digging additional egg pockets upstream, which also results in protecting the earlier nest site by covering it with additional gravel (Montgomery et al. 1996). In two streams in southwestern Alaska, such spawning behavior by Sockeye Salmon (*Oncorhynchus nerka*) has consistently disturbed more than 5000 m^2 of streambed (constituting about 30% of the available streambed) every summer over the last 50 years, and the effect would almost double in the absence of commercial harvesting. Nest digging displaces fine sediments, resulting in a coarser stream bottom in the spawning area, the deposition of fine materials downstream in pools and sandbars, and an increase in the concentration of suspended particulate materials in the water column. The impact of Sockeye Salmon excavations is detrimental to algal and invertebrate biomass. Above a density of 0.1 salmon m^{-2}, algal and benthic invertebrate biomass decreased by an average of 75–85% during the approximately month-long spawning season (Moore and Schindler 2008). Recovery of algal biomass to presalmon levels only occurred in streams with low salmon densities (< 0.1 salmon m^{-2}), and within the same season benthic invertebrate biomass did not recover. Although salmon spawning also increased the availability of nutrients in streams, there was no evidence that salmon positively impacted algal biomass on short-term (same year) or long-term (subsequent year) time frames. Overall, anadromous Pacific Salmon are an important component of stream disturbance regimes, with their impacts roughly similar to erosive flood events (Moore 2006).

The activities associated with nest construction by salmon also have a positive role in the formation and maintenance of salmon spawning habitats and on embryo survival. Survival of the developing embryos is dependent on the eggs being buried below the level of bed-scour during the period of incubation. In a small stream entering Puget Sound, Washington, and in a stream near Juneau, Alaska, mass spawning activity by Chum Salmon (*O. keta*) reduced the vulnerability of eggs to bed-scour by removing fine sediments from the spawning area, although the effect of increased coarseness on reducing bed-scour was somewhat lessened by the more loosely packed bed materials. Overall, spawning activity decreases streambed mobility at a time when the developing embryos are most susceptible to displacement by high stream flow and thus may enhance the likelihood of offspring survival (Montgomery et al. 1996).

CYPRINID PITS AND MOUNDS Males of several genera of minnows excavate nest pits surrounded by pebbles and then attract females to spawn. For instance, Bluehead Chub, as well as other *Nocomis* species, build pebble mounds by excavating a pit in a shallow, gravel-bottomed section of a stream, then creating a mound by picking up stones with their mouths and filling the pit with gravel (Figure 13.3). Mounds can be group projects with 1–5 males involved, and the stones may be transported to the nest from as much as four meters away. When the mound is complete, smaller depressions are dug into the top or upstream edge, which serve as the spawning sites. The mounds are located in flowing water (ca. 11–25 cm s^{-1}) and can be of substantial size, averaging 14,500 stones, 50–69 cm in diameter, and 14–24 cm in height (Wallin 1989; Johnston and Page 1992; Maurakis et al. 1992; Johnston and Kleiner 1994). Nests of the closely related Hornyhead Chub (*N. biguttatus*) can be up to 90 cm across and 30 cm tall and, in Wisconsin streams, averaged 40 × 50 cm and 6.4 cm above the substratum (Hankinson 1932; Robison and Buchanan 1988; Vives 1990).

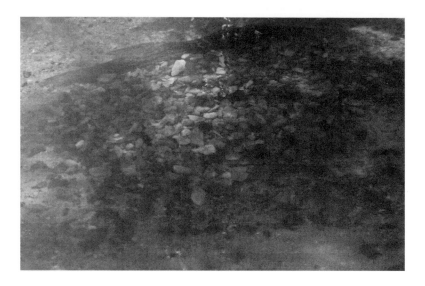

FIGURE 13.3. Pebble nest of a Bluehead Chub (*Nocomis leptocephalus*) from a small stream in Mississippi. The nest is 66 cm × 40 cm. Photo courtesy of Mollie Cashner. See also corresponding figure in color insert.

Male Creek Chubs construct a "pit-ridge" nest by initially removing stones with their mouths, or by shoving larger stones with their snouts, to dig a pit in the stream bottom. The stones from the pit are piled immediately upstream and nests are usually located in shallow channels. Males spawn multiple times, both with the same and with different females, and after each spawning event the male extends the pit farther downstream and covers the newly laid eggs with pebbles from the extended pit. Ultimately, the pit is preceded upstream by a longitudinal ridge (which fills in the trench) that may reach lengths of 5.5 m, and averages 69 cm long, 22 cm wide, and 4 cm high (Reighard 1910; M. R. Ross 1977; Maurakis et al. 1990; Johnston and Page 1992).

Male stonerollers (*Campostoma* spp.) also construct pebble nests by picking up pebbles in their mouths and moving them from the center to the edge of the nest pit. Pebble size is determined primarily by mouth size, although the male may remove larger stones from the nest area by pushing them with his head. This behavior also loosens sand and gravel, which is then carried away by the current. The nests vary in size, but most nest activity is focused in an area 15–30 cm in diameter and nests are generally 7.6 cm or less in depth (R. J. Miller 1962).

SUNFISH AND BASS NESTS Male centrarchids build and defend oval to circular-shaped nests. Most of the nest-building work is done by downward thrusts of the large caudal fin, although sometimes the nest area is cleared by sweeps of the pectoral fin, by carrying stones in the mouth or by pushing stones with the head (Warren 2009). For example, male White Crappie (*Pomoxis annularis*), occasionally with help from the female, build nests in the shallow water of streams or ponds. The nests average 30 cm in diameter and are somewhat irregular in shape (Hansen 1965; Siefert 1968). Largemouth Bass (*Micropterus salmoides*) males also build their nests in shallow water, usually around a meter deep or less. Nests are generally at least 2 m apart, but sometimes closer if underwater objects shield nests from view by other males (Clugston 1966; Heidinger 1976). During nest construction, the male places his head at the center of the nest and, by undulating the body and sweeping the caudal fin, clears out the area. Because of this activity, the nest diameter is usually equal to twice the body length of the male (M. H. Carr 1942). After spawning, the male bass (and rarely the female as well) cares for the eggs by fanning them day and night and by chasing away potential predators (Kramer and Smith 1960a). Nest guarding continues for several weeks after hatching (M. H. Carr 1942; Heidinger 1976).

Like other sunfish species, Longear Sunfish males fan out a depression in shallow water over areas with a gravel bottom and slow current flow using vigorous motion of the caudal fin. As with Largemouth Bass, the nest diameter is about twice the length of the fish, and nests are often constructed close together so that there are large aggregations of territorial, nesting males (Huck and Gunning 1967; Bietz 1981).

Mutualisms

In addition to asymmetrical (+/o) relationships, facilitative interactions may be symmetrical (mutualistic) when both species benefit from the association (Box 13.1) (Boucher et al. 1982). Mutualisms can be direct or indirect, the latter being the case of no direct contact but both parties benefit from the other's presence. Direct mutualisms can be further subdivided into symbiotic and nonsymbiotic mutualisms with the distinction between them based on the level of physiological integration. Although exceptions are frequent, symbiotic mutualisms are usually coevolved and obligate with a strong physiological connection. In contrast, nonsymbiotic mutualisms tend to be facultative and not coevolved (Boucher et al. 1982). Much of the earlier literature on species associations treated mutualism as an obligatory response. However, numerous studies indicate that mutualisms are often context dependent and may be obligatory in one area but not another and may even change to antagonistic interactions (Gomulkiewicz et al. 2003; Hay et al. 2004).

Direct Mutualisms

Cleaning behavior is an example of a direct, nonsymbiotic mutualism. Although much more common in tropical fish assemblages, such behavior has been documented in Atlantic drainages of North America for coastal marsh or estuarine species, including the Rainwater Killifish (*Lucania parva*), Sheepshead Minnow (*Cyprinodon variegatus*), and Striped Killifish (*Fundulus majalis*) (Tyler 1963; Able 1976). Along the British Columbia coast, Threespine Stickleback are opportunistic cleaners, removing ectoparasitic copepods from juvenile Pink Salmon (*Oncorhynchus gorbuscha*) (Losos et al. 2010).

Indirect Mutualisms

In addition to direct physical contact, species can benefit each other in a variety of ways even though physical contact does not occur (Boucher et al. 1982). Benefits could occur through structures and other alterations of the physical habitat (ecosystem engineering) or by the importation of nutrients.

PACIFIC SALMON MIGRATIONS AND MARINE-DERIVED NUTRIENTS Historically, the annual migrations of Pacific salmon from the ocean into freshwater streams and lakes, where they spawn and die, constituted the largest cross-ecosystem movement of biomass in the North Pacific region (Hocking and Reynolds 2011). Although the degree of ecosystem effects are dependent on the salmon species and the environmental context, the annual migrations of species of Pacific salmon can influence virtually all ecosystem components, from primary producers, invertebrates, and fishes in lakes and streams, to bears, eagles, and old-growth spruce trees (Piccolo et al. 2009). Nutrient subsidies to riparian ecosystems are greater for salmonid species that spawn in dense aggregations, such as Chum Salmon, Pink Salmon, and Sockeye Salmon, rather than at low densities, such as anadromous Rainbow Trout (Steelhead) and Coho Salmon (*O. kisutch*) (Figure 13.4) (Bilby et al. 2003).

An intriguing example of a potential indirect mutualism is the linkage between Pacific salmon and the forests surrounding their natal streams (Hay et al. 2004). The basic tenets are

FIGURE 13.4. Pacific salmon migrations import large quantities of marine-derived nutrients into freshwater ecosystems. The picture shows a decomposing head of a Chum Salmon (*Oncorhynchus keta*) along the shore of the Skokomish River, Washington. Photograph courtesy of Peter Bisson. See also corresponding figure in color insert.

first that forested streams result in greater survival of embryos and newly emerged salmon, and thus support greater densities of young fish when compared with nonforested streams. This effect occurs through the input of nutrients from leaf litter and increased habitat complexity resulting from instream wood. Second, spawning runs of adult salmon import large quantities of marine-derived nutrients (nitrogen, carbon, and phosphorus) into headwater streams and thus subsidize the growth of riparian trees. By their instream carcasses they also benefit benthic invertebrates that are preyed on by juvenile salmon, as well as directly benefiting juvenile salmon who feed on the adult carcasses (Figure 13.5) (Helfield and Naiman 2001; Piccolo et al. 2009). Instream nutrient subsidies by salmon are, however, somewhat counteracted by their physical disturbance of the streambed (Moore and Schindler 2008; Hocking and Reynolds 2011).

The positive effect of salmon on the riparian forest is further supported by a study using salmon carcasses placed in the floodplain and another study using a nitrogen tracer released in the riparian zone of a salmon spawning river (Drake et al. 2006; Gende et al. 2007). The study using salmon carcasses showed that they were providing nitrogen to soils adjacent to carcasses over the course of several months at concentrations capable of influencing plant growth (Gende et al. 2007). The tracer study showed that nitrogen amounts equivalent to those derived from salmon carcasses are taken up in the fall by the roots of riparian trees, especially Western Red Cedar, and allocated to leaves and stems the following spring (Drake et al. 2006).

The high densities of predators and scavengers that are attracted to salmon rivers for a few months each year provide a vector for moving the salmon-derived nutrients and energy upslope. In portions of Alaska, British Columbia, and Washington, primary movement of carcasses to the riparian zone occurs mainly by Brown Bears (*Ursus arctos*), Black Bears (*U. americanus*), and Bald Eagles (*Haliaeetus leucocephalus*); in more altered landscapes, such as the wine country of central California, domestic animals (dogs and cats) and also Raccoons (*Procyon lotor*), Opossums (*Didelphis virginiana*),

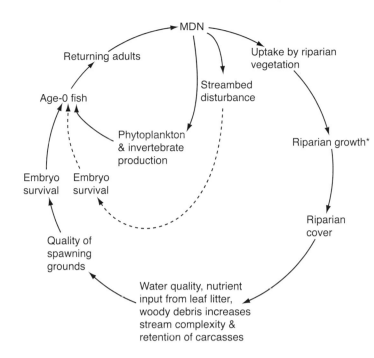

FIGURE 13.5. A possible indirect mutualism between Pacific salmon and components of the aquatic and riparian systems. Solid arrows show positive effects or increases; the dashed arrows show negative effects. *Some types of riparian vegetation do respond to marine-derived nutrients (MDN) by showing increased growth but other species do not. Based on data from Drake et al. (2006) and Piccolo et al. (2009).

and Turkey Vultures (*Cathartes aura*) play a similar role (Merz and Moyle 2006; Gende et al. 2007).

Although it is established that marine-derived nutrients from Pacific salmon species can subsidize riparian vegetation, the importance of the subsidies in fostering plant growth and/or species composition is both species and site dependent (Wipfli and Baxter 2010). Where an effect occurs, it generally results in shifts in plant species composition away from those typical of nutrient-poor habitats to those typical of nutrient-rich habitats, with a concomitant reduction in species richness. However, the influence of salmon is always associated with other factors, including the canopy plant community, the size of the watershed, and the slope of the site. The greatest effects from salmon on riparian vegetation occur in small drainages having generally lower productivity (Hocking and Reynolds 2011). Impacts of salmon on watersheds where nitrogen and phosphorus are not limiting are no doubt less than in those where these nutrients are limiting. In addition, stream gradient and geomorphology can affect flushing rates, and increased flushing could reduce the residency time of salmon-derived

nutrients. Because of their lower flushing rates, lakes may show a much longer response time over which marine-derived nutrients can be incorporated into the ecosystem (Piccolo et al. 2009).

The available information suggests that, at least in some instances, the fitness of Pacific salmon species is increased by a mutualistic feedback from riparian growth on stream habitat quality. In addition, the previously mentioned studies make it clear that conservation management of salmon species provides benefits that extend way beyond a catchable harvest for anglers (Merz and Moyle 2006).

MINNOW AND SUNFISH NESTS AND NEST ASSOCIATES Even though minnow and sunfish engineers are not subsidized by resources outside of their immediate ecosystem as is the case with salmon, their engineering products can be substantial. Occupied nests can be used as spawning sites by other species of fishes, especially minnows, and in fact, may be a requirement for the successful reproduction of many species (Wallin 1989). In addition, a single nest may be used by several or more associate species at the same time. Within the family Cyprinidae, at least 35 species are known to be nest associates (Johnston and Page 1992; Wisenden 1999).

Nest associates can gain improved reproductive success by spawning over host nests. Depending on the host species, eggs/larvae of nest associates benefit by being in the clean gravel afforded by the constructed nests, by protection against predation if the host guards the nest, and by possible aeration of the eggs by the host (Johnston and Page 1992; Johnston 1994a). Associates respond both to the nest and the presence of the host. In an experiment that consisted of constructing in two streams artificial gravel-mound nests that mimicked the actual nests of Bluehead Chub and Green Sunfish (*Lepomis cyanellus*), the artificial nests were not used by any of the numerous minnow species (including Central Stoneroller; Rosyside Dace; and Greenhead Shiner, *Notropis chlorocephalus*) that were observed spawning over nearby natural nests. In fact, once a host nest is constructed, it seems that the parental care by the host and not the physical environment of the nest is the major benefit to nest associates (Johnston 1994a).

Potential nest-building hosts can provide different services and risks to the associates, depending on the species. Minnow hosts do not defend the nest site once they are through spawning but also are not a predatory threat to the spawning associates. Even sunfish (*Lepomis*) hosts are generally not large enough to be a major threat to the associate species and also provide defense of the nest site for several weeks (Katula and Page 1998; Warren 2009). In contrast, nest associates of large piscivores such as Black Basses (*Micropterus*) and Bowfin (*Amia calva*) balance the apparent gain in defense of the nest by the host against the very real chance of being eaten by the host. For instance, Golden Shiner are apparently facultative nest associates of Bowfin, a large predator that routinely feeds on small fishes, including Golden Shiner, but which also offers aggressive and extended (ca. 9 days) defense of the nest site (Reighard 1903; Katula and Page 1998). In addition, at least three species of minnows, Golden Shiner, Common Shiner, and Taillight Shiner (*Notropis maculatus*), are known to spawn over nests of bass and would clearly be susceptible to bass predation but also gain extended protection of the nest by the host (Kramer and Smith 1960b; Johnston and Page 1992; Katula and Page 1998). Finally, there is an example of a large piscivore, Longnose Gar (*Lepisosteus osseus*), laying eggs over the nests of another piscivore, Smallmouth Bass, most likely during the night when the bass was inactive (Goff 1984). The gar eggs and larvae were vigorously defended by the resident Smallmouth Bass and young of both species seemed to have higher survival when together rather than alone. Thus nest associates and hosts represent a wide range of relative sizes, potential benefits, and potential risks to the spawning associates by the host species. Apparently no studies have rigorously examined the gain in offspring survival of the associates against the risk of predation by the host.

In some cases the relationship between associate and host is facultative. For instance, Topeka Shiner (*Notropis topeka*) is a nest associate with sunfish, including Green Sunfish and Orangespotted Sunfish (*L. humilis*). However, it will also spawn in the absence of sunfish nests, as shown by males selecting and defending spawning sites over sand substrata (Witte et al. 2009). In other cases the relationship between host and associate is obligate—Yellowfin Shiner do not reproduce in the absence of Bluehead Chub nests (Wallin 1989, 1992).

The relationship between host and associate is generally thought to be mutualistic, with a few exceptions where it is perhaps parasitic (Wisenden 1999). However, very few interactions have been adequately tested for benefits (or costs) to both hosts and associates. In one of the few studies to do this, Carol Johnston (1994a, b) first looked at the survival of the larvae of nest associates spawning over host nests, using Striped Shiner, which are nest associates of Hornyhead Chub, and Redfin Shiner (*Lythrurus umbratilis*), which are nest associates of Green Sunfish. The results clearly showed increased

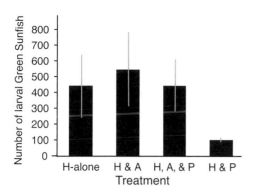

FIGURE 13.6. The benefit of a nest associate (A; Redfin Shiner, *Lythrurus umbratilis*) to its host (H; Green Sunfish, *Lepomis cyanellus*), as shown by mean larval survival of Green Sunfish in the presence of an egg/larval predator (P; Longear Sunfish, *L. megalotis*). Vertical lines are 95% confidence intervals. Based on data from Johnston (1994b).

brood survivorship for both associates when in the presence of the host fish. Next, using a replicated mesocosm experiment, she tested the potential benefit that the host might gain from the associate, using the host Green Sunfish, the nest-associate Redfin Shiner, and an egg/larval predator (Longear Sunfish, *Lepomis megalotis*) (Figure 13.6). Somewhat surprisingly, the mean number of surviving Green Sunfish larvae in treatments with the host, associate, and predator did not differ from treatments with the host alone or with the host and associate. However, the number of surviving Green Sunfish larvae was reduced by 80% in experimental treatments of the host and the egg/larval predator in the absence of the associate. Hosts with associates had significantly higher reproductive success than those without associates, providing convincing support for a mutualistic relationship. The proposed mechanism for protection against predation was the selfish herd effect (see Chapter 12), the dilution of the host eggs by those of the associate. In fact, the number of host eggs was diluted 15% by eggs of the associate.

A possible example of a parasitic relationship between nest associate and host occurs with Dusky Shiner (*Notropis cummingsae*) and Redbreast Sunfish (Fletcher 1993). Dusky Shiners spawn over nests of Redbreast Sunfish, which can form large nesting colonies and actively defend their nests against intrusion by large schools of Dusky Shiner. Dusky Shiner schools generally entered a nest when the host was absent and tended to be near the edges of the sunfish colony where the likelihood of finding an unguarded nest was greater compared to nests in the middle of the colony. Dusky Shiner readily consumed eggs and larvae of Redbreast Sunfish, and an examination of gut contents of Dusky Shiner taken from over nests showed that almost all identifiable larvae were sunfish. Based on the usual school size and the number of larvae consumed by a single Dusky Shiner, there could be approximately 1000 sunfish larvae removed from the nest during one feeding episode. Dusky Shiner most likely were feeding opportunistically and may have selected sunfish larvae because of their larger size. Although the apparent decrease in reproductive success of the host due to Dusky Shiner predation on eggs and larvae strongly suggest a parasitic relationship, actual parasitism cannot be definitively shown until the overall reproductive cost to the host is determined. Although unlikely, the decreased survival of sunfish eggs/larvae from shiner predation could be compensated for by the presence of the associate, perhaps via the selfish herd effect.

Species Associations and the Potential for Other Forms of Facilitation

Proposed benefits from interspecific aggregation include increased foraging efficiency and enhanced detection and avoidance of predators. Early warning benefits in mixed groups are well known in some organisms such as birds (Boucher et al. 1982; Thompson and Thompson 1985) and also fishes (Chapter 12). Fishes in monospecific shoals are able to quickly transmit information about a predator. In an experiment with European Minnows (*Phoxinus phoxinus*), one group of minnows (the transmitter minnows) was exposed to a model of a Northern Pike (*Esox lucius*) while a second group of minnows (the receiver minnows) could only see the transmitter minnows but not the predator. When the transmitter minnows were exposed to the predator model, fish in the receiver group responded by hiding and/or reduced activity—behaviors associated with predator avoidance (Magurran and Higham 1988). For early warning to be effective in mixed-species shoals, the information must also be transmitted rapidly (Godin et al. 1988), something that may be less likely among different species. However, by joining mixed-species groups, fishes could reap the benefits of being a member of a larger group (increased vigilance and reduced probability of predation) yet reduce the cost of intraspecific competition to less than it would be in an equally large group of conspecifics, assuming that predator evasion maneuvers would not be impeded by heterospecifics during an actual predator attack (Allan 1986; Pitcher 1986). Also, rare species might persist by virtue of hiding within schools of more abundant heterospecifics (Moyle and Li 1979). Based on studies done in an artificial stream, three sympatric European cyprinids that frequently associate in shoals, Dace (*Leuciscus leuciscus*), European Minnow, and Gudgeon (*Gobio gobio*), showed a balance between avoiding identical resources while remaining close enough to the other species to gain antipredation benefits from the appearance of a large shoal (Allan 1986). If all species in a mixed-species shoal would benefit to some degree, the interaction could be termed facultatively mutualistic; however, such fitness benefits have not yet been shown.

Suggestions of potentially facilitative or mutualistic interactions among mixed species of freshwater fishes are not new. Often observing fishes with field glasses while sitting on a railroad trestle over a stream, Jacob Reighard (1920) provided careful descriptions of feeding interactions between minnows and both the Northern Hogsucker (*Hypentelium nigricans*) and White Sucker. He also commented that he had observed White Sucker to be much less easily startled when accompanied by a group of Logperch (*Percina caprodes*). Although based only on visual observations, Reighard suggested enhanced feeding and safety as benefits from mixed-species groupings. Reighard's observations have since been extended by underwater observations of Blacktail Shiners and Northern Hog Sucker in a clear upland stream in Mississippi. Blacktail Shiners commonly followed closely behind large Northern Hog Suckers, and were observed feeding on invertebrates suspended by benthic disturbance caused by feeding actions of the Northern Hog Sucker (Baker and Foster 1994). In this relationship, the larger, benthic feeding sucker might also benefit by receiving early warnings of a threat if the smaller shiners responded to a potential predator. If it turns out that the interaction is beneficial to both participants, then the sucker-minnow tandem could represent a true mutualism.

Similar feeding associations have been observed among pairs of minnows and also darters. For instance, Bigeye Shiners (*Notropis boops*) swim above schools of benthic-foraging Ozark Minnows (*N. nubilus*), apparently feeding on items that are suspended in the water column by the benthic forager. The same pattern occurs with Hornyhead Chubs following the benthic algivore, Central Stoneroller (Gorman 1988a). Similarly, Gilt Darter (*Percina evides*) commonly follow two other darters, Logperch and Blotchside Logperch (*P. burtoni*), feeding on items exposed when the logperches flip over stones (Greenberg 1991). Overall, mixed feeding groups of fishes are apparently common. Various minnow species frequently form mixed schools, with individuals of one species following and responding to individuals of another species (Mendelson 1975; Copes 1983). Shoals of mixed species are also known among centrarchids, including species that can also interact as predator and prey as the size differential between them increases. In Florida canals, Largemouth Bass ≤ 30 cm TL and similarly sized Bluegill form mixed-species foraging groups of around five individuals that interact in hunting small poeciliids and cichlids (Annett 1998).

COEVOLUTION: DIFFUSE AND DIRECT

Thus far, this chapter has covered forms of facilitation, mutually positive or positive-neutral interactions among species. The topic of this section relates to a set of mechanisms, coevolution, potentially responsible for the interactions. Coevolution is not synonymous with mutualisms, species interactions, or symbiosis (Janzen 1980). Most basically, it constitutes "an evolutionary change in a trait of the individuals in one population in response to a trait of the individuals of a second population, followed by an evolutionary response by the second population to the change in the first" (Janzen 1980). As such, coevolutionary responses are widespread in evolution and potentially could include various forms of facilitation as well as resource divergence between species and predator-prey (including parasite-host) interactions; however, such interactions cannot be assumed due to coevolution (Janzen 1980; Thompson 2005, 2009). In fact, coevolution is difficult to prove and, even if interactions among species are coevolutionary responses, they could be a consequence of coevolution with organisms no longer present in their habitat (recall the disparate ages of major groups of North American freshwater fishes; Chapter 2) (Janzen 1980).

Forms of Coevolution and Eco-Evolution

Coevolutionary interactions can occur between tightly coupled species pairs (i.e., pairwise coevolution), where the evolution of a given trait in one species produces a subsequent evolution of a trait in the other, resulting in selective pressure producing further modification of the trait in the first species, and so on (Ehrlich and Raven 1964). Coevolution occurring in multispecies interactions was termed diffuse coevolution by Daniel Janzen (1980), and the fact that communities can evolve by virtue of overall contacts among species, without involving reciprocal evolutionary steps by all species, is presently considered an important aspect of coevolution (Inouye and Stinchcombe 2001; Thompson 2009).

The understanding that most species are collections of genetically distinct populations, that levels of selection often vary among the populations, and that consequently, the type and strengths of interactions among species should also be expected to vary, has led to the development of the geographic mosaic theory of coevolution (Thompson 1994, 1999a, 1999b, 2005, 2009; Gomulkiewicz et al. 2000; Nuismer et al. 2003). This much more realistic appreciation of species' populations leads to the view that interactions coevolve as geographic mosaics and that such interactions are continually modified across ecosystems (Thompson 2009). Because of this, one population within a species might be coevolved with another species, yet other populations of the same species would not necessarily show the same relationship.

It is well understood that natural selection resulting from ecological changes affects evolutionary changes in organisms over long and short time scales (Gould 2002; Schoener 2011). It is also well documented that evolutionary changes over long time periods (i.e., evolutionary time scale) have important consequences for ecological communities (Schoener 2011). However, studies of community ecology, such as resource separation among species (Chapter 11) or food web interactions (Chapter 12), typically consider species as static units that are not composed of locally evolving populations. The appreciation that evolutionary dynamics are affecting communities in "ecological time" is much more recent and has occurred because of the understanding that evolution can occur over ecological time scales. The interaction between ecology and evolution within an ecological time frame is referred to as eco-evolutionary dynamics, with the major precept that both directions of effect, ecology to evolution and evolution to ecology, are substantial (Johnson et al. 2009; Schoener 2011). The recognition of the interplay of short- and long-term evolutionary processes with a matrix of ecological interactions among species is currently leading to an exciting link between community ecology and evolutionary biology, although few studies as yet have documented the strength of the evolution to ecology pathway (Johnson and Stinchcombe 2007; Schoener 2011).

Obviously, one important requirement for such interaction to occur is that there is rapid evolution—evolutionary change that occurs on an ecological time scale (Schoener 2011). Other things being equal, rates of evolutionary change should increase as a function of shorter generation time. For instance, some livebearing fishes such as Least Killifish (*Heterandria formosa*) and Western Mosquitofish (*Gambusia affinis*) can produce several generations per year, and Red Shiner (*Cyprinella lutrensis*), as well as various other small cyprinids, can reproduce in their first summer of life, resulting in two generations per year (Henrich 1988; Haynes and Cashner 1995; Marsh-Matthews et al. 2002). Many other small-bodied fishes (minnows, topminnows, and some sunfish) reproduce when they are a year old (Ross 2001). Another set of requirements for eco-evolutionary interactions to occur is that a population shows genetic variation in traits, that traits respond to directional selection, and that the selection has an effect on a community variable (Johnson et al. 2009). As an example of rapid genetic change to directional selection, Red Shiner populations near cold-water releases from dams showed increased heterozygosity compared to populations in unaltered environments within a span of 50 years (Zimmerman and Richmond 1981). Mosquitofishes (*Gambusia* spp.), introduced to Hawaii in 1905, showed heritable differences in life-history traits in populations from thermally stable versus thermally variable reservoirs over a span of 70 years and approximately 140 generations (Stearns 1983). In a small northern California stream, individuals from an anadromous population of Rainbow Trout (Steelhead) were transplanted by humans above a waterfall some 100 years ago. The population above the falls shows a strong tendency for a resident life-history pattern and has diverged genetically from the source population below the waterfall to the extent of showing reproductive isolation from the source population (Pearse et al. 2009).

Whether or not the suggested mutualisms treated earlier in this chapter or in previous chapters have been shaped by coevolution or are simply fortuitous interactions is difficult to discern. The well-studied divergence of benthic and limnetic forms of Pumpkinseed (*Lepomis gibbosus*) and Threespine Sticklebacks (*Gasterosteus aculeatus*) (see Chapter 11) provides an example of pairwise coevolution via resource competition (Robinson et al. 1993; Schluter 2010). However, resource differences among species in many systems, especially in small streams, is likely due more to phylogeny or to environmental disturbances such as droughts and floods than by closely evolved interactions among species (Grossman and Freeman 1987; Grossman and Ratajczak 1998).

As yet, surprisingly little is understood about the importance of coevolution in shaping how ecosystems function. In a landmark study using Trinidadian populations of Giant Rivulus (Aplocheilidae: *Rivulus hartii*; mean body mass 0.75 g) and Guppy (Poeciliidae: *Poecilia reticulata*; mean body mass 0.15 g), Palkovacs et al. (2009) compared the effects of Guppy introduction, Guppy evolution (using populations that evolved with low and high predation risks), and Giant Rivulus-Guppy coevolution on food webs, including invertebrate biomass. Locally coevolved populations were more efficient at exploiting invertebrates compared to populations of Giant Rivulus and Guppy that were not coevolved. This was perhaps due to selection for enhanced competitive ability in the coevolved populations for exploitation of a shared resource. The results from this study show that coevolution of local populations may play a critical role in shaping ecological processes over contemporary time scales.

Longevity of Species Associations

Material in Part 1 would indicate that species composing local assemblages are often of widely different evolutionary ages and origins and that fish assemblages have gained and lost species over evolutionary time scales.

Consequently, the probability of a suite of species remaining together over long periods of time (i.e., thousands of years) is no doubt low, a conclusion also supported by others (Matthews 1998, Chapter 9). Especially for pairwise coevolution to occur, the interacting species need to remain associated for a substantial period of time, should have high encounter rates, and have strong interspecific interactions (Thompson 1982; Futuyma and Slatkin 1983; Price 1984; Farrell and Mitter 1993; Brown 1995). Of course, what constitutes a substantial period of time, a strong interaction, or a physically close association likely varies widely among taxa. Habitat availability would seem an important factor in fostering close association of species. Certain habitats, like deep pools or lakes, can be critical to the stability of a fish community (Schlosser 1987; Fausch and Bramblett 1991; Lohr and Fausch 1997). Whether fishes congregate consistently in a particular kind of habitat because of past coevolution of traits or because of independently acquired traits, the likelihood of future trait modification by coevolutionary interactions is increased if they now co-occur in regular, direct contact. What is generally absent in discussions of coevolution is actual evidence on how long (months, years, or millennia) patterns of direct contact among mobile species in real-world communities must persist to provide a consistent template within which interspecific genetic adjustments of traits, hence local coevolution, can occur (Ross and Matthews, in press). Among other plant and animal taxa, certain coevolved relationships are thought to be quite old or to have developed over millions of years of intimate contact, as in the case of the obligate, mutualistic relationship between fungus-growing ants and the cultivated fungus (McNaughton 1984; Currie et al. 2003). However, other studies indicate that coevolutionary interactions among species can develop over much shorter time periods (e.g., decades) (Thompson 2005). Consequently, observations of persistent contacts among fish species within an ecological time scale can be pertinent to evolutionary processes for species and communities (Chapter 6, Persistence stability of local associations).

SUMMARY

Species interactions that are mutually positive, or asymmetrically positive and neutral, are important in fish assemblages and aquatic ecosystems in general. Close associations of fish species in various kinds of habitats suggest the potential for such interactions to develop, although our understanding of how common positive or positive/neutral associations occur is still limited.

Facilitative interactions can occur through the presence of foundation species or through the activity of ecosystem engineers, although activities of ecosystem engineers can benefit some species more than others. Beavers represent important ecosystem engineers responsible for creating extensive lentic habitat patches in upland streams. The decline of Beavers as a consequence of trapping, likely had major impacts on fish habitats and the occurrence of fish species in low-order streams. Fish species that construct various types of gravel or pebble nests are also ecosystem engineers, and their nests provide nesting habitat for nest associates. Some nest hosts also contribute to the survival of young of the nest associates through nest guarding.

Mutualisms occur at various levels in fish assemblages and aquatic systems. Pacific salmon import huge quantities of marine-derived nutrients into streams, and such nutrients can benefit instream productivity and also riparian vegetation. A mutualistic response occurs if increased riparian vegetation improves stream conditions for the Pacific salmon. Interactions of nest associates and hosts also can be mutualistic if the associates increase the fitness of the hosts. One mechanism would be that the young of the associates dilute the impact of predation on the young of the hosts. Mixed associations of fishes can also be mutualistic, such as when the collective group vigilance increases the time available for foraging by individual members rather than sacrificing feeding time to vigilance.

The understanding that evolutionary changes in local fish populations can occur on an ecological time scale has contributed to the development of eco-evolutionary dynamics, as has the geographic mosaic theory of coevolution. Especially in populations with short generation times and thus high turnover, rapid evolutionary responses are likely and could lead to shifts in behaviors with consequent impacts on aquatic ecosystems.

SUPPLEMENTAL READING

Boucher, D. H., S. James, and K. H. Keeler. 1982. The ecology of mutualism. *Annual Review of Ecology and Systematics* 13: 315–47. A historical account of mutualism with suggestions for appropriate use of terms.

Johnston, C. E. 1994. Nest association in fishes: evidence for mutualism. *Behavioral Ecology* 35:379–83. One of the few studies to clearly demonstrate the benefit of nest associates to the nest-building host species.

Palkovacs, E. P., M. C. Marshall, B. A. Lamphere, B. R. Lynch, D. J. Weese, D. F. Fraser, D. N. Reznick, C. M. Pringle, and M. T. Kinnison. 2009. Experimental evaluation of evolution and coevolution as agents of ecosystem change in Trinidadian streams. *Philosophical Transactions of the Royal Society B* 364:1617–28. A landmark paper, based on years of supportive research, that shows how locally coevolved populations have different impacts on ecosystems than non-coevolved populations of the same species.

Pollock, M. M., M. Heim, and D. Werner. 2003. Hydrologic and geomorphic effects of Beaver dams and their influence on fishes, 213–33. In *The ecology and management of wood in world rivers*. S. Gregory, K. Boyer, and A. Gurnell (eds.). American Fisheries Society Symposium 37, Bethesda, Maryland. An excellent review of the past and present impacts of Beaver dams, and their widespread removal.

Power, M. E., D. Tilman, J. A. Estes, B. A. Menge, W. J. Bond, L. S. Mills, G. Daily, J. C. Castilla, J. Lubchenco, and R. Paine. 1996. Challenges in the quest for keystones. *BioScience* 46:609–20. An important paper clarifying the relationship between foundation species (or dominant species) and ecosystem engineers (or keystone species).

Schoener, T. W. 2011. The newest synthesis: Understanding the interplay of evolutionary and ecological dynamics. *Science* 331:426–29. A current synopsis of the eco-evolutionary interactions.

Wipfli, M. S., and C. V. Baxter. 2010. Linking ecosystems, food webs, and fish production: subsidies in salmonid watersheds. *Fisheries* 35:373–87. A broad look at food sources, including marine-derived nutrients, and food webs in streams.

PART FIVE

Issues in Conservation

THE FOUR PREVIOUS PARTS of this book have covered geological and climatic events shaping the North American freshwater fish fauna, how fish assemblages are formed and maintained, the functional morphology and life-history characteristics of fishes, and ways in which fish populations and species interact. This part integrates many of the previous topics by focusing on several areas of how fishes interact with their environments, how human-caused impacts have altered these environments and the responses of fishes, and how the relatively new field of conservation biology has contributed to the restoration of habitats and native fish populations. I have selected only several key issues because of the vast and rapidly expanding literature on fish conservation. This is exemplified by the excellent and recent book on fish conservation by Gene Helfman, which provides a much more expansive coverage of the topic (Helfman 2007).

To a great extent, conservation biology is a "crisis discipline" that takes a group of traditionally academic disciplines and applies the knowledge to the real-world challenge of maintaining functional biodiversity (Soulé

1985; Meffe and Carroll 1997). That is, not just keeping extant taxa alive, but also maintaining landscapes and ecosystems in which their populations and assemblages interact and evolve. In addition, conservation biology frequently includes the added challenge of immediate need—the necessity of proposing management actions based on currently available knowledge, without the luxury of additional time for studies. Ideally, the goal is to shift conservation biology from a reactive to a proactive discipline, aimed at avoiding or minimizing problems rather than attempting to restore severely damaged populations and ecosystems (Meffe and Carroll 1997).

Conservation of North American native fish diversity is approaching a crisis situation. In the twentieth century, 40 North American fish species became extinct, reflecting an extinction rate 1,000 times higher than the estimated background rate for fishes. Overall, the extinction rate for North American freshwater species is five times greater than for terrestrial species and is equivalent to the estimated extinction rate of tropical rainforest communities (Ricciardi and Rasmussen 1999).

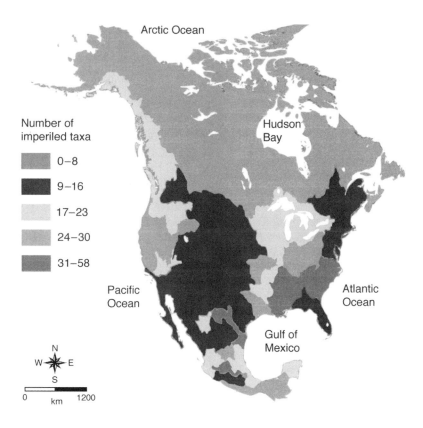

The distribution of imperilment (extinct, endangered, threatened, or vulnerable) of North American freshwater fishes. Degrees of imperilment are indicated for ecoregions. Redrawn from Jelks et al. (2008) with permission of the American Fisheries Society. See also corresponding figure in color insert.

In spite of conservation efforts for North American freshwater and diadromous fishes, the number of species considered to be imperiled increased from 1989 to 2008. At present, 39% of the described fish species or subspecies are imperiled and 21 taxa are considered to be extinct from wild habitats, being maintained only in captive breeding programs. Not surprisingly, the current North American map of imperiled species (see figure in Part 5) is quite similar to the map (Figure 1.4) of North American species diversity—the richest faunas have the most species to lose (Jelks et al. 2008).

As a matter of convenience, Part 5 is divided into two chapters—Chapter 14 emphasizes conservation biology of lotic systems, including fringing floodplains as used primarily by riverine fishes, and Chapter 15 emphasizes conservation biology of lentic systems, including floodplain lakes as used primarily by lentic specialists. This division is somewhat arbitrary for several reasons. First, man-made impoundments, reservoirs, are treated in both chapters. Chapter 14 deals primarily with the impact of dams on disrupting movement patterns of fishes in streams and in affecting downstream water quality, whereas the nature of the impounded water and its effects on the tributary streams is primarily covered in Chapter 15. Second, rather than considering streams, reservoirs, and natural lakes as discrete entities, the degrees of similarity in ecological structure and function are largely related to water residence times and flow characteristics. Consequently a more useful approach to viewing lentic-lotic distinctions is that of a continuum ordered by water flow and hydrologic flushing; lotic habitats are at one extreme with unidirectional flow and high flushing rates, whereas large lentic systems are at the other extreme with turbulent mixing and low flushing rates (Soballe and Kimmel 1987; Essington and Carpenter 2000; Wetzel 2001). Some of the factors affected by water movement include the time available for attached or suspended biota to interact with transported materials, abrasion and resuspension of materials, spatial and temporal variation, turbidity, and nutrient supply (Soballe and Kimmel 1987). Over time, lakes and streams can converge—streams change in character with age and/or longitudinally, becoming more lake-like with lower gradients and greater phytoplankton and zooplankton development, and lakes fill and gradually return to a riverine, wetland, or terrestrial habitat (Wetzel 2001). Finally, and perhaps most important, the separation of lotic and lentic systems runs the risk of de-emphasizing the important conclusion that ecosystems are rarely closed and that virtually all ecosystems receive benefits from or provide resources to other ecosystems—without which such ecosystems will generally suffer (Lamberti et al. 2010).

FOURTEEN

Streams Large and Small

CONTENTS

STREAMS AND RIVERS CONSTITUTE a major feature of the North American landscape—indeed, much of North American topography has been shaped by the impacts of flowing water on the erosion and deposition of materials (Leopold et al. 1964; Leopold 1994; Mount 1995). North American streams and rivers collectively discharge approximately 8,200 km³ (580 billion gallons) each year, which is about 17% of the world's total. Although there is a general relationship between the area of the watershed and the annual mean discharge (Leopold 1994), rivers in arid lands have discharges much lower than would be predicted by watershed area, whereas those in regions with high precipitation have annual discharges that are much greater (Figure 14.1). For instance, of 28 of the largest North American rivers, those with watersheds in Alaska, Canada, the Pacific Northwest, and subtropical Mexico have mean annual virgin discharges (i.e., the estimated discharge prior to human modifications) close to or above the regression line. In contrast, those in the arid central and western United States have discharges much lower than predicted by drainage area, with the most extreme being the Rio Grande, which has the sixth largest area but the lowest discharge among the 28 large rivers. Rivers in arid regions are particularly subject to droughts as well as periodic flooding with concomitant challenges to fishes and other aquatic organisms.

No matter where they occur, large rivers in North America and elsewhere in the world are now highly modified by humans through damming, water diversions, introduction of

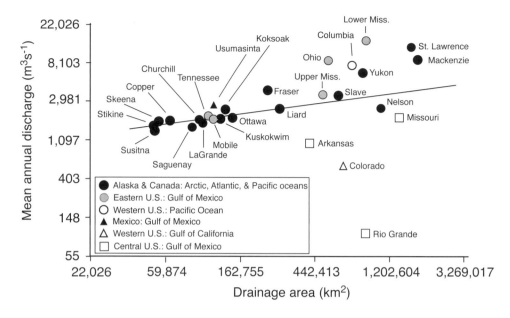

FIGURE 14.1. A log-log plot of the relationship between mean annual virgin discharge and drainage area for 28 large North American rivers. The angled line is a regression fit to a log-log plot (r = 0.31); for clarity the numbers on the x and y axes are shown as untransformed values. Based on data from Benke and Cushing (2005).

nonindigenous species, and various forms of water pollution. Over half of the world's large rivers (i.e., those with virgin mean annual discharge of at least 350 m³s⁻¹) are strongly affected by dams (Nilsson et al. 2005). Of the 74 large North American rivers, 39 (53%) are moderately to strongly affected by impoundments or flow regulation (Dynesius and Nilsson 1994). In North America, rivers that are not strongly impacted by human alterations are disproportionately located in areas of low human-population density—extreme northern areas of Alaska and Canada. In fact, only one of the 35 rivers not strongly affected by human alterations is located in temperate North America—the Pascagoula River located in Mississippi and Alabama (see also Jackson 2012). At the other extreme, the Colorado River is one of the world's most regulated rivers, with impoundments capable of storing and releasing more than 2.5 times the annual flow (Nilsson et al. 2005). Even in regions of North America where water has long been considered an abundant resource, such as the eastern United States, increasingly rapid human population growth is now exerting stressful demands on the available water (Freeman and Marcinek 2006).

The future of aquatic habitats and organisms in North America, as is true elsewhere, is closely linked to patterns of human population growth, and particularly to per capita demands on water use as well as the release of greenhouse gases (Meffe and Carroll 1997). There are a number of threats to aquatic biodiversity, including increasing demands for water and associated corollaries of impoundments and withdrawals, declines in water quality, channelization, levees, the introduction of nonindigenous species, and impacts of global climate change. Although all are important, increasing demands for water and changes to natural flow patterns are certainly major concerns and are the topic of this chapter. Impacts of nonindigenous species are covered in the following chapter.

HYDROLOGIC CYCLES AND FISH ASSEMBLAGES • Natural Patterns

The development of aquatic and riparian ecosystems is closely tied to the hydrologic regime, and daily or seasonal variations in stream discharge can have major impacts on

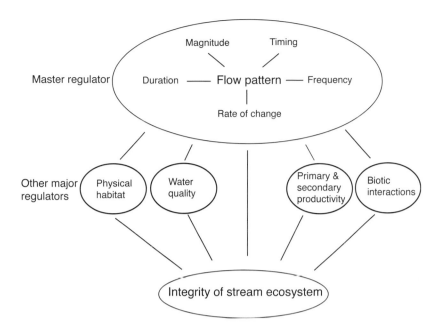

FIGURE 14.2. The flow pattern of streams is the master regulator of the ecological integrity of the stream ecosystem. Patterns of flow can act directly on ecological integrity or indirectly through other primary regulators. Based on Poff et al. (1997).

the behavior and ecology of aquatic organisms (Richter et al. 1996; Poff et al. 1997; Bunn and Arthington 2002). Patterns of flow can vary in their timing, magnitude, duration, frequency, and rate of change, and this variation has numerous direct and indirect effects on the integrity of the stream ecosystem (Box 14.1). The magnitude of flow, among other things, controls access to floodplain resources, and in western rivers that are driven by snowmelt, magnitude changes drastically over time. Flow variation can act directly on ecological integrity, or indirectly via primary regulators of ecological integrity, including physical habitat, water quality, primary and secondary productivity, and biotic interactions (Figure 14.2). In spite of its importance to aquatic communities, the extent of short-term variations in stream depth, flow velocity, and water quality (including temperature) often surprise many biologists accustomed to thinking of stream characteristics on a monthly or seasonal basis coincident with fish or invertebrate sampling, especially when dealing with small streams.

Importance of a Variable Flow Regime

Rises in stream discharge of sufficient magnitude and duration can trigger upstream or downstream movement of fishes, provide potential access to new habitats, rescue stranded populations in cut-off pools, or provide access to refuges during high flood events (Albanese et al. 2004). In addition, seasonally variable flow patterns provide increased food resources in the channel and also provide access to food and habitat resources on river floodplains, those areas that are periodically inundated by lateral overflow of rivers, lakes or reservoirs, by groundwater, or by direct precipitation. Importantly, streams and rivers and their associated floodplains should be considered as functional units and not as separate entities because they are tightly linked in terms of water, sediment, and organic matter exchanges (Junk et al. 1989; Junk and Wantzen 2004).

The river-floodplain system includes the main channel, permanent lakes and ponds on the floodplain, and the periodically inundated floodplain habitats. The temporally variable patterns in magnitude and frequency create

BOX 14.1 • Defining the Master Variable

Knowing basic hydrologic characteristics of a stream is fundamental to an understanding of fish life histories, fish-habitat relationships, and fish assemblage structure and persistence. This "master variable" of stream flow can be described by five main components: magnitude, frequency of occurrence, timing, flashiness, and duration, each of which can include several or more different measurements (Richter et al. 1996; Poff et al. 1997; Lytle and Poff 2004).

MAGNITUDE A measure of water condition, such as the amount of water moving past a fixed point per unit time (e.g., cubic meters per second), or some other measure of amount, such as water depth or a relative measure such as the amount of water relative to bank-full depth. Magnitude provides a measure of habitat availability and is often given in monthly averages of daily values. In the following figure, magnitude could be expressed as the 16-day mean, or simply by the line showing water depth.

FREQUENCY OF OCCURRENCE A measure of the pulsing behavior of flow—that is, how often a flow above or below a given magnitude occurs over a given time interval. In the figure below, there are three pulses above overbank flow within 16 days. Frequency usually shows an

inverse relationship with magnitude, indicating the increasing rarity of high magnitude flows.

TIMING The date when the highest and lowest flows (or intermediate flow events) occur; usually expressed on an annual cycle. Over annual cycles, the timing of flow events (either high or low) can occur during the same months (high predictability) or over different months (low predictability).

FLASHINESS A measure referring to the rate of change in flow conditions from one magnitude to another. The example shows a flashy stream, with water depth changing from 39 to 40 cm on the morning of April 30 to a peak of 151 cm by noon the following day. Also, the first two high-flow events have an interval of only nine days.

DURATION A measure of how long specific flow conditions last. Duration can be expressed as the number of days a particular flow is exceeded (such as the number of days per month or year of overbank flow) or as the length of time a particular flow event occurs. In the example figure, the overbank flows lasted from 1 to 1.5 days. On an annual cycle, overbank flows in Beaverdam Creek from 1992 to 1994 averaged 43 hours (Slack 1996).

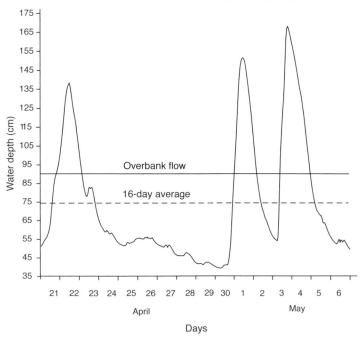

Short-term flow variation (water depth) in Beaverdam Creek, a small headwater stream in southeastern Mississippi. Based on unpublished data from W. T. Slack and S. T. Ross.

pulses of overbank flow. The flood-pulse concept (FPC), proposed by Junk et al. (1989), considers annual flooding events as the principal factor responsible for the "existence, productivity, and interactions of the major biota in river-floodplain systems." The FPC is a general concept for large river-floodplain systems in temperate as well as tropical habitats, and a major tenet is that much of the primary and secondary production of these systems takes place on the floodplain, with the river channel viewed as a transport mechanism for water and dissolved/suspended matter, a corridor for the movement of aquatic organisms, and as a refuge for aquatic organisms during periods of low flow (Junk and Wantzen 2004).

Flow, Temperature, and Movement Patterns

The timing of life-history events, such as spawning migrations, actual spawning, and embryo, larval, and adult growth and survival, are related to seasonal environmental changes, in particular day length acting in conjunction with changes in water temperature and rising or falling water levels (Leggett 1977; Helfman et al. 2009). For instance, in California's Sacramento-San Joaquin River system, most Steelhead (*Oncorhynchus mykiss*) time their migration into fresh water with the high flows generated by fall and winter rains, and also take advantage of the associated low water temperatures for spawning. For both Chinook Salmon (*O. tshawytscha*) and Steelhead in this system, water temperatures have profound impacts on growth and survival, with very narrow temperature tolerances, especially in the embryonic and early juvenile stages. In general, different stocks of Steelhead and Pacific salmon along the Pacific coast are adapted to different water temperature regimes (Myrick and Cech 2004). In the Columbia River, the timing of upstream migration of native Sockeye (*O. nerka*) and Chinook salmon, as well as the nonindigenous American Shad (*Alosa sapidissima*), are correlated primarily with water temperature and perhaps secondarily to spring discharge (Quinn and Adams 1996).

In tributaries of the Wabash River, Indiana, nine species of suckers in the genera *Carpiodes*, *Catostomus*, *Erimyzon*, *Hypentelium*, and *Moxostoma* show a seasonal progression of occurrence on spawning bars that is related to changes in water temperature and stream discharge (Curry and Spacie 1984). Similarly, the appearance of larvae of Flannelmouth (*Catostomus latipinnis*), Bluehead (*C. discobolus*), and Razorback (*Xyrauchen texanus*) suckers; Colorado Pikeminnow (*Ptychocheilus lucius*); and Speckled Dace (*Rhinichthys osculus*) in the San Juan River of the Colorado River system occurs progressively during the year in relation to rising water temperatures and a rising and then falling spring hydrograph (Figure 14.3). Of course, actual spawning and hatching dates of fishes predate the occurrence of protolarvae or mesolarvae (the two earliest larval stages) by a matter of days or weeks (Box 14.2). In the case of Razorback Sucker, over a 12-year period, hatching and emergence from gravel spawning bars included periods prior to spring runoff or a portion of the ascending limb of the hydrograph. One-third of the time, hatch dates also extended into the descending limb of spring runoff. In general, the native suckers initiate spawning earlier in the season, usually during a rising hydrograph and rising water temperatures, compared to the native minnows (Speckled Dace and Colorado Pikeminnow), which spawn during a falling hydrograph but still rising water temperatures (Brandenburg and Farrington 2011). Colorado Pikeminnow in the free-flowing Yampa River, Colorado, also spawned as flows were decreasing and as temperatures reached 22–25° C (Tyus 1990).

In a study of movement of eight fish species (families Cyprinidae, Catostomidae, Ictaluridae, and Cottidae) in small streams in the James River drainage, Virginia, five species showed increased movement rates in association with high flow events (Figure 14.4) (Albanese et al. 2004). Movement was based on the number of fish moving past bidirectional fish traps (e.g., traps that captured fishes moving both upstream and downstream). Although more species showed upstream movements,

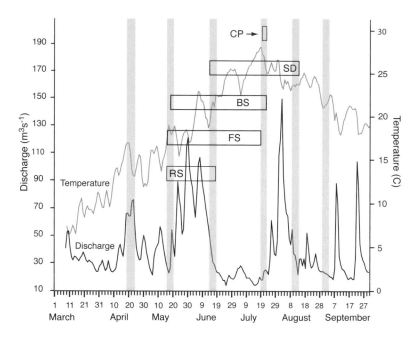

FIGURE 14.3. River discharge, temperature, and the occurrence of native fish larvae in the San Juan River in 2010. Temperature and discharge data are based on USGS gauge 09379500, near Bluff, Utah. Rectangles show the occurrence of larvae for five native fish species. The light gray bars are sampling trips. RS = Razorback Sucker (*Xyrauchen texanus*); FS = Flannelmouth Sucker (*Catostomus latipinnis*); BS = Bluehead Sucker (*C. discobolus*); SD = Speckled Dace (*Rhinichthys osculus*); CP = Colorado Pikeminnow (*Ptychocheilus lucius*). Based on data from Brandenburg and Farrington (2011) and unpublished data from Howard Brandenburg, Mike Farrington, and Steve Platania.

downstream movement occurred as well. Voluntary flow-mediated movement within stream channels (in contrast to passive displacement by flood waters or lateral movement onto floodplains) is extremely widespread among North American freshwater fish families, including fishes in the families Acipenseridae, Polyodontidae, Cyprinidae, Catostomidae, Ictaluridae, Salmonidae, Gasterosteidae, and Centrarchidae (Leggett 1977; Schlosser 1995b; Albanese et al. 2004; Heise et al. 2005; Gerken and Paukert 2009).

The Pascagoula River, Mississippi, supports a population of Gulf Sturgeon (*Acipenser oxyrinchus desotoi*), a large anadromous fish that feeds in coastal waters of the Gulf of Mexico and which spawns in coastal rivers along the northern and eastern Gulf of Mexico. The return migration into fresh water occurs over a narrow time period in February and March, with fish reaching their spawning site approximately 250

river km upstream from the Gulf Coast by mid-April (Heise et al. 2004; Ross et al. 2009). Similar patterns of movement occur in other populations of Gulf Sturgeon (Fox et al. 2000). Rising water temperature is no doubt an important cue, but upstream movement is also related to rises in stream discharge (Heise et al. 2004).

Gulf Sturgeon leave fresh water in the fall and the timing of the out-migration is associated with seasonal cues, including shorter day lengths and declining water temperatures, but once these conditions are appropriate, the actual trigger for movement seems to be spikes in river discharge. In a drought year, Gulf Sturgeon out-migration from the Pascagoula River, Mississippi, occurred much later than in years with higher fall flows (Heise et al. 2005). On the Pacific coast, Green Sturgeon (*A. medirostris*) also move out of fresh water as a function of dropping water temperature and increases in discharge (Erickson et al. 2002).

BOX 14.2 • Larval Occurrence and Spawning or Hatching Dates

Environmental conditions during fish spawning, embryo development, and larval emergence (generally coinciding with hatching followed by first feeding) are highly important in the understanding of fish recruitment patterns and the relationship of spawning and survival to environmental conditions (see also Chapter 9). However, in most instances, the actual dates of spawning, emergence, or hatching are not directly observable. Although the occurrence of eggs or larvae gives an approximation of spawning and hatching dates, egg capture or larval presence can postdate spawning and hatching by days or even weeks, depending on the species and environmental conditions. For instance, Bloater larvae (*Coregonus hoyi*) in Lake Michigan generally occur in the hypolimnion 10 or fewer days following the initiation of exogenous feeding (which occurs around three days after hatching), but in the epilimnion, some larvae were 55 days beyond the initiation of first feeding (Rice et al. 1987).

Dates of hatching of many fishes, and thus environmental conditions, such as temperature or stream discharge, near the time of spawning and at the time of emergence from spawning areas, can be determined by back-calculations based on the size or age of fishes at known capture dates. The most straightforward approach is to determine the actual age (in days) of fish—which, combined with a known date of capture, readily provides the age at hatching or first feeding. For example, this approach was used in a northern population of Brook Silverside (*Labidesthes sicculus*) based on the determination of daily growth bands in otoliths of juvenile and adult fish (Powles and Sandeman

2008). In another case, spawning dates for White Sturgeon (*Acipenser transmontanus*) in the Kootenai River, Idaho, were back-calculated from the dates of capture of fertilized eggs on spawning mats (White Sturgeon spawn in the water column and have demersal adhesive eggs), determining the developmental stage of the viable eggs and knowing, from other studies, the relationship of embryo stage and developmental time (Paragamian et al. 2002).

In the San Juan River of Colorado, New Mexico, and Utah, Razorback Sucker (*Xyrauchen texanus*) appeared in the ichthyoplankton from May 17 to June 20, 2010 (see Figure 14.3). Larvae hatch 6–7 days after fertilization at 18° C and begin initial feeding nine days after hatching. At 16.5° C, the growth rate of larval Razorback Sucker is around 0.35 mm per day. Consequently, hatching dates can be back-calculated by subtracting the average length of larvae at hatching (8.0 mm TL) from the length at capture, and then dividing by the average growth rate at an appropriate water temperature (i.e., 0.35 mm per day at 16.5° C). Using this approach, in 2010, Razorback Sucker hatching dates ranged from April 28 to June 11. Furthermore, because the larvae hatch 6–7 days postfertilization (Bestgen 2008), actual spawning likely occurred from April 21 to June 4. Because river discharge and temperature during spawning, hatching, and larval occurrence vary considerably, being able to back-calculate more exact dates is important for understanding spawning requirements and success (Brandenburg and Farrington 2011; Howard Brandenburg, Mike Farrington, and Steve Platania, unpublished data 2011).

Upstream spawning migration of the Splittail (*Pogonichthys macrolepidotus*), a cyprinid endemic to the San Francisco Estuary and associated rivers, is triggered by spikes in river flow during the winter. High flows are also required to attract fish to floodplain habitats that are used for spawning and for nursery areas (see also Chapter 6) (Feyrer et al. 2006). Similarly, upstream spawning migration of Colorado Pikeminnow in the free-flowing Yampa River, Colorado and Utah, is apparently triggered by discharge and perhaps temperature, occurring about a month following the

highest spring flows at water temperatures above 9° C (Tyus 1990).

Flow characteristics, especially timing, magnitude, and duration, can have major impacts on the survival of eggs and larvae and thus drive population success or failure (see also Chapter 5). For instance, newly emerged fry of Rainbow Trout (*O. mykiss*) are highly vulnerable to downstream displacement and mortality caused by flooding. In the Pacific Northwest, their native range, Rainbow Trout spawn in late winter to early spring and the fry emerge from the redds during late spring to early summer. This timing

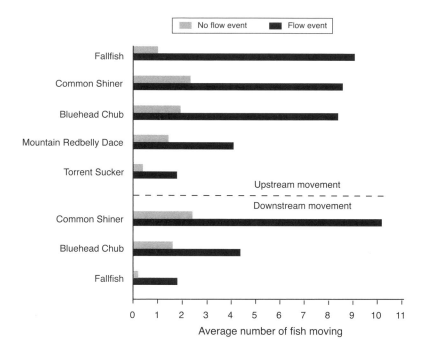

FIGURE 14.4. The effect of short-term flow events on the movement of stream fishes in the James River drainage, Virginia. Based on data from Albanese et al. (2004).

of fry emergence generally follows months with high likelihoods of floods—illustrating the importance of the timing and magnitude of flow on population success (Figure 14.5A) (Fausch et al. 2001). However, Rainbow Trout are one of the most widely transplanted fish species, and their invasion success is often hard to predict (MacCrimmon 1971, 1972). Fausch et al. (2001) compared flow regimes among five Holarctic regions where Rainbow Trout are native or have been introduced, including regions where introductions have been highly successful (Southern Appalachians and Colorado—prior to the whirling disease epidemic), moderately successful (Hokkaido Island, Japan), and unsuccessful (Honshu Island, Japan). The principal hypothesis was that Rainbow Trout invasions would be successful when flow regimes most closely matched those in the native range (i.e., emergence of fry following high flows), and least successful when high flows overlapped with or followed the period of fry emergence. The results supported the hypothesis that invasion success related to the match between the timing of fry emergence and months with low probability

of flood disturbance. In the Pacific Northwest and in the southern Appalachians, floods are caused by winter rains and flows drop to base levels by early summer (Figure 14.5A, B). Fry emergence occurs during a period of declining flows. Streams on Honshu Island, where invasion of Rainbow Trout has been unsuccessful, have flooding caused by summer monsoonal rains followed by typhoons in the fall. Consequently, there is a high likelihood of flooding from April through October. Fry emerge in May and June, preceding rather than following months with the highest percentage of flooding (Figure 14.5C). Flooding of Hokkaido streams is a factor of melting snow in late spring and the likelihood of typhoons in early fall. In contrast to the Honshu Island trout, the emergence of fry occurs in a two-month slot after the end of snowmelt and before increased chances of typhoons (Figure 14.5D). Finally, Colorado streams are driven primarily by snowmelt and have highly predictable flows. Rainbow Trout emergence extends from June through late July when the likelihood of flooding is rapidly declining (Figure 14.5E).

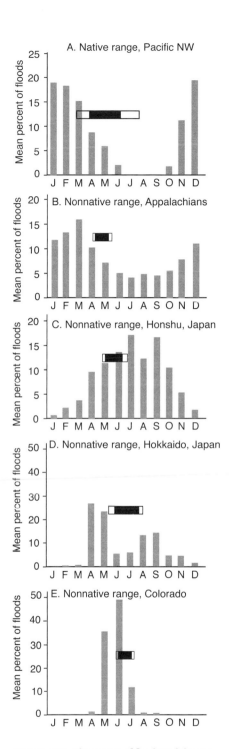

FIGURE 14.5. The timing of floods and the successful survival or colonization of Rainbow Trout (*Oncorhynchus mykiss*) in five Holarctic regions. Open horizontal rectangles show the range of dates for emergence of fry from spawning redds; the closed horizontal rectangles indicate the central 80% of emergence times. Based on Fausch et al. (2001).

Flow and Access to Resources

Although flood events in higher gradient streams can have catastrophic impacts on aquatic and floodplain biota (see Chapter 6), in lowland rivers with fringing floodplains, increases in discharge can cue movement of fishes out of the stream channel and onto the inundated floodplain for purposes of feeding, reproduction, or for refuge from high main-channel flows. The most extreme examples of such movement occur with tropical freshwater systems, such as the Amazon Basin; however, similar patterns occur, or occurred before levee construction and channelization, in many North American streams (Welcomme 1979; Ross and Baker 1983; Junk et al. 1989).

The importance of access to inundated floodplains by fishes has been recognized by biologists for over 100 years. Stephen Forbes, one of the first presidents of the Ecological Society of America and one of the founding fathers of North American ecology, clearly understood the great importance of floodplains, as is evident from an address given to the American Fisheries Society in 1910, stating, "Nothing can be more dangerous to the continued productiveness of these waters [the Illinois River] than a shutting of the river into its main channel and the drainage of the bottom-land lakes for agricultural purposes" (Schneider 2000). Forbes further emphasized the importance of natural access to fringing floodplains by fishes and other aquatic organisms in his classic paper, "The Lake as a Microcosm" (Forbes 1925). Sadly, in spite of the knowledge and long efforts of Forbes and his colleagues to influence law makers, some of the major habitats on the Illinois River floodplain, including those that had formed the nucleus of their research, were ultimately levied and drained for agricultural purposes (Schneider 2000).

The periodic inundations of forested bottomlands along many low-gradient streams provide fishes with direct access to terrestrial food resources. Beaverdam Creek, a tributary of Black Creek in the Pascagoula River drainage of southeastern Mississippi, provides an excellent example of connectivity between a low-order stream (second order at the study site) and its floodplain. The flow pattern can be highly flashy (see the figure in Box 14.1) and is driven by rainfall events. Periods of overbank flow are of short duration, typically 80 hours or less, and occur primarily in the winter and early spring (November–April), although there is considerable annual variation. Fishes move onto the inundated floodplain quickly even as the water is rising; species occurring in 60% or more of the floods were Striped Shiner (*Luxilus chrysocephalus*), Tadpole Madtom (*Noturus gyrinus*), Blackspotted Topminnow (*Fundulus olivaceus*), Longear Sunfish (*Lepomis megalotis*), and Pirate Perch (*Aphredoderus sayanus*) (Slack 1996). Another 14 species occurred in 20–60% of the floods, and 13 species rarely moved out of the stream channel, occurring in < 20% of all floods. However, of the 36 fish species documented from the study site (including both the stream channel and the floodplain), only four, Speckled Darter (*Etheostoma stigmaeum*), Southern Brook Lamprey (*Ichthyomyzon gagei*), Spotted Bass (*Micropterus punctulatus*), and Flagfin Shiner (*Pteronotropis signipinnis*), were never collected on the floodplain. Not surprisingly, the number of species on the floodplain increased as a function of flood duration, although species composition varied substantially among floods (i.e., species similarity was not related to flood duration). Because of the short duration of inundation and the generally low frequency of duration during the spawning periods for most of the species, fishes were not using the inundated floodplain for spawning but instead to take advantage of a pulse of terrestrial prey items (Slack 1996; S. T. Ross unpublished data). Also, given the short duration of flooding, "the occurrence of fishes on the floodplain represents opportunistic exploitation of additional habitat, and not floodplain habitat, per se" (Slack 1996).

The hypothesis that the floodplain of this low-order stream provides significant food resources was tested using enclosures in the stream channel and on the inundated floodplain, which were stocked with Cherryfin Shiner (*Lythrurus roseipinnis*), a shoaling, drift-feeding minnow (O'Connell 2003). Of course, enclosures could not be placed in the stream channel during periods of floodplain inundation because of water depth and current speed, so the comparison is between the low-flow stream and the high-flow, inundated floodplain. Overall, there were significantly more items in the drift on the inundated floodplain compared to the stream channel. Cherryfin Shiner, held at natural densities in floodplain enclosures, consumed more than twice the mass of prey compared to those in stream enclosures. Although the actual diversity of prey items did not differ between the two habitats for either availability or presence in fish gut contents, the taxonomic composition of the diet of floodplain fish differed significantly from those in the stream. Most of the difference was due to the high number of terrestrial arthropods, especially Springtails (Collembola) and terrestrial mites that were eaten by fish on the floodplain. The study showed that the inundated floodplain offered more food than the low-water stream, with the mean density of drift items never below eight per m^3, whereas the mean densities of drift items in the low-water stream were never more than six per m^3, and floodplain drift densities can be 2–3 times higher than the maximum drift density expected for the stream. Consequently, even for this low-order stream, periodic flooding allows fishes access to abundant prey sources, and because there were not differences in drift densities among sampling periods, the inundated floodplain can consistently provide high densities of drift (O'Connell 2003). These results are somewhat counter to predictions of the flood-pulse concept for small streams in that fishes are actively moving onto the inundated floodplain and feeding directly on available resources rather than using resources indirectly as they are washed into the stream channel.

Black Creek, a larger stream in the same Mississippi watershed (annual mean discharge of 6.8 m^3s^{-1}), has winter and spring overbank floods of longer duration (although stream discharge or stage were not continually monitored at the site) and the occurrence of fishes moving onto the floodplain, versus those that did not, was more distinct (Ross and Baker 1983). Species moving onto the floodplain, termed flood exploitative, were hypothesized to have increased reproductive success, as shown by population size, because of increased food availability prior to spawning. A second suite, termed flood quiescent, did not venture onto the floodplain and were hypothesized to decline following years with more flooding. A test of the hypothesis with Weed Shiner (*Notropis texanus*), a flood exploitative species, and Blackbanded Darter (*Percina nigrofasciata*), a flood quiescent species, showed that Weed Shiner abundance was significantly correlated with high-flow years. In contrast, Blackbanded Darter population size did not increase or decrease in relation to spring discharge prior to spawning (Ross and Baker 1983).

In progressively larger rivers, the importance of periodic inundation of floodplains changes to include spawning and nursery habitat in addition to increased food availability. The Tallahatchie River of Mississippi, a large tributary of the Yazoo River (a major tributary of the Mississippi River), has an average monthly discharge of 1,525 to over 3,600 m³s⁻¹. The Yazoo River floodplain, dominated by agricultural land and associated second-growth hardwood and sparse cypress, is inundated during winter and spring. The floodplain also supports some permanent bodies of water. Based on light-trap sampling, more larval and juvenile fishes were collected in floodplain habitats than in the river channel. From early spring to midsummer, larvae of spring-spawning fishes were dominant in floodplain habitats, suggesting the importance of the floodplain, even though partially modified as agricultural land, as a spawning and nursery ground for riverine fishes (Turner et al. 1994). Numerically abundant taxa included Gizzard Shad (*Dorosoma cepedianum*), Crappie (*Pomoxis* spp.), and various darters (Percidae).

The upper Mississippi River watershed includes the states of Missouri, Illinois, Iowa, Wisconsin, and Minnesota and, although a highly modified system, has floodplain habitats primarily associated with impounded sections. Fishes associated with littoral zone habitats in the impounded river sections, such as Largemouth Bass (*Micropterus salmoides*) and Bluegill (*Lepomis macrochirus*), showed increased growth responses following extreme flooding compared to lower-flow periods. Such results, albeit only shown by certain size classes, are consistent with predictions of the flood-pulse concept for large river-floodplain systems (Gutreuter et al. 1999). Black Crappie (*Pomoxis nigromaculatus*), which are less associated with the littoral zone, showed an ambiguous response to flooding, and the even more lotic species, White Bass (*Morone chrysops*), did not show an immediate growth response to flooding, again consistent with predictions of the flood-pulse concept (Junk et al. 1989; Gutreuter et al. 1999).

Although many North American freshwater fishes do show spawning, hatching, and growth patterns that are related to high river discharge and access to floodplain habitats, there are also species that do not respond to high flow in terms of growth (e.g., White Bass in the upper Mississippi River) or reproduction (e.g., Blackbanded Darter in Black Creek, Mississippi) (Ross and Baker 1983; Gutreuter et al. 1999). Consequently, flow regimes in rivers are not a "one size fits all" situation, especially when freshwater faunas are considered on a worldwide scale (King et al. 2003). Particularly in arid regions of the world where river flows are unpredictable, such as in parts of Australia, at least some groups of native fishes successfully recruit during summer low-flow conditions (King et al. 2003). This life-history pattern fits predictions of the low-flow recruitment hypothesis (LFR) (Humphries et al. 1999). In the low-gradient Brazos River, Texas, where overbank flooding is highly irregular and unpredictable, oxbow lakes did support higher numbers of juvenile fishes relative to the main channel and were particularly important for equilibrium strategists such as Bluegill and White Crappie (*Pomoxis annularis*)—species that build nests and defend young. Small species with prolonged reproductive periods, such as Red Shiner (*Cyprinella lutrensis*), and large, long-lived species, such as Longnose Gar (*Lepisosteus osseus*), were abundant in the river channel. Fishes with a periodic strategy, such as gar, are able to retain their reproductive potential for long periods in the river channel until high flows allow access to oxbow lakes for spawning. In fact, Longnose Gar was the only species in the Brazos River study showing greater abundances in the wet year. In contrast, the low-flow, dry year favored Red Shiner, Western Mosquitofish (*Gambusia affinis*), and Gizzard Shad. Overall, hydrologic connectivity did not increase juvenile production for most species, suggesting that recruitment dynamics of these species were similar to predictions of the LFR (Zeug and Winemiller 2008b).

HYDROLOGIC CYCLES AND FISH ASSEMBLAGES • *Altered Patterns*

Perhaps one of the oldest of human impacts on aquatic systems is the building of dams (Baxter 1977). Dams have major impacts on riverine ecosystem function by changing downstream flow patterns and sediment supply, altering water temperature, and blocking the upstream and downstream movement of nutrients and organisms (see Chapter 5 and Figure 5.5). Most dams can be categorized into two major functional groups—those built to store water and those built to raise water levels. Storage dams generally have a large storage volume (e.g., some large western impoundments can retain almost four years of average annual runoff), a large hydraulic head, long retention times, and major control over the downstream release of water. Run-of-river dams are characterized by low storage volumes, short hydraulic residence times, a small hydraulic head, and little or no control over water releases. Although they do trap sediment, they tend not to alter

downstream water temperatures (Poff and Hart 2002). In addition to blocking passage, specific impacts of dams depend on the type and purpose of the impoundment—storage dams built for flood control are usually emptied as soon as possible following a flood; storage dams built for other purposes retain water during high stream inflow and then gradually release water over time, or retain most water until additional reservoir capacity is needed for the next high flow (Baxter 1977).

What is retained by the dam (water, sediment, nutrients, and heat) is typically lost to the stream. Typical impacts of large to medium-sized dams on flow patterns include creating a lag in peak flows and reducing the overall magnitude of flows, and either increasing low-flow discharge (in dams associated with hydropower generation or in response to downstream irrigation needs), or totally drying downstream channels (Baxter 1977; Mount 1995). For instance, Navajo Dam, which impounded the San Juan River in 1962, reduced the magnitude and duration of spring flows while increasing low-water discharge (Figure 14.6). However, generalizing on the specific hydrological impacts of dams, especially dams on smaller watersheds, is challenging because variation caused by dam operations are superimposed on the already great variation in flow patterns caused by geographical and climatic regions, land use patterns, and position in the stream network (Poff et al. 2006).

In a broad-based study of the effects of dams on small watersheds (< 282 km²) within the United States, dams reduced peak flows while elevating low flows, similar to effects of large dams. Dams on small watersheds also caused increases in flow duration, except for those in the southeastern United States, which did not show any differences from pre-dam conditions. Finally, small dams, as with large dams, tended to reduce flow variability, with the exception of hydropower dams whose flow characteristics are often driven by distant electrical demands (Poff et al. 2006).

In addition to altering flow and associated sediment dynamics, dams can also have a major

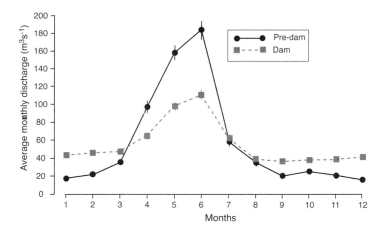

FIGURE 14.6. The effects of dams on river hydrology, based on discharge of the San Juan River downstream of the Navajo Dam. Data are means and 95% confidence intervals of monthly flows during pre-dam (1942–1961) and dam conditions (1962–1991). Discharge data are from the USGS Gauge 09368000 in Shiprock, New Mexico.

impact on water temperatures at certain times of the year, especially summer and fall. Dams, and especially those with discharges originating below the thermocline (hypolimnial discharge), gain heat in the epilimnion during the summer as solar radiation on the impounded water is converted to thermal energy. Because this has little effect on the cold hypolimnion, hypolimnial discharges are colder compared to pre-dam conditions. When reservoir stratification breaks down in late fall and winter, some of the stored heat can enter a hypolimnial outflow, warming downstream water relative to pre-dam conditions (Baxter 1977). In a detailed study of the effects of four dams in the Pacific Northwest, summer water releases were cooler compared to temperatures before the dams were built, and generally were warmer in the fall and winter. However, even though two of the dams released water from the hypolimnion, one released water from the epilimnion, and one released water from various levels, all four had similar effects on water temperatures. Two dams were located in the Willamette River drainage, Oregon, one on the Cowlitz River in Washington and one on the Rogue River in southern Oregon (Angilletta et al. 2008).

Chinook Salmon used all four of the dammed streams studied by Angilletta et al., and numerous aspects of salmon biology are keyed to temperature (Brett 1971). Spawning of Pacific salmon species, the emergence of fry, and the migration of smolts all are related to appropriate thermal cues in natal lakes and streams (Brett 1971; Angilletta et al. 2008). For instance, downstream migration of Chinook Salmon smolts begins at water temperatures above 10° C and reaches a maximum between 12.5 and 15° C; higher average spring temperatures result in earlier emigration (Roper and Scarnecchia 1999). Because dams maintain higher stream temperatures into fall and winter, they should impact late spawners more than early spawners. Consequently, there is a very real possibility that dams and other anthropogenic changes can result in mismatches between life histories and environmental conditions or increased mortality because of temperature stress. Elevated water temperatures increase the rate of embryonic development and thus accelerate the timing of emergence of alevin from the redd. The early emergence can potentially expose young fish to peak flows, higher predation rates, or lowered resource abundance (Angilletta et al. 2008).

One of the more obvious effects of dams is that of blocked passage caused by large dams as well as by numerous smaller obstacles such as low dams and culverts. In the Columbia River basin, the largest river basin in the Pacific Northwest, large, impassable dams are responsible for the loss of one-third of the historical Pacific salmon and Steelhead habitats, including 87% of mainstem Columbia River spawning habitat for Chinook Salmon (Sheer and Steel 2006). In addition, almost one-half of the historical populations of Sockeye Salmon are extinct (Waples et al. 2011).

There is no doubt that large dams have had profound impacts on aquatic systems in the Columbia River basin, in particular on migratory fishes. However, quite often the declines in migratory species are due to multiple causes rather than a single factor. An excellent example of this point is the case of the Snake River-Redfish Lake population of anadromous Sockeye Salmon. The Salmon River, named for the abundance of Pacific salmon it used to support, and Redfish Lake, named for the large numbers of Sockeye Salmon that formerly spawned in the lake tributaries (redfish is a local name for anadromous Sockeye Salmon), historically supported large numbers of Sockeye Salmon that were a significant component of the overall Columbia River salmon production (Bjornn et al. 1968). Redfish Lake, and other nearby lakes in the Sawtooth Valley, Idaho, represent the farthest upstream spawning sites known for Sockeye Salmon—the distance from Redfish Lake to the mouth of the Columbia River is some 1400 km and represents an elevation gain of almost 2000 m. Fish reach Redfish Lake from the Pacific Ocean via the Columbia, Snake, and Salmon rivers, with a final ascent up Redfish Lake Creek (Figure 14.7). Spawning of Chinook Salmon, anadromous Sockeye Salmon, and Kokanee (nonmigratory form of *O. nerka*) in these lakes was documented by fisheries scientists from 1894 to 1896 (Evermann and Meek 1898). However, the complex genetic structure of *O. nerka* populations in Redfish Lake has only been understood much more recently. Both resident (Kokanee) and anadromous Sockeye Salmon in Redfish Lake are more similar to each other than to Sockeye Salmon populations in the upper Columbia River, and Kokanee from Redfish Lake and other lakes in the Sawtooth Valley are genetically distinct from other Columbia River populations. Furthermore, Redfish Lake populations of Kokanee and anadromous Sockeye Salmon are also distinct, but with an additional level of complexity. There are two quite distinct gene pools of *O. nerka* in

Redfish Lake, one being the resident Kokanee and the other comprising the anadromous Sockeye Salmon plus a generally nonmigratory or "residual" form that matures at a smaller body size than the anadromous form (Waples et al. 2011). Given the amazingly long migration route of the Snake River-Redfish Lake anadromous Sockeye Salmon, perhaps the "residual" form represents a type of "bet-hedging" to survive times when migration is less successful or unsuccessful, much as variation in age at maturity does for populations of Chinook Salmon (Waples et al. 2010).

The natural harvest of Columbia River salmon has occurred for around 10,000 years, although population-altering levels of fishing pressure did not occur until large-scale commercial harvests began in the 1860s, resulting in obvious declines in all salmon species. Even though extensive declines of Pacific salmon were occurring before the added impact of blocked river passage by dams, declines of the Snake River salmon were accentuated by the construction in 1913 of the Sunbeam Dam, located on the main channel of the Salmon River about 32 km downstream from the mouth of Redfish Lake Creek (Figure 14.7). The dam was essentially a complete barrier to fish passage until it was partially removed in 1934, although recent genetic evidence indicates that the original anadromous plus "residual" Sockeye Salmon population was not extirpated (Bjornn et al. 1968; Waples et al. 2011). The 1930s also saw the start of major development of downstream hydroelectric dams and the continued addition of main-channel dams on the Columbia and Snake rivers into the 1970s (Selbie et al. 2007; Waples et al. 2007). At present, only about 5% of the historical spawning habitat for Sockeye Salmon remains accessible in the Columbia River basin and the number of Snake River Sockeye Salmon returning to lakes in central Idaho has dropped by over 99% (Selbie et al. 2007). The Snake River Sockeye Salmon was listed as endangered under the U.S. Endangered Species Act in 1991 and there are major efforts to restore the stock, although it still

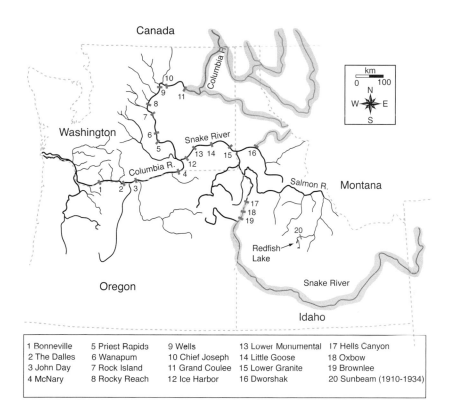

FIGURE 14.7. Major rivers and associated dams in the Columbia River basin that are referred to in the text. Light gray shading indicates upstream reaches blocked to migratory salmonids by impassable dams. Based on Bjornn et al. (1968), Selbie et al. (2007), and Waples et al. (2007, 2011).

1 Bonneville	5 Priest Rapids	9 Wells	13 Lower Monumental	17 Hells Canyon
2 The Dalles	6 Wanapum	10 Chief Joseph	14 Little Goose	18 Oxbow
3 John Day	7 Rock Island	11 Grand Coulee	15 Lower Granite	19 Brownlee
4 McNary	8 Rocky Reach	12 Ice Harbor	16 Dworshak	20 Sunbeam (1910-1934)

remains near the brink of extinction (Selbie et al. 2007; Keefer et al. 2008a).

Overfishing and blocked passage are not the only factors involved in the decline of Snake River Sockeye Salmon. There have also been major changes in the ecology of its natal streams and lakes, particularly Redfish Lake. Fish introductions were common in Redfish Lake beginning in 1921. Even though the lake supported an apparently native population of Kokanee Salmon, juvenile Kokanee were introduced from multiple sources, as well as Rainbow Trout and millions of Sockeye Salmon eggs, again from unknown sources in the United States and Canada. Overall, the introductions increased predation pressure on zooplankton because Kokanee Salmon remained planktivorous and remained in the lake throughout their life cycle. Consequently, juvenile anadromous Sockeye Salmon were competing directly for

zooplankton resources with juvenile and adult Kokanee. The change in predation pressure resulted in the loss of large zooplankton (i.e., *Daphnia*), which were replaced by smaller, less nutritious forms (see also Chapter 13). In addition, nutrient enrichment began in the 1950s, perhaps due to enhanced atmospheric nitrogen deposition as well as purposeful fertilization with the goal of enhancing the forage base for Sockeye Salmon, but with the unintended consequences of further altering the phytoplankton and zooplankton composition of the lake (Selbie et al. 2007). The Snake River Sockeye Salmon are also the most southerly of all Sockeye Salmon populations, and the currently high mortality of migrating fish is due in part to water temperatures near their upper tolerance levels (21–24° C) (Keefer et al. 2008a).

Fishes with complex life-history patterns, such as migratory fishes traversing a variety of habitats, can have a life-history pattern that is no longer optimal or even possible given current environmental conditions (see Chapter 9). The overall low productivity of the rearing system, the extraordinary migration demands on juvenile and adult fish even under natural conditions, their current problems with blocked passage, their sensitivity to regional climate warming, and currently high parasite loads, all suggest that Snake River Sockeye Salmon may be naturally vulnerable to the novel environmental stresses of twentieth century (Selbie et al. 2007; Keefer et al. 2008a).

Smaller dams and diversions have also contributed to declines of Pacific salmon in the Columbia River basin. The Willamette and Lower Columbia river basins include all Columbia River tributaries downstream of the Dalles Dam, located 308 river km inland from the river mouth (Figure 14.7). Within this region there are at least 1,491 anthropogenic barriers to fish passage resulting in the blocked access to 14,931 km of streams (a 42% loss). This includes a loss of 10,142 km of prime Steelhead habitat, based on river gradient, and an 8,380 km loss of prime Chinook Salmon habitat. Habitat loss disproportionately includes high-quality streams characterized by intact coniferous forests that provide greater shading, more woody debris, and cooler water temperatures. It is not surprising that the loss of large amounts of habitat is significantly correlated with the decline in salmon populations (Sheer and Steel 2006).

As is obvious from the Columbia River, in many instances a single river has been impacted by multiple dams—the Colorado River has 12 major reservoirs for purposes of water supply, flood control, and hydropower, and the upper Mississippi River, between Minneapolis, Minnesota, and Cairo, Illinois, has 27 low-head navigation dams (Christensen et al. 2004; Zigler et al. 2004). Effects of high dams on the Colorado River, including blocked fish passage and altered flow and sediment regimes, are related to the severe population declines of Colorado Pikeminnow, Humpback Chub (*Gila cypha*), Bonytail

Chub (*G. elegans*), and Razorback Sucker, all of which have evolved long-distance migration patterns as part of their life histories (Minckley et al. 2003). Low-head navigation dams on the upper Mississippi River can act as semipermeable barriers to fish movement. Paddlefish (*Polyodon spathula*) have evolved long-distance migrations for access to prime growth areas, such as oxbow lakes, which differ from suitable spawning areas generally characterized by clean gravel substrata in flowing water (Purkett 1963; Jennings and Zigler 2009). Based on a radio-tagging study, Paddlefish populations in the upper Mississippi River are generally fragmented in an upstream direction. Because gates of navigation dams are raised off the bottom as river flow increases, the best opportunity for upstream passage is during high flows when the gates are raised out of the water resulting in open river conditions. Overall, upstream passage is more affected than downstream passage (Zigler et al. 2004).

The Fox River, a sixth-order tributary of the Illinois River draining parts of Wisconsin and Illinois, has a 171 km reach that is fragmented by 15 low-head dams. The dams are all run-of-river structures that are less than 15 m high, with surface spillways and small, shallow impoundments (Santucci et al. 2005). The distributions of at least 30 fish species are affected by the dams, either by having upstream movements blocked or reduced or by having fragmented populations. For instance, the 15 species limited in their upstream distributions by dams dropped out sequentially in an upstream progression, with only five passing the first dam in the series and only four passing the second. Two species, Gizzard Shad and River Redhorse (*Moxostoma carinatum*), made it past the third, fourth, and fifth dams, and River Redhorse passed the sixth dam in the series before being blocked by the seventh. The 15 species with truncated distributions occurred in the upper and lower river, but usually not in the more central, urbanized area that also had a high density of dams (eight dams in 22 river km) compared to other parts of the river (dams every 15.3 km). Limiting factors likely included low water quality in the impoundments as well as effects of barriers; for instance, species richness, abundance, and measures of biotic integrity were all reduced in impoundments versus the free-flowing sections of the river. Overall, the study shows the strong impact that even low-head dams have on fishes and other aquatic organisms.

In eastern North America, including the southeastern United States but especially in the northeastern United States, mill ponds and other low-head dams are or were common and appeared in large numbers beginning in the seventeenth century and extending into the early twentieth century (Walter and Merritts 2008). Mill ponds essentially replaced Beaver ponds, except that they represented much more of a barrier to fishes. In these regions, most small to medium streams (first–third order) were impacted, often by multiple dam sites. Milldams (usually 2.5–3.7 m high) were the primary power source for industry, especially in the nineteenth century, and as early as 1840 there were more than 65,000 water-powered mills in the eastern United States. Not surprisingly, impacts on fish populations were noticed early on. For instance, a milldam on the Conestoga River, Pennsylvania, was torn down in 1731 because it was damaging the local fishing industry, and a 1763 petition concerning the same river cited milldams as responsible for destroying shad, salmon, rockfish (Striped Bass), and trout fisheries (Walter and Merritts 2008).

Although many milldams are now breached (a much reduced zone of impoundment and with spillway open or with other damage to dam) or relict (free-flowing; evidence of dam often reduced to bank-side structures), intact and even breached dams continue to have impacts on fish assemblages. Currently there are more than 10,000 low-head dams, including milldams, in Alabama. In a study of 20 streams, nine having intact dams, six with breached dams, and five with relict dams, fish species richness in the flowing stream above the impoundment was lower in streams with intact or breached dams compared to assemblages immediately downstream of the dams.

Overall, the fish fauna comprised 88 species in 12 families, with minnows (55%), darters (20%) and sunfishes (14%) being the most abundant groups. Recovery of fish assemblages, once effects of the barrier were largely removed (relict dams), was evidenced by no differences between upstream and downstream reaches (Helms et al. 2011).

In species-rich, small upland streams of the southeastern United States, movements of fishes are associated with feeding, reproduction, and recolonization of areas that might be periodically dewatered and are important for maintenance of viable populations (Chapter 5). In these small streams with rich faunas of relatively small-bodied fishes, barriers to movement need not be large dams or even low-head dams such as mill ponds. For instance, road crossings on national forest streams can act as potential barriers to fish movement. In a study in the Ouachita National Forest, Arkansas, on small tributaries of the Ouachita River, the potential of road crossings as barriers to fish movement was studied on nine crossings on seven streams (Warren and Pardew 1998). Four types of crossings, fords, open-box, culvert, and slab, were compared in terms of fish passage to natural reaches. Ford crossings were submerged roadbeds with a compacted gravel substratum; open-box crossings had 1–3 bays, each 3–4 m wide and 24–30 m long, topped with a concrete roadbed and with a concrete or gravel substratum; culvert crossings were two to four 1 m–diameter pipes placed on a concrete pad and covered by a concrete or earth/gravel roadbed and with a concrete apron extending downstream for 3–4 m; concrete slab crossings were low dams with a 25 cm vertical drop off the downstream edge to the pool below (Warren and Pardew 1998).

Both culvert and slab crossings reduced fish movement overall, and had a lower diversity of fishes moving through them compared to natural reaches. Current velocity through the crossing was closely related to the degree a crossing acted as a barrier—culverts had the highest water velocity and the lowest passage, whereas open box crossings had the lowest velocities and the highest level of fish passage. A key point from this study is that it doesn't require a dam tens of meters tall to act as a barrier to small fishes—relative size matters. Small-stream fishes need to have water velocities through crossings that are much lower than maximum velocities suggested for migratory fishes such as salmonids (Warren and Pardew 1998).

DAM REMOVAL, FISH PASSAGES, ENVIRONMENTAL FLOWS, AND OTHER REMEDIAL APPROACHES

Given the previously mentioned examples and the various ways in which the behavior and life cycles of fishes and other aquatic organisms are closely tied to patterns of flow and water temperature, it stands to reason that even apparently minor changes in these characteristics can have wide-ranging impacts (Poff et al. 1997). As shown previously, fishes whose movements are strongly related to flow events may be more vulnerable to extirpation when faced with reduced flow patterns or be blocked from movement by dams, including small structures (Albanese et al. 2004). Successful spawning of native fishes can be reduced if high reservoir releases coincide with larval emergence, or nonindigenous fishes can be favored if release patterns favor their survival (Harvey 1987; Fausch et al. 2001). Based on currently available knowledge, protecting or restoring natural flow regimes is one of the most prudent approaches for promoting the colonization and persistence of fish populations (Bain et al. 1988; Poff et al. 1997, 2003; Baron et al. 2002; Albanese et al. 2004; Poff and Zimmerman 2010).

Regaining Natural Flow Characteristics

Historically, management of riverine ecosystems, including efforts of water allocation to sustain ecosystems and meet human demands, has not favored environmental flow protection. This is particularly true for the protection of high-discharge events that are necessary for

ecological purposes such as channel maintenance and the maintenance and creation of floodplain habitats (Poff et al. 2003; Richter 2010). In addition, even where environmental flow protection that mimics natural flow variability has been agreed upon, the actual implementation has proven difficult for a variety of reasons, including natural changes (i.e., droughts and floods), political will, other societal demands, and hydrological engineering (Poff et al. 2003; Richter 2010).

Although seeming to make good ecological sense, ecological impacts of environmental flows from dams, achieved through modifying dam operation or dam structure, are rarely evaluated (Souchon et al. 2008; Bradford et al. 2011). One of the few long-term studies focused on the Bridge River, a tributary of the Fraser River in British Columbia, which was dammed in 1960 for hydropower production (see also Chapter 6). The dam is located 41 km upstream from the confluence with the Fraser River. Because flow was diverted into another watershed, a 3 km stretch of the river immediately below the dam was dewatered. Below this reach, five small tributaries and groundwater provide low but continuous stream flow, although the inflow from tributaries is < 1% of the preimpoundment flow. The Bridge River receives an unregulated tributary 15 km downstream from the dam (Bradford et al. 2011). Consequently, there were three different dam-altered habitats downstream from the dam: 3 km of dry channel, 12 km of consistent but greatly reduced flow, and 30 km of flow augmented by an unregulated tributary, albeit still with flows much less than during pre-dam conditions (Bradford et al. 2011).

A 13-year experiment of flow releases from the Bridge River dam was designed to test the efficacy of environmental flows to increase salmonid production (primarily Rainbow Trout; Coho Salmon, *O. kisutch*; and Chinook Salmon). The environmental flow has been equivalent to a mean annual discharge of 3 m^3s^{-1} and is varied seasonally to provide a summertime peak and improved winter flow. Even though the environmental flow is < 1% of the pre-dam flow (approximately 100 m³s⁻¹), greater flows were not possible given hydropower demands and because much of the prime salmonid habitat was located in the now-impounded area above the dam—the first section of the river below the dam was likely most important for fish passage to those upstream habitats. The monetary cost of providing the environmental flow, in terms of lost hydropower revenue, is approximately $5–8 million Canadian dollars (Bradford et al. 2011).

The total number of juvenile salmonids did increase significantly following implementation of the environmental flow, with most of the increase as a result of repopulation of the formerly dry 3 km section immediately below the dam. In the next two reaches of river that were already flowing, although at much reduced levels prior to the environmental release, there was not a detectable change in total salmonid abundance. Consequently, the overarching hypothesis that fish production in all three sections would increase as a consequence of increased discharge (or more water equals more fish) was not supported by the results outside of the section that was dry before managed flows resumed (Bradford et al. 2011). Importantly, another lesson from the Bridge River study is that impacts of environmental flows are extremely context dependent—the outcomes being driven by the amount and quality of downstream habitats. Although the restored flow only averaged 3 m³s⁻¹, it was managed to reflect the snowmelt-driven flow pattern of the natural river. One result of this was the restoration of a narrow riparian zone that included young cottonwood trees (Hall et al. 2009). In fact, because it is often not possible to regain the magnitude of natural flows present prior to dam construction, the concept of river downsizing becomes an important alternative. By creating flows that follow a natural progression, even though much less in magnitude, much of the natural function of at least some rivers can be restored (Trush et al. 2000).

As with the Bridge River system, managed flows are often designed for particular species, especially in the northwestern United States and western Canada, where species diversity is low and a major emphasis is on the recovery of migratory salmonids. The Skagit River, in northwestern Washington, includes prime spawning and rearing habitat for anadromous salmonids. However, the survival of eggs and fry was compromised by operation of the Skagit hydroelectric project, which was initiated in the early 1900s with additional dams added into the 1960s and which supplies electricity to Seattle. As with other hydroelectric facilities, discharge varied with power demands and created significant problems of egg and fry mortality, especially for Pink (*O. gorbuscha*), Chum (*O. keta*), and Chinook salmon (Graybill et al. 1978). Beginning in 1991, discharges from the Skagit project were changed to improve habitat quality for the three salmon species by reducing mean and maximum flows during the period of snowmelt and increasing minimum and mean flows during the fall and winter. Rate changes were also modified because the rapid declines in flow (downramping) during the day had caused considerable stranding mortality of fry. Under the managed flows, downramping was reduced and daytime downramping nearly eliminated. Overall compliance with the flow recommendations has been 99% (Connor and Pflug 2004).

Studies during the period of managed released (1991–2001) compared to preagreement conditions show that the environmental flow measures resulted in substantially increased Pink and Chum salmon abundances and in a sustainable healthy population of Chinook Salmon in the upper Skagit River. The study area now supports the greatest percentage of spawning adults of the three salmon species in the Skagit River basin, and the Skagit River basin has the largest runs of Pink and Chum salmon in the contiguous United States (Connor and Pflug 2004).

Determining appropriate releases from dams can be more challenging in species-rich eastern and southeastern U.S. streams, especially when the magnitude of a pre-dam hydrograph cannot be attained. As with the Bridge River study, substantial benefits can sometimes be obtained by changing releases to guarantee a small minimum flow (fish really do need water). Thurlow Dam, a hydroelectric facility, was constructed on the Tallapoosa River, Alabama, in 1930. Water was released from the dam based on power demands—averaging 230 m^3s^{-1} when generating electricity or no release when there was no power generation. Not surprisingly, fishes were affected by the rapidly pulsating release pattern, with those typified as fluvial specialists absent or rare in the sampling area below the dam but gradually recovering in abundance farther downstream where other tributaries resulted in a more normal hydrograph (Kinsolving and Bain 1993).

Starting in February 1991, a minimum flow of at least 34 m^3s^{-1} was established as part of a relicensing agreement, although the high flows still occurred during power generation—so the overall change was from a highly pulsating flow regime that dropped to zero discharge between the peaks, to a highly pulsating flow regime that maintained a continuous base flow. Before the establishment of a base flow, the shoreline fish abundance and diversity 3 km below the dam was low and dominated by generalist species, such as Emerald Shiner (*Notropis atherinoides*), Weed Shiner, Blacktail Shiner (*Cyprinella venusta*), Redbreast Sunfish (*Lepomis auritus*), Longear Sunfish, Redear Sunfish (*L. microlophus*), Spotted Bass, and Largemouth Bass (Travnichek et al. 1995). Approximately one year after the initiation of a minimum low flow, species richness more than doubled to 19 and included a majority of fluvial specialists, such as Speckled Chub (*Macrhybopsis* sp. cf. *M. aestivalis*), Fluvial Shiner (*N. edwardraneyi*), Skygazer Shiner (*N. uranoscopus*), Bronze Darter (*Percina palmaris*), Speckled Darter, and Banded Sculpin (*Cottus carolinae*).

Reduced stream discharge, as a consequence of impoundments, is one of the major factors in breaking the connection between fringing floodplains and the river (Poff et al. 2010). The other is the construction of levees to block a river from access to its historic floodplains, generally for the development of agriculture on the rich floodplain soils. Levees, however, also isolate a river from its natural reservoir (its floodplain) so that the intensity of otherwise smaller floods is increased (Belt 1975). For instance, although the annual discharge of the Mississippi River at Vicksburg has not changed appreciably in 170 years, the Mississippi River and its major tributaries now have extensive development of lateral levees, with continuous levees along both banks south of Vicksburg, Mississippi, and the channel of the Mississippi River has lost about one-third of its volume since 1873 (Belt 1975; Turner and Rabalais 2003).

Reconnecting Floodplains and Streams

Because of the known importance of floodplain habitats as spawning and/or nursery areas for many native fish species, the restoration or creation of floodplains is often a major goal of recovery efforts (Poff et al. 2010). Such restoration requires hydrological connectivity between the floodplain and river, a flow regime that has sufficient variability, and a magnitude of scale that is adequate for biological processes to occur and for the restoration to have meaningful ecological benefits. In addition, floods need to occur for sufficient duration and season to coincide with life-history patterns of floodplain associated organisms (Opperman et al. 2010). The Sacramento-San Joaquin river system in California's Central Valley has undergone extensive modification from its natural state through dam building, water extraction, channelization, bank armoring, channel incision, and levee construction, all of which have resulted in major negative impacts on the aquatic and riparian biota (Opperman et al. 2010). Two case studies show the importance of floodplain restoration at differing scales—one affecting 0.4 km^2 and providing benefits at a local level (Cosumnes River), and the other affecting 240 km^2 and providing benefits at the population level (Yolo Bypass).

The Cosumnes River is unique among streams in the region in that it lacks mainstream dams and still has a fairly normal hydrograph. However, it does have levees along most of the lower reaches, denying juvenile Chinook Salmon and other fishes access to fringing floodplains

(Moyle et al. 2007). The Nature Conservancy (TNC) acquired land along the lower Cosumnes River in 1984 and began restoring native vegetation. In 1985, a levee broke around an agricultural field, and although soon repaired, the freshly deposited sediment on the field soon was colonized by native cottonwoods and willows. When the farm property was acquired by TNC in 1987, the success of the accidental breach in fostering rapid growth of riparian vegetation led to the ultimate intentional breeching of levees starting in 1995 (Swenson et al. 2003).

A study of larval fishes showed that the inundated floodplain was used early in the season by native fishes such as Splittail, Sacramento Blackfish (*Orthodon microlepis*), and Chinook Salmon, as a feeding, spawning, and rearing area and also by the nonindigenous Common Carp (*Cyprinus carpio*) as a spawning and rearing area. By summer, only nonindigenous fishes, especially Mississippi Silverside (*Menidia audens*), Mosquitofishes (*Gambusia* spp.), and centrarchids were present (Crain et al. 2004; Moyle et al. 2007). Studies of floodplain use by native and nonindigenous fishes show that management of flows can favor native fishes by having extensive, early season flooding, followed by complete dewatering of the floodplain by the end of the flooding season (Figure 14.8) (Moyle et al. 2007).

The restored floodplain is also important for growth and retention of juvenile Chinook Salmon. Based on a series of replicated channel and floodplain enclosures stocked with young Chinook Salmon, the off-channel floodplain habitats provided a significantly better rearing habitat because of higher temperatures and productivity compared to the river (Jeffres et al. 2008). Interestingly, the increases of native fish retention and growth have been achieved while still retaining almost 90% of the preserve in active agricultural production, including grazing, rice farming, and other annual crops (Swenson et al. 2003).

The Yolo Bypass is a large, constructed wetland on the lower Sacramento River that has been in continuous use since the 1930s. Although originally engineered strictly to let

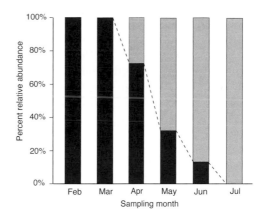

FIGURE 14.8. The progression of numerically dominant native and nonindigenous fishes from the onset of floodplain access to the end of the flooding season in the restored floodplain of the Cosumnes River, California. Black bars indicate native species; gray bars show nonindigenous species; the dashed line shows the division between native and nonindigenous species. Based on Moyle et al. (2007).

floodwater from the Sacramento Valley overflow onto a broad floodplain, it is also important as wildlife and fish habitat. Among fishes, it is particularly important as a nursery area for Chinook Salmon and as a spawning, nursery, and feeding area for Splittail (Sommer et al. 2001). Splittail are endemic to the upper San Francisco estuary and its lower watershed, and have annual spawning migrations from the brackish estuary to upstream floodplains and tributaries. Larval and juvenile Splittail remain in floodplain habitats, until the high spring flows recede and the floodplain begins to dry, before moving back into the estuary. Splittail show increased growth and spawning success in wet years that are significantly related to the magnitude and duration of inundation of the constructed floodplain during their spawning and juvenile rearing periods. Winter flood pulses are also important to initiate upstream spawning migrations and to attract fish onto the floodplain (Feyrer et al. 2006).

Restoring Connectivity of Watersheds

One of the most obvious impacts of dams is the break in connectivity of aquatic ecosystems, and especially the loss of upstream habitat to migratory fishes. In fact, the need to maintain connectivity by using fish passages was recognized as early as the twelfth century during the reign of King Richard I in England with a statute instructing that "English rivers be kept free of obstructions so that a well-fed three-year-old pig could stand sideways in the stream without touching either side," which was referred to as the King's gap (Montgomery 2003). Although fish passages have a long history of use, studies of the efficiency of different types of passages for various species of fishes have only recently become common. In a broad review of research assessing the success of fish passages for the years 1965 to 2008, including 50 North American studies, three-fourths of the studies were published from 1999 to 2008. Although North American studies more often included postpassage effects compared to studies in other countries, the number of such studies was still only 11. Postpassage studies often showed serious negative impacts on fishes that successfully traversed a fish ladder, including delayed mortality or failure to reach spawning sites (Roscoe and Hinch 2010). Because of downstream effects of dams, and issues of upstream and downstream passage of fishes and other organisms, dams have contributed to significant declines in anadromous salmon, as well as other migratory species, throughout North America (Cooke et al. 2005; Roscoe et al. 2010).

Passages around low-head dams can be achieved by a relatively natural flow of water around the structure (termed nature-like passages). These passages typically have low gradients and mimic natural side channels. In contrast, fish ladders around large dams are highly engineered structures—generally a step-like series of compartments containing downstream-flowing water, requiring a fish to swim against the current and then jump to the next compartment (Volpato et al. 2009; Roscoe and Hinch 2010). The success of fish ladders is varied, especially when success is measured by the passage of multiple species of fishes rather than a single target group, such as anadromous salmonids.

Most of the attention on fish passage problems has been on economically valuable species, such as migratory shad or salmonids. However, other important components of native fish assemblages are also migratory (see Chapter 5). For instance, many catostomids show upstream spawning migrations, and the removal of migration barriers, either by the construction of fish passages or by actual physical removal. would benefit numerous species of North American catostomids (Cooke et al. 2005). As a case in point, a weir approximately 1 m high was constructed in 1971 to divert water to a power generating station, but which blocked upstream passage of fishes in the San Juan River, near Farmington, New Mexico, except during high flows. Construction of a nature-like fish passage in 2003 around the weir has successfully provided upstream access for native suckers, including the endangered Razorback Sucker, as well as Flannelmouth and Bluehead suckers. The passage has also been used by the endangered Colorado Pikeminnow. Because a grate holds the fish in an enclosure after they ascend the passage, nonindigenous fishes, such as Channel Catfish (*Ictalurus punctatus*) and Common Carp, can be removed from the system. In 2010, 28,292 native fishes moved through the passage and 785 nonindigenous fishes were removed from the system (Figure 14.9) (Morel 2011; J. Morel, pers. comm., 2012).

Landsburg Dam, on the Cedar River, near Seattle, Washington, provides water to Seattle but had blocked fish passage for 103 years. In 2003, the installation of a fish passage facility allowed migratory and resident fishes the opportunity to regain access to upstream habitats. Once the fish passage became operational, the recolonization above the dam by adult salmonids, including Chinook Salmon, Steelhead, and Coho Salmon, was rapid and shows that the barrier bypass was successful in allowing migratory species to reestablish populations, leading to range expansion and potentially increasing population sizes (Kiffney et al. 2009).

Bonneville Dam, on the lower Columbia River (Figure 14.7), has the most complex

FIGURE 14.9. The nature-like passage stream around a weir on the San Juan River near Farmington, New Mexico. Photo courtesy of James Morel. See also corresponding figure in color insert.

passage facilities of the Columbia River dams, with two powerhouses, a spillway, and four lower-gradient passages that service two fish ladders (Keefer et al. 2008b). Based on radio-tagged adult fishes, passage efficiencies ranged from 92.1% for fall Chinook Salmon, 98.5% for spring-summer Chinook Salmon, 97.7% for Steelhead, and 98.8% for Sockeye salmon. Long passage times can be a source of stress to migrating salmonids. Median passage times were shortest for Sockeye Salmon (15 h), intermediate for Steelhead (17–24 h), and longest for spring-summer Chinook salmon (19–41 h). Overall, passage times were fairly similar for other large dams on the Columbia River (Keefer et al. 2008b).

In the Seton River, a tributary of the Fraser River of southwest British Columbia, after traveling 320 km from the mouth of the Fraser River, Sockeye Salmon must get around a 7.2 m diversion dam via a fish ladder to travel the final 55 or so km to their spawning grounds. The fish ladder is 107 m long, with 32 pools and a grade of 6.9%. Eighty-seven fish that had successfully ascended the fish ladder were radio-tagged and, using blood samples, assayed for indications of stress before being released either above the fish passage (28 fish) or below the dam (59 fish). Of

those released downstream of the dam, 85% successfully returned to the tailrace below the dam, 44 entered the fish way, and 41 successfully ascended the fish ladder for the second time, for an overall success rate of 80% once fish entered the tailrace area. Postpassage mortality differed significantly between the fish initially released above the fish ladder and those forced to ascend the ladder a second time. Survival to the spawning grounds was only 73% for the second climbers and 93% for the fish that only ascended the fish ladder once. Although there could be positive (the test fish had found the fish ladder once already) and negative (some of the fish had to climb the fish ladder a second time, although physiological studies showed little indication of stress or exhaustion) biases by using fish that had successfully found and ascended the fish ladder, the important points from the study are that success in finding the fish ladder is no more than 85% and perhaps even less, and second, that some level of postpassage mortality likely occurs (Roscoe et al. 2010).

A second approach to restoring connectivity is, of course, the removal of the obstruction. Although dam removal at first seems like an elegantly simple solution, the removal of a dam can carry with it a number of concerns and trade-offs (Box 14.3). All dams retain sediment, but depending on the geomorphology of the region and age of the dam, sediment buildup in some dams can be extensive. The rapid downstream movement of sediment during and after dam removal (often a 3–5 order of magnitude increase) can cause major ecological impacts, especially because some sediments harbor decades-long accumulations of toxins or nutrients. Riverine fishes generally benefit from dam removal, although instances of downstream mortality because of high sediment loads have occurred. Consequently, dam removal should, at least in the short-term, be viewed as a disturbance (Stanley and Doyle 2003). Between 1982 and 2002, over 500 dams were removed in the United States. By 2020, a vast majority of large U.S. dams will have reached or passed their intended life expectancy either requiring extensive repairs or removal (Bednarek 2001; Stanley and Doyle 2003). Although dam removal is becoming increasingly common, there are still relatively few studies of the stream channel and faunal responses that ensue (Hart et al. 2002; Stanley and Doyle 2003; Gardner et al. 2011).

In Wisconsin, the removal of small dams started in the 1960s and accelerated into the 1990s. In 1988, the low-head (4.3 m high) Woolen Mills Dam on the Milwaukee River was removed after approximately 80 years of existence. Within five years of removal, habitat quality had improved in the formerly impounded section from a rating of fair to ratings of good or excellent. The abundance and biomass of Smallmouth Bass (*Micropterus dolomieu*), a popular sportfish, had increased substantially in the restored river as well as in upstream sites, as had populations of native darters and suckers. At the same time, Common Carp abundance declined. Below the former dam site, habitat quality initially decreased, as did desirable fish abundances, but then recovered. Overall, removal of the Woolen Mills Dam has been viewed as beneficial to ecological integrity of the Milwaukee River (Kanehl et al. 1997).

In a synthesis of removal studies of small Wisconsin dams, Doyle et al. (2005) found that ecosystems affected by the dams responded in different ways and at different rates to removal. Macroinvertebrates tended to recover quickly to pre-dam conditions, mussels only slightly or not at all, and fishes at an intermediate level, often showing rapid upstream movement following dam removal. Two general recovery trajectories were apparent: some ecosystems fully recovered to pre-dam conditions, as suggested by the Woolen Mills Dam, whereas others, the majority, only partially recovered because of enduring impacts of the dam or because of other watershed changes.

In a small tributary, Sedgeunkedunk Stream, of the Penobscot River in coastal Maine, anadromous fishes declined or disappeared following the building of multiple dams. In 2008, a middle dam on Sedgeunkedunk Stream was bypassed by a rock-ramp fish passage, and in 2009, the lowest dam was removed so that habitat connectivity has been restored for most of the watershed—something that likely had not occurred for over 100 years (Gardner et al. 2011). Fishes responded almost immediately after dam removal by recolonizing upstream areas—fish density increased five-fold and fish biomass eight-fold, although initially this was caused by movement of fishes formerly found downstream of the dam.

Impacts of large dam removals are still poorly known, given that their removal has greatly lagged behind that of small dams with heights typically under 6 m. However, in 2011 the Condit Dam was removed from the White Salmon River, a Columbia River tributary, and the process of removing two large hydroelectric dams from the Elwha River, Washington, was finally started after years of study. The complete removal will occur over a three-year period. Removal of the dams will restore access to extensive upstream habitat and should provide unprecedented opportunity for recovery of 10 anadromous fish populations, especially because the upstream habitats are located in the relatively undisturbed wilderness of Olympic

BOX 14.3 • The Ethics of Dam Removal

Derek G. Ross, PhD, Auburn University,
Master of Technical and Professional Communication Program

We build dams because we think we need what they provide, thus dam building often proceeds in deliberate and cognizant opposition to nature. In 1946, for example, the Bureau of Reclamation published a 292-page book titled *The Colorado River: A Natural Menace Becomes a National Resource* (Billington and Jackson 2006). Dams are visually impressive, and they speak eloquently of a government's ability to wield power while also standing as physical representations of progress. To decry dams as obscene, unnecessary, or even environmentally unsound appears, at first blush, to be flagrantly un-American. After all, through production of hydroelectric power, which makes up 20% of global electric supply (Marks 2007), dams provide a clear alternative to the burning of fossil fuels for energy needs. Calling for a stop to damming, or calling for dam removal, is not likely to sit well with community members who have been taught to appreciate progress and growth.

Growth, however, is often misrepresented. Killingsworth and Palmer (1992) point out that growth is a myth rarely discussed in relation to its original denotation—the idea of something springing forth and developing to maturity. As writer Edward Abbey often noted, for example, "Growth for the sake of growth is the ideology of the cancer cell" (Cahalan 2001). If we stop considering growth as the exigency for culture and instead consider progress toward sustainability, then, at some point, some dams need to come down to begin to attempt to restore a healthy coexistence between humankind and nature.

Unfortunately, we can't just rip dams out of the earth and expect the environment as it was to suddenly reappear. Authors draw distinction between "big" and "small" dams—big dams are massive constructions used to bring natural forces under total human control, smaller dams are less complex constructions used for the same purposes but on a smaller scale. A big dam may serve a city of hundreds of thousands, while a small dam may be built for an individual, an industry, or a village's use. It is the "big" dams with which we are most familiar and which cause the most environmental impact. Between 1900 and 2004, the number of big dams in the world rose from around 600 to

45,000 (Khagram 2004). These big dams, the Glen Canyon Dam, for example, have been built to provide water and power for hundreds of thousands of humans—their removal creates a resource hole that must somehow be addressed.

Given the emotional appeal that environmentalism has for many people, indeed, some view "ecology as a replacement for religion" (Killingsworth and Palmer 1992), debates about dam removal often become heated. Despite the arguments, however, dams do come down. As Marks (2007) writes, "In the U.S., where hydropower dams must be relicensed every 30 to 50 years, the rate of dam removal has exceeded the rate of construction for the past decade or so. In the previous two years alone [2005–2006], about 80 dams have fallen, and researchers following the trend expect that dams will continue to come down, especially small ones."

Just like fixing global warming is not as simple as "stop driving cars," deciding to take down a dam is not simply about figuring out from where else to get water.

While building dams has significant and long-lasting ecological effects (Babbit 2002), removing dams is as potentially problematic as building them in the first place. Removing a dam can release large amounts of sediment, which may then affect important habitats (Bednarek 2001), may wash away habitat formed since the erection of the dam (Meretsky et al. 2000), and may cause major changes in species distributions (Gregory et al. 2002). Additionally, sediment held back by the dam may be heavily polluted, causing massive problems when released into the river ecosystem (Marks 2007). Ecological responses to dam removal may occur over extended periods of time (Hart et al. 2002; Pizzuto 2002). Dam removal does not mean a quick fix or an instant return to natural conditions. Additionally, federal, state, and local laws must be taken into account when removing dams, often resulting in complex interactions between environmental protection and environmental restoration legislation (Bowman 2002).

Removal of large dams also introduces issues of repopulation of previously water-covered areas.

(continued)

Massive dam-building projects have often resulted in the transplantation of native peoples from their historic lands to other, potentially less suitable, areas (Khagram 2004). Between 1947 and 1992, for example, one estimate has the number of people displaced by dam construction at 20 million (Lowry 2003). This displacement has historically occurred by force (Khagram 2004; Leslie 2005), though also by mutual agreement. Glen Canyon National Recreational Area, for example, exists partially on lands belonging to the Navajo Indian Reservation (Begay 2001). The town of Page, Arizona, which was originally created for the many workers and their families involved with the construction of the dam, exists where over 3,000 Navajos once lived, but was obtained with permission from the Navajo Tribal Council in 1956 (Martin 1999).

As dams come down, decisions—not easy ones—need to be made about what will be done with the reclaimed land. Such decisions should consider not only the environment and reintroduction of native species but also cultural rights and expectations. Land destroyed by dam building may not have the same cultural value it once held when the dam comes down or regular river flows are restored (such as the more than 250 acres of aquatic habitat along the Colorado River restored by the Colorado River Indian Tribes' use of water allotments [Cohn 2001]). For example, Robert Begay (2001), director of the Navajo Nation Historic Preservation Department, notes that "these places [culturally significant areas flooded when the dams went up] can never be recreated. Even reclamation activities, such as trying to recreate the original landscape and planting native grasses, have destroyed the sacred quality of places and thereby destroyed the homes of deities or holy people who lived at many of these places."

Dam removal is an important and vital part of our attempts to reclaim environmentally and culturally important land: unsafe and unnecessary dams need to go. Removal, however, must be considered in light of historical use, environmental history, and cultural expectations so that benefit to both humans and the environment is the end result.

National Park. Currently, the anadromous populations of salmonids are limited to 7.5 river km below the Elwha Dam—a dam that lacked any fish passage (Brenkman et al. 2012).

Darwinian Debt and Other Issues

As emphasized in Chapter 13, populations of fishes and other aquatic populations are not static elements but are, to varying degrees, responding phenotypically and genetically to changing selective pressures brought on by altered habitats and/or by human exploitation. This has been recognized as an important issue in the management of commercial fisheries, where removal of larger individuals has driven down body sizes and ages at maturity of exploited species (Dunlop et al. 2009). Darwinian debt refers to the expectation, based on numerous genetic models, that the evolutionary recovery from harmful genetic changes caused by various anthropogenic impacts takes longer than the time required to induce the changes. It is a debt, because the results of anthropogenic selection may result in evolutionary costs for future generations of the species (Loder 2005; Allendorf et al. 2008; Dunlop et al. 2009).

Although recognized as important in freshwater and marine fisheries management, such evolutionary responses are also important in conservation issues, including discussions of dams, fish passages, and dam removal (Loder 2005; Williams et al. 2008). For example, fish ladders can be considered artificial selection devices because phenotypes successful in traversing them can differ from phenotypes best suited for long-distance migration. In a South American study of fish ladder passage by a migratory characiform, fish collected from the top of the fish ladder differed significantly in their morphological and physiological profiles from those collected at the bottom of the fish ladder, thus strongly suggesting that selection caused by differential fish ladder passage could contribute to evolutionary change (Volpato et al. 2009).

Because of the real possibility of eco-evolutionary interactions (Chapter 13) and the apparent decrease in fitness of many fish species, such as Chinook Salmon, caused by the presence of dams, there could well be strong selection for regaining fitness in the dam-altered habitats. With Chinook Salmon, the increased stream temperatures caused by dam releases during late fall and winter can increase the rate of development of embryos such that emergence occurs too early, at a time when high spring flows are still prevalent. Hence selection to regain an optimal emergence time by slowing embryo development or delaying spawning may be ongoing. Based on the high heritability of spawning time in salmonids, a modeling exercise indicates that an evolutionary change in the time of spawning of 0.7–1.1 days per generation is theoretically possible, although not yet demonstrated, and could result in reaching a new optimum spawning time in around 20 generations (85 years) (Angilletta et al. 2008).

The changes in many rivers brought on by dams also alter the fitness consequences of adaptations for long distance migrations. The alteration of the Columbia River into a series of slow-moving reservoirs has relaxed selection for adults capable of swimming long distances upstream against strong river currents. Instead, conditions now favor fish that are able to navigate through lakes and locate and traverse fish ladders (Waples et al. 2007). Rather than generally passive downstream transport of smolts, they must now actively swim through reservoirs, often encountering nonindigenous predators along the way. Juveniles must also survive passage through multiple dams or the collection and transportation around dams. Overall, dams have increased the cost of migration, reducing energy available for sexual selection, and perhaps favoring a nonmigratory life history. Waples et al. (2007) have raised the possibility that were the Columbia River ever returned to a quasi-natural, free-flowing state, remaining salmonid populations might face a Darwinian debt (and temporarily reduced fitness) as they struggle to reevolve historical

adaptations. As further evidence of this, migration tactics of Snake River fall-run Chinook Salmon have changed since the advent of large dams on the Columbia River. Historically, fish undertook the downstream migration as sub-yearlings, but now many individuals migrate as yearlings. Furthermore, the yearling migration tactic has made a large contribution to adult returns over the last decade. Such a change is most likely a consequence of rapid genetic change and raises important concerns in the conservation attempts to recover federally listed salmon populations (Williams et al. 2008).

As salmon and other fish populations first declined in Europe and North America in the eighteenth and nineteenth centuries, the use of hatcheries in augmenting overfished stocks came into common practice. In instances where overfishing was the principal cause of declines, hatcheries showed some success in rebuilding populations, especially once the localized nature of populations came to be appreciated (Moring 2000; Montgomery 2003). Successes generally did not occur when the declines were caused by factors other than overfishing, such as habitat degradation—a lesson that took a very long time to be fully appreciated. During the advent of dam building in North America, and well into the 1970s and 1980s, hatchery production of Pacific salmon continued to be seen as the countermeasure to the impacts of dams (Montgomery 2003). In spite of good intentions, hatcheries have "failed to mitigate the effects of habitat loss and damage in the Columbia River basin," a major reason being the impact of other negative factors on salmon life history, as well as an early lack of appreciation for local genetic stocks (Montgomery 2003).

Hatcheries select for different traits, such as fast growth and aggressive feeding, compared to natural environments. Genetic studies have shown that hatchery fish typically have lower survival in nature compared to wild fish. For instance, by using genetic markers, Chilcote et al. (1986) were able to compare the reproductive success of hatchery-produced and wild-produced summer Steelhead spawning under natural conditions in the Kalama River of southwestern Washington. Over a four-year period, survival of progeny from hatchery spawners averaged only 28% of that of progeny from wild spawners. In spite of this, most of the smolts emigrating to sea were produced by naturally spawning hatchery fish, simply because hatchery spawners greatly outnumbered wild spawners. Even a few generations of domestication in a hatchery can result in negative effects on natural reproduction in the wild. In a study of the lifetime reproductive success of the first two generations of Steelhead that were initially reared in a hatchery but which bred in the wild after their release, the genetic effects of domestication reduced their subsequent reproductive capabilities by around 40% per captive-reared generation (Araki et al. 2007). In conjunction with efforts to restore environmental quality and manage levels of exploitation, hatcheries can be critical for restoring fish populations in cases where natural populations have been eliminated or extremely low. They are not the answer to restoring populations in the absence of other restorative approaches.

Finally, responses of fishes, in streams or other habitats, to recovery efforts can occur over much longer time scales than generally assumed. In a study of 24 tributary watersheds in western North Carolina, the overall best predictor of present-day diversity of fishes and macroinvertebrates was not watershed and land use in the decade of the study (1990s) but rather the land and watershed use in the 1950s. Past land use, especially long-term agriculture, may result in long-term reductions to aquatic biodiversity, perhaps requiring many decades to recover (Harding et al. 1998). Consequently, even though a number of recovery efforts listed previously showed some positive results, other responses could potentially require much longer time periods.

SUMMARY

Flowing water (lotic) systems of North America support a greater diversity of fishes compared to natural lakes but are now highly altered by dams,

diversions, changes in land use and pollution. In North America, most of the rivers that have little evidence of human impacts are distant from areas of high human population density, such as high-latitude areas of Canada and Alaska—regions with naturally low fish diversity.

A key feature of lotic systems is the hydrologic regime—the frequency, magnitude, duration, timing, and rate of change in flow. Many fish species, as well as other aquatic organisms, have major life-history events (e.g., migration, spawning) that are closely tied in with variations in flow patterns. In addition, overbank flow allows fishes direct access to resources on inundated floodplains and also transports floodplain nutrients back into river systems—a key point of the flood-pulse concept. Flow patterns, especially magnitude and timing, also have major impacts on the survival of embryos and larvae of species that construct nests or broadcast their eggs along stream bottoms.

Dams and diversions alter flow patterns and block movements of fishes. Alteration of flow can have strong negative impacts on embryo and larval survival, if, for instance, egg emergence is timed to coincide with natural low flows but releases from dams result in unnatural high flows. Dams also alter downstream water temperatures, generally providing cooler water in the summer and warmer water in the late fall, resulting in impacts on developmental rates of embryos and larvae. Blocked passage can eliminate large areas of habitat, especially for migratory species, but the impacts of dams are not necessarily related to their sizes. Dams associated with mill ponds no doubt had major impacts on movement patterns of fishes beginning in the nineteenth century. For small fishes that are not strong swimmers, even culverts can reduce or eliminate fish movement.

Virtually all efforts to recover fishes and ecosystems at risk involve some measure of restoration of natural flow patterns, even if only timing or frequency, and maintenance of appropriate minimal flows to prevent dangerous physical conditions (e.g., high temperatures and low oxygen) or total dewatering. Restoring connectivity to floodplains is an important action of many recovery plans. Fish passages around dams can contribute to recovery efforts but still have their limitations, especially because only a portion of fishes successfully traverse most fish passages, and some fish passages (e.g., fish ladders) can impose novel selective regimes on fishes. Dam removal is also a recovery tool that is being increasingly considered, especially for structures that are no longer needed or which might be compromised structurally.

Important in all recovery issues is awareness that fish populations are dynamic and continually responding to changes in their environments. Some of these are phenotypic shifts in behavior or morphology and others are genetically based responses to altered or new environments. For instance, well-intentioned as stocking may be, hatchery-raised fishes typically have lower survival rates than do wild fishes, in part because hatcheries offer a very different selective regime. Recovery of imperiled native fishes and other aquatic organisms will require identifying and considering environmental flow requirements to allow streams and rivers to function on a more natural basis, resulting in their being able to provide a wide range of ecosystem services.

SUPPLEMENTAL READING

Benke, A. C., and C. E. Cushing (eds.). 2005. *Rivers of North America*. Elsevier Academic Press, Boston, Massachusetts. Thorough coverage of physical and biological characteristics of North American rivers.

Jelks, H. L., S. J. Walsh, N. M. Burkhead, S. Contreras-Balderas, E. Díaz-Pardo, D. A. Hendrickson, J. Lyons, N. E. Mandrak, J. S. Nelson, S. P. Platania, B. A. Porter, C. B. Renaud, J. J. Schmitter-Soto, E. B. Taylor, M. L. Warren, Jr. 2008. Conservation status of imperiled North American freshwater and diadromous fishes. *Fisheries* 33:372–407. The current compilation of imperiled North American freshwater fishes.

Montgomery, D. R. 2003. *King of fish, the thousand-year run of salmon*. Westview Press, Boulder,

Colorado. An excellent account of the multitude of factors responsible for the declines of anadromous salmon species, as well as the challenges faced in recovering populations.

Schneider, D. W. 2000. Local knowledge, environmental politics, and the founding of ecology in the United States, Stephen Forbes and "The lake as a microcosm" (1887). *Isis* 91:681–705. A fascinating account of early research on floodplains and their ecological importance, and an early conservation battle (albeit largely unsuccessful in the short-term) to protect such habitats.

Sommer, T., B. Harrell, M. Nobriga, R. Brown, P. Moyle, W. Kimmerer, and L. Schemel. 2001. California's Yolo Bypass: Evidence that flood control can be compatible with fisheries, wetlands, wildlife, and agriculture. *Fisheries* 26:6–16. A synthesis of studies showing that a large constructed wetland can achieve flood control and provide important spawning and nursery habitat for fishes and habitat for other wildlife.

Stanley, E. H., and M. W. Doyle. 2003. Trading off: The ecological effects of dam removal. *Frontiers in Ecology and the Environment* 1:15–22. An overview of important considerations surrounding the removal of dams.

Waples, R. S., R. W. Zabel, M. D. Scheuerell, and B. L. Sanderson. 2007. Evolutionary responses by native species to major anthropogenic changes to their ecosystems: Pacific salmon in the Columbia River hydropower system. *Molecular Ecology* 17:84–96. An in-depth account of the issues and impacts of hydropower dams on anadromous salmonids.

WEB SOURCES

Dam removal. Time-Lapse Video of 98-year-old Condit dam removal on the White Salmon River, Washington. Removal. http://news .nationalgeographic.com/news/2011/10/111028 -condit-dam-removal-video/

Dams on the Colorado River. http://environment .nationalgeographic.com/environment/ freshwater/colorado-river-map/

Elwha River, Washington, dam removal webcams. http://www.nps.gov/olym/photosmultimedia/ elwha-river-webcams.htm

Federal listing and status of Pacific salmon populations in the Columbia River basin. http://www .nwr.noaa.gov/ESA-Salmon-Listings/Salmon -Populations/index.cfm

FIFTEEN

Ponds, Lakes, and Impoundments

CONTENTS

IN CONTRAST TO STREAMS in the Northern Hemisphere, lakes are generally recent geological phenomena being primarily post-Pleistocene (approximately 10,000 years old or less). Lakes can be categorized by their nature of origin. For instance, the eminent limnologist G. Evelyn Hutchinson (1957b) divided lake origins into 10 main categories with numerous subcategories, and another eminent limnologist, Robert Wetzel (2001), considered nine main groups as follows. Lake origins are typically catastrophic, such as being formed as a result of tectonic activity (faulting), volcanic activity, landslides, glacial activity, dissolving soluble rock, river activity, wind action, shoreline activity, and damming by the buildup of organic materials. In addition, two mammals, humans and Beaver, are adept at building barriers across rivers. Probably the majority of lakes in central and northern North America have been formed by glacial activity (see Figure 3.4 on the extent of Pleistocene glaciation). Lakes in the southeastern United States, especially in Mississippi, Alabama, and Louisiana, are mainly oxbow lakes, formed by river activity through the cutting-off of meanders on mature floodplains. Florida's Lake Okeechobee, one of the largest lakes completely in the United States, was formed partly as a result of tectonic activity, as a minor depression in the sea floor was uplifted during the Pliocene (Wetzel 2001). Both natural lakes and man-made impoundments are geologically transitory structures, with sediment capture starting as soon as they are formed.

Chapters in Parts 1 and 2 emphasized the hierarchical role of large-scale biogeographic processes, such as plate tectonics; climatic and more localized tectonic factors, such as glaciation and volcanism; as well as regional and local factors, such as species interactions, on the formation and maintenance of fish assemblages in lakes as well as streams. The emphasis in this chapter is to take a closer look at species occurrences and interactions in lakes and ponds, and then to examine how such occurrences and interactions have been altered by several major types of perturbations, especially shoreline development and the introduction of nonindigenous species. In the latter case, because of the similarity of issues, examples are also included from flowing water systems.

CONNECTIVITY (AGAIN)

Historically, limnologists tended to focus on lakes more as individual entities, perhaps because lakes are encompassed by a natural shoreline. However, just as streams show a continuum of processes and faunal change from low to high orders, lakes also need to be considered in terms of their position in the landscape—their connectivity. Somewhat surprisingly given their importance, studies of linkages between lentic and lotic systems, as well as among ecosystems of all kinds, are relatively rare although increasing in number. Linkages can occur in different ways, and are defined as "any persistent or recurring process or attribute that connects different ecosystems in some manner" (Lamberti et al. 2010). For instance, in terms of chemical and physical properties, water in lakes that are lower in the landscape tends to have greater inputs of landscape-derived materials because of the longer contact time with soils or agricultural nutrients (Soranno et al. 1999). Just as inflowing streams deliver sediment, nutrients, and organic matter to lentic ecosystems, lentic systems can deliver dissolved and particulate materials to outlet streams, potentially affecting the streams as well as lakes or reservoirs located downstream.

As also emphasized in Chapter 14, the connectivity between stream and floodplain habitats is particularly important for riverine fishes as well as fishes adapted to the lentic habitats of floodplains. Floodplain lakes, formed by the cut-off of a river meander, channel braiding, or other processes, are particularly abundant within the Mississippi Alluvial Plain, which extends south from Illinois to the Gulf Coast. For instance, the map of the Yazoo River drainage, in the Mississippi portion of the alluvial plain (i.e., the Mississippi Delta), shows the extensive complex of river channels (Figure 15.1). Before human intervention, approximately 13–15 new lakes formed per century in the Mississippi Alluvial Plain. However, beginning in 1735, levee construction had already started on the lower Mississippi River above and below New Orleans, and by the late 1930s all of the lower Mississippi River was bounded by a coordinated system of levees. Although many of these lakes still exist, the formation of new oxbow lakes is now essentially halted by the presence of levees separating the rivers from their former floodplains (Baker et al. 1991; Miranda 2005).

If left undisturbed, oxbow lakes progress through stages from the active meander bend, to the cutoff from the river, to a lacustrine phase, and finally a terrestrial phase. In a study of sediment deposits of Sky Lake, an old oxbow lake formed by the Mississippi River, an initial sedimentation stage occurred during the time the lake was still part of the flowing river, approximately 3,800–5,000 years ago. After it was abandoned by the river, perhaps 3,600 years ago, there was initially relatively high sedimentation caused by seasonal flooding, but sedimentation rates declined as the river migrated farther away and flooding was less frequent. During this period, sedimentation rates remained fairly low and consistent, with a total sediment deposition of approximately 120 cm. Within the last 110–120 years, as the surrounding floodplain was cleared for agriculture, another 120 cm of sediment was deposited—the same amount that was deposited in the

FIGURE 15.1. Portions of the Mississippi Alluvial Plain, including the Mississippi Delta, showing place names referred to in the text. The wide gray line surrounds the Mississippi Alluvial Plain; the dashed line shows state boundaries; black lines show major river channels. Map based on Chapman et al. (2004).

previous 3,600 years, or a 50-fold increase in the rate of sedimentation (Wren et al. 2008).

Floodplain lakes naturally separated from the floodplain will gradually fill with sediment, becoming progressively shallower and more turbid over time. The conditions of lakes that are artificially disconnected from the river by levee construction initially reflect their status (i.e., depth and water clarity) at the time of disconnection. However, given that such lakes are frequently surrounded by agricultural land, they tend to suffer from an accelerated rate of sediment accumulation, as shown by Sky Lake. Overall, floodplain lakes separated from the river will senesce over time, and without the capability of the river to create new floodplain lakes, such lakes will gradually disappear unless direct management efforts are used to maintain them (Baker et al. 1991; Miranda 2005; Wren et al. 2008; Miyazono et al. 2010).

The degree of connectivity between a floodplain lake and river can be an important determinant of the diversity and kinds of fishes found in the lake. In a study of 11 oxbow lakes along a 500-km section of the Mississippi River between Memphis, Tennessee, and Vicksburg, Mississippi (Figure 15.1), Miranda (2005) studied the relationship between fish composition and river-lake connectivity. Only large-bodied fishes were included in the analyses. Because all sampling was based on boat electrofishing along shorelines, smaller species, such as numerous minnows, topminnows, livebearers, silversides, and darters, were not effectively captured. There were strong interactions between connectivity, lake area, and lake depth, but lakes with high connectivity were generally closer to the active river channel, deeper, and clearer compared to lakes with lower or no connectivity, which were typically farther from the active channel, smaller, and more turbid. Large lakes with high connectivity to the Mississippi River had fish faunas that included species that were also characteristic of rivers, such as River Carpsucker (*Carpiodes carpio*) (Table 15.1), and small lakes with high connectivity had species such as gar. Both large and small lakes with low connectivity supported fish faunas that were more typical of lentic habitats, such as Yellow Bass (*Morone mississippiensis*) in large lakes, and numerous centrarchid species, such as Orangespotted Sunfish (*Lepomis humilis*), in small lakes. Fishes typical of intermediate-sized lakes included Skipjack Herring (*Alosa chrysochloris*) in highly connected lakes, and Freshwater Drum (*Aplodinotus grunniens*) in poorly connected lakes. Although connectivity had a strong effect on species composition, it was not related to the number of species found in a lake.

In another study of fishes and floodplain lakes on the Mississippi Alluvial Plain, within the White River National Wildlife Refuge of Arkansas (Figure 15.1), Lubinski et al. (2008) sampled 16 randomly selected lakes. Again, floodplain lakes located farthest from the main channel tended to be smaller, shallower, and more turbid, although because of complex connections to the river, the more distant lakes were not necessarily the least connected with the river channel. Although connectivity, as measured by the degree of flooding, did not have a strong impact on the species composition of the lakes, it was positively correlated with the number of fish species captured. However, the lack of appropriate gauges could have biased the use of flood frequency as a measure of connectivity and affected the apparent strength of the relationship between connectivity and species composition. In support of this, more typical riverine species, such as Skipjack Herring, Sauger (*Sander canadensis*), Longnose Gar (*Lepisosteus osseus*), Smallmouth Buffalo (*Ictiobus bubalus*), and White Bass (*Morone chrysops*), showed varying degrees of positive association with lakes that had higher degrees of flooding. Likewise, lentic species such as Yellow Bullhead (*Ameiurus natalis*), Spotted Gar (*Lepisosteus oculatus*), and Bowfin (*Amia calva*) were more associated with lakes having lower degrees of flooding.

The use of three different types of sampling gear (fyke nets; gill nets; and nighttime, boat-mounted electrofishing) in the White River study resulted in a more complete assessment of the fish assemblages found in the floodplain lakes. These lakes supported a high species richness, which averaged 37 species per lake. To compare abundances across the three gear types (catch-per-unit-effort would not be comparable among gears), average rank abundance was used instead. Based on this, the most abundant species were Bluegill (*Lepomis macrochirus*), Gizzard Shad (*Dorosoma cepedianum*), Spotted Gar, Orangespotted Sunfish, Warmouth (*L. gulosus*); Largemouth Bass (*Micropterus salmoides*), White Crappie (*Pomoxis annularis*), Freshwater Drum, Smallmouth Buffalo, and Yellow Bass (*Morone mississippiensis*) (Lubinski et al. 2008).

The Yazoo River, also part of the Mississippi Alluvial Valley, includes hundreds of floodplain lakes in its basin (Figure 15.1). In a study of 17 of these lakes, Miyazono et al. (2010) assessed the role of habitat connectivity and local environmental factors on fish functional groups. The groups were based on the three life-history strategies—opportunistic, equilibrium, and periodic—as proposed by Winemiller and Rose (1992) and discussed in Chapter 9 (Figure 9.3). Periodic strategists, which colonized floodplain lakes from the rivers, had large body sizes, high annual fecundity, and high variation among years in recruitment, as determined by lake-river connectivity. The common feature of floodplain lakes having faunas dominated by periodic strategists was river-lake connectivity. All such lakes had similar fish assemblages in spite of great variation in lake morphometry and overall size. As the connectivity between river and floodplain lakes decreased, water retention times increased with overall limnological conditions shifting from lotic to lentic, and opportunistic and lentic-adapted equilibrium species, such as Bluegill, increased. Opportunistic strategists with slower swimming speeds and equilibrium species with greater reproduction allocation to parental care became the dominant fish species. In lakes surrounded by better developed forested wetlands, fishes including gar, Bowfin, Black Crappie (*Pomoxis nigromaculatus*), and Buffalo (*Ictiobus* spp.) increased.

TABLE 15.1

Large-Bodied Fishes Typical of Floodplain Lakes of Different Sizes and Levels of Connectivity to the Mississippi River

Lake type	Fishes
Large and connected	River Carpsucker, White Bass
Large and disconnected	Yellow Bass, Smallmouth Buffalo, Catfish species, Threadfin Shad
Small and connected	Gar species
Small and disconnected	Orangespotted Sunfish, White Crappie, Green Sunfish, Warmouth, Bluegill

SOURCE: Based on Miranda (2005).

Finally, in a broader-scale study of 54 floodplain lakes in the Mississippi Alluvial Valley region of Arkansas and Mississippi, Dembkowski and Miranda (2012) found that measures of fish diversity were positively correlated with lake depth and inversely correlated with agricultural development, so that fish diversity was greatest in large, deep lakes surrounded with lower proportions of agricultural land. Connectivity was only weakly correlated with fish diversity measures, but likely affected species composition.

The four studies of fishes in floodplain lakes, all located within the broad region of the Mississippi Alluvial Plain, provide similar messages of the importance of river-lake interactions to maintaining diverse floodplain lake and channel faunas, although direct comparisons are complicated because of different study designs, analyses, and sampling gears. The role of connectivity or distance from the river channel clearly impacted the physical nature of the lakes. Lakes closer to the channel were generally larger, deeper, and with lower turbidity compared to more distant lakes, which were smaller, shallower, and more turbid (Miranda 2005; Lubinski et al. 2008). Measures of connectivity or distance from the river channel can have a strong impact on fish species composition and species richness. Not surprisingly, more highly connected lakes are characterized by fish faunas that include more riverine species (Miranda 2005; Miyazono et al. 2010; Dembkowski and Miranda 2012). Even though the White River study did not show an overall

strong relationship between connectivity and species composition, there was support for the role of inundation frequency in affecting species composition (Lubinski et al. 2008). Two studies showed strong positive correlations of connectivity with various measures of species diversity and richness (Lubinski et al. 2008; Miyazono et al. 2010), but the other two did not (Miranda 2005; Dembkowski and Miranda 2012). This suggests that the primary impact of connectivity is with species composition rather than with species richness—species richness can remain unchanged even though faunal composition totally changes. Finally, deeper lakes likely offer greater environmental stability and overall habitat persistence, compared to shallow lakes, which could periodically suffer declines in water quality or even drying during drought years (Lubinski et al. 2008; Dembkowski and Miranda 2012).

PHYSICAL VARIABLES AND SPECIES RICHNESS IN LENTIC SYSTEMS

Because of the relatively recent age of extant lakes in North America, fish assemblages in lakes tend to reflect riverine faunas, and because of the fluvial origin of lake fishes, one would expect the greatest area of interaction between fishes and their habitats to be in the littoral zone.

The abundance of fishes in the littoral zone of lakes is typically greater than in pelagic or deep benthic zones, and fishes in the littoral zone are frequently important as prey to larger pelagic species (Werner et al. 1977; Lyons and Magnuson

1987; Pierce et al. 1994). Even fishes that are primarily pelagic often use nearshore habitats in some part of their life cycles. For instance, approximately 80% of fishes from the Laurentian Great Lakes use such areas in some way (e.g., feeding, spawning, or nursery area) for at least part of the year (Reid and Mandrak 2009).

Fish assemblages in lentic ecosystems are influenced by a number of interacting factors, but several key physical variables stand out as being particularly important. The number of different kinds of species is generally related to the size of the water body, determined by depth or area (see also Chapter 4). For example the number of species found in North American lakes, inclusive of Mexico, the United States, and Canada, increases by a power of 0.13 as lake area increases (Figure 15.2). Both lower-latitude (Mexico) and higher-latitude (Canada) lakes tend to fall below the predicted relationship, likely because of the reduced overall species richness of habitats in low and high latitudes of North America (see Figure 1.4) (Barbour and Brown 1974). A more recent analysis of factors related to species richness in lakes of the world also showed that in North America (inclusive of the United States and Canada), lake area was the primary determinant of species richness in natural lakes, with species richness increasing by a power of 0.20 with area. Area was followed by lake depth, pH, and, to a lesser extent, latitude as predictors of fish species richness (Figure 15.3) (Amarasinghe and Welcomme 2002).

At a regional level, fish species richness and composition of North American lakes are also related to lake area and water depth, especially as it relates to winterkill of fishes, but also to lake productivity, water quality (such as pH), and temperature (Rahel 1986; Matuszek and Beggs 1988; Pierce et al. 1994). In addition, regional species richness in formerly glaciated areas is related to time since deglaciation and distance from glacial refugia (Chapter 3) (Eadie and Keast 1984).

Much of Canada is underlain by the Precambrian Shield, a massive formation of crystalline rocks such as granites, and with typically shallow, nutrient-poor soils with low buffering capacity (Clerk et al. 2000; Natural Resources Canada 2006). In a study of 75 lakes in Ontario, all located on the Precambrian Shield, the fish assemblages in individual lakes, comprising 21 species overall, were strongly related to lake depth and particularly whether the lakes showed dimictic or polymictic circulation patterns. Dimictic lakes typically undergo thermal stratification and mix twice each year in the spring and fall, whereas polymictic lakes have relatively continuous periods of mixing throughout the ice-free period (Marshall and Ryan 1987; Wetzel 2001). Lakes greater than 6.3 m in average depth were dimictic and supported a "salmonid assemblage," comprising Lake Trout (*Salvelinus namaycush*), Burbot (*Lota lota*), Lake Whitefish (*Coregonus clupeaformis*), and Cisco (formerly referred to as Lake Herring, *C. artedi*). Lakes shallower than 6.3 m were polymictic and support a "percid assemblage," comprising Walleye (*Sander vitreus*), Northern Pike (*Esox lucius*), White Sucker (*Catostomus commersonii*), and Yellow Perch (*Perca flavescens*) (Marshall and Ryan 1987).

Other local factors, such as habitat complexity and the diversity of vegetation, also contribute to species richness. In fact, the strong relationship of species richness to lake area or depth is likely due in part to increased habitat diversity of larger lakes (Eadie and Keast 1984; Matuszek and Beggs 1988; Amarasinghe and Welcomme 2002). Based on a broad review of fish-habitat relationships in both lotic and lentic systems, abundant evidence indicates that decreases in structural habitat complexity are detrimental to fish assemblage diversity and composition (Smokorowski and Pratt 2007). Lakes in northern Ontario support fewer common littoral-zone fish species than those in southeastern Ontario (7 versus 12), likely because of the more recent colonization of the northern lakes (Eadie and Keast 1984). In both groups of lakes, littoral fish species richness was strongly correlated with lake area and with the level of shoreline complexity, a measure independent of lake size. At a finer spatial scale, that of species diversity among littoral habitats within lakes, diversity

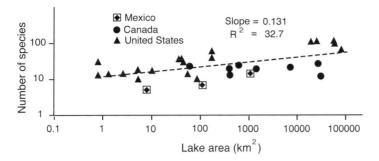

FIGURE 15.2. The effect of habitat area on the number of fish species in North American lakes. The dashed line shows the least square regression. Based on data from Barbour and Brown (1974) and used with permission from Ross and Matthews (in press), Johns Hopkins University Press.

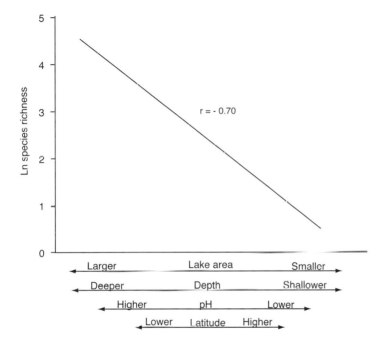

FIGURE 15.3. The relationship of fish species richness in North American lakes to lake area, depth, pH, and latitude. The length of the arrows indicates the relative contributions of the different variables to species richness. Ln = natural logarithm. Based on Amarasinghe and Welcomme (2002).

was correlated with the diversity of benthic prey in northern Ontario lakes, but with measures of habitat complexity, especially foliage height and substratum diversity, in southeastern Ontario lakes. Overall, fish species richness in Ontario lakes is strongly related to resource heterogeneity (Eadie and Keast 1984).

In a survey of 18 lakes in northern Wisconsin, the summer fish species richness was particularly related to the diversity of vegetation (Tonn and Magnuson 1982). In addition, species composition was strongly affected by winter oxygen concentrations and the availability of refuges from severe physical conditions, such as connecting streams. Species composition was also affected by the availability of high

densities of rooted, emergent macrophytes, such as Arrowhead (*Sagittaria*), to provide small fishes a refuge from large piscivores (e.g., Northern Pike). Because of the reduced risk of predation on young fishes provided by littoral-zone macrophytes, such areas often represent important foraging areas (Mittelbach 1988).

Complex littoral-zone habitats, both as a result of a range of substratum sizes as well as submerged and emergent macrophytes, are also important spawning areas for many North American fish species (see also Chapter 9). In addition to clean sand used by various nest-building species, such as many centrarchids, numerous species rely on aquatic macrophytes for spawning (Balon 1975; Reed and Pereira

2009; Warren 2009). In a meta-analysis of papers that presented experimental results from direct habitat manipulations in lakes and also streams, Smokorowski and Pratt (2007) found strong correlational evidence that complex substrata support higher fish abundances, richness, and diversity than fine substratum areas, and that sites with submerged aquatic macrophytes support even higher levels.

The pelagic zone of lakes or reservoirs is typically occupied by fewer species compared to the littoral zone, but is still an important component of lentic systems. Although specialized pelagic and deep-pelagic species are often considered to be characteristic of only the largest of North American lakes, in fact, examples of pelagic specialization are also found for small lakes. For example, consider the specialized limnetic forms of Threespine Stickleback (*Gasterosteus aculeatus*) and Pumpkinseed (*Lepomis gibbosus*) (see Chapters 7 and 11), as well as benthic and limnetic pairs of Lake Whitefish in eastern Canada and Maine, and another pair in the Yukon Territory and Alaska (Schluter 1996). In addition, in smaller lakes many species readily move between the littoral and pelagic zones, sometimes on a diel basis, but also as a function of age and/or size (Reid and Mandrak 2009).

Exploitation of pelagic habitats of large lakes, and especially of deep water benthic or epibenthic habitats is much more limited, in part because of the recent age of large North American lakes (see also Chapter 1). In large lakes, such as the Laurentian Great Lakes, deepwater habitats refer to those within the hypolimnion, characterized by high hydrostatic pressures, low ambient light, and relatively unvarying cold temperatures (Janssen et al. 2007). Several different groups of fishes have radiated to exploit these cold, deep environments. Ciscoes (Salmonidae, *Coregonus*) have differentiated into multiple species/morphotypes in the North American Great Lakes and other large northern lakes, and occupy pelagic habitats in the epilimnion as well the hypolimnion (Turgeon et al. 1999; Clemens and Crawford 2009). The widespread Lake Trout, which generally inhabits relatively shallow water, has radiated into two deepwater morphotypes in some of the largest lakes of northern North America. As adults, the so-called Humper form is a smaller-bodied feeding specialist on Opossum Shrimp (*Mysis relicta*); adults of the larger, more robust Siscowet are piscivores, feeding particularly on deepwater ciscoes (Eshenroder 2008).

Deepwater fish assemblages in the Great Lakes historically were composed of Lake Trout, Burbot, Shortjaw Cisco (*Coregonus zenithicus*), Kiyi (*C. kiyi*), Bloater (*C. hoyi*), Blackfin Cisco (*C. nigripinnis*), Round Whitefish (*Prosopium cylindraceum*), Pygmy Whitefish (*P. coulterii*), Deepwater Sculpin (*Myoxocephalus thompsonii*), and Spoonhead Sculpin (*Cottus ricei*). However, entire fish assemblages of deepwater fishes were totally or partially eliminated from nearly all five of the Great Lakes by the mid-twentieth century (Zimmerman and Krueger 2009; Favé and Turgeon 2008). For instance, the current-day shallow and deep pelagic fauna of Lake Superior consists of Cisco, Kiyi, and Bloater, plus the nonindigenous Rainbow Smelt (*Osmerus mordax*) (T. B. Johnson et al. 2004).

ALTERED FAUNAS AND NOVEL HABITATS

Lentic systems, as is true also of streams and rivers, have been highly altered by human activities, either directly or indirectly. Direct actions include alteration of shoreline and littoral-zone habitats, exploitation of recreational and commercial fisheries, and the introduction of nonindigenous organisms. Indirect actions include changes in land use in the watershed, alteration of connectivity of rivers and lakes, reductions in inflow to lakes, lowering of water tables by urban and agricultural pumping, and changes in climate related to the accelerated production of greenhouse gases.

Quantifying the effects of human and natural impacts on fish assemblages and ecosystems is often difficult because of the usual absence of data prior to the onset of major land use and associated ecosystem changes and the long history of impacts from human activity (Harig and Bain 1998; D. Foster et al. 2003). Impacts of indigenous people on fish species and populations are difficult to discern unless faunas can be partially reconstructed by unique isotopic ratios (such as the importation of marine-derived nutrients by anadromous salmonids), archeological evidence, or paleolimnological studies (Garrison and Wakeman 2000; Selbie et al. 2007). Even faunal consequences of more recent impacts occurring in the middle to late nineteenth century are problematic unless there are museum records or other historical data available to reconstruct species occurrences. Cases where historical data are available support the reality of the concern over the early loss of native fishes—the fish fauna of Spirit Lake, Iowa, has declined about 25% in the last 70 years (Pierce et al. 2001). In addition, the effects of agriculture in the nineteenth century can have lingering effects on ecosystems, with the potential of continuing impacts on the occurrence and abundance of species. In Harvard Forest, Massachusetts, legacies of past agricultural practices continue to influence aquatic ecosystems through sediment accumulation rates and organic content, even 100–150 years after cessation of agriculture and reforestation (D. Foster et al. 2003). Especially vexing is that experimental studies of habitat changes on fish assemblages necessarily deal with the contemporary flora and fauna, so the degree of effect may be lessened because more sensitive species have already been eliminated (Taillon and Fox 2004).

Littoral-Zone Impacts

Changes in the littoral zone, not surprisingly, can have major impacts on fishes, and the alteration of littoral habitats and adjacent shorelines is increasing in many North American lakes. In Minnesota, and most likely elsewhere, small, often seasonal cabins are being replaced by larger, permanent homes, with a concomitant increase in highly managed shoreline and beach areas (Radomski and Goeman 2001; Reed and Pereira 2009). One consequence of the increasing domestication of lakes is the loss, through direct removal or other factors, of emergent and floating-leaf vegetation, which are more impacted by human activities than submerged vegetation. In a study of 44 small, clear Minnesota lakes, there was a 66% average reduction in vegetation coverage in developed shoreline plots compared to undeveloped shorelines. When this is projected to some 531 similar lakes in Minnesota, the estimated total loss of emergent and floating-leaf vegetation from human development is approximately 20–28%. Impacts of vegetation losses on recreational fisheries is likely—both the relative biomass and mean size of Northern Pike, Bluegill, and Pumpkinseed were positively correlated with the amounts of emergent and floating-leaf vegetation (Radomski and Goeman 2001). A larger-scale study used aerial imagery taken in the summer and fall, spanning the period from 1939 to 2003. One-hundred randomly selected Minnesota lakes were selected from an initial pool of over 3,000 lakes greater than 0.04 km². Similar to other studies, shoreline development negatively impacted emergent and floating-leaf plant species—the loss of plant cover averaged 17% for shorelines in the highest development category (Radomski 2006).

The loss of macrophytes can have negative impacts on fish assemblages, especially juvenile life-history stages. In a study of shoreline development on fishes in Spirit Lake, Iowa, the same lake for which Pierce et al. (2001) showed a 25% decline in species number over 70 years, approximately half of the species sampled occurred in the limnetic zone as larvae before moving into the nearshore littoral zone as juveniles. Species richness and total abundance of juvenile fishes were greater in the shallow depth zones (1–3 m) of natural littoral areas than in developed sites, where nearshore beds of submerged and emergent vegetation had been removed to provide better recreational access. Only juvenile Small-mouth Bass (*Micropterus dolomieu*) did not show major differences between developed and undeveloped shorelines. At the deepest zone studied (2–3 m), there were few differences in fish richness or abundance between developed and undeveloped sites. In this study, the loss of submerged and emergent vegetation in shallow, nearshore areas as a consequence of shoreline development had a negative impact on the fish assemblage (Bryan and Scarnecchia 1992).

However, negative responses of fishes and other organisms to shoreline development are not always apparent. For instance, Pigeon Lake, a large, shallow Ontario Lake surrounded by vacation cottages, did not show reductions in fish species richness in response to altered shoreline development. Fish species occurrences in three types of natural habitats and three levels of shoreline development were assessed by underwater visual transects (Taillon and Fox 2004). Both the mean species richness and the life-stage richness did not differ among the three levels of shoreline development (undeveloped, moderately developed, or highly developed). Highly developed sites showed such alterations as removal of terrestrial shoreline vegetation, dredged areas, imported beach sand, seawalls, as well as docks, boathouses, and lifts. However, moderately developed sites actually supported more life stages of species compared to either undisturbed or highly disturbed sites. Overall, the mean species and life-stage richness did not vary among the three habitat categories (rocky with low vegetation diversity, a combination of sand and cobble with moderately dense vegetation, and silt substratum with relatively dense and diverse macrophytes). The influence of habitat on within-taxon abundance was greater than influence of development. Adult Blackchin (*Notropis heterodon*) and Spottail (*N. hudsonius*) shiners, juvenile and adult Yellow Perch, Pumpkinseed, and young-of-year and adult Largemouth Bass (*Micropterus salmoides*) had higher abundances in the densely vegetated, silt-bottom habitats. Smallmouth Bass, adult Rock Bass (*Ambloplites rupestris*), and Logperch (*Percina caprodes*) showed greater abundances in the lightly vegetated, rocky habitat, and the moderately vegetated, sandy habitats supported the highest densities of young-of-year Yellow

Perch and adult Walleye. The absence of a negative response to shoreline development could have been due to the general lack of disturbance of nearshore macrophyte species richness and density, such as occurred in the Iowa and Minnesota studies just discussed. In addition, some shoreline developments, such as pilings associated with piers or boathouses, can offset reductions of natural habitat complexity. Finally, as shown previously for stream biotas, long-term land use may well have already eliminated the more sensitive species (Harding et al. 1998). In support of this hypothesis, six littoral-zone cyprinids that were collected in the 1972 fish surveys were not seen in visual surveys nor collected in the studies from 2000 or later (Taillon and Fox 2004).

A major factor in the long-term impacts of shoreline development is the impact on spawning habitats of various fish species. Nest-site locations of Black Crappie (478 nests) and Largemouth Bass (119 nests) were compared among littoral-zone habitats in three small, shallow Minnesota lakes. Black Crappie nests were somewhat shallower (0.25–1.5 m) than those of Largemouth Bass (0.3–3.0 m), and overall, Black Crappie were more sensitive to shoreline development than were Largemouth Bass. In all three lakes, Black Crappie were more likely to nest adjacent to undeveloped shoreline compared to developed sites. Most nest sites were over hard sand substrata with stands of Hardstem Bulrush (*Scirpus acutus*); nest sites were also associated with shorelines having developed canopy cover, understory vegetation, and little submerged vegetation. The nests that were adjacent to developed shorelines were deeper than those in undisturbed areas. Largemouth Bass also nested primarily near undeveloped shorelines, but were less dependent on Bulrush. As with Black Crappie, nests adjacent to developed shorelines tended to be deeper. Because the hard sand bottoms suitable for Hardstem Bulrush are also prime areas for recreational development, and development usually is associated with the removal of emergent vegetation, continued loss of such areas will progressively limit prime spawning locations for Black Crappie in these lakes, with the potential for increased recruitment failure (Reed and Pereira 2009).

The impact of human activities on littoral-zone habitats of lakes and reservoirs is not necessarily associated with losses of vegetation. Indirect effects, such as the introduction of nonindigenous plants can result in excessive plant densities, which are detrimental to fishes, other aquatic organisms, and recreational users. Excessive growth of macrophytes is often shown by such exotic plants as Eurasian Milfoil (*Myriophyllum spicatum*), Water Hyacinth (*Eichornia crassipes*), and Hydrilla (*Hydrilla verticillata*). As a consequence, purposeful removal of vegetation in ponds, lakes, and reservoirs can be important for improvement of fish populations, including recreational fisheries (Olson et al. 1998).

Another example of the complexity of human impacts and intervention on aquatic habitats is provided by fishes inhabiting small springs and streams in intermountain and hot desert regions of the western United States. Because of extreme competition for water from human development and agriculture, many such habitats, and their generally unique faunas, are at great risk of extirpation and extinction (Minckley and Douglas 1991). Although some areas are now protected from most sources of damage, the distant groundwater pumping in arid-land metropolitan areas is a continual threat. For example, continued groundwater pumping by Las Vegas is causing a gradual decline of the water table in such unique areas as the Ash Meadows spring complex, in the Death Valley region of Nevada (Deacon et al. 2007). In 1984, much of the Ash Meadows area was designated as a National Wildlife Refuge, and on-site efforts at recovery of the native flora and fauna accelerated (see Deacon and Williams 1991 for an account of the tremendous environmental battles that made this possible). One of the management actions was to remove feral livestock by fencing the preserve. Although this was done

with the best of intensions, both aquatic and riparian vegetation around all of the approximately 20 springs greatly increased and open-water habitat for fishes declined. This was especially severe with small springs (Figure 15.4). Through their work on the springs of Ash Meadows, and in a similar Australian spring system, Kodric-Brown and Brown (2007) proposed the hypothesis that the natural ecological structure and function of desert springs relied on continual and high levels of physical disturbance by large mammals as they used springs as watering holes and feeding areas. In the late Pleistocene, the mammal fauna of the desert surrounding Ash Meadows included species of the now-extinct megafauna—mammoths, camels, llamas, horses, bison, mountain deer, giant mountain sheep, and giant ground sloths. As this megafauna declined, aboriginal humans took over the role of creating disturbance by such actions as burning the emergent vegetation to reduce cover and provide access to the spring pools and to maintain open water. As indigenous people were removed, the primary impact of European settlers was through their introduction of domestic livestock such as cattle and horses. To counter the loss of disturbance and its impact on riparian vegetation, the management plans now include manual removal of vegetation, although this is time-consuming and expensive (Kodric-Brown and Brown 2007). The challenge, of course, is finding an appropriate balance between too little and too much disturbance—uncontrolled and even moderate access of livestock is almost always damaging to riparian and aquatic systems in the western United States (Belsky et al. 1999). Especially because of the annual migratory pattern of some of the native large herbivores, such as bison, impacts on particular areas and ponds were lessened (D. Foster et al. 2003).

Littoral zones in very large lakes show some similarities with those discussed previously, but there are also important distinctions. In the Great Lakes, approximately 80% of the fish fauna use nearshore areas for at least part of the

FIGURE 15.4. A small spring in Ash Meadows, Nevada. The right side of the picture shows recent efforts to clear emergent and shoreline vegetation. See also corresponding figure in color insert.

FIGURE 15.5. A sandy shoreline along the northern shore of Lake Michigan, near the Straits of Mackinac. See also corresponding figure in color insert.

year for such purposes as feeding, reproduction, and nursery habitats. The value of such habitats to fishes is potentially lessened or eliminated by shoreline structures such as jetties, groins (spur breakwaters), and breakwaters (Reid and Mandrak 2009). Sandy coastal areas may support fewer species compared to more structurally complex shorelines, but they constitute important fish and invertebrate habitats (Figure 15.5). Fishes characteristic of high-energy shorelines of the Great Lakes are more typical of stream species rather than species prevalent in smaller lakes. For instance, based on a large multi-disciplinary study of shoreline areas in Lakes Erie and Michigan, typical lake shoreline species, such as sunfishes, were generally absent. Instead, lotic species such as mayfly nymphs (Heptageniidae), Longnose Dace (*Rhinichthys cataractae*), and Mottled Sculpin (*Cottus bairdi*) were typical (Meadows et al. 2005).

Shoreline areas of many Great Lakes coastlines have been extensively modified as a consequence of large structures for erosion and wave control to protect harbors and lakeshore property and the dredging of channels for navigation. Over time, such modifications can alter sediment dynamics via longshore currents, lead to the reduction or elimination of beaches and barrier systems, cause down-cutting of the lake bed resulting in greater nearshore water depths, and increase the amount of hard substrata. Such altered shorelines increase the chances of successful colonization for lithophilic, invasive species, such as Zebra Mussel (*Dreissena polymorpha*) and Round Goby (*Neogobius melanostomus*). This in turn can cause changes in trophic dynamics and productivity in the Great Lakes by shifting energy flow from predominantly pelagic communities to benthic communities in nearshore areas, with the coastal changes also affecting offshore regions via upwelling/downwelling cycles. The bottom line is that nearshore structures can have cumulative impacts on habitat quality, and that such modifications can enhance the colonization success of nuisance species, thus facilitating much larger-scale changes in biological community composition, trophic structure, and ecosystem function. As a consequence, numerous ecosystem services, including fisheries production, can be impacted (Meadows et al. 2005).

Biotic Homogenization and Differentiation

Over 50 years ago, the ecologist Charles Elton noted that the breakdown in the isolation of continents and islands through accidental and purposeful transplantation of plants and animals by humans was resulting in "ecological explosions" such as pest outbreaks (Elton 1958).

The rate of introduction of nonindigenous species is not at all reflective of natural dispersal rates. Echoing comments by Elton (1958), Ricciardi (2007) considers that "the current mass invasion event is without precedent and should be regarded as a unique form of global change." In reference to the filter analogy of Part 1 and its effect on faunal assembly (see the figure in Part 1), human introductions serve to bypass natural filters and thus greatly increase the rate of introductions while also bypassing the selective process of natural filters, with the consequence of species combinations that would be highly unlikely under natural conditions (Figure 15.6) (Rahel 2002; Ricciardi 2007).

One result from the breakdown of barriers that have in many instances existed since the early and middle Triassic (235–250 mya) is the increasing level of similarity among local, regional, and continental faunas and floras—a process referred to as biotic homogenization (Boxes 15.1 and 15.2). Based on data published in 2000, pairs of states within the United States were 7.2% more similar and shared an average of 15.4 more species compared to pre-European settlement conditions (i.e., similarities based only on native species prior to any extirpations or extinctions) (Rahel 2000).

Proportionately fewer introductions of nonindigenous fishes have occurred in Canada, but 64% of pairwise comparisons showed increases in similarity among regions, and faunas have significantly increased in similarity by an average of 1.2%. The homogenization has been driven by relatively few species compared to the United States, namely Goldfish (*Carassius auratus*), Rainbow Trout (*Oncorhynchus mykiss*), Brown Trout (*Salmo trutta*), and Smallmouth Bass. As with the United States, introductions were much more common than extirpations (Taylor 2004). Concern over biotic homogenization has led to a greatly accelerated rate of publications on the topic since 2000 (Olden 2006).

Another way of looking at faunal homogenization is to compare two measures of diversity. The level of species-packing in a single location, such as a lake, is determined by α-diversity. In

Freshwater fish faunas of different regions of
North America, and of the different biogeographic regions of the world, are becoming
increasingly similar—a process termed faunal homogenization by Vitousek et al. (1997)
and brought about by invasions and extinctions stemming from human-caused habitat
alterations and the bypassing of natural biogeographic barriers through the purposeful or
accidental introductions of organisms outside
of their natural ranges (Rahel 2000, 2002;
Rosenzweig 2001). The last time that such
biotic connectedness took place was when the
earth's landmasses were joined in the supercontinent Pangea, some 250 mya (see Chapter 2). Harold Mooney, of Stanford University,
suggested the term "Neo-Pangea" to describe
our newly and artificially connected world.
Neo-Pangea is equivalent to the world's largest biogeographical province, being formed
by the progressive shrinking of all other biotic
provinces. Gordon Orians, of the University of
Washington, suggested that the time period in
which we are now living is appropriately called
the Homogocene (Rosenzweig 2001).

Even though introductions are often done
for specific reasons, such as improving a sport
fishery, unintended effects are widespread. The
occurrence of negative and often broad-scale
unintended consequences caused by having too
narrow a focus has been termed the Frankenstein Effect—good intentions leading to greatly
unintended consequences (Moyle et al. 1986).

species). All of these are facilitated by habitat
alteration (Rahel 2000, 2002). For instance,
watershed deforestation frequently results in
elevated water temperatures and in increased
sediment input to streams, with the latter a
particular issue in the higher gradient areas of
upland streams. As a consequence, cool upland
streams, characterized by allochthonous nutrient inputs and complex substrata including
coarse and fine materials, are functionally
transformed into lowland streams characterized by warmer water temperatures, finer substrata, and increased autochthonous production. This habitat homogenization can result
in the gradual replacement of more sensitive
upland species by more tolerant native, lowland
species (Scott and Helfman 2001).

In a smaller-scale study of biotic homogenization, Radomski and Goeman (1995) quanti-
fied changes in 300 natural lakes in central
Minnesota over a 50-year period. During this
time the lakes were actively managed for recreational fishing and 70% had been stocked
with game fishes. As expected, β-diversity
declined with increased fish stocking, whereas
local diversity, measured as species richness,
increased. Six species became significantly
more common over the study period (Walleye;
Black Bullhead, *Ameiurus melas*; Yellow Bullhead; Brown Bullhead, *A. nebulosus*; Bluegill;
and Common Carp, *Cyprinus carpio*), and one
species (Bowfin) declined. However, the effects
of fish management were entangled with those
of habitat changes and unrecorded bait-bucket
introductions over the period of study.

In the study of faunal homogenization of fish
faunas in the United States, by far the greatest
influence on increasing biotic similarity among
regions was the introduction of nonindigenous
species, with only slight changes caused by the
loss of native species (Rahel 2000). The species that was lost from the largest number of
states is the now extinct Harelip Sucker (*Moxostoma lacerum*; formerly placed in the monotypic
genus *Lagochila*) that was known to occur in
eight states. Food and game species make up the
majority of the introduced species, with almost

contrast, the change in species composition
among different locations (i.e., among lakes)
can be expressed by β-diversity (Box 15.2) (Whittaker 1972; Willig et al. 2003; Jurasinski et al.
2009). Declines in regional distinctiveness can
be caused by the loss of local, often endemic
species, and/or by the addition and establishment of more widespread species. The species
can come from within the same drainage, such
as would be caused by upstream movement of
more tolerant native species, or by species introduced outside of their range (nonindigenous

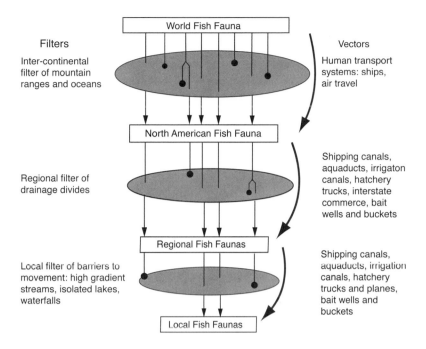

FIGURE 15.6. The unintentional and intentional introductions of nonindigenous species through the bypassing (thick curved arrows) of natural biogeographic filters as a consequence of human assistance. Small arrows show successful passage of taxon through natural filter; arrows ending in closed circles indicate blockage by a natural filter. Based in part on Rahel (2007) and Ross and Matthews (in press).

all the introductions following the east-to-west trajectory of European colonization. Only one species, Rainbow Trout, has been widely introduced in eastern states. States with relatively low numbers of native species show the greatest degree of faunal change; more than half of the fish species occurring in Nevada, Utah, and Arizona are introduced from other regions. Broader-scale comparisons of faunal similarity also show the effects of homogenization (Rahel 2000).

Similarly, in a comparison of historic and current fish assemblages in California, most watersheds showed net gains in species richness, and faunal similarity (i.e., homogenization) at the broad scale of California's six zoogeographic regions has increased since the 1850s. As with the broader-scale study of the United States, most of the change in similarity was caused by introductions and not extinctions (Marchetti et al. 2001).

The introduction of nonindigenous species does not necessarily result in biotic

homogenization; it could also result in greater differentiation among sampled habitats. Assessment of such changes is greatly influenced by the spatial and temporal scale of a study. For instance, a study with a broad geographical scale might be less likely to record species as extinct and more likely to record introductions of species, due to the greater number of communities and habitats sampled. Similarly, the time period over which comparisons are made has a potential effect on the outcome. A greater time interval increases the opportunity for introduction of species as well as the opportunity for extinction to occur (Olden and Poff 2003). This leads to difficulties in comparing the degrees of homogenization among regions. As an example, the study of Minnesota lakes showed a 9% increase in biotic similarity compared to a 20% increase in aquatic habitats of California. Although it might be tempting to interpret this as a greater rate of homogenization of California fish faunas, the Minnesota

BOX 15.2 • Basic Measures of Biotic Homogenization

The homogenization of faunas is expressed in two basic ways: indices of faunal resemblance and comparison of alpha and beta diversity. Although there are numerous measures of faunal resemblance, one of the oldest, simplest, and most widely used measures is the Jaccard coefficient (C_j). C_j is based on species presence or absence, which is determined as

$$C_j = \frac{a}{(a + b + c)} \times 100 \qquad \text{(Eq. 15.1)}$$

where a is the number of species common to both communities, b is the number of species unique to the first area, and c is the number of species unique to the second area (Pielou 1984; Brower et al. 1998).

Three forms of diversity are typically recognized: alpha, or local, diversity (α); beta diversity (β); and gamma, or regional, diversity (γ) (Whittaker 1972). Alpha diversity refers to the local diversity of a particular area, however it is defined (e.g., pond, stream, or watershed). Alpha diversity could simply be the number of species or some other measure of species diversity, such as Shannon's index (Poole 1974). Regional diversity is determined in the same manner but on a larger spatial scale (Jurasinski et al. 2009). Beta diversity is a measure of the difference in species composition either between two or more local assemblages, such as along an environmental gradient, or between local and regional assemblages. It provides a useful measure of how species have partitioned habitats (Wilson and Shmida 1984; Koleff et al. 2003). In pioneering works on communities, Robert Whittaker recognized two basic approaches for expressing β-diversity. For samples taken periodically over time, or along a habitat gradient, a measure of species turnover between paired samples is often used. This would be based on coefficients that express the change in species composition between samples, such as dissimilarity measures (e.g., $100 - C_j$). When samples are taken from different communities within a landscape, β-diversity (B_w) can be expressed most simply by the ratio of the total number of species represented in the samples to the mean number of species per sample (Whittaker 1972; Jurasinski et al. 2009). B_w is determined as

$$B_w = \frac{S}{\alpha} - 1 \qquad \text{(Eq. 15.2)}$$

where S is the total number of species in the system being studied, and is the average number of species found within the samples (i.e., the mean of the α-diversities) (Whittaker 1972).

The effect of two nonindigenous species introductions on the two measures of faunal homogenization shows that faunal similarity increases and β-diversity decreases.

SPECIES OCCURRENCES IN PREINTRODUCTION SAMPLES

Pond 1	Pond 2	Pond 3
Blackchin Shiner	Golden Shiner	Golden Shiner
Smallmouth Bass	Pumpkinseed	Largemouth Bass
Bluegill	Channel Catfish	Bluegill
		Channel Catfish

SPECIES OCCURRENCES IN POSTINTRODUCTION SAMPLES

Pond 1	Pond 2	Pond 3
Mosquitofish	Mosquitofish	Mosquitofish
Common Carp	Common Carp	Common Carp
Blackchin Shiner	Golden Shiner	Golden Shiner
Smallmouth Bass	Pumpkinseed	Largemouth Bass
Bluegill	Channel Catfish	Bluegill
		Channel Catfish

Examples of preintroduction calculations:

$$C_{j\,[1,2]} = \frac{0}{6} \times 100 = 0\% \qquad \text{(Eq. 15.3)}$$

$$C_{j\,[2,3]} = \frac{2}{5} \times 100 = 40\%$$

$$B_w = \frac{7}{3.33} - 1 = 1.12$$

FAUNAL SIMILARITY

Comparison	Pre-introduction	Post-introduction
Ponds 1 and 2	0	25%
Ponds 1 and 3	17%	37%
Ponds 2 and 3	40%	57%

β-DIVERSITY

Preintroduction	Postintroduction
1.12	0.69

study was of lakes all located within central Minnesota, and change was determined over approximately 50 years, whereas the California study involved statewide, zoogeographic provinces and compared change over approximately 150 years (Radomski and Goeman 1995; Marchetti et al. 2001; Olden and Poff 2003).

Patterns of biotic homogenization and differentiation are also a function of the type and number of winning (i.e., increasing populations size and range) and losing species (i.e., decreasing population size, shrinking range, or extirpation), the historical degree of similarity among the communities, and to a lesser degree the richness of the recipient communities. Because of the uneven nature of fish introductions, if nonindigenous species are primarily introduced into a single watershed there can be increased differentiation of species (i.e., less similarity) among watersheds (Olden and Poff 2003).

California fishes provide an example of the complexity of faunal change over time. Fish faunas in California historically had a high level of regional distinctness. Even though there are only 66 native freshwater fish species, over one-third (24) are endemic to specific watersheds. This is attributed to the high level of faunal isolation by mountains and deserts, and California's impressive physical diversity (Moyle 2002). Since pre-1850 to present, there have also been introductions of 50 nonindigenous species, the statewide extinction of seven fish species, and the recognition that almost half of the native fish fauna is in danger of becoming extinct. Although faunal homogenization increased in the state's zoogeographic regions, as previously shown, this is not necessarily true of comparisons made at smaller spatial scales such as watersheds (Marchetti et al. 2006).

In California, and likely elsewhere, urbanization is strongly related to the level of endangerment of native fishes and the likelihood of invasion by nonindigenous fishes. Contrary to the pattern of the large-scale zoogeographic regions, the introductions of nonindigenous fishes and the extinction of native fishes have caused fish faunas of watersheds to differentiate over time, a process also related to the level of urbanization. The reason for the differentiation, rather than homogenization, is a reflection of the haphazard manner in which nonindigenous fishes have been introduced, the degree of isolation of many watersheds that makes range expansion unlikely, and the historic distinctness of many of the regional faunas. The increased differentiation at the scale of watershed is also likely ephemeral and could disappear with the continued introduction of nonindigenous species and extinctions of native fishes (Marchetti et al. 2006).

Most of the documentation of homogenization has focused on taxonomic homogenization, the increase in phylogenetic similarity of biotas over time as a consequence of establishment of cosmopolitan species and extirpation of endemic species—indeed the terms taxonomic and biotic homogenization are often used interchangeably. Olden and Rooney (2006) argue that biotic homogenization should be defined more broadly as the "overarching ecological process by which formerly disparate biotas lose biological distinctiveness at any level of organization, including their genetic, taxonomic, and functional characteristics."

Genetic homogenization is of increasing concern in many fish populations, especially where natural populations have been supplemented by hatchery-produced individuals (Dann et al. 2010). Genetic homogenization results in greater genetic similarity of individuals and can be caused by interspecific hybridization or intraspecific hybridization caused by the mixing of formerly isolated populations. Genetic homogenization through the hybridization of individuals from local and distant populations of a species is a concern because it can lower the ability of a population to respond to environmental change and also decrease fitness (termed outbreeding depression) under existing environmental conditions (Olden et al. 2004; Olden 2006; Dann et al. 2010). The decrease in fitness from outbreeding depression can occur because of the mismatch between the average gene effects and local habitats, or from the disruption of coadapted gene complexes (Dann et al. 2010). For example, hatchery introductions since the 1960s have resulted in fall-run Chinook Salmon (*Oncorhynchus tshawytscha*) in California's Central Valley comprising essentially a single population—one that includes hatchery fish. The extreme level of genetic homogenization of these populations is in stark contrast to populations of Chinook Salmon throughout the Pacific Northwest, where hatchery introductions have been less prevalent (Williamson and May 2005).

Functional homogenization is an increase in the functional similarity of biotas over time because of the establishment of species with similar roles in the ecosystem and by the loss of species possessing unique functional roles. Although changes in functional similarity can be related to taxonomic similarity, they do not necessarily follow the same temporal or spatial patterns. The different rates of change in taxonomic and functional similarities are evident in a study assessing 100 years of change in the distribution of native and nonindigenous fishes in the Lower Colorado River basin (Pool and Olden 2011). Functional characteristics included two ecological traits, trophic guild and water temperature preferences, and six life-history traits. The latter included adult body size, age at maturity, fecundity, egg size, parental care, and reproductive guild (see Chapter 9). Examined at intervals of decades, there was initially a weak relationship between functional and taxonomic homogenization that gradually became stronger over time. Functional homogenization became evident by the early 1960s, but taxonomic homogenization was not evident until the 1990s. From 1900 to 1999, the fish faunas in watersheds of the Lower Colorado River basin have increased only slightly in taxonomic similarity from 30.8% to 31.9%. In contrast, functional homogenization increased from 6.2% to 45.2% over the same time period (Pool and Olden 2011). The loss of specialized functional groups of fishes has likely impacted the broad-scale functioning of aquatic ecosystems.

Nonindigenous Species Background and Terminology

The earliest introduction of a nonindigenous fish species to North America seems to have been the Goldfish, which arrived in the 1600s from its native Asia via Europe. As an early harbinger to the release of nonindigenous aquarium fishes, Goldfish were introduced as ornamentals to be "kept in glass vases as an ornament to the parlor and drawing room" (De Kay 1842). This was followed, apparently in 1831, by the introduction of Common Carp from France (De Kay 1842), followed by more introductions later in the 1800s, which included Brown Trout (Courtenay 2007). Movement of North American fishes outside of their native drainages was also widespread in the nineteenth century. For instance, in a chapter giving instructions and encouragement for the transplantation of black basses (genus *Micropterus*), Henshall (1881) stated, "there is no fish more worthy of cultivation; none that can be so easily transplanted, and none that is so well adapted to the various waters of our country." By 1999, Fuller et al. reported the occurrence of 536 fish taxa outside of their native ranges in U.S. waters; of these, 316 were moved beyond their native ranges in the United States and 185 were of foreign origin, with 75 of these being established (Fuller et al. 1999; Courtenay 2007). Since then, an additional 118 species have been collected outside of their native ranges in the United States, and another 73 species native to other countries have been recorded (Courtenay 2007). Tracking numbers of introduced species is problematic because new introductions are continually happening. Many of the introductions occur by moving species native to North American watersheds to watersheds where they are not native, and whether or not the introductions have resulted in established, breeding populations can be difficult to determine (Fuller et al. 1999). The most commonly introduced fishes in the contiguous United States include five Eurasian forms, Common Carp, Goldfish, Brown Trout, Tilapia (*Oreochromis* spp.), and Grass Carp (*Ctenopharyngodon idella*),

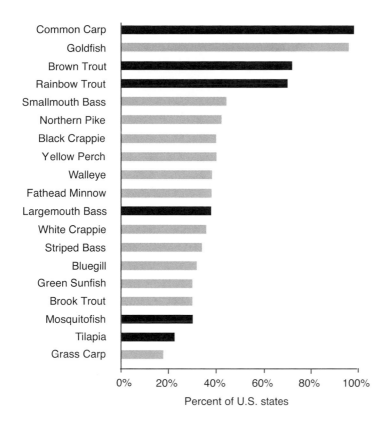

FIGURE 15.7. Eighteen of the most commonly introduced nonindigenous fishes in the United States, ranked by the number of states where nonindigenous populations are established. The black bars refer to species that are also included in the World Conservation Union's (IUCN) database of "100 of the World's Worst Invasive Alien Species." Based on Fuller et al. (1999), Rahel (2000), Florida Fish and Wildlife Conservation Commission (2012), International Nonindigenous Species Database Network (2012), and Lowe et al. (2004).

plus 15 species native to the United States but introduced into other regions, such as Rainbow Trout, Smallmouth Bass, Northern Pike, and Mosquitofish (includes two species, Western Mosquitofish [*Gambusia affinis*] and Eastern Mosquitofish [*G. holbrooki*] but many records of introductions do not distinguish between the two) (Figure 15.7). The list also includes six species (out of a total of eight) that have the dubious distinction of being listed in the World Conservation Union's (IUCN) database of "100 of the World's Worst Invasive Alien Species" (Lowe et al. 2004). They are so listed because of the damage that they have caused to native species and ecosystems in many parts of the world.

The terminology of nonindigenous species often distinguishes between species transplanted from their native range to another watershed within the same political boundary (sometimes also referred to as nonnative) versus exotic species—those of foreign origin (sometimes referred to as alien species) (Shelton and Smitherman 1984; Courtenay et al. 1986; Fuller

et al. 1999). There are certain instances when political boundaries are important in dealing with introduced species, such as legislation controlling species' introductions. Moreover, negative ecological responses can occur when species are moved from their native ecosystems to ecosystems where they do not naturally occur—something that is independent of political boundaries (Helfman 2007; Gozlan et al. 2010). Based on data for plant and fish introductions in 12 U.S. states, species introduced from within the United States (i.e., transplanted species) had greater homogenizing effects compared to species introduced from outside the United States (i.e., exotic species) (McKinney 2005). Consequently, it is much preferable to refer to all species found outside of their native ranges as nonindigenous species (Helfman 2007).

Introductions of nonindigenous fish species occur for many different reasons. In the United States, the largest number of introductions has involved purposeful introductions from government or private groups to improve recreational fishing. The effect of this has been further magnified by the emptying of bait buckets or fish wells in boats at the end of a fishing trip—irrespective of the watershed. This is followed by releases of aquarium fishes and by escape of fishes from aquaculture facilities (Courtenay 2007). Indeed, the ultimate escape of nonindigenous fishes from aquaculture or other facilities is considered "virtually inevitable" (Shelton and Smitherman 1984). Species are also intentionally introduced to improve or recover wild stocks, and unintentionally released by discharge of ballast water from ships (Claudi and Leach 1999; Gozlan et al. 2010). Finally, some recent introductions, such as the Northern Snakehead (*Channa argus*), are thought to have occurred as a consequence of prayer animal release, the purposeful release of animals to favor some deity (Courtenay 2007).

Reservoirs, Nonindigenous Species, and Biotic Homogenization

Reservoir construction can be a major contributor to the homogenization of North American fish assemblages. There are over 80,000 large impoundments and 2.5 million smaller impoundments in the United States alone, resulting in a major shift in the occurrence of standing water, especially in regions where lakes and ponds were naturally absent or rare. For instance, the construction of reservoirs has led to a 228-fold increase in the area of standing water in Oklahoma, putting this historically arid state almost on a par in terms of standing water with lake-rich states such as Wisconsin (Havel et al. 2005; P. T. Johnson et al. 2008). Reservoirs are among the most homogeneous of human-created habitats, and in California the warm lentic waters of reservoirs and ditches offer ideal habitats for many nonindigenous fishes. It is no surprise that faunal similarity of reservoir fish faunas is much greater compared to the fish faunas in general (Marchetti et al. 2001). Overall, reservoirs result in local riverine species, such as riffle-inhabiting darters in the southeastern United States, being replaced by a widespread suite of lentic species that are often not native to the region (Rahel 2002).

A major question is how reservoirs contribute to the spread of nonindigenous species, especially whether species common in reservoirs readily invade streams flowing into reservoirs.

The presence of reservoirs is related to the number of nonindigenous species, as shown by a study of 125 North American drainages. There were positive correlations to the number of nonindigenous species with the number of impoundments and the size of the watersheds, and a negative correlation to the number of native fish species (Gido and Brown 1999). Based on a study of five invasive species, including one fish species (Eurasian Water Milfoil, Spiny Water Flea, Zebra Mussel, Rainbow Smelt, and Rusty Crayfish), in lakes and impoundments of Wisconsin and the upper peninsula of Michigan, impoundments were significantly more likely to be invaded than natural lakes. The strong association between impoundments and nonindigenous species is a consequence of the young age of reservoirs (generally < 60–100 years), the different habitat offered by impoundments compared to natural habitats, and the high disturbance regime characteristic of most impoundments (P. T. Johnson et al. 2008). Also, impoundments are more likely to be accessible to humans compared to natural lakes, more likely to have hydrological connections via stream networks, and are often stocked by similar suites of species. Examples include catfishes; pikes and pickerels; salmon, trout, and whitefishes; temperate basses; and bass and sunfishes, many of which are among the most widely introduced species (Figure 15.7). In addition, movement of anglers from reservoirs to natural lakes increases the opportunity of dispersal. All of this can lead to reservoirs acting as stepping-stone habitats for the spread of nonindigenous species (Crossman and Cudmore 1999; Havel et al. 2005; P. T. Johnson et al. 2008).

The level of upstream colonization of nonindigenous fishes from reservoirs varies, in part because of biogeographic differences. Based on two studies done in Oklahoma and Kansas and spanning the interval between 1997 and 2003, reservoirs had only a very localized effect on the upstream occurrence of nonindigenous fishes. In part, this was influenced by the number of native species, such as Bluegill, Largemouth Bass, Channel Catfish (*Ictalurus punctatus*), and White Bass, that survive well in reservoirs (Gido 2004; Falke and Gido 2006). The most important factor in predicting the occurrence of nonindigenous fishes in streams was not proximity to reservoirs but instead was human population density in a watershed. This likely occurs both because of more extreme habitat alteration as well as propagule pressure—the frequent release of nonindigenous fishes (Gido 2004).

Detecting Impacts of Nonindigenous Species

Successful establishment of nonindigenous fishes in an ecosystem will always constitute risk, and, worldwide, fishes are among the most widely introduced of all taxa (Li et al. 1999; Gozlan et al. 2010). However, there are many challenges to actually assessing negative, or positive, impacts, especially given that introduction and establishment of nonindigenous species is usually coincident with other potential drivers of change, such as habitat alteration. Measures such as species richness or diversity often show little evidence of impact of nonindigenous species, and in the short-term may even show increases at more localized scales of analysis, especially if the nonindigenous species are included in the analysis. If the latter occurs, species richness can actually increase when the establishment of nonindigenous species has outpaced the extinction of native species (Gido and Brown 1999; Rahel 2002; Brown and Sax 2007). Clear Lake, California, provides an example of the paradox of gaining species locally and losing diversity globally. Clear Lake, a large, natural lake located in central California, had a native fish fauna of 12 species, including three endemics. To improve angling, species such as sunfishes, black basses, and catfishes were introduced, increasing the fauna

by 16 widespread, nonindigenous species. As a consequence of the nonindigenous introductions coupled with habitat changes, two native species, Clear Lake Splittail (*Pogonichthys ciscoides*) and Thicktail Chub (*Gila crassicauda*), became extinct. The result is that local diversity increases to 26 and global diversity declines by two. At the level of the United States, 39 nonindigenous species have become established, all common elsewhere in the world, and 19 species found only in the United States have been lost (Rahel 2002). Clearly, assessing impacts of nonindigenous species on measures of local (alpha) diversity only tells a small part of the story!

Key issues in understanding impacts of nonindigenous species include the metrics used to detect changes and, equally important, the time scale over which changes are assessed—some species introductions show a lag phenomenon, with species remaining innocuous for years before they suddenly become pests (Mooney and Cleland 2001; Simberloff 2011). Changes in alpha diversity alone are poor indicators of impact, and certainly the inclusion of nonindigenous species in diversity calculations (although appropriate for studies of taxonomic homogenization) further clouds actual impacts (Angermeier 1994; Scott and Helfman 2001).

Although it is perhaps likely that introductions of many nonindigenous species (including plants and animals) pose minimal risk, generalizations about the lack of impacts of nonindigenous species are strongly biased by the overall paucity of rigorous evaluations of impacts for both ecosystems and species (Gozlan 2008; Vitule et al. 2009). The "rule of tens" suggests that approximately 10% of introduced species become established and approximately only 10% of those that become established become pests (e.g., out of 100 introductions, one species will have a negative impact) (Williamson and Fitter 1996). However, negative impacts from fish introductions on native freshwater fishes seem to be an exception to this rule. In seven studies covering the introductions of 352 nonindigenous freshwater fishes worldwide, on average 55% (range = 38–77%) of the introductions resulted in established populations. Because many introductions go unreported or undetected, this is perhaps an overestimation. Furthermore, of 26 freshwater systems with established populations of nonindigenous fishes, 77% of the cases showed a reduction or elimination of native fishes (e.g., out of 100 introductions, 55 would become established and 42 would have negative impacts) (Ross 1991). A more recent study, using FishBase, a large international database on fish biology, Ruesink (2005) found a similar high level of establishment of 64%. In contrast, based on a more restricted Food and Agricultural Organization of the United Nations (FAO) database on 103 aquaculture species, Gozlan (2008) found that 48% of introduced freshwater fishes had a negative ecological impact on the native ecosystem (the likelihood of establishment was not reported separately). This is rather similar to the resultant 42% shown by primarily nonaquaculture species surveyed by Ross (1991). However, Gozlan (2008) also considered that the risk of ecological impact for the majority of fish species was less than 10%. The take-home message seems to be that fish species vary greatly in terms of their potential threat following introduction into habitats where they are not native, but that the "rule of tens" is likely not appropriate for fishes. The challenge, of course, is determining what the threat of a particular species might be. Contrary to the suggestion by Gozlan (2009) that "on a global scale, the majority of freshwater fish introductions are not identified as having an ecological impact while having great societal benefits," Vitule et al. (2009) point out that the absence of impacts is often an artifact of the lack of rigorous data on impacts or on findings from superficial studies.

Potential impacts of nonindigenous species need to be assessed at multiple scales, and especially with the overarching goal of protecting ecological integrity, of which biodiversity is only one component (Angermeier 1994; Meffe and Carroll 1997). Importantly, diversity, biodiversity, and ecological integrity (used interchangeably with biological integrity) are all different measures but are also nested elements. Diversity (species diversity), is a measure of the number of species in an area. This can be assessed using various indices of species richness or diversity (Peet 1974). Biological diversity (biodiversity) encompasses multiple levels of organization, including genes, populations, species, communities, ecosystems, and landscapes, and is thus more comprehensive than species diversity, although species diversity is part of biodiversity. Theoretically, an estimate of biodiversity would include all taxonomic elements present, all genetic elements present, and all the ecological elements present—a daunting task at best! Both diversity and biodiversity are concerned with describing the number of elements. Biological integrity is a synthetic (in contrast to a collective) property and most simply refers to a system's wholeness—"the presence of all appropriate elements and the occurrence of all processes at appropriate rates" (Angermeier and Karr 1994). Unlike diversity, biological integrity encompasses the elements as well as the processes responsible for generating and maintaining the elements. A general definition of biological integrity is a "system's ability to generate and maintain adaptive biotic elements through natural evolutionary processes" (Angermeier and Karr 1994). Biological integrity refers to natural conditions with little or no human impacts, and a system with high biological integrity is the outcome of natural evolutionary and biogeographic processes. Examples of important interactions and processes included in measures of biological integrity would be the transfer of nutrients and energy through multiple trophic levels via food webs as well as top-down effects of predators on lower trophic groups (see Chapters 11 and 12). Simberloff (2011) argues that when ecosystem impacts of nonindigenous species are assessed at multiple scales, as explained previously, negative impacts would be associated with many introduced species. Finally, the absence of an immediate impact from introduced species does not preclude later impacts. As environmental alterations continue in land use and global climate change, species that were once considered low risk may suddenly become high risk (Gozlan 2009). The ecological and financial costs of nonindigenous species that have caused impacts, such as the Sea Lamprey (*Petromyzon marinus*) in the Great Lakes, led Simberloff (2007) to argue that the appropriate position in dealing with deliberate introductions must be "guilty until proven innocent."

Examples of Impacts

Predicting potential impacts of species' introductions is clearly difficult, although various studies suggest that species that are ecosystem engineers (see Chapter 13), or species that are near the top of food webs (see Chapter 12), are perhaps more likely to have negative effects (Helfman 2007). Common Carp can have impacts at multiple levels within an ecosystem and are one of the most widely introduced fish species worldwide. In North America they occur in over 90% of U.S. states and are found in some 95% of Mexican water bodies (Figure 15.7; Weber and Brown 2009). Carp are included in the IUCN database of "100 of the World's Worst Invasive Alien Species" (Lowe et al. 2004). The spread and consequent impacts of Common Carp are related to their being ecological generalists and in having rapid growth, high fecundity, and a prolonged spawning season (Weber and Brown 2009). Carp are omnivorous, feeding on algae, seeds of vascular plants, organic detritus, zooplankton, and benthic invertebrates. When feeding, they take in a mouthful of bottom materials, expel it into the water, and then pick out food items, with large carp being able to penetrate about 13 cm into the bottom in search of food. Consequences of this are the removal of submerged plants, resuspension of bottom materials, and increased turbidity (Walburg and Nelson 1966; Eder and Carlson 1977; Panek 1987). Based on a broad survey of literature on the impacts of Common Carp, more than 90% of the studies found decreases in macrophytes and benthic invertebrates; approximately 70–90% of the studies reported increases in nutrient levels,

turbidity, chlorophyll *a*, and phytoplankton; and approximately 60% of the studies found that Common Carp were responsible for increases in small-bodied zooplankton and declines in fish populations. As a result of these broad ecosystem impacts, the introduction of Common Carp triggers a phase-shift from clear to turbid water in lakes (Weber and Brown 2009). In addition to devastating impacts on native fishes, thousands of hectares of waterfowl habitat have been lost (Simberloff 2011).

The introduction of nonindigenous predators can also have dramatic effects on aquatic systems by direct predation, such as the impact on Lake Trout and large coregonids following the invasion of the North American Great Lakes by Sea Lamprey. Predation impacts have also resulted from the widespread stocking of Smallmouth Bass in lakes where they were not native, and the transplantation of Flathead Catfish (*Pylodictis olivaris*), a voracious predator, into streams in the eastern and western United States, to name only a few examples (R. R. Miller et al. 1989; Whittier and Kincaid 1999; Helfman 2007).

Sea Lamprey occur on both sides of the North Atlantic Ocean and are anadromous, migrating into freshwater streams to spawn. Along the North American coast, Sea Lamprey ascend coastal streams from Greenland to Florida and are also known to form landlocked populations in some lakes. Landlocked Sea Lamprey tend to have smaller body sizes than the sea-run populations (Hubbs and Pope 1937; Scott and Crossman 1973). The occurrence of Sea Lampreys in the Great Lakes is of particular concern because of the impact that they have had on native fishes such as Lake Trout and large ciscoes. Although still debated, Sea Lamprey were likely native to Lake Ontario and the Finger Lakes of New York as a dwarf, landlocked form (Hubbs and Pope 1937; Lawrie 1970). Other hypotheses for the occurrence of Sea Lamprey in Lake Ontario are that they traveled via nineteenth-century navigation canals (viewed as unlikely by Daniels 2001), or that they could have been stocked— they were actually a popular food fish in the early nineteenth century and the ammocoetes (larval lampreys) were also used widely as bait (Daniels 2001). Sea Lamprey were blocked from entry into the upper Great Lakes (lakes Erie, Huron, Michigan, and Superior) by the impressive and impassable barrier of Niagara Falls. The means by which Sea Lamprey, and also the introduced Alewife (*Alosa pseudoharengus*), reached the upper Great Lakes was most likely the Welland Ship Canal connecting Lakes Ontario and Erie (Lawrie 1970; Daniels 2001). The canal was first constructed in 1881 but had little water flow and likely offered poor habitat for Sea Lamprey because of their normal propensity to ascend rapidly flowing streams for spawning. The design of the Welland Canal changed over time, with the fourth version completed in 1930. This was more suitable for fish passage, having fewer locks, greater water volume and water flow, and took water directly from Lake Erie rather than from the Grand River, which was not in the Lake Erie watershed (Mills et al. 1999; Daniels 2001). Sea Lamprey were first noted in Lake Erie in 1921, but the first documented spawning was in 1932 shortly after the last modification of the Welland Canal. Sea Lamprey were first reported from Lakes Michigan and Huron in 1936 and 1937 and from Lake Superior in 1946. By 1947, established populations of Sea Lamprey existed in all the upper Great Lakes (Hubbs and Pope 1937; Lawrie 1970; Daniels 2001).

Sea Lamprey were a factor in, albeit not the primary cause of, the decline of large, open-water fishes in the Great Lakes, such as Lake Trout and ciscoes (*Coregonus* spp.). Initially, those populations declined because of overfishing, but in the 1940s the already declining stocks were subjected to direct predation by the invasion of the Sea Lamprey. Although Lake Trout have somewhat recovered as a consequence of continuing efforts to control Sea Lampreys, three species of ciscoes (Deepwater Cisco, *C. johannae*; Blackfin Cisco; Longjaw Cisco, *C. alpenae*) became extinct as the remnant populations hybridized with more common ciscoes (Miller et al. 1989).

Introduced predators can also have indirect effects on native fishes and aquatic ecosystems through the alteration of food webs (Brown et al. 1999). Examples of food web effects include the declines in native Lake Trout populations following the introduction of Smallmouth Bass (Vander Zanden et al. 2004), and the declines in Yellowstone Cutthroat Trout (*Oncorhynchus clarkii bouvieri*) subsequent to the introduction of Lake Trout (Chapter 12).

Smallmouth Bass is a popular sportfish native to northeastern North America. Through planned and unplanned introductions, the range of Smallmouth Bass has expanded northward and westward, with often unintended consequences on resident fishes (Whittier and Kincaid 1999). One such impact has been the decline in Lake Trout populations in certain lakes following Smallmouth Bass introductions. Lake Trout primarily consume pelagic prey fishes, such as various coregonids; however, in lakes without pelagic fishes, Lake Trout switch to zooplankton and to moving into the littoral zone to feed on various species of minnows. Smallmouth Bass are a more efficient littoral predator than Lake Trout and are able to deplete littoral populations of prey fishes. Consequently, in lakes without a natural pelagic fish forage base, the introduction of Smallmouth Bass causes Lake Trout to switch to a lower-quality diet with consequences for growth and survival (Vander Zanden et al. 2004). By modeling the potential for Smallmouth Bass impacts, Vander Zanden et al. (2004) were able to identify a suite of lakes across Quebec whose Lake Trout populations would be most vulnerable to Smallmouth Bass introductions. Identifying where to focus conservation efforts is important, given that the spread of Smallmouth Bass will likely increase as the impact of global climate change increases water temperatures in relatively small, shallow lakes (Stefan et al. 2001).

In contrast to predation, there are fewer examples of species declines or extirpations as a consequence of competitive interactions between native and nonnative species. The

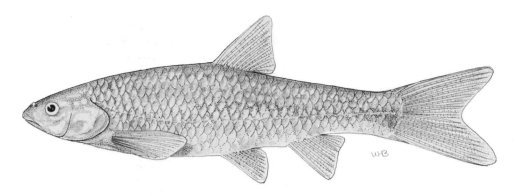

FIGURE 15.8. The Rio Grande Silvery Minnow (*Hybognathus amarus*). Picture courtesy of Howard Brandenburg, courtesy of the New Mexico Game and Fish Commission. See also corresponding figure in color insert.

few examples of extinctions associated with competitive interactions relative to predation impacts may simply be caused by competition-induced extinction generally being a slower process than predation-induced extinction (Ross 1991; Mooney and Cleland 2001; Marchetti et al. 2006). One of the strongest examples of species loss through apparent competitive exclusion is the loss of the native Rio Grande Silvery Minnow (*Hybognathus amarus*) from the Pecos River that occurred in the surprisingly short time of 10 years following the introduction of the Plains Minnow (*H. placitus*) (Figure 15.8) (see also Chapter 11). Because of the lack of support for hybridization as a factor in the loss of the Rio Grande Silvery Minnow, exploitation competition from the Plains Minnow remains as the most likely cause of the species replacement (Moyer et al. 2005).

The role of nonindigenous species as vectors of parasites and diseases is also of great concern (Vitule et al. 2009). The devastating impact of whirling disease (caused by the parasite *Myxobolus cerebralis*) on North American salmonid populations is a case in point and also shows the circuitous route that invasions can take. Whirling disease is native to Europe where it is carried by Brown Trout, although the trout are resistant to the disease. The parasite causes inflammation of the cerebrospinal fluid, deforms the brain stem, causes other nerve damage, and results first in swimming in circles and then death of infected,

nonresistant trout (Helfman 2007; Vitule et al. 2009). Rainbow Trout were introduced into Europe in the nineteenth century and acquired whirling disease from the native Brown Trout. In 1956, when infected, frozen Rainbow Trout were shipped back to the United States, part of the shipment ended up in a Pennsylvania fish hatchery where it resulted in infection of trout fingerlings. The fingerlings were disseminated to various regions of the United States, and in 1965, whirling disease was first detected west of the Mississippi River in both Nevada and California. Especially in the western United States, whirling disease has decimated many trout populations, including populations of imperiled species. The spread of the disease is most likely due to the transfer of infected fish (Bergersen and Anderson 1997; Helfman 2007; Vitule et al. 2009). As a final twist, the inadvertent exportation of infected hatchery fish, as well as frozen fish, has sent the disease back to Europe—this time as a form pathogenic to Brown Trout (Helfman 2007).

AVENUES FOR RECOVERY?

Examples of adverse impacts of nonindigenous species on ecosystems abound, but economic drivers, such as aquaculture expansion, will continue to push for further introductions, especially aquaculture and aquarium species (Rahel 2007; Gozlan 2009). The rate of introduction of sport fishes has declined because the demands for sport-fish expansion have largely been met, along with the increased awareness of state and federal agencies of the potential risk of introductions, and the establishment of regulations by federal agencies and many state agencies controlling the release of nonindigenous species. The risk of establishment of fishes used in biological control methods has also declined because of the now common practice of using sterile fishes (although because sterilization is not 100% certain, a small risk remains) (Rahel 2007). Introductions from the release of ballast water are still a concern because of the level of international shipping, but there have been major advances in ways to control releases of organisms from ballast water using either port-based or ship-based systems (Crossman and Cudmore 1999; Bronte et al. 2003; Tsolaki and Diamadopoulos 2010). Other unauthorized introductions, including live-bait releases and the accidental release of species present as contaminants in authorized fish stocking (e.g., small sunfishes; Brook Stickleback, *Culaea inconstans*; and various catfishes) will no doubt continue to be a problem. Interbasin transfer of species via water transport systems and shipping canals also continues to be a problem. For instance, consider the current concern and efforts to stop the movement of Bighead (*Hypophthalmichthys nobilis*) and Silver (*H. molitrix*) carp from the Chicago Sanitary and Shipping Canal into Lake Michigan and beyond. Although originally constructed in 1900, the water quality of the canal was initially so poor that fishes could not survive. Paradoxically, only as water quality has drastically improved has the man-made linkage between the Mississippi River basin and the Great Lakes system become a potential avenue of dispersal (Rahel 2007). To counter this, an elaborate electric barrier has been constructed by the Army Corps of Engineers; however, in 2010 a Bighead Carp was captured north of the barrier and only six miles from Lake Michigan (Vroman 2011). Finally, there is also the real concern over the expansion of warm-water-species' ranges as a consequence of global warming (Stefan et al. 2001).

Preventing the introduction of nonindigenous species is clearly challenging and an ongoing issue, but prevention is almost always preferable to attempting the removal of such species once they are established. The overall success rate of removing unwanted fishes from small lentic waters is generally < 50%, and removal of nonindigenous fishes from large lakes is virtually impossible unless there is a vulnerable life-history stage (e.g., spawning migrations of Sea Lamprey into streams) (Meronek et al. 1996; Bronte et al. 2003). There are some moderate success stories for removal of nonindigenous fishes from small lakes. Common Carp removal has often been successful in small lakes where aggregations can be targeted by chemical removal methods (Weber and Brown 2009). Smallmouth Bass removal efforts over a 6-year period in a 271 ha lake in the Adirondacks resulted in a 90% population reduction based on underwater visual surveys. Native fish abundances increased coincident with Smallmouth Bass removal, although the removal also triggered a compensatory recruitment response in the remaining bass contributing to a shift in Smallmouth Bass size structure to smaller individuals (Weidel et al. 2007). Efforts to remove nonnative fishes, especially predators, from the Colorado River have proven difficult at best (Tyus and Saunders 2000). In the upper Colorado River, nonnative removal efforts exceeded $4 million by 2005, with estimated costs of $2–$6 per removal of each nonindigenous fish (Mueller 2005). Rigorous studies of positive effects from removal efforts are limited, although removal efforts generally are successful in reducing (but not eliminating) unwanted populations, but perhaps not on a scale required for native fish recovery. In some cases, removal efforts also are effective in shifting the size structure to smaller fishes. However, a shift to smaller predators could actually increase predatory impacts on early life stages of native fishes (Mueller 2005). In the lower Colorado River, in the Grand Canyon, a 4-year removal project captured 36,500 nonindigenous fishes, including Rainbow Trout, Fathead Minnow (*Pimephales promelas*), Common Carp, and Brown Trout. The percentage of nonindigenous fishes in the removal reach fell from > 95% in 2003 to < 50% in 2005; although declines of nonindigenous fishes also occurred in an upstream control reach as a consequence of reduced immigration of Rainbow Trout from upstream spawning areas. Declines in nonindigenous fishes were accompanied by some evidence of increases in native fish abundance in the removal reach (Coggins et al. 2011). Importantly, because nonindigenous species are rarely completely extirpated, ongoing, or at least periodic, removal efforts will generally be required (Tyus and Saunders 2000).

Although hopefully lessened in terms of rates of introduction, it is obvious that the problems of nonindigenous species and biotic homogenization are not going to disappear (Rahel 2007; Gozlan 2009). Ideally, nonindigenous organisms could be ranked by their potential risk so that limited resources for control could be focused on those with the greatest potential impact on species, communities, and ecosystems. Considerable work has been done on the characteristics of species that increase their likelihood of becoming established once they are introduced. Key findings are the similarity of the native habitat to the habitat into which species are introduced (i.e., lowered environmental resistance; see Chapter 5) and the trophic level of the nonindigenous species as discussed previously. Fishes at both low and high trophic levels seem to be more likely to become established than those in other trophic groups. Top predators are able to take advantage of naïve prey, and omnivores/detritivores are less apt to be resource limited. Also, rapidly reproducing generalists tend to be better invaders than specialists (Moyle and Light 1996a, b; Brown et al. 1999; Ruesink 2005). In spite of this, successful predictions of establishment are still limited and often specific to certain habitats or regions, although there are some promising advances (see Ruesink

2005). Prediction of potential risk is an even more challenging task. Although predators and ecosystem engineers are known to carry a high level of risk, in general our knowledge of ecological impacts is usually too limited to reliably predict high-risk species (Brown and Sax 2007; Gozlan 2009).

SUMMARY

In North America, natural lakes are all of fairly recent origin and their characteristics are influenced by their size, age, position in a watershed, underlying geological strata, and the level of connectivity with flowing water or other lakes. The strong effect of size (area and depth) is likely related to increased levels of habitat diversity in larger lakes. Connectivity is particularly important in influencing physical and biological characteristics, such as fish species composition, of lakes. This is especially true of oxbow lakes, which are prevalent in the eastern and southeastern United States.

The littoral zones of lakes and impoundments generally support greater fish diversity and abundance compared to pelagic or deep benthic zones. Especially in smaller bodies of water, nearshore and shoreline vegetation influences species occurrence and composition, but even in large lakes early life-history stages of many fish species use nearshore areas for development. In the North American Great Lakes, the pelagic and deep pelagic regions were occupied by a group of specialized salmonids, including Lake Trout and species of ciscoes. Sadly, many of these have been totally extirpated or have greatly reduced populations as a consequence of overfishing, the introduction of nonindigenous predators, and hybridization.

Losses of native fish species from North America began early in the nineteenth century as a consequence of stream alterations, damming, land use practices, and movement of fish taxa outside of their native ranges. The rate of loss and the introduction of nonindigenous species accelerated through the twentieth century. A consequence of species loss and the bypassing of natural barriers to dispersal by purposeful or accidental transport by humans is resulting in the biotic homogenization of the North American fish fauna. Regions typically share more species, and are thus more similar, compared to historic conditions. Homogenization not only occurs by increased taxonomic similarity but also at the genetic level through increased similarity among individuals in a taxon. Functional homogenization also is of concern as the number of functional groups (i.e., how species make a living) has been reduced. The creation of large impoundments, especially in areas where lentic ecosystems were rare or absent, has contributed to the homogenization of faunas. Reservoirs tend to offer similar habitats to fishes and other aquatic organisms and tend to be stocked with similar suites of species across broad geographical regions.

Nonindigenous species can impact native faunas in various ways and their impacts do not necessarily happen immediately. Impacts must be assessed at multiple scales, with the emphasis on the sustained ecological integrity of natural aquatic systems. Compared to other taxonomic groups, data on fishes suggest that the frequency of impacts are greater, with perhaps 40% or more of introductions resulting in negative impacts on native fish populations or aquatic ecosystems. The successful prediction of the occurrence and magnitude of impacts of nonindigenous species is still limited, so that any potential introduction should be considered to carry some risk. Preventing introductions of nonindigenous taxa is far better than attempting to remove unwanted species once they are established. Removal or control of nonindigenous species is difficult, time-consuming, and expensive, and likely will require continual or at least periodic control measures.

SUPPLEMENTAL READING

Angermeier, P. L. 1994. Does biodiversity include artificial diversity? *Conservation Biology* 8:600–602. Makes the important point that ecological integrity should be the goal of conservation efforts.

Deacon, J. E., and C. D. Williams. 1991. Ash Meadows and the legacy of the Devils Hole Pupfish, 69–87. In *Battle against extinction, native fish management in the American west.* W. L. Minckley and J. E. Deacon (eds.). The University of Arizona Press, Tucson. Describes the heroic efforts and dedication required to gain protection for endangered fishes and aquatic habitats in the Death Valley region.

Rahel, F. J. 2000. Homogenization of fish faunas across the United States. *Science* 288:854–56. A landmark paper on the homogenization of fish faunas.

Rahel, F. J. 2007. Biogeographic barriers, connectivity and homogenization of freshwater faunas: It's a small world after all. *Freshwater Biology* 52:696–710. The impact of human transportation systems, commerce, and culture on the bridging of natural barriers to dispersal.

Scott, M. C., and G. S. Helfman. 2001. Native invasions, homogenization, and the mismeasure of integrity of fish assemblages. *Fisheries* 26:6–15. Explains why the impacts of habitat homogenization cannot be assessed by determining alpha diversity and why coarser-scale studies can miss important signs of damage to fish assemblages within a river system.

Vitule, J. R. S., C. A. Freire, and D. Simberloff. 2009. Introduction of non-native freshwater fish can certainly be bad. *Fish and Fisheries* 10:98–108. An important documentation of the actual problems and threats posed by the introduction of nonindigenous species.

WEB SOURCES

Canada physiographic regions, Precambrian Shield. http://atlas.nrcan.gc.ca/sites/english/maps/reference/anniversary_maps/physiographic regions/map.pdf.

Global invasive species database. http://www.issg.org/database/welcome/.

International Nonindigenous Species Database Network. http://www.nisbase.org/nisbase/index.jsp.

GLOSSARY

ALEVIN Life-history stage characteristic of catfishes and trouts that is equivalent to yolk-sac larva.

ALLOGENIC ECOSYSTEM ENGINEERS Organisms that alter the environment by transforming living or nonliving materials from one physical state to another and are not necessarily part of the modified physical structure.

ALLOPATRIC SPECIATION Formation of species through geographic isolation.

ALLOZYMES Variant forms of an enzyme that vary in their amino acid sequences and that are coded by different alleles at the same genetic locus.

AMMOCOETE The larval stage of lampreys.

AMPHIDROMOUS A type of diadromy in which the adult feeding and growing habitat and the spawning habitat are in the same biome (e.g., fresh water or sea). In freshwater amphidromy, movement from fresh water to the sea occurs during the larval stage, followed by a return migration to fresh water by early juveniles.

ANADROMOUS Type of diadromous life-history pattern characterized by movement from the sea to fresh water for purposes of spawning. After spawning, the young fishes move back to the sea where most of their feeding and growth occurs.

AREA CLADOGRAM Essentially taking a diagram of evolutionary relationships of taxa (i.e., cladogram) and replacing the taxa with their geographical areas; a way of reconstructing ancestral

communities and determining causes of trait change.

AUTECOLOGY The study of how a single species interacts with its biotic and physical environments.

AUTOGENIC ECOSYSTEM ENGINEERS Organisms that transform their environment through endogenous processes, such as tree growth, and remain part of the altered environment.

BASIOCCIPITAL BONE The posterior-most bone of the neurocranium that contributes to the bottom and sides of the braincase and articulates with the first vertebra.

BET HEDGING Adaptations that minimize risk, such as reproducing over multiple years in a highly variable environment.

BUCCAL CAVITY Space enclosed by the cheeks and mouth and located anterior to the opercular chamber.

CATADROMOUS Type of diadromous life-history pattern characterized by movement from fresh water to the sea for purposes of spawning. After spawning, the young fishes move into fresh water, where most of their feeding and growth occurs.

CERATOBRANCHIAL BONES Paired bones forming part of the ventral gill arches.

CHARACTER DISPLACEMENT Enhanced differences between two species where they occur together compared to locations where they occur apart.

CLADOGRAM A branching diagram showing the evolutionary relationships among taxa and based on cladistic methodology.

CONGENERIC Referring to species in the same genus.

COPEPODS Small crustaceans (class Copepoda), characterized by a cylindrical, segmented body; two terminal processes on the abdomen; and a head with a single eye and five pairs of jointed appendages.

CRATON A continental nucleus; a stable part of the continental lithosphere.

CRITICAL THERMAL MAXIMUM (CTM) A measure of tolerance to elevated water temperature in which water temperatures are steadily raised over a time period that is short enough to keep the fish from physiologically adapting yet long enough so that the core body temperature is equal to the water temperature. The endpoint is usually judged as the point at which the fish loses equilibrium.

CRUSTACEA Subphylum (sometimes considered a class) of primarily aquatic arthropods that includes crabs, shrimp, lobsters, crayfishes, cladocerans, and copepods.

CRYPSIS Traits reducing the risk of detection of an organism by a potential prey or predator when it is potentially perceivable to an observer.

COEFFICIENT OF VARIATION (CV) A relative measure of variation within a data set, calculated by dividing the standard deviation by the mean and usually multiplied by 100.

DARWINIAN DEBT Refers to the expectation that evolutionary recovery of a population from harmful genetic changes caused by various anthropogenic impacts takes longer than the amount of time required to cause the changes; it is a debt because the results of anthropogenic selection may result in evolutionary costs for future generations of the population.

DEME A local population united by interbreeding.

DEMERSAL Referring to organisms or objects that have negative buoyancy and occur on the bottom of water bodies.

DETRITUS Organic debris, such as accumulations of leaf particles, on a stream or lake bottom.

DIADROMOUS Life-history pattern characterized by periodic, physiologically mediated movement between the sea and fresh water; includes amphidromous, anadromous, and catadromous patterns.

DIATOMS Unicellular or colonial algae of the phylum Chrysophyta.

DYNAMIC LIFT Hydrodynamic lift; the lift generated by the movement of water over a hydrofoil.

ENDEMIC Occurring in a restricted area and nowhere else.

ENVELOPE OF DETECTION Distance away from a prey that a predator can locate it; its radius equals maximum visibility.

EPAXIAL Elements originating from above the axis of the notochord or vertebral column.

EPILIMNION In a thermally stratified water body, the warm, mixed layer of water located above the thermocline.

EPINEURAL Above the spinal cord; the dorsal surface of the vertebral centrum.

EPIPLEURAL Above the pleural ribs (the ribs above the body cavity).

ETHMOID BONES A series of small paired and median bones located anteriorly on the skull above the upper jaw.

EXAPTATION A character that enhances fitness but was not built by natural selection for its current role.

FITNESS The genetic contribution of an organism to subsequent generations.

FRY General term for a young fish at the age when the yolk sac has been absorbed and the fish is actively feeding. In salmonids the fry stage follows the alevin stage.

GENUS (PL. GENERA) Taxonomic category below the level of family that contains one or more species.

GEOGRAPHIC INFORMATION SYSTEMS (GIS) A system of computer programs that stores, edits, analyzes, and presents data that can be linked to a location, such as species locations, elevation, rainfall, soil types, water bodies, human population densities, and so on.

GNATHOSTOMES Referring to the Gnathostomata, the jawed vertebrates.

GONOCHORISTIC Having separate sexes.

HETEROTROPHY Obtaining nutrients by the consumption of organic matter.

HYDRAULIC RETENTION Measure of relative complexity of a stream reach as determined by the time required for a dye to travel a specified distance through the reach.

HYOID APPARATUS Bony elements supporting the tongue and gill rakers and including the branchiostegal rays.

HYPAXIAL Elements originating from below the axis of the notochord or vertebral column.

HYPOLIMNION In a thermally stratified water body, the dense, cold layer of water located below the thermocline.

ICHTHYOLOGY Scientific study of fishes.

IMPOUNDMENT A body of water formed by damming a stream either by natural means (e.g., Beaver dams or landslides) or by human efforts (e.g., a reservoir).

INCLUSIVE FITNESS A measure of fitness that includes an individual's own genetic fitness as well as its influence on the fitness of its kin.

INDUCIBLE DEFENSES Phenotypic responses by prey organisms induced by exposure to predators or to chemical cues from predators.

INFERIOR MOUTH The mouth is directed downward and the upper jaw and snout project beyond the lower jaw when the mouth is closed.

INSOLATION A measure of the amount of sunlight reaching the surface of an area.

INTRODUCED Used in reference to a nonnative species.

INVERTIVORES Predators of invertebrates.

IPSILATERAL On the same side of the body.

ITEROPARITY Breeding multiple times throughout the reproductive life span.

LACUSTRINE Referring to a lake or lake-like habitat.

LARVA (PL. LARVAE) The immature stage of fishes (or other taxa) prior to metamorphosis or transformation into the juvenile or adult form.

LEPTOKURTIC Frequency distribution in which there are more low values, fewer intermediate values, and more high values than would be predicted from a normal distribution.

MESOLARVA The second larval period characterized by the initial development of principal rays in the median fins.

METACENTRIC HEIGHT The difference between the center of mass and the center of buoyancy.

METALARVA The third larval period characterized by the formation of pelvic fin buds (when pelvic fins are present) and lasting until the fin-fold is completely resorbed.

MHC GENES Genes associated with the major histocompatability complex (MHC) that are important in the immune system of vertebrates.

MYOSEPTA Sheets of connective tissue that separate blocks of axial muscles (myomeres) and onto which the muscle fibers insert.

NEUROCRANIUM The fairly rigid bony "box" surrounding the brain.

OOCYTE The developing egg cell in the female's ovary that is formed from oogonia and is initially surrounded by follicular cells until ovulation.

OOGONIA Female reproductive cells that arise from primordial sex cells in or near the (peritoneal) germinal epithelium and are the precursor to oocytes.

PALATAL ORGAN An expansive patch of tissue on the roof of the pharynx containing numerous taste buds; in some suspension-feeding fishes it is papillate and covered in mucous.

PERIPHYTON A composite organic film found on aquatic substrata made up of algae, cyanobacteria, organic detritus, and other microbes.

PARR The third juvenile stage of salmonids, following the alevin and fry stages, that is characterized by species-specific patterns of vertical bars.

PHARYNGEAL JAWS Modified gill-arch elements, often bearing tooth pads, that are used in processing prey.

PHYSOSTOMES Fishes having a connection between the swimbladder and the mouth through a pneumatic duct.

POTAMODROMOUS Regular migration occurring wholly within fresh water.

PROSTAGLANDINS A group of unsaturated, polyoxygenated fatty acids; F prostaglandins are important in fish ovulation, the initiation of female sexual behavior, and also as a pheromone that attracts males to females.

PROTOCERCAL CAUDAL FIN A relatively undifferentiated caudal fin typical of lampreys and the larvae of most modern fishes.

PROTOLARVAE The initial larval period following hatching or parturition and which is characterized by the absence of distinct rays or spines in the median fins.

REALIZED NICHE The postinteractive niche, or the portion of the multidimensional niche that is actually occupied by a species, in contrast to the fundamental niche, which includes all areas of multidimensional space that a species might potentially occupy over all life stages.

REDD The gravel spawning depression excavated by salmonids.

SEMELPARITY Breeding only once in a lifetime.

SETIFORM TEETH Small, sometimes sharp, and usually very numerous teeth that are shaped like bristles or setae.

SMOLT The silvery, juvenile stage of anadromous salmonids that follows the freshwater-restricted parr stage and occurs prior to transitioning from fresh to salt water.

SPATULATE TEETH Rounded and flattened distally with a narrow shaft, much in the shape of a spatula.

STANDARD LENGTH (SL) Measured as the straight-line distance from the most anterior part of the snout to the base of the caudal fin (the hypural plate).

STATIC LIFT Buoyant lift, the lift produced by the inclusion in the body of low-density lipids or air bladders.

SUPERIOR MOUTH The mouth is directed upward and the lower jaw projects beyond the upper jaw when the mouth is closed.

SUSPENSORIUM A series of bones extending from the neurocranium, which forms the lateral wall of the mouth cavity and supports the jaw joint. It includes the quadrate, hyomandibula, pterygoids, symplectic, and preopercle bones.

SYMPATRIC Occurring within the normal range of one or more additional populations or species.

SYNECOLOGY The study of two or more species interacting in a community.

SYSTEMATICS The study of the evolutionary relationships among organisms.

TAXONOMY The branch of systematic biology concerned with the recognition, description, and hierarchical placement of taxa.

TERETE A streamlined body shape; cylindrical and tapering anteriorly and posteriorly.

TERMINAL MOUTH The mouth is directed forward with the upper and lower jaws approximately even.

THERMOCLINE In a thermally stratified body of water, the boundary between the upper (mixed) layer of warm water and the cold, dense (unmixed) layer of water.

TERRANE A discrete, fault-bounded crustal element that is added to a craton through plate movement.

VAGILITY Ability of an organism to move or disperse in an environment.

VISCOSITY Resistance of a fluid to deformation because of internal friction.

WEBERIAN APPARATUS Series of bony ossicles, formed from modified anterior vertebrae, that connect the swimbladder to the inner ear in the Otophysi (a large group including minnows, suckers, characins, and catfishes).

YOY Abbreviation of young-of-year in reference to fishes in their first calendar year of life (i.e., before January 1 of the year following hatching).

LITERATURE CITED

Aberhan, M. 1999. Terrane history of the Canadian Cordillera: Estimating amounts of latitudinal displacement and rotation of Wrangellia and Stikinia. *Geological Magazine* 136:481–92.

Able, K. W. 1976. Cleaning behavior in the cyprinodontid fishes: *Fundulus majalis, Cyprinodon variegatus,* and *Lucania parva. Chesapeake Science* 17:35–39.

Adams, S. B., and M. L. Warren, Jr. 2005. Recolonization by warmwater fishes and crayfishes after severe drought in upper coastal plain hill streams. *Transactions of the American Fisheries Society* 134:1173–92.

Adams, S. R., G. L. Adams, and J. J. Hoover. 2003. Oral grasping: A distinctive behavior of cyprinids for maintaining station in flowing water. *Copeia* 2003:851–57.

Adams, S. R., J. J. Hoover, and K. J. Killgore. 2000. Swimming performance of the Topeka shiner (*Notropis topeka*) an endangered midwestern minnow. *American Midland Naturalist* 144:178–86.

Albanese, B., P. L. Angermeier, and S. Dorai-Raj. 2004. Ecological correlates of fish movement in a network of Virginia streams. *Canadian Journal of Fisheries and Aquatic Sciences* 61:857–69.

Albanese, B. W., P. L. Angermeier, and C. Gowan. 2003. Designing mark-recapture studies to reduce effects of distance weighting on movement distance distributions of stream fishes. *Transactions of the American Fisheries Society* 132:925–39.

Albanese, B. W., P. L. Angermeier, and J. T. Peterson. 2009. Does mobility explain variation in colonization and population recovery among stream fishes? *Freshwater Biology* 54:1444–60.

Aleev, Y. G. 1969. *Function and gross morphology in fish.* Translated from Russian. Israel Program for Scientific Translations, Jerusalem.

Alexander, R. McN. 1966. The functions and mechanisms of the protrusible upper jaws of two species of cyprinid fish. *Journal of Zoology, London* 149:288–96.

———. 1967a. *Functional design in fishes.* Hutchinson University Library, London, United Kingdom.

———. 1967b. The functions and mechanisms of the protrusible upper jaws of some acanthopterygian fish. *Journal of Zoology, London* 151:43–64.

———. 1967c. Mechanisms of the jaws of some atheriniform fish. *Journal of Zoology, London* 151:233–55.

———. 1969. Mechanics of the feeding action of a cyprinid fish. *Journal of Zoology, London* 159:1–15.

Alfaro, M. E., J. Janovetz, and M. W. Westneat. 2001. Motor control across trophic strategies: Muscle activity of biting and suction feeding fishes. *American Zoologist* 41:1266–79.

Alford, J. B., and D. C. Beckett. 2006. Dietary specialization by the Speckled Darter, *Etheostoma stigmaeum,* on chironomid larvae in a Mississippi stream. *Journal of Freshwater Ecology* 21:543–51.

———. 2007. Selective predation by four darter (Percidae) species on larval chironomids (Diptera) from a Mississippi stream. *Environmental Biology of Fishes* 78:353–64.

Ali, M., and R. J. Wootton. 1999. Effect of variable food levels on reproductive performance of breeding female Three-Spined Sticklebacks. *Journal of Fish Biology* 55:1040–53.

Allan, J. R. 1986. The influence of species composition on behaviour in mixed-species cyprinid shoals. *Journal of Fish Biology* 29:97–106.

Allendorf, F. W, P. R. England, G. Luikart, P. A. Ritchie, and N. Ryman. 2008. Genetic effects of harvest on wild animal populations. *Trends in Ecology and Evolution* 23:327–37.

Alò, D., and T. F. Turner. 2005. Effects of habitat fragmentation on effective population size in the endangered Rio Grande Silvery Minnow. *Conservation Biology* 19:1138–48.

Amarasinghe, U. S., and R. L. Welcomme. 2002. An analysis of fish species richness in natural lakes. *Environmental Biology of Fishes* 65:327–39.

Ames, B. N. 1983. Dietary carcinogens and anticarcinogens. Oxygen radicals and degenerative diseases. *Science* 221:1256–64.

Amorim, M. C. P. 2006. Diversity of sound production in fish, 71–104. In *Communication in fishes*. Vol. 1. F. Ladich, S. P. Collin, P. Moller, and B. G. Kapoor (eds.). Science Publishers, Enfield, New Hampshire.

Andrew, J. H., N. Jonsson, B. Jonsson, K. Hindar, and T. G. Northcote. 1992. Changes in use of lake habitat by experimentally segregated populations of Cutthroat Trout and Dolly Varden Char. *Ecography* 15:245–52.

Angermeier, P. L. 1992. Predation by Rock Bass on other stream fishes: Experimental effects of depth and cover. *Environmental Biology of Fishes* 34:171–80.

———. 1994. Does biodiversity include artificial diversity? *Conservation Biology* 8:600–602.

Angermeier, P. L., and J. R. Karr. 1994. Biological integrity versus biological diversity as policy directives. *BioScience* 44:690–97.

Angermeier, P. L., and M. R. Winston. 1998. Local vs. regional influences on local diversity in stream fish communities. *Ecology* 79:911–27.

Angilletta, M. J., Jr., E. A. Steel, K. K. Bartz, J. G. Kingsolver, M. D. Scheuerell, B. R. Beckman, and L. G. Crozier. 2008. Big dams and salmon evolution: Changes in thermal regimes and their potential evolutionary consequences. *Evolutionary Applications* 1:286–99.

Anker, G. Ch. 1974. Morphology and kinetics of the head of the stickleback, *Gasterosteus aculeatus*. *Transactions of the Zoological Society of London* 32:311–416.

Annett, C. A. 1998. Hunting behavior of Florida Largemouth Bass, *Micropterus salmoides floridanus*, in a channelized river. *Environmental Biology of Fishes* 53:75–87.

Araki, H., B. Cooper, and M. S. Blouin. 2007. Genetic effects of captive breeding cause a rapid, cumulative fitness decline in the wild. *Science* 318:100–103.

Archard, G. A., I. C. Cuthill, J. C. Partridge, and C. van Oosterhout. 2008. Female Guppies (*Poecilia reticulata*) show no preference for conspecific chemosensory cues in the field or an artifical flow chamber. *Behaviour* 145:1329–46.

Armbruster, J. W., and L. M. Page. 1996. Convergence of a cryptic saddle pattern in benthic freshwater fishes. *Environmental Biology of Fishes* 45:249–57.

Arnold, G. P., P. W. Webb, and B. H. Holford. 1991. The role of the pectoral fins in station-holding of Atlantic Salmon parr (*Salmo salar* L.). *Journal of Experimental Biology* 156:625–29.

Arnold, G. P., and D. Weihs. 1978. The hydrodynamics of rheotaxis in the Plaice (*Pleuronectes platessa* L.). *Journal of Experimental Biology* 75:147–69.

Auer, N. A., and E. A. Baker. 2002. Duration and drift of larval lake sturgeon in the Sturgeon River, Michigan. *Journal of Applied Ichthyology* 18:557–64.

Avila, V. L. 1976. A field study of nesting behavior of male Bluegill sunfish (*Lepomis macrochirus* Rafinesque). *American Midland Naturalist* 96:159–206.

Avise, J. C. 2004. *Molecular markers, natural history, and evolution*. 2nd ed. Sinauer Associates, Inc. Publisher, Sunderland, Maryland.

Avise, J. C., A. G. Jones, D. Walker, J. A. DeWoody, et seq. 2002. Genetic mating systems and reproductive natural histories of fishes: Lessons for ecology and evolution. *Annual Review of Genetics* 36:19–45.

Avise, J. C., and J. E. Mank. 2009. Evolutionary perspectives on hermaphroditism in fishes. *Sexual Development* 3:152–63.

Avise, J. C., J. M. Quattro, and R. J. Vrijenhoek. 1992. Molecular clones within organismal clones: Mitochondrial DNA phylogenies and the evolutionary histories of unisexual vertebrates. *Evolutionary Biology* 26:225–46.

Babbit, B. 2002. What goes up, may come down. *BioScience* 52:656–58.

Bagamian, K. H., D. C. Heins, and J. A. Baker. 2004. Body condition and reproductive capacity of Three-spined Stickleback infected with the cestode *Schistocephalus solidus*. *Journal of Fish Biology* 64:1568–76.

Bain, M. B., and L. A. Helfrich. 1983. Role of male parental care in survival of larval Bluegills.

Transactions of the American Fisheries Society 112:47–52.

Bain, M. B., J. T. Finn, and H. E. Booke. 1988. Streamflow regulation and fish community structure. *Ecology* 69:382–92.

Baker, J. A., W. A. Cresko, S. A. Foster, and D. C. Heins. 2005. Life-history differentiation of benthic and limnetic ecotypes in a polytypic population of Threespine Stickleback (*Gasterosteus aculeatus*). *Evolutionary Ecology Research* 7:121–31.

Baker, J. A., and D. C. Heins. 1994. Reproductive life history of the North American madtom catfish *Noturus hildebrandi* (Bailey and Taylor 1950), with a review of data for the genus. *Ecology of Freshwater Fishes* 3:167–75.

Baker, J. A., D. C. Heins, S. A. Foster, and R. W. King. 2008. An overview of life-history variation in female Threespine Stickleback. *Behaviour* 145:579–602.

Baker, J. A., K. J. Killgore, and R. L. Kasul. 1991. Aquatic habitats and fish communities in the lower Mississippi River. *Reviews in Aquatic Sciences* 3:313–56.

Baker, J. A., and S. T. Ross. 1981. Spatial and temporal resource utilization by southeastern cyprinids. *Copeia* 1981:178–89.

Ballantyne, P. K., and P. W. Colgan. 1978. Sound production during agonistic and reproductive behaviour in the Pumpkinseed (*Lepomis gibbosus*), the Bluegill (*L. macrochirus*), and their hybrid sunfish. I. Context. *Biology of Behaviour* 3:113–35.

Balon, E. K. 1975. Reproductive guilds of fishes: A proposal and definition. *Journal of the Fisheries Research Board of Canada* 32:821–64.

———. 1981. Additions and amendments to the classification of reproductive styles in fishes. *Environmental Biology of Fishes* 6:377–89.

———. 1986. Types of feeding in the ontogeny of fishes and the life-history model. *Environmental Biology of Fishes* 16:11–24.

Barber, I., P. Walker, and P. A. Svensson. 2004. Behavioural responses to simulated avian predation in Threespined Sticklebacks: The effect of experimental *Schistocephalus solidus* infections. *Behaviour* 141:1425–40.

Barbour, C. D. 1973. A biogeographical history of *Chirostoma* (Pisces: Atherinidae): A species flock from the Mexican Plateau. *Copeia* 1973:533–56.

Barbour, C. D., and J. H. Brown. 1974. Fish species diversity in lakes. *The American Naturalist* 108:473–89.

Bardack, D., and R. Zangerl. 1968. First fossil lamprey: A record from the Pennsylvanian of Illinois. *Science* 162:1265–67.

Baron, J. S., N. L. Poff, P. L. Angermeier, C. N. Dahm, P. H. Gleick, N. G. Hairston, Jr., R. B. Jackson, C. A. Johnston, B. D. Richter, and A. D Steinman. 2002. Meeting ecological and societal needs for freshwater. *Ecological Applications* 12:1247–60.

Bart, H. L., and L. M. Page. 1991. Morphology and adaptive significance of fin knobs in egg-clustering darters. *Copeia* 1991:80–86.

Baxter, R. M. 1977. Environmental effects of dams and impoundments. *Annual Review of Ecology and Systematics* 8:255–83.

Baylis, J. R. 1981. The evolution of parental care in fishes, with reference to Darwin's rule of male sexual selection. *Environmental Biology of Fishes* 6:223–51.

———. 1982. Unusual escape response by two cyprinodontiform fishes, and a Bluegill predator's counter-strategy. *Copeia* 1982:455–57.

Beaudoin, C. P., W. M. Tonn, E. E. Prepas, and L. I. Wassenaar. 1999. Individual specialization and trophic adaptability of Northern Pike (*Esox lucius*): An isotope and dietary analysis. *Oecologia* 120:386–96.

Bednarek, A. T. 2001. Undamming rivers: A review of the ecological impacts of dam removal. *Environmental Management* 27:803–14.

Begay, R. 2001. Doo dilzin da: Abuse of the natural world. *American Indian Quarterly* 25:21–27.

Begon, M., J. L. Harper, and C. R. Townsend. 1996. *Ecology, individuals, populations and communities.* 3rd ed. Blackwell Science, Cambridge, Massachusetts.

Beisner B. E., P. R. Peres-Neto, E. S. Lindström, A. Barnett, and M. L. Longhi. 2006. The role of environmental and spatial processes in structuring lake communities from bacteria to fish. *Ecology* 87:2985–91.

Bell, M. A., W. E. Aguirre, and N. J. Buck. 2004. Twelve years of contemporary armor evolution in a Threespine Stickleback population. *Evolution* 58:814–24.

Bell, M. A., and S. A. Foster. 1994. Introduction to the evolutionary biology of the Threespine Stickleback, 1–27. In *The evolutionary biology of the Threespine Stickleback.* M. A. Bell and S. A. Foster (eds.). Oxford University Press, Oxford, United Kingdom.

Belsky, A. J., A. Matzke, and S. Uselman. 1999. Survey of livestock influences on stream and riparian ecosystems in the western United States. *Journal of Soil and Water Conservation* 54:419–31.

Belt, C. B., Jr. 1975. The 1973 flood and man's constriction of the Mississippi River. *Science* 189:681–84.

Belyea, L. R., and J. Lancaster. 1999. Assembly rules within a contingent ecology. *Oikos* 86:402–16.

Bemis, W. E., E. K. Findeis, and L. Grande. 1997. An overview of Acipenseriformes. *Environmental Biology of Fishes* 48:25–71.

Bemis, W. E., and B. Kynard. 1997. Sturgeon rivers: An introduction to acipenseriform biogeography and life history. *Environmental Biology of Fishes* 48:167–83.

Bender, E. A., T. J. Case, and M. E. Gilpin. 1984. Perturbation experiments in community ecology: Theory and practice. *Ecology* 65:1–13.

Benke, A. C., and C. E. Cushing (eds.). 2005. *Rivers of North America*. Elsevier Academic Press, Boston, Massachusetts.

Bennett, W. A., and T. L. Beitinger. 1997. Temperature tolerance of the Sheepshead Minnow, *Cyprinodon variegatus*. *Copeia* 1997:77–87.

Benson, B. J., and J. J. Magnuson. 1992. Spatial heterogeneity of littoral fish assemblages in lakes: Relation to species diversity and habitat structure. *Canadian Journal of Fisheries and Aquatic Sciences* 49:1493–500.

Bentzen P., and J. D. McPhail. 1984. Ecology and evolution of sympatric sticklebacks (*Gasterosteus*): Specialization for alternative trophic niches in the Enos Lake species pair. *Canadian Journal of Zoology* 62:2280–86.

Berendzen, P. B., A. M. Simons, and R. M. Wood. 2003. Phylogeography of the Northern Hogsucker, *Hypentelium nigricans* (Teleostei: Cypriniformes): Genetic evidence for the existence of the ancient Teays River. *Journal of Biogeography* 30:1139–52.

Bergersen, E. P., and D. E. Anderson. 1997. The distribution and spread of *Myxobolus cerebralis* in the United States. *Fisheries* 22:6–7.

Bergstedt, L. C., and E. P. Bergersen. 1997. Health and movements of fish in response to sediment sluicing in the Wind River, Wyoming. *Canadian Journal of Fisheries and Aquatic Sciences* 54:312–19.

Bernardi, G. 1997. Molecular phylogeny of the Fundulidae (Teleostei, Cyprinodontiformes) based on the cytochrome b gene, 189–97. In *Molecular systematics of fishes*. T. D. Kocher and C. A. Stepien (eds.). Academic Press, San Diego, California.

Berra, T. M. 2001. *Freshwater fish distribution*. Academic Press, San Diego, California.

Bertschy, K. A., and M. G. Fox. 1999. The influence of age-specific survivorship on Pumpkinseed sunfish life histories. *Ecology* 80:2299–313.

Bestgen, K. R. 2008. Effects of water temperature on growth of Razorback Sucker larvae. *Western North American Naturalist* 68:15–20.

Bestgen, K. R., D. W. Beyers, J. A. Rice, and G. B. Haines. 2006. Factors affecting recruitment of young Colorado Pikeminnow: Synthesis of predation experiments, field studies, and individual-based modeling. *Transactions of the American Fisheries Society* 135:1722–42.

Bestgen, K. R., and S. P. Platania. 1990. Extirpation of *Notropis simus simus* (Cope) and *Notropis orca* Woolman (Pisces: Cyprinidae) from the Rio Grande in New Mexico, with notes on their life history. *Occasional Papers of the Museum of Southwestern Biology* 6:1–8.

Biehl, C. C., and W. J. Matthews. 1984. Small fish community structure in Ozark streams: Improvements in the statistical analysis of presence-absence data. *American Midland Naturalist* 111:371–82.

Bietz, B. F. 1981. Habitat availability, social attraction and nest patterns in Longear Sunfish (*Lepomis megalotis peltastes*). *Environmental Biology of Fishes* 6:193–200.

Bilby, R. E., E. W. Beach, B. R. Fransen, J. K. Walter, and P. A. Bisson. 2003. Transfer of nutrients from spawning salmon to riparian vegetation in western Washington. *Transactions of the American Fisheries Society* 132:733–45.

Bilkovic, D. M., J. E. Olney, and C. H. Hershner. 2002. Spawning of American Shad (*Alosa sapidissima*) and Striped Bass (*Morone saxatilis*) in the Mattaponi and Pamunkey rivers, Virginia. *Fishery Bulletin* 100:632–40.

Billington, D. P., and D. C. Jackson. 2006. *Big dams and the New Deal era: A confluence of engineering and politics*. University of Oklahoma Press, Norman.

Bjerselius, R., W. Li, J. H. Teeter, J. G. Seelye, P. B. Johnsen, P. J. Maniak, G. C. Grant, C. N. Polkinghorne, and P. W. Sorensen. 2000. Direct behavioral evidence that unique bile acids released by larval Sea Lamprey (*Petromyzon marinus*) function as a migratory pheromone. *Canadian Journal of Fisheries and Aquatic Sciences* 57:557–69.

Bjornn, T. C., D. R. Craddock, and D. R. Corley. 1968. Migration and survival of Redfish Lake, Idaho, Sockeye Salmon, *Oncorhynchus nerka*. *Transactions of the American Fisheries Society* 97:360–73.

Blais, J., C. Rico, and L. Bernatchez. 2004. Nonlinear effects of female mate choice in wild Threespine Sticklebacks. *Evolution* 58:2498–510.

Blake, R. W. 1980. Undulatory median fin propulsion of two teleosts with different modes of life. *Canadian Journal of Zoology* 58:2116–19.

———. 1983a. *Fish locomotion*. Cambridge University Press, United Kingdom.

———. 1983b. Functional design and burst-and-coast swimming in fishes. *Canadian Journal of Zoology* 61:2491–94.

———. 2004. Fish functional design and swimming performance. *Journal of Fish Biology* 65:1193–222.

Blanchfield, P. J., and M. S. Ridgway. 1999. The cost of peripheral males in a Brook Trout mating system. *Animal Behaviour* 57:537–44.

Blinn, D. W., and N. L. Poff. 2005. Colorado River Basin, 483–538. In *Rivers of North America*. A. C. Benke and C. E. Cushing (eds.). Elsevier Academic Press, San Diego, California.

Blum, M. J., D. A. Neely, P. M. Harris, and R. L. Mayden. 2008. Molecular systematics of the cyprinid genus *Campostoma* (Actinopterygii: Cypriniformes): Disassociation between morphological and mitochondrial differentiation. *Copeia* 2008:360–69.

Blumer, L. S. 1979. Male parental care in the bony fishes. *The Quarterly Review of Biology* 54:149–61.

———. 1982. A bibliography and categorization of bony fishes exhibiting parental care. *Zoological Journal of the Linnean Society* 76:1–22.

Bolnick, D. I. 2004. Can intraspecific competition drive disruptive selection? An experimental test in natural populations of Sticklebacks. *Evolution* 58:608–18.

Bond, C. E. 1996. *Biology of fishes*. 2nd ed. Saunders College Publishing, Harcourt Brace College Publishing, Fort Worth, Texas.

Boschung, H. T., Jr., and R. L. Mayden. 2004. *Fishes of Alabama*. Smithsonian Books, Washington, DC.

Boucher, D. H., S. James, and K. H. Keeler. 1982. *The ecology of mutualism. Annual Review of Ecology and Systematics* 13:315–47.

Bowen, S. H. 1996. Quantitative description of the diet, 513–29. In *Fisheries techniques*. 2nd ed. B. R. Murphy and D. W. Willis (eds.). American Fisheries Society, Bethesda, Maryland.

Bowman, M. B. 2002. Legal perspectives on dam removal. *BioScience* 52:739–46.

Bradford, M. J., P. S. Higgins, J. Korman, and J. Sneep. 2011. Test of an environmental flow release in a British Columbia river: Does more water mean more fish? *Freshwater Biology* 56:2119–34.

Brainerd, E. L. 2001. Caught in the crossflow. *Nature* 412:387–88.

Brainerd, E. L., B. N. Page, and F. E. Fish. 1997. Opercular jetting during fast-starts by flatfishes. *Journal of Experimental Biology* 200:1179–88.

Bramblett, R. G., and K. D. Fausch. 1991. Variable fish communities and the index of biotic integrity. *Transactions of the American Fisheries Society* 120:452–769.

Brandenburg, W. H., and M. A. Farrington. 2011. *Colorado Pikeminnow and Razorback Sucker larval fish survey in the San Juan River during 2010*. The San Juan River Basin Recovery Implementation Program, U.S. Fish and Wildlife Service, Albuquerque, New Mexico.

Breder, C. M., Jr. 1926. The locomotion of fishes. *Zoologica* (New York) 4:159–297.

———. 1976. Fish schools as operational structures. *Fishery Bulletin* 74:471–502.

Brenkman, S. J., J. J. Duda, C. E. Torgersen, E. Welty, G. R. Pess, R. Peters, and M. L. Henry. 2012. A riverscape perspective of Pacific salmonids and aquatic habitats prior to large-scale dam removal in the Elwha River, Washington, USA. *Fisheries Management and Ecology* 19:36–53.

Brett, B. L. H., and D. J. Grosse 1982. A reproductive pheromone in the Mexican poeciliid fish *Poecilia chica*. *Copeia* 1982:219–23.

Brett, R. 1971. Energetic responses of salmon to temperature. A study of some thermal relations in the physiology and freshwater ecology of Sockeye Salmon (*Oncorhynchus nerka*). *American Zoologist* 11:99–113.

Brezner, J. 1958. Food habits of the Northern River Carpsucker in Missouri. *Progressive Fish Culturist* 20:491–94.

Briggs, J. C. 1974. Operation of zoogeographic barriers. *Systematic Zoology* 23:248–56.

———. 1986. Introduction to the zoogeography of North American fishes, 1–16. In *The zoogeography of North American freshwater fishes*. C. H. Hocutt and E. O. Wiley (eds.). John Wiley and Sons, New York, New York.

———. 1989. The historic biogeography of India: Isolation or contact? *Systematic Zoology* 38:322–32.

Briggs, J. C. (ed.). 1987. Biogeography and plate tectonics. *Developments in Palaeontology and Stratigraphy* 10.

———. 1995. Global biogeography. *Developments in Palaeontology and Stratigraphy* 14.

Britten, R. J. 1986. Rates of DNA sequence evolution differ between taxonomic groups. *Science* 231:1393–98.

Britz, R., K. W. Conway, and L. Rüber. 2009. Spectacular morphological novelty in a miniature cyprinid fish, *Danionella dracula* n. sp. *Proceedings of the Royal Society B* 276:2179–86.

Brönmark, C., and L-A. Hansson. 2000. Chemical communication in aquatic systems: An introduction. *Oikos* 88:103–9.

Bronte, C. R., M. P. Ebener, D. R. Schreiner, D. S. DeVault, M. M. Petzold, D. A Jensen, C. Richards, and S. J. Lozano. 2003. Fish community change in Lake Superior, 1970–2000. *Canadian Journal of Fisheries and Aquatic Sciences* 60:1552–74.

Brooks, D. R., and D. A. McLennan. 1991. *Phylogeny, ecology, and behavior, a research program in comparative biology.* University of Chicago Press, Chicago, Illinois.

Brooks, J. L. 1968. The effects of prey size selection by lake planktivores. *Systematic Zoology* 17:273–91.

Brooks, J. L., and S. I. Dodson. 1965. Predation, body size, and composition of plankton. *Science* 150:28–35.

Brower, J. E., J. H. Zar, and C. N. von Ende. 1998. *Field and laboratory methods for general ecology.* 4th ed. WCB, McGraw-Hill, Boston, Massachusetts.

Browman, H. I., and W. J. O'Brien. 1992. Foraging and prey search behaviour of Golden Shiner (*Notemigonus crysoleucas*) larvae. *Canadian Journal of Fisheries and Aquatic Sciences* 49:813–19.

Brown, A. V., and M. L. Armstrong. 1985. Propensity to drift downstream among various species of fish. *Journal of Freshwater Ecology* 3:3–17.

Brown, G. E., and J. A. Brown. 1996. Kin discrimination in salmonids. *Reviews in Fish Biology and Fisheries* 6:201–19.

Brown, G. E., D. P. Chivers, and R. J. F. Smith. 1995. Localized defecation by pike: A response to labelling by cyprinid alarm pheromone? *Behavioral Ecology and Sociobiology* 36:105–10.

Brown, J. A., and P. W. Colgan. 1986. Individual and species recognition in centrarchid fishes: Evidence and hypotheses. *Behavioral Ecology* 19:5:373–79.

Brown, J. H. 1995. *Macroecology.* University of Chicago Press, Chicago, Illinois.

Brown, J. H., and C. R. Feldmeth. 1971. Evolution in constant and fluctuating environments: Thermal tolerances of desert pupfish (*Cyprinodon*). *Evolution* 25:390–98.

Brown, J. H., B. J. Fox, and D. A. Kelt. 2000. Assembly rules: Desert rodent communities are structured at scales from local to continental. *The American Naturalist* 156:314–21.

Brown, J. H., D. A. Kelt, and B. J. Fox. 2002. Assembly rules and competition in desert rodents. *The American Naturalist* 160:815–18.

Brown, J. H., and D. F. Sax. 2007. Do biological invasions decrease biodiversity? *Conservation Magazine* 8:16–17.

Brown, L. R., A. M. Brasher, B. C. Harvey, and M. Matthews. 1999. Success and failure of nonindigenous aquatic species in stream ecosystems: Case studies from California and Hawaii, 415–30. In

Nonindigenous freshwater organisms, vectors, biology, and impacts. R. Claudi and J. H. Leach (eds.). Lewis Publishers, Boca Raton, Florida.

Brown, L. R., and P. B. Moyle. 1997. Invading species in the Eel River, California: Successes, failures, and relationships with resident species. *Environmental Biology of Fishes* 49:271–91.

Brown, W. L., Jr., and E. O. Wilson. 1956. Character displacement. *Systematic Zoology* 5:49–64.

Brown-Peterson, N. J., and D. C. Heins. 2009. Interspawning interval of wild female Three-Spined Stickleback *Gasterosteus aculeatus* in Alaska. *Journal of Fish Biology* 74:2299–312.

Brown-Peterson, N. J., S. Lowerre-Barbieri, B. Macewicz, F. Saborido-Rey, J. Tomkiewicz, and D. Wyanski. 2007. *An improved and simplified terminology for reproductive classification in fishes.* http://digital.csic.es/handle/10261/11844.

Bruno, J. F., J. J. Stachowicz, and M. D. Bertness. 2003. Inclusion of facilitation into ecological theory. *Trends in Ecology and Evolution* 18:119–25.

Bryan, M. D., and D. L. Scarnecchia. 1992. Species richness, composition, and abundance of fish larvae and juveniles inhabiting natural and developed shorelines of a glacial Iowa lake. *Environmental Biology of Fishes* 35:329–41.

Buckland, J. 1880. *Natural history of British fishes.* Unwin, London, United Kingdom (not seen; cited in Hara, T. J. 1993. Role of olfaction in fish behaviour, 173–99. In *Behaviour of teleost fishes.* 2nd ed. Chapman and Hall, New York, New York).

Bunn, S. E., and A. H. Arthington. 2002. Basic principles and ecological consequences of altered flow regimes for aquatic biodiversity. *Environmental Management* 30:492–507.

Burgess, G. H. 1980. *Trinectes maculatus* (Block and Schneider), Hogchoker, 831. In *Atlas of North American freshwater fishes.* D. S. Lee, C. R. Gilbert, C. H. Hocutt, R. E. Jenkins, D. E. McAllister, and J. R. Stauffer, Jr., et al. 1980. North Carolina State Museum of Natural History, Raleigh.

Burggren, W. W., and W. E. Bemis. 1992. Metabolism and ram gill ventilation in juvenile Paddlefish, *Polyodon spathula* (Chondrostei: Polyodontidae). *Physiological Zoology* 65:515–39.

Burnard, D., R. E. Gozlan, and S. W. Griffiths. 2008. The role of pheromones in freshwater fishes. *Journal of Fish Biology* 73:1–16.

Burr, B. M., and R. L. Mayden. 1992. Phylogenetics and North American freshwater fishes, 18–75. In *Systematics, historical ecology, and North American freshwater fishes.* R. L. Mayden (ed.). Stanford University Press, Stanford, California.

Burt, A., D. L. Kramer, K. Nakatsuru, and C. Spry. 1988. The tempo of reproduction in *Hyphessobrycon pulchripinnis* (Characidae), with a discussion on the biology of "multiple spawning" in fishes. *Environmental Biology of Fishes* 22:15–27.

Butler, D. R., and G. P. Malanson. 2005. The geomorphic influences of beaver dams and failures of beaver dams. *Geomorphology* 71:48–60.

Cabana, G., and J. B. Rasmussen. 1996. Comparison of aquatic food chains using nitrogen isotopes. *Proceedings of the National Academy of Sciences* 93:10844–47.

Cahalan, J. M. 2001. *Edward Abbey: A life.* University of Arizona Press, Tucson.

Callan, W. T., and S. L. Sanderson. 2003. Feeding mechanisms in Carp: Crossflow filtration, palatal protrusions and flow reversals. *Journal of Experimental Biology* 206:883–92.

Campbell Grant, E. H., W. H. Lowe, and W. F. Fagan. 2007. Living in the branches: Population dynamics and ecological processes in dendritic networks. *Ecology Letters* 10:165–75.

Cardinale, B. J., M. A. Palmer, and S. L. Collins. 2002. Species diversity enhances ecosystem functioning through interspecific facilitation. *Nature* 41:426–29.

Carlson, R. L., P. C. Wainwright, and T. J. Near. 2009. Relationship between species co-occurrence and rate of morphological change in *Percina* darters (Percidae: Etheostomatinae). *Evolution* 63:767–78.

Carney, J. P., and T. A. Dick. 2001. The historical ecology of Yellow Perch (*Perca flavescens* [Mitchill]) and their parasites. *Journal of Biogeography* 27:1337–47.

Carpenter, J. 2005. Competition for food between an introduced crayfish and two fishes endemic to the Colorado River basin. *Environmental Biology of Fishes* 72:335–42.

Carpenter, S. R., J. F. Kitchell, and J. R. Hodgson. 1985. Cascading trophic interactions and lake productivity. *BioScience* 35:634–39.

Carr, M. G., and J. E. Carr. 1985. Individual recognition in the juvenile Brown Bullhead (*Ictalurus nebulosus*). *Copeia* 1985:1060–62.

———. 1986. Characterization of an aggression supressing pheromone in the juvenile Brown Bullhead (*Ictalurus nebulosus*). *Copeia* 1986:540–44.

Carr, M. H. 1942. The breeding habits, embryology and larval development of the large-mouthed black bass in Florida. *Proceedings of the New England Zoological Club* 20:43–77.

Carranza, J., and H. E. Winn. 1954. Reproductive behavior of the Blackstripe Topminnow, *Fundulus notatus*. *Copeia* 1954:273–78.

Carroll, A. M., and P. C. Wainwright. 2009. Energetic limitations on suction feeding performance in centrarchid fishes. *Journal of Experimental Biology* 212:3241–51.

Carroll, A. M., P. C. Wainwright, S. H. Huskey, D. C. Collar, and R. G. Turingan. 2004. Morphology predicts suction feeding performance in centrarchid fishes. *Journal of Experimental Biology* 207:3873–81.

Cashner, R. C., W. J. Matthews, E. Marsh-Matthews, P. J. Unmack, and F. M. Cashner. 2010. Recognition and redescription of a distinctive stoneroller from the southern Interior Highlands. *Copeia* 2010:300–311.

Cattermole, P. 2000. *Building planet Earth.* Cambridge University Press, Cambridge, United Kingdom.

Cavender, T. M. 1986. Review of the fossil history of North American freshwater fishes, 699–724. In *The zoogeography of North American freshwater fishes.* C. H. Hocutt and E. O. Wiley (eds.). John Wiley and Sons, New York, New York.

———. 1991. The fossil record of the Cyprinidae, 34–54. In *Cyprinid fishes, systematics, biology and exploitation.* I. J. Winfield and J. S. Nelson (eds.). Chapman and Hall, New York, New York.

Cavender, T. M., and M. M. Coburn. 1992. Phylogenetic relationships of North American Cyprinidae, 293–327. In *Systematics, historical ecology, and North American freshwater fishes.* R. L. Mayden (ed.). Stanford University Press, Stanford, California.

Chakrabarty, P. 2004. Cichlid biogeography: Comment and review. *Fish and Fisheries* 5:97–119.

Chapman, L. J., and C. A. Chapman. 1993. Fish populations in tropical floodplain pools: A re-evaluation of Holden's data on the River Sokoto. *Ecology of Freshwater Fish* 2:23–30.

Chapman, S. S., B. A. Kleiss, J. M. Omernik, T. L. Foti, and E. O. Murray. 2004. *Ecoregions of the Mississippi Alluvial Plain.* U.S. Geological Survey, Reston, Virginia.

Charnov, E. L., J. M. Smith, and J. J. Bull. 1976. Why be an hermaphrodite? *Nature* 263:125–26.

Charnov, E. L., T. F. Turner, and K. O. Winemiller. 2001. Reproductive constraints and the evolution of life histories with indeterminate growth. *Proceedings of the National Academy of Sciences* 98:9460–64.

Charnov, E. L., and W. Zuo. In press. Growth, mortality, and life-history scaling across species. *Evolutionary Ecology Research.*

Chen, P., E. O. Wiley, and K. M. McNyset. 2007. Ecological niche modeling as a predictive tool: Silver and Bighead carps in North America. *Biological Invasions* 9:43–51.

Chilcote, M. W., S. A. Leider, and J. J. Loch. 1986. Differential reproductive success of hatchery and wild summer-run Steelhead under natural conditions. *Transactions of the American Fisheries Society* 115:726–35.

Childs, M. R., R. W. Clarkson, and A. T. Robinson. 1998. Resource use by larval and early juvenile native fishes in the Little Colorado River, Grand Canyon, Arizona. *Transactions of the American Fisheries Society* 127:620–29.

Chivers, D. P., B. D. Wisenden, C. J. Hindman, T. A. Michalak, R. C. Kusch, S. G. W. Kaminskyj, K. L. Jack, et al. 2007. Epidermal "alarm substance" cells of fishes maintained by non-alarm functions: Possible defence against pathogens, parasites and UVB radiation. *Proceedings of the Royal Society B* 274:2611–19.

Christensen, B. 1996. Predator foraging capabilities and prey antipredator behaviours: Pre- versus postcapture constraints on size-dependent predator-prey interactions. *Oikos* 76:368–80.

Christensen, N. S., A. W. Wood, N. Voisin, D. P. Lettenmaier, and R. N. Palmer. 2004. The effects of climate change on the hydrology and water resources of the Colorado River basin. *Climatic Change* 62:337–63.

Clack, J. A. 2002. *Gaining ground: The origin and evolution of tetrapods.* Indiana University Press, Bloomington.

Clague, J. J., and T. S. James. 2002. History and isostatic effects of the last ice sheet in southern British Columbia. *Quaternary Science Reviews* 21:71–87.

Clark, P. U., J. M. Licciardi, D. R. MacAyeal, and J. W. Jenson. 1996. Numerical reconstruction of a soft-bedded Laurentide Ice Sheet during the last glacial maximum. *Geology* 24:679–82.

Claudi, R., and J. H. Leach (eds.). 1999. *Nonindigenous freshwater organisms, vectors, biology, and impacts.* Lewis Publishers, Boca Raton, Florida.

Clemens, B. J., and S. S. Crawford. 2009. The ecology of body size and depth use by Bloater (*Coregonus hoyi* Gill) in the Laurentian Great Lakes: Patterns and hypotheses. *Reviews in Fisheries Science* 17:174–76.

Clerk, S., R. Hall, R. Quinlan, and J. P. Smol. 2000. Quantitative inferences of past hypolimnetic anoxia and nutrient levels from a Canadian Precambrian Shield lake. *Journal of Paleolimnology* 23:319–36.

Cloutman, D. G., and R. D. Harrell. 1987. Life history notes on the Whitefin Shiner, *Notropis niveus* (Pisces: Cyprinidae), in the Broad River, South Carolina. *Copeia* 1987:1037–40.

Clugston, J. P. 1966. Centrarchid spawning in the Florida Everglades. *Quarterly Journal of the Florida Academy of Science* 29:37–143.

Coggins, L. G., Jr., M. D. Yard, and W. E. Pine, III. 2011. Nonnative fish control in the Colorado River in Grand Canyon, Arizona: An effective program or serendipitous timing? *Transactions of the American Fisheries Society* 140:456–70.

Cohen, D. M. 1970. How many recent fishes are there? Festchrift for George Sprague Myers. *Proceedings of the California Academy of Science* 38:341–46.

Cohn, J. P. 2001. Resurrecting the dammed: A look at Colorado River restoration. *BioScience* 51:998–1003.

Cole, K. S., and R. J. F. Smith. 1987. Release of chemicals by prostaglandin-treated female Fathead Minnows, *Pimephales promelas*, that stimulate male courtship. *Hormones and Behavior* 21:440–56.

———. 1992. Attraction of female Fathead Minnows, *Pimephales promelas*, to chemical stimuli from breeding males. *Journal of Chemical Ecology* 18:1269–84.

Collar, D. C., B. C. O'Meara, P. C. Wainwright, and T. J. Near. 2009. Piscivory limits diversification of feeding morphology in centrarchid fishes. *Evolution* 63:1557–73.

Collette, B. B. 1961. Correlations between ecology and morphology in anoline lizards from Havana, Cuba and southern Florida. *Bulletin of the Museum of Comparative Zoology at Harvard College* 125:137–62.

Collette, B. B., and P. Banarescu. 1977. Systematics and zoogeography of the fishes of the family Percidae. *Journal of the Fisheries Research Board of Canada* 34:1450–63.

Collin, S. P., and A. E. O. Trezise. 2006. Evolution of colour discrimination in vertebrates and its implications for visual communication, 303–35. In *Communication in fishes.* Vol. 2. F. Ladich, S. P. Collin, P. Moller, and B. G. Kapoor (eds.). Science Publishers, Enfield, New Hampshire.

Colwell, R. K., and D. W. Winkler. 1984. A null model for null models in biogeography, 344–59. In *Ecological communities: Conceptual issues and the evidence.* D. R. Strong, Jr., D. Simberloff, L. G. Abele, and A. B. Thistle (eds.). Princeton University Press, Princeton, New Jersey

Connell, J. H. 1978. Diversity in tropical rain forests and coral reefs. *Science* 199:1302–10.

———. 1983. On the prevalence and relative importance of interspecific competition: Evidence from field experiments. *The American Naturalist* 122:661–96.

Connell, J. H., and W. P. Sousa. 1983. On the evidence needed to judge ecological stability or persistence. *The American Naturalist* 121:789–824.

Conner, J. V., and R. D. Suttkus. 1986. Zoogeography of freshwater fishes of the Western Gulf Slope, 413–56. In *The zoogeography of North American freshwater fishes.* C. H. Hocutt and E. O. Wiley (eds.). John Wiley and Sons, New York, New York.

Connor, E. F., and D. Simberloff. 1979. The assembly of species communities: Chance or competition? *Ecology* 60:1132–40.

———. 1986. Competition, scientific method, and null models in ecology. *American Scientist* 74:155–62.

Connor, E. J., and D. E. Pflug. 2004. Changes in the distribution and density of Pink, Chum, and Chinook salmon spawnings in the upper Skagit River in response to flow management measures. *North American Journal of Fisheries Management* 24:835–52.

Conover, D. O. 1985. Field and laboratory assessment of patterns in fecundity of a multiple spawning fish: The Atlantic Silverside, *Menidia menidia. Fishery Bulletin*, U.S. 83:331–41.

Conover, D. O., and S. W. Heins. 1987. The environmental and genetic components of sex ratio in *Menidia menidia* (Pisces: Atherinidae). *Copeia* 1987:732–43.

Cook, R. E. 1977. Raymond Lindeman and the trophic-dynamic concept of ecology. *Science* 198:22–26.

Cooke, S. J., C. M. Bunt, S. J. Hamilton, C. A. Jennings, M. P. Pearson, M. S. Cooperman, and D. F. Markle. 2005. Threats, conservation strategies, and prognosis for suckers (Catostomidae) in North America: Insights from regional case studies of a diverse family of non-game fishes. *Biological Conservation* 121:317–31.

Cooke, S. J., D. P. Philipp, D. H. Wahl, and P. J. Weatherhead. 2006. Energetics of parental care in six syntopic centrarchid fishes. *Oecologia* 148:235–49.

Cooke, S. J., D. P. Phillipp, and P. J. Weatherhead. 2002. Parental care patterns and energetics of Smallmouth Bass (*Micropterus dolomieu*) and Largemouth Bass (*Micropterus salmoides*) monitored with activity transmitters. *Canadian Journal of Zoology* 80:756–70.

Cooke, S. J., P. J. Weatherhead, D. H. Wahl, and D. P. Philipp. 2008. Parental care in response to natural variation in nest predation in six sunfish (Centrarchidae: Teleostei) species. *Ecology of Freshwater Fish* 17:628–38.

Cooley, W. W., and P. R. Lohnes. 1971. *Multivariate data analysis.* John Wiley and Sons, Inc. New York, New York.

Copes, F. A. 1983. The Longnose Dace *Rhinichthys cataractae* Valenciennes in Wisconsin and Wyoming waters. *Museum of Natural History* 19:1–11.

Corbett, B. W., and P. M. Powles. 1986. Spawning and larval drift of sympatric Walleyes and White Suckers in an Ontario stream. *Transactions of the American Fisheries Society* 115:41–46.

Cornell, H. V., and J. H. Lawton. 1992. Species interactions, local and regional processes, and limits to the richness of ecological communities: A theoretical perspective. *Journal of Animal Ecology* 61:1–12.

Courtenay, W. R., Jr. 2007. Introduced species: What species do you have and how do you know? *Transactions of the American Fisheries Society* 136:1160–64.

Courtenay, W. R., Jr., D. A. Hensley, J. N. Taylor, and J. A. McCann. 1986. Distribution of exotic fishes in North America, 675–98. In *The zoogeography of North American freshwater fishes.* C. H. Hocutt and E. O. Wiley (eds.). John Wiley and Sons, New York, New York.

Cox, C. B. 2001. The biogeographic regions reconsidered. *Journal of Biogeography* 28:511–23.

Cox, C. B., and P. D. Moore. 2005. *Biogeography, an ecological and evolutionary approach.* 7th ed. Blackwell Publishing, Malden, Massachusetts.

Cox, D. L. 1980. A note on the queer history of "Niche." *Bulletin of the Ecological Society of America* 61:201–2.

Cox, S., S. Chandler, C. Barron, and K. Work. 2009. Benthic fish exhibit more plastic crypsis than non-benthic species in a freshwater spring. *Journal of Ethology* 27:497–505.

Cracraft, J. 1974. Continental drift and vertebrate distribution. *Annual Review of Ecology and Systematics* 4077:215–61.

Crain, P. K., K. Whitener, and P. B. Moyle. 2004. Use of a restored central California floodplain by larvae of native and alien fishes, 125–40. In *Early life history of fishes in the San Francisco Estuary and watershed.* F. Feyrer, L. R. Brown, R. L. Brown, and J. J. Orsi (eds.). *American Fisheries Society Symposium* 39, Bethesda, Maryland.

Crane, A. L., D. Woods, and A. Mathis. 2009. Behavioural responses to alarm cues by free-ranging Rainbow Darters (*Etheostoma caeruleum*). *Behaviour* 146:1565–72.

Crespi, B. J., and R. Teo. 2002. Comparative phylogenetic analysis of the evolution of semelparity and life history in salmonid fishes. *Evolution* 56:1008–20.

Crisp, D. T. 1988. Prediction, from temperature, of eyeing, hatching and "swim-up" times for salmonid embryos. *Freshwater Biology* 19:41–48.

Croizat, L. 1958. *Panbiogeography, or, an introductory synthesis of zoogeography, phytogeography, and geology, with notes on evolution, systematics, ecology, anthropology, etc.* Published by the author, Caracas, Venezuela.

Croizat, L., G. Nelson, and D. E. Rosen. 1974. Centers of origin and related concepts. *Systematic Zoology* 23:265–87.

Cross, F. B. 1970. Fishes as indicators of Pleistocene and Recent environments in the Central Plains, 241–58. In *Pleistocene and Recent environments of the central Great Plains.* Dort, W., Jr., and J. K. Jones, Jr. (eds.). University of Kansas, Special Publication 3.

Cross, F. B., R. L. Mayden, and J. D. Stewart. 1986. Fishes in the western Mississippi Basin (Missouri, Arkansas and Red Rivers), 363–412. In *The zoogeography of North American freshwater fishes.* C. H. Hocutt and E. O. Wiley (eds.). John Wiley and Sons, New York, New York.

Crossman, E. J., and B. C. Cudmore. 1999. Summary of fishes intentionally introduced in North America, 99–111. In *Nonindigenous freshwater organisms, vectors, biology, and impacts.* R. Claudi and J. H. Leach (eds.). Lewis Publishers, Boca Raton, Florida.

Crossman, E. J., and D. E. McAllister. 1986. Zoogeography of freshwater fishes of the Hudson Bay drainage, Ungava Bay and the Arctic Archipelago, 53–104. In *The zoogeography of North American freshwater fishes.* C. H. Hocutt and E. O. Wiley (eds.). John Wiley and Sons, New York, New York.

Cudmore-Vokey, B., and E. J. Crossman. 2000. Checklists of the fish fauna of the Laurentian Great Lakes and their connecting channels. *Canadian Manuscript Report of Fisheries and Aquatic Science* 2550:1–39.

Curio, E. 1976. *The ethology of predation.* Springer-Verlag, Berlin, Germany.

Currie, C. R., B. Wong, A. E. Stuart, T. R. Schulltz, S. A. Rehner, U. G. Mueller, G. Sung, J. W. Spatafora, and N. A. Straus. 2003. *Ancient tripartite coevolution in the attine ant microbe symbiosis. Science* 299:386–88.

Curry, D. R. 1990. Quaternary palaeolakes in the evolution of semidesert basins, with special emphasis on Lake Bonneville and the Great Basin, U.S.A. *Palaeogeography, Palaeoclimatology, Palaeoecology* 76:189–214.

Curry, K. D., and A. Spacie. 1984. Differential use of stream habitat by spawning catostomids. *American Midland Naturalist* 111:267–69.

Daeschler, E. B. 2000. An early actinopterygian fish from the Catskill Formation (Late Devonian, Famennian) in Pennsylvania, U.S.A. *Proceedings of the Academy of Natural Sciences of Philadelphia* 150:181–92.

Dahlberg, M. D. 1979. A review of survival rates of fish eggs and larvae in relation to impact assessments. *Marine Fisheries Review* 41:1–12.

Dale, G., and A. Pappantoniou. 1986. Eye-picking behavior of the Cutlips Minnow, *Exoglossum maxillingua*: Applications to studies of eyespot mimicry. *Annals of the New York Academy of Sciences* 463:177–78.

Dalerum, F., and A. Angerbjörn. 2005. Resolving temporal variation in vertebrate diets using naturally occurring stable isotopes. *Oecologia* 144:647–58.

Dalton, B. E., T. W. Cronin, N. J. Marshall, and K. L. Carleton. 2010. The fish eye view: Are cichlids conspicuous? *Journal of Experimental Biology* 213:2243–55.

Daniels, R. A. 2001. Untested assumptions: The role of canals in the dispersal of Sea Lamprey, Alewife, and other fishes in the eastern United States. *Environmental Biology of Fishes* 60:309–29.

Dann, T. H., W. W. Smoker, J. J. Hard, and A. J. Gharrett. 2010. Outbreeding depression after two generations of hybridizing southeast Alaska Coho Salmon populations? *Transactions of the American Fisheries Society* 139:1292–305.

Danos, N., N. Fisch, and S. Gemballa. 2008. The musculotendinous system of an anguilliform swimmer: Muscles, myosepta, dermis, and the interconnections in *Anguilla rostrata. Journal of Morphology* 269:29–44.

Darlington, P. J., Jr. 1957. *Zoogeography: The geographical distribution of animals.* John Wiley, New York, New York.

Davis, B. J., and R. J. Miller. 1967. Brain patterns in minnows of the genus *Hybopsis* in relation to feeding habits and habitat. *Copeia* 1967:1–39.

Davis, M. B. 1983. Quaternary history of deciduous forests of eastern North America and Europe. *Annals of the Missouri Botanical Garden* 70:550–63.

Day, S. W., T. E. Higham, A. Y. Cheer, and P. C. Wainwright. 2005. Spatial and temporal patterns of water flow generated by suction-feeding Bluegill sunfish *Lepomis macrochirus* resolved by

Particle Image Velocimetry. *Journal of Experimental Biology* 208:2661–71.

De Kay, J. E. 1842. *Zoology of New-York, or the New-York fauna.* Part IV. Fishes. W. and A. White and J. Visscherr. Albany, New York.

De Silva, S. S. 1980. Biology of the juvenile Grey Mullet: A short review. *Aquaculture* 19:21–36.

Deacon, J. E., and C. D. Williams. 1991. Ash Meadows and the legacy of the Devils Hole Pupfish, 69–87. In *Battle against extinction, native fish management in the American west.* W. L. Minckley and J. E. Deacon (eds.). The University of Arizona Press, Tucson.

Deacon, J. E., A. E. Williams, C. D. Williams, and J. E. Williams. 2007. *Fueling population growth in Las Vegas: How large-scale groundwater withdrawal could burn regional biodiversity. BioScience* 57:688–98.

Decker, A. S., M. J. Bradford, and P. S. Higgins. 2008. Rate of biotic colonization following restoration below a diversion dam in the Bridge River, British Columbia. *River Research and Applications* 24:876–83.

Deevey, E. S., Jr. 1947. Life tables for natural populations of animals. *The Quarterly Review of Biology* 22:283–314.

Delbeek, J. C., and D. D. Williams. 1987. Food resource partitioning between sympatric populations of brackishwater sticklebacks. *Journal of Animal Ecology* 56:949–67.

Delco, E. A., Jr. 1960. Sound discrimination by males of two cyprinid fishes. *Texas Journal of Science* 12:48–54.

Delcourt, H. R., and P. A. Delcourt. 1988. Quaternary landscape ecology: Relevant scales in space and time. *Landscape Ecology* 2:23–44.

Deliagina, T. G., F. Ullén, M.-J. Gonzalez, H. Ersson, G. N. Orlovsky, and S. Grillner. 1995. Initiation of locomotion by lateral line photoreceptors in lamprey: Behavioural and neurophysiological studies. *Journal of Experimental Biology* 198:2581–91.

del Rio, C. M., N. Wolf, S. A. Carleton, and L. Z. Gannes. 2009. Isotopic ecology ten years after a call for more laboratory experiments. *Biological Reviews* 84:91–111.

Dembkowski, D. J., and L. E. Miranda. 2012. Hierarchy in factors affecting fish biodiversity in floodplain lakes of the Mississippi alluvial valley. *Environmental Biology of Fishes* 93:357–68.

Desert Fishes Council. 2012. Species tracking. http://www.desertfishes.org/?page_id=327.

Detenbeck N. E., P. W. DeVore, G. J. Niemi, and A. Lima. 1992. Recovery of temperate-stream fish communities from disturbance: A review of case

studies and synthesis of theory. *Environmental Management* 16:33–53.

DeWoody, J. A., and J. C. Avise. 2001. Genetic perspectives on the natural history of fish mating systems. *Journal of Heredity* 92:167–72.

DeWoody, J. A., D. E. Fletcher, M. Mackiewicz, S. D. Wilkins, and J. C. Avise. 2000a. The genetic mating system of Spotted Sunfish (*Lepomis punctatus*): Mate numbers and the influence of male reproductive parasites. *Molecular Ecology* 9:2119–28.

DeWoody, J. A., D. E. Fletcher, S. D. Wilkins, and J. C. Avise. 2000b. Parentage and nest guarding in the Tessellated Darter (*Etheostoma olmstedi*) assayed by microsatellite markers (Perciformes: Percidae). *Copeia* 2000:740–47.

DeWoody, J. A., D. E. Fletcher, S. D. Wilkins, W. S. Nelson, and J. C. Avise. 1998. Molecular genetic dissection of spawning, parentage, and reproductive tactics in a population of Redbreast Sunfish, *Lepomis auritus. Evolution* 52:1802–10.

———. 2000c. Genetic monogamy and biparental care in an externally fertilizing fish, the Largemouth Bass (*Micropterus salmoides*). *Proceedings of the Royal Society of London B* 267:2431–37

Diamond, J. N. 1975. Assembly of species communities, 342–444. In *Ecology and evolution of communities.* M. L. Cody and J. M. Diamond (eds.). Belknap Press of Harvard University Press, Cambridge, Massachusetts.

Dickinson, W. R. 2004. Evolution of the North American Cordillera. *Annual Review of Earth and Planetary Science* 32:13–45.

Dillon, J. T., and P. L. Ehlig. 1993. Displacement on the southern San Andreas Fault, 199–216. In *The San Andreas fault system: Displacement, palinspastic reconstruction, and geologic evolution.* R. E. Powell, R. J. Weldon, II, and J. C. Matti (eds.). *Geological Society of America*, memoir 178, Boulder, Colorado.

Dingle, H. 1996. *Migration, the biology of life on the move.* Oxford University Press, New York, New York.

Dittman, A. H., and T. P. Quinn. 1996. Homing in Pacific salmon: Mechanisms and ecological basis. *Journal of Experimental Biology* 199:83–91.

Doadrio, I., and O. Domínguez. 2004. Phylogenetic relationships within the fish family Goodeidae based on cytochrome b sequence data. *Molecular Phylogenetics and Evolution* 31:416–30.

Dobzhansky, T. 1950. Evolution in the tropics. *American Scientist* 38:209–21.

Dodds, W. K., K. Gido, M. R. Whiles, K. M. Fritz, and W. J. Matthews. 2004. Life on the edge: The ecology of Great Plains prairie streams. *BioScience* 54:205–16.

Dodson, J. J. 1997. Fish migration: An evolutionary perspective, 10–36. In *Behavioural ecology of teleost fishes*. J-G. J. Godin (ed.). Oxford University Press, New York, New York.

Dodson, S. 1989. Predator-induced reaction norms. *BioScience* 39:447–52.

Dominey, W. J. 1980. Female mimicry in male Bluegill sunfish—a genetic polymorphism? *Nature* 284:546–48.

———. 1981. Maintenance of female mimicry as a reproductive strategy in Bluegill sunfish (*Lepomis macrochirus*). *Environmental Biology of Fishes* 6:59–64.

———. 1984. Alternative mating tactics and evolutionarily stable strategies. *American Zoologist* 24:385–96.

Domínguez-Domínguez, O., E. Martínez-Meyer, L. Zambrano, and G. Pérez-Ponce de León. 2006. Using ecological-niche modeling as a conservation tool for freshwater species: Livebearing fishes in Central Mexico. *Conservation Biology* 20:1730–39.

Douglas, M. E. 1987. An ecomorphological analysis of niche packing and niche dispersion in stream-fish clades, 144–49. In *Community and evolutionary ecology of North American stream fishes*. W. J. Matthews and D. C. Heins (eds.). University of Oklahoma Press, Norman.

Douglas, M. E., P. C. Marsh, and W. L. Minckley. 1994. Indigenous fishes of western North America and the hypothesis of competitive displacement: *Meda fulgida* (Cyprinidae) as a case study. *Copeia* 1994:9–19.

Douglas, M. E., and W. J. Matthews. 1992. Does morphology predict ecology? Hypothesis testing within a freshwater stream fish assemblage. *Oikos* 65:213–24.

Douglas, M. R., P. C. Brunner, and M. E. Douglas. 2003. Drought in an evolutionary context: Molecular variability in Flannelmouth Sucker (*Catostomus latipinnis*) from the Colorado River Basin of western North America. *Freshwater Biology* 48:1254–73.

Dovel, W. L., J. A. Mihursky, and A. J. McErlean. 1969. Life history aspects of the Hogchoker, *Trinectes maculatus*, in the Patuxent River estuary, Maryland. *Chesapeake Science* 110:104–19.

Dowling, T. E., and W. S. Moore. 1986. Absence of population subdivision in the Common Shiner, *Notropis cornutus* (Cyprinidae). *Environmental Biology of Fishes* 15:151–55.

Dowling, T. E., C. A. Tibbets, W. L. Minckley, and G. R. Smith. 2002. Evolutionary relationships of the plagopterins (Teleostei: Cyprinidae) from cytochrome *b* sequences. *Copeia* 2002:665–78.

Doyle, M. W., E. H. Stanley, C. H. Orr, A. R. Selle, S. A. Sethi, and J. M. Harbor. 2005. Stream ecosystem response to small dam removal: Lessons from the Heartland. *Geomorphology* 71:227–44.

Drake, D. C., R. J. Naiman, and J. S. Bechtold. 2006. Fate of nitrogen in riparian forest soils and trees: An ^{15}N tracer study simulating salmon decay. *Ecology* 87:1256–66.

Drake, J. A., T. E. Flum., G. J. Witteman, T. Voskuil, A. M. Hoylman, C. Creson, D. A. Kenny, G. R. Huxel, C. S. Larue, and J. R. Duncan. 1993. The construction and assembly of an ecological landscape. *Journal of Animal Ecology* 62:9117–30.

Drenner, R. W., F. deNoyelles, Jr., and D. Kettle. 1982a. Selective impact of filter-feeding Gizzard Shad on zooplankton community structure. *Limnology and Oceanography* 27:965–68.

Drenner, R. W., J. R. Mummert, F. deNoyelles, Jr., and D. Kettle. 1984. Selective particle ingestion by a filter-feeding fish and its impact on phytoplankton community structure. *Limnology and Oceanography* 29:941–48.

Drenner, R. W., W. J. O'Brien, and J. R. Mummert. 1982b. Filter-feeding rates of Gizzard Shad. *Transactions of the American Fisheries Society* 111:210–15.

Drucker, E. G., and G. V. Lauder. 2000. A hydrodynamic analysis of fish swimming speed: Wake structure and locomotor force in slow and fast labriform swimmers. *Journal of Experimental Biology* 203:2379–93.

———. 2003. Function of pectoral fins in Rainbow Trout: Behavioral repertoire and hydrodynamic forces. *Journal of Experimental Biology* 206:813–26.

Duarte, C. M., and M. Alcaraz. 1989. To produce many small or few large eggs: A size-independent reproductive tactic of fish. *Oecologia* 80:401–4.

Dudgeon, D., A. H. Arthington, M. O. Gessner, Z.-I. Kawabata, D. J. Knowler, C. Lévêque, R. J. Naiman, et al. 2006. Freshwater biodiversity: Importance, threats, status and conservation challenges. *Biological Reviews* 81:163–82.

Dudley, R. K., and W. J. Matter. 1999. Effects of a record flood on fishes in Sabino Creek, Arizona. *The Southwestern Naturalist* 44:218–21.

Dudley, R. K., and S. P. Platania. 2007. Flow regulation and fragmentation imperil pelagic-spawning riverine fishes. *Ecological Applications* 17:2074–86.

Dunlop, E. S., K. Enberg, C. Jorgensen, and M. Heino. 2009. Toward Darwinian fisheries management. *Evolutionary Applications* 2:245–59.

Dyke, A. S., J. T. Andrews, P. U. Clark, J. H. England, G. H. Miller, J. Shaw, and J. J. Veillette. 2002. The Laurentide and Innuitian ice sheets during the last glacial maximum. *Quaternary Science Reviews* 21:9–31.

Dynesius, M., and C. Nilsson. 1994. Fragmentation and flow regulation of river systems in the northern third of the world. *Science* 266:753–62.

Eadie, J. McA., and A. Keast. 1984. Resource heterogeneity and fish species diversity in lakes. *Canadian Journal of Zoology* 62:1689–95.

Eberhardt, L. L., and J. M. Thomas. 1991. Designing environmental field studies. *Ecological Monographs* 61:53–73.

Echelle, A. A., E. W. Carson, A. F. Echelle, R. A. van den Bussche, T. E. Dowling, and A. Meyer. 2005. Historical biogeography of the New-World pupfish genus *Cyprinodon* (Teleostei: Cyprinodontidae). *Copeia* 2005:320–39.

Echelle, A. A., and A. F. Echelle. 1992. Mode and pattern of speciation in the evolution of inland pupfishes in the *Cyprinodon variegatus* complex (Teleostei: Cyprinodontidae): An ancestor-descendant hypothesis, 691–709. In *Systematics, historical ecology, and North American freshwater fishes*. R. L. Mayden (ed.). Stanford University Press, Stanford, California.

———. 1993. Allozyme variation and systematics of the New World cyprinodontines (Teleostei: Cyprinodontidae). *Biochemical Systematics and Ecology* 21:583–90.

———. 1997. Patterns of abundance and distribution among members of a unisexual-bisexual complex of fishes (Atherinidae: *Menidia*). *Copeia* 1997:249–59.

Echelle, A. A., and I. Kornfield. 1984. *Evolution of fish species flocks*. University of Maine at Orono Press, Orono.

Eck, G. W., and L. Wells. 1987. Recent changes in Lake Michigan's fish community and their probable causes, with emphasis on the role of the Alewife (*Alosa pseudoharengus*). *Canadian Journal of Fisheries and Aquatic Sciences* 44:53–60.

Edds, D. R., W. J. Matthews, and F. P. Gelwick. 2002. Resource use by large catfishes in a reservoir: Is there evidence for interactive segregation and innate differences? *Journal of Fish Biology* 60:739–50.

Eder, S., and C. A. Carlson. 1977. Food habits of Carp and White Suckers in the South Platte and St. Vrain rivers and Goosequill Pond, Weld County, Colorado. *Transactions of the American Fisheries Society* 106:339–46

Eggers, D. M. 1977. The nature of prey selection by planktivorous fish. *Ecology* 58:46–59.

Eggold, B. T., and P. J. Motta. 1992. Ontogenetic dietary shifts and morphological correlates in Striped Mullet, *Mugil cephalus. Environmental Biology of Fishes* 34:139–58.

Ehlers, J. 1996. *Quaternary and glacial geology*. John Wiley and Sons, New York, New York.

Ehrlich, P. R., and P. H. Raven. 1964. Butterflies and plants: A study in coevolution. *Evolution* 18:586–608.

Eidietis, L., T. L. Forrester, and P. W. Webb. 2003. Relative abilities to correct rolling disturbances of three morphologically different fish. *Canadian Journal of Zoology* 80:2156–63.

Eldredge, N., and J. Cracraft. 1980. *Phylogenetic patterns and the evolutionary process*. Columbia University Press, New York, New York.

Elgar, M. A. 1990. Evolutionary compromise between a few large and many small eggs: Comparative evidence in teleost fish. *Oikos* 59:283–87.

Elton, C. 1927. *Animal ecology*. The Macmillan Company, New York, New York.

Elton, C. S. 1958. *The ecology of invasions by animals and plants*. Methuen, London.

Emlen, J. M. 1966. *The role of time and energy in food preference. The American Naturalist* 100:611–17.

Endler, J. A. 1978. A predator's view of animal color patterns. *Evolutionary Biology* 11:319–64.

———. 1986. Defense against predators, 109–34. In *Predator-prey relationships*. M. E. Feder and G. V. Lauder (eds.). The University of Chicago Press, Chicago, Illinois.

———. 1988. Frequency-dependent predation, crypsis and aposematic coloration. *Philosophical Transactions of the Royal Society B* 319:505–23.

———. 1992. Signals, signal conditions, and the direction of evolution. *The American Naturalist* 139:S125–S153.

Endler, J. A., A. Basolo, S. Glowacki, and J. Zerr. 2001. Variation in response to artificial selection for light sensitivity in Guppies (*Poecilia reticulata*). *The American Naturalist* 158:36–48.

Erickson, D. L., J. A. North, J. E. Hightower, J. Weber, and L. Lauck. 2002. Movement and habitat use of Green Sturgeon *Acipenser medirostris* in the Rogue River, Oregon, USA. *Journal of Applied Ichthyology* 18:565–69.

Eschmeyer, W. N. (ed.). 1998. *Catalog of fishes*. Vol. 3. California Academy of Sciences. San Francisco.

Eshenroder, R. L. 2008. Differentiation of deep-water Lake Charr *Salvelinus namaycush* in North

American lakes. *Environmental Biology of Fishes* 83:77–90.

Essington, T. E., and S. R. Carpenter. 2000. Nutrient cycling in lakes and streams: Insights from a comparative analysis. *Ecosystems* 3:131–43.

Etnier, D. A., and C. E. Skelton. 2003. Analysis of three cisco forms (*Coregonus*, Salmonidae) from Lake Saganaga and adjacent lakes near the Minnesota/Ontario border. *Copeia* 2003:739–49.

Etnier, D. A., and W. C. Starnes. 1993. *The Fishes of Tennessee.* University of Tennessee Press, Knoxville.

Evans, H. E., and E. E. Deubler, Jr. 1955. Pharyngeal tooth replacement in *Semotilus atromaculatus* and *Clinostomus elongatus* two species of cyprinid fishes. *Copeia* 1955:31–41.

Evermann, B. W., and S. E. Meek. 1898. A report upon salmon investigations in the Columbia River basin and elsewhere on the Pacific Coast in 1896. *Bulletin of the United States Fish Commission* 17(1897):15–84.

Fagan, W. F. 2002. Connectivity, fragmentation, and extinction risk in dendritic metapopulations. *Ecology* 83:3243–49.

Fairchild, G. W., R. J. Horwitz, D. A. Nieman, M. R. Boyer, and D. F. Knorr. 1998. Spatial variation and historical change in fish communities of the Schuylkill River drainage, southeast Pennsylvania. *American Midland Naturalist* 139:282–95.

Falke, J. A., and K. B. Gido. 2006. Effects of reservoir connectivity on stream fish assemblages in the Great Plains. *Canadian Journal of Fisheries and Aquatic Sciences* 63:480–93.

Farina, A. 2006. *Principles and methods in landscape ecology. Toward a science of landscape.* Springer, A. A. Dordrecht, The Netherlands.

Farrell, B. C., and C. Mitter. 1993. Phylogenetic determinants of insect/plant community diversity, 253–66. In *Species diversity in ecological communities.* R. E. Ricklefs and D. Schluter (eds.). University of Chicago Press, Chicago, Illinois.

Fausch, K. D. 1988. Tests of competition between native and introduced salmonids in streams: What have we learned? *Canadian Journal of Fisheries and Aquatic Sciences* 45:2238–46.

———. 1998. Interspecific competition and juvenile Atlantic Salmon (*Salmo salar*): On testing effects and evaluating the evidence across scales. *Canadian Journal of Fisheries and Aquatic Sciences* 55:218–31.

Fausch, K. D., and R. G. Bramblett. 1991. Disturbance and fish communities in intermittent tributaries of a western Great Plains river. *Copeia* 1991:659–74.

Fausch, K. D., Y. Taniguchi, S. Nakano, G. D. Grossman, and C. R. Townsend. 2001. Flood disturbance regimes influence Rainbow Trout invasion success among five Holarctic regions. *Ecological Applications* 11:1438–55.

Fausch, K. D., C. E. Torgersen, C. V. Baxter, and H. W. Li. 2002. Landscapes to riverscapes: Bridging the gap between research and conservation of stream fishes. *BioScience* 52:483–98.

Fausch, K. D., and R. J. White. 1981. Competition between Brook Trout (*Salvelinus fontinalis*) and Brown Trout (*Salmo trutta*) for positions in a Michigan stream. *Canadian Journal of Fisheries and Aquatic Sciences* 38:1220–27.

———. 1986. Competition among juveniles of Coho Salmon, Brook Trout, and Brown Trout in a laboratory stream, and implications for Great Lakes tributaries. *Transactions of the American Fisheries Society* 115:363–81.

Favé, M.-J., and J. Turgeon. 2008. Patterns of genetic diversity in Great Lakes Bloaters (*Coregonus hoyi*) with a view to future reintroduction in Lake Ontario. *Conservation Genetics* 9:281–93.

Feldmeth, C. R. 1981. The evolution of thermal tolerance in desert pupfish (genus *Cyprinodon*), 357–84. In *Fishes in North American deserts.* R. J. Naiman and D. L. Soltz (eds.). John Wiley and Sons, New York, New York.

Felley, J. D. 1984. Multivariate identification of morphological-environmental relationships within the Cyprinidae (Pisces). *Copeia* 1981:442–55.

Ferry-Graham, L. A., A. C. Gibb, and L. P. Hernandez. 2008. Premaxillary movements in cyprinodontiform fishes: An unusual protrusion mechanism facilitates "picking" prey capture. *Zoology* 111:455–66.

Ferry-Graham, L. A., and G. V. Lauder. 2001. Aquatic prey capture in ray-finned fishes: A century of progress and new directions. *Journal of Morphology* 248:99–119.

Ferry-Graham, L. A., P. C. Wainwright, and G. V. Lauder. 2003. Quantification of flow during suction feeding in Bluegill sunfish. *Zoology* 106:159–68.

Feyrer, F., T. Sommer, and W. Harrell. 2006. Managing floodplain inundation for native fish: Production dynamics of age-0 Splittail (*Pogonichthys macrolepidotus*) in California's Yolo Bypass. *Hydrobiologia* 573:213–26.

Findeis, E. K. 1997. Osteology and phylogenetic interrelationships of sturgeons (Acipenseridae). *Environmental Biology of Fishes* 48:73–126.

Findley, S. S. 1976. The structure of bat communities. *The American Naturalist* 110:129–39.

Fine, J. M., L. A. Vrieze, and P. W. Sorensen. 2004. Evidence that petromyzontid lampreys employ a common migratory pheromone that is partially comprised of bile acids. *Journal of Chemical Ecology* 30:2091–110.

Fine, M. L., J. P. Friel, D. McElroy, C. B. King, K. E. Loesser, and S. Newton. 1997. Pectoral spine locking and sound production in the Channel Catfish *Ictalurus punctatus. Copeia* 1997:777–90.

Fine, M. L., D. McElroy, J. Rafi, C. B. King, K. E. Loesser, and S. Newton. 1996. Lateralization of pectoral stridulation sound production in the Channel Catfish. *Physiology and Behavior* 60:753–57.

Fine, M. L., H. E. Winn, and B. L. Olla. 1977. Communication in fishes, 472–518. In *How animals communicate.* T. A. Sebeok (ed.). Indiana University Press, Bloomington.

Fisher, D. L. 1984. Central capacity limits consistent mapping, visual search tasks: Four channels or more? *Cognitive Psychology* 16:449–84.

Fisher, W. L. 1990. Life history and ecology of the Orangefin Darter *Etheostoma bellum* (Pisces: Percidae). *American Midland Naturalist* 123:268–81.

Fitzgerald, G. J. 1993. Seeing red, turning red. *Reviews in Fish Biology and Fisheries* 3:286–92.

Fitzgerald, G. J., M. Fournier, and J. Morrissette. 1994. Sexual selection in an anadromous population of Threespine Sticklebacks—no role for parasites. *Evolutionary Ecology* 8:348–56.

Fiumera, A. C., B. A. Porter, G. D. Grossman, and J. C. Avise. 2002. Intensive genetic assessment of the mating system and reproductive success in a semi-closed population of the Mottled Sculpin, *Cottus bairdi. Molecular Ecology* 11:2367–77.

Fleming, I. A. 1998. Pattern and variability in the breeding system of Atlantic Salmon (*Salmo salar*), with comparisons to other salmonids. *Canadian Journal of Fisheries and Aquatic Science* 55, supplement 1:59–76.

Fleming, I. A., and M. R. Gross. 1990. Latitudinal clines: A trade-off between egg number and size in Pacific salmon. *Ecology* 71:1–11.

Fleming, I. A., and J. D. Reynolds. 2004. Salmonid breeding systems, 264–94. In *Evolution illuminated, salmon and their relatives.* A. P. Hendry and S. C. Stearns (eds.). Oxford University Press, New York, New York.

Fletcher, D. E. 1993. Nest association of Dusky Shiners (*Notropis cummingsae*) and Redbreast Sunfish (*Lepomis auritus*), a potentially parasitic relationship. *Copeia* 1993:159–67.

Fletcher, D. E., E. E. Dakin, B. A. Porter, and J. C. Avise. 2004. Spawning behavior and genetic parentage in the Pirate Perch (*Aphredoderus sayanus*), a fish with an enigmatic reproductive morphology. *Copeia* 2004:1–10.

Flint, R. F. 1971. *Glacial and Quaternary geology.* John Wiley and Sons, New York, New York.

Florida Fish and Wildlife Conservation Commission. 2012. *Nonnative freshwater fish.* http://myfwc.com/wildlifehabitats/nonnatives/freshwater-fish/.

Floyd, K. B., R. D. Hoyt, and S. Timbrook. 1984. Chronology of appearance and habitat partitioning by stream larval fishes. *Transactions of the American Fisheries Society* 113:217–23.

Forbes, S. A. 1925. The lake as a microcosm. *Illinois Natural History Survey Bulletin* 15:537–50.

Forman, R. T. T. 1995. Some general principles of landscape and regional ecology. *Landscape Ecology* 10:133–42.

Forman, R. T. T., and M. Godron. 1981. Patches and structural components for a landscape ecology. *BioScience* 31:733–40.

Foster, D., F. Swanson, J. Aber, I. Burke, N. Brokaw, D. Tilman, and A. Knapp. 2003. The importance of land-use legacies to ecology and conservation. *BioScience* 53:77–88.

Foster, N. R. 1967. *Comparative studies on the biology of killifishes (Pisces, Cyprinodontidae).* PhD dissertation, Cornell University, Ithaca, New York.

Foster, S. A. 1994. Inference of evolutionary pattern: Diversionary displays of Three-Spined Sticklebacks. *Behavioral Ecology* 5:114–21.

———. 1995. Understanding the evolution of behavior in Threespine Stickleback: The value of geographic variation. *Behaviour* 132:1107–29.

Foster, S. A., and J. A. Baker. 2004. Evolution in parallel: New insights from a classic system. *Trends in Ecology and Evolution* 19:456–59.

Foster, S. A., J. A. Baker, and M. A. Bell. 2003. The case for conserving Threespine Stickleback populations. *Fisheries* 28:10–18.

Fox, B. J. 2001. The genesis and development of guild assembly rules, 23–57. In *Ecological assembly rules.* E. Weiher and P. Keddy (eds.). Cambridge University Press, Cambridge, United Kingdom.

Fox, D. A., J. E. Hightower, and F. M. Parauka. 2000. Gulf Sturgeon spawning migration and habitat in the Choctawhatchee River system, Alabama-Florida. *Transactions of the American Fisheries Society* 129:811–26.

Fox, M. G., and A. J. Crivelli. 1998. Body size and reproductive allocation in a multiple spawning centrarchid. *Canadian Journal of Fisheries and Aquatic Sciences* 55:737–48.

Franssen, N. R., and K. B. Gido. 2006. Use of stable isotopes to test literature-based trophic classifications of small-bodied stream fishes. *American Midland Naturalist* 156:1–10.

Fraser, D. J., P. Duchesne, and L. Bernatchez. 2005. Migratory charr schools exhibit population and kin associations beyond juvenile stages. *Molecular Ecology* 14:3133–46.

Freeman, M. C., M. K. Crawford, J. C. Barrett, D. E. Facey, M. G. Flood, J. Hill, D. J. Stouder, and G. D. Grossman. 1988. Fish assemblage stability in a southern Appalachian stream. *Canadian Journal of Fisheries and Aquatic Sciences* 45:1949–58.

Freeman, M. C., and G. D. Grossman. 1992. A field test for competitive interactions among foraging stream fishes. *Copeia* 1992:898–902.

Freeman, M. C., and P. A. Marcinek. 2006. Fish assemblage responses to water withdrawals and water supply reservoirs in Piedmont streams. *Environmental Management* 38:435–50.

Freeman, M. C., and D. J. Stouder. 1989. Intraspecific interactions influence size specific depth distribution in *Cottus bairdi*. *Environmental Biology of Fishes* 24:231–36.

Friedman, M., and H. Blom. 2006. A new Actinopterygian from the Famennian of East Greenland and the interrelationships of Devonian ray-finned fishes. *Journal of Paleontology* 80:1186–204.

Frietsche, R. A., R. D. Miracle, and R. W. McFarlane. 1979. Larvae and juveniles of the Brook Silverside, *Labidesthes sicculus*, 187–97. In *Proceedings of a workshop on freshwater larval fishes*. R. Wallus and C. W. Voigtlander (eds.). Tennessee Valley Authority, Norris, Tennessee.

Fritts, A. L., and T. N. Pearsons. 2006. Effects of predation by nonnative Smallmouth Bass on native salmonid prey: The role of predator and prey size. *Transactions of the American Fisheries Society* 135:853–60.

Frommen, J. G., and T. C. M. Bakker. 2004. Adult Three-Spined Sticklebacks prefer to shoal with familiar kin. *Behaviour* 141:1401–9.

Fry, B., and E. B. Sherr. 1984. $\delta^{13}C$ measurements as indicators of carbon flow in marine and freshwater ecosystems. *Contributions in Marine Science* 27:13–47.

Frye, J. C., H. B. Willman, R. F. Black. 1965. Outline of glacial geology of Illinois and Wisconsin, 43–61. In *The Quaternary of the United States*.

H. E. Wright, Jr., and D. G. Frey (eds.). Princeton University Press, New Jersey.

Fryer, G., and T. D. Iles. 1972. *The cichlid fishes of the great lakes of Africa: Their biology and evolution*. Oliver Boyd, Edinburgh, United Kingdom.

Fuiman, L. A. 2002. Special considerations of fish eggs and larvae, 1–32. In *Fishery science, the unique contributions of early life stages*. L. A. Fuiman and R. G. Werner (eds.). Blackwell Science, Oxford, United Kingdom.

Fuller, P. L., L. G. Nico, and J. D. Williams. 1999. *Nonindigenous fishes introduced into inland waters of the United States*. American Fisheries Society, Special Publication 27, Bethesda, Maryland.

Fuller, R. C. 2001. Patterns in male breeding behaviors in the Bluefin Killifish, *Lucania goodei*: A field study. *Copeia* 2001:823–28.

———. 2002. Lighting environment predicts the relative abundance of male colour morphs in Bluefin Killifish (*Lucania goodei*) populations. *Proceedings of the Royal Society B* 269:1457–65.

———. 2003. Disentangling female mate choice and male competition in the Rainbow Darter, *Etheostoma caeruleum*. *Copeia* 2003:138–48.

Fuller, R. C., and A. M. Johnson. 2009. A test for negative frequency-dependent mating success as a function of male colour pattern in the Bluefin Killifish. *Biological Journal of the Linnean Society* 98:489–500.

Fuller, R. C., and L. A. Noa. 2010. Female mating preferences, lighting environment, and a test of the sensory bias hypothesis in the Bluefin Killifish. *Animal Behaviour* 80:23–35.

Funk, J. L. 1955. Movement of stream fishes in Missouri. *Transactions of the American Fisheries Society* 85:39–57.

Futuyma, D. J., and G. Moreno. 1988. The evolution of ecological specialization. *Annual Review of Ecology and Systematics* 19:207–33.

Futuyma, D. J., and M. Slatkin, editors. 1983. *Coevolution*. Sinauer Associates, Sunderland, Massachusetts.

Gadomski, D. M., and C. A. Barfoot. 1998. Diel and distributional abundance patterns of fish embryos and larvae in the lower Columbia and Deschutes rivers. *Environmental Biology of Fishes* 51:353–68.

Gaffney, P. M. 1975. Roots of the niche concept. *The American Naturalist* 109:490.

Galbraith, M. G., Jr. 1967. Size-selective predation on *Daphnia* by Rainbow Trout and Yellow Perch. *Transactions of the American Fisheries Society* 96:1–10.

Gale, W. F. 1983. Fecundity and spawning frequency of caged Bluntnose

Minnows—fractional spawners. *Transactions of the American Fisheries Society* 112:398–402.

———. 1986. Indeterminate fecundity and spawning behavior of captive Red Shiners—fractional, crevice, spawners. *Transactions of the American Fisheries Society* 115:429–37.

Gale, W. F., and G. L. Buynak. 1978. Spawning frequency and fecundity of Satinfin Shiner (*Notropis analostanus*)—a fractional, crevice spawner. *Transactions of the American Fisheries Society* 107:460–63.

———. 1982. Fecundity and spawning frequency of the Fathead Minnow—a fractional spawner. *Transactions of the American Fisheries Society* 111:35–40.

Gale, W. F., and W. G. Deutsch. 1985. Fecundity and spawning frequency of captive Tessellated Darters—fractional spawners. *Transactions of the American Fisheries Society* 114:220–29.

Gale, W. F., and H. W. Mohr, Jr. 1978. Larval fish drift in a large river with comparison of sampling methods. *Transactions of the American Fisheries Society* 107:46–55.

Gallagher, R. P., and J. V. Conner. 1980. Spatio-temporal distribution of ichthyoplankton in the lower Mississippi River, Louisiana, 101–15. In *Proceedings of the fourth annual larval fish conference, University of Mississippi*. L. A. Fuiman (ed.). United States Fish and Wildlife Service, Biological Service Program, National Power Plant Team, Ann Arbor, Michigan.

Gans, C. 1986. Functional morphology of predator-prey relationships, 6–23. In *Predator-prey relationships*. M. E. Feder and G. V. Lauder (eds.). The University of Chicago Press, Illinois.

Gard, R., and G. A. Flittner. 1974. Distribution and abundance of fishes in Sagehen Creek, California. *Journal of Wildlife Management* 38:347–58.

Gardner, C., Jr., S. J. Coghlan, J. Zydlewski, and R Saunders. 2011. Distribution and abundance of stream fishes in relation to barriers: Implications for monitoring stream recovery after barrier removal. *River Research and Applications*, doi:10.1002/rra.1572.

Garrison, P. J., and R. S. Wakeman. 2000. Use of paleolimnology to document the effect of lake shoreland development on water quality. *Journal of Paleolimnology* 24:369–93.

Garrison, T. 2009. *Essentials of oceanography*. 5th ed. Brooks/Cole Cengage Learning, Belmont, California.

Gatz, A. J., Jr. 1979a. Ecological morphology of freshwater stream fishes. *Tulane Studies in Zoology and Botany* 21:91–124.

———. 1979b. Community organization in fishes as indicated by morphological features. *Ecology* 60:711–18.

———. 1981. Morphologically inferred niche differentiation in stream fishes. *American Midland Naturalist* 106:10–21.

Gee, J. H. 1983. Ecologic implications of buoyancy control in fish, 140–76. In *Fish biomechanics*. P. W. Webb and D. Weihs (eds.). Praeger Publishers, CBS Educational and Professional Publishing, New York, New York.

Gelwick, F. P. 1990. Longitudinal and temporal comparisons of riffle and pool fish assemblages in a northeastern Oklahoma stream. *Copeia* 1990:1072–82.

Gelwick, F. P., and W. J. Matthews. 1992. Effects of an algivorous minnow on temperate stream ecosystem properties. *Ecology* 73:1630–45.

Gende, S. M., A. E. Miller, and E. Hood. 2007. The effects of salmon carcasses on soil nitrogen pools in a riparian forest of southeastern Alaska. *Canadian Journal of Forest Research* 37:1194–202.

Gerald, J. W. 1971. Sound production during courtship in six species of sunfish (Centrarchidae). *Evolution* 25:75–87.

Gerken J. E., and C. P. Paukert. 2009. Threats to Paddlefish habitat: Implications for conservation, 173–83. In *Paddlefish management, propagation, and conservation in the 21st century*. C. P. Paukert and G. D. Scholten (eds.). American Fisheries Society, Symposium 66. Bethesda, Maryland.

Gerking, S. D. 1959. *The restricted movement of fish populations. Biological Reviews* 34:221–42.

Gibb, A. C., M. A. Ashley-Ross, C. M. Pace, and J. H. Long, Jr. 2011. Fish out of water: Terrestrial jumping by fully aquatic fishes. *Journal of Experimental Biology* 313A:1–5.

Gibb, A. C., and L. Ferry-Graham. 2005. Cranial movements during suction feeding in teleost fishes: Are they modified to enhance suction production? *Zoology* 108:141–53.

Gibb, A. C., L. A. Ferry-Graham, L. P. Hernandez, R. Romansco, and J. Blanton. 2008. Functional significance of intramandibular bending in poeciliid fishes. *Environmental Biology of Fishes* 83:507–19.

Gido, K. 2004. Patterns of fish invasions in the Great Plains of North America. *Biological Conservation* 118:121–31.

Gido, K. B., and J. H. Brown. 1999. Invasion of North American drainages by alien fish species. *Freshwater Biology* 42:387–99.

Gido, K. B., and N. R. Franssen. 2007. Invasion of stream fishes into low trophic positions. *Ecology of Freshwater Fish* 16:457–64.

Gido, K. B., N. R. Franssen, and D. L. Propst. 2006. Spatial variation in δ¹⁵N and δ¹³C isotopes in the San Juan River, New Mexico and Utah: Implications for the conservation of native fishes. *Environmental Biology of Fishes* 75:197–207.

Gido, K. B., C. S. Guy, T. R. Strakosh, R. J. Bernot, K. J. Hase, and M. A. Shaw. 2002. Long-term changes in the fish assemblages of the Big Blue River Basin 40 years after the construction of Tuttle Creek Reservoir. *Transactions of the Kansas Academy of Science* 105:193–208.

Gido, K. B., and W. J. Matthews. 2001. Ecosystem effects of water column minnows in experimental streams. *Oecologia* 126:247–53.

Gido, K. B., W. J. Matthews, and W. C. Wolfinbarger. 2000. Long-term changes in a reservoir fish assemblage: Stability in an unpredictable environment. *Ecological Applications* 10:1517–29.

Gido, K. B., and D. L. Propst. 1999. Habitat use and association of native and nonnative fishes in the San Juan River, New Mexico and Utah. *Copeia* 1999:321–32.

Gilbert, C. R. 1976. Composition and derivation of the North American freshwater fish fauna. *Florida Scientist* 39:104–11.

Gilbert, C. R., and G. H. Burgess. 1980. *Notropis galacturus* (Cope), Whitetail Shiner, 266. In *Atlas of North American freshwater fishes*. D. S. Lee, C. R. Gilbert, C. H. Hocutt, R. E. Jenkins, D. E. McAllister, and J. R. Stauffer, Jr., et al. 1980. North Carolina State Museum of Natural History, Raleigh.

Giles, N. 1987. Predation risk and reduced foraging activity in fish: Experiments with parasitized and non-parasitized Three-Spined Sticklebacks, *Gasterosteus aculeatus* L. *Journal of Fish Biology* 31:37–44.

Gill, A. B. 2003. The dynamics of prey choice in fish: The importance of prey size and satiation. *Journal of Fish Biology* 63:105–16.

Gill, H. S., C. B. Renaud, F. Chapleau, R. L. Mayden, and I. C. Potter. 2003. Phylogeny of living parasitic lampreys (Petromyzontiformes) based on morphological data. *Copeia* 2003:687–703.

Gilliam, J. F., and D. F. Fraser. 1987. Habitat selection under predation hazard: Test of a model with foraging minnows. *Ecology* 68:1856–62.

Gillis, G. B., and G. V. Lauder. 1995. Kinematics of feeding in Bluegill sunfish: Is there a general distinction between aquatic capture and transport mechanisms? *Journal of Experimental Biology* 198:709–20.

Gliwicz, Z. M., E. Szymanska, and D. Wrzosek. 2010. Body size distribution in *Daphnia* populations as an effect of prey selectivity by planktivorous fish. *Hydrobiologia* 643:5–19.

Godin, J.-G. J. 1997. Evading predators, 191–236. In *Behavioural ecology of teleost fishes*. J-G. J. Godin (ed.). Oxford University Press, New York, New York.

Godin, J.-G. J., L. J. Classon, and M. V. Abrahams. 1988. Group vigilance and shoal size in a small characin fish. *Behaviour* 104:29–40.

Godsoe, W. 2010. I can't define the niche but I know it when I see it: A formal link between statistical theory and the ecological niche. *Oikos* 119:53–60.

Goff, G. P. 1984. Brood care of Longnose Gar (*Lepisosteus osseus*) by Smallmouth Bass (*Micropterus dolomieui*). *Copeia* 1984:149–52.

Goldstein, R. M., and M. R. Meador. 2004. Comparisons of fish species traits from small streams to large rivers. *Transactions of the American Fisheries Society* 133:971–83.

Goldstein, R. M., and T. P. Simon. 1999. Toward a unified definition of guild structure for feeding ecology of North American freshwater fishes, 123–220. In *Assessing the sustainability and biological integrity of water resources using fish communities*. T. P. Simon (ed.). CRC Press, Boca Raton, Florida.

Goldthwait, R. P., A. Dreimanis, J. L. Forsyth, P. F. Karrow, and G. W. White. 1965. Pleistocene deposits of the Erie Lobe, 85–97. In *The Quaternary of the United States*. H. E. Wright, Jr., and D. G. Frey (eds.). Princeton University Press, New Jersey.

Gomahr, A., M. Palzenberger, and K. Kotrschal. 1992. Density and distribution of external taste buds in cyprinids. *Environmental Biology of Fishes* 33:125–34.

Gomulkiewicz, R., S. L. Nuismer, and J. N. Thompson. 2003. Coevolution in variable mutualisms. *The American Naturalist* 162:S80–S93.

Gomulkiewicz, R., J. N. Thompson, R. D. Holt, S. L. Nuismer, and M. E. Hochberg. 2000. Hot spots, cold spots, and the geographic mosaic theory of coevolution. *The American Naturalist* 156:156–74.

Goodrich, J. S., S. L. Sanderson, I. E. Batjakas, and L. S. Kaufman. 2000. Branchial arches of suspension-feeding *Oreochromis esculentus*: Sieve or sticky filter? *Journal of Fish Biology* 56:858–75.

Gorman, O. T. 1986. Assemblage organization of stream fishes: The effect of rivers on adventitious streams. *The American Naturalist* 128:611–16.

———. 1988a. The dynamics of habitat use in a guild of Ozark minnows. *Ecological Monographs* 58:1–18.

———. 1988b. An experimental study of habitat use in an assemblage of Ozark minnows. *Ecology* 69:1239–50.

———. 1992. Evolutionary ecology and historical ecology: Assembly, structure, and organization of stream fish communities, 659–88. In *Systematics, historical ecology, and North American freshwater fishes*. R. L. Mayden (ed.). Stanford University Press, Stanford, California.

Gosline, W. A. 1971. *Functional morphology and classification of teleostean fishes*. The University Press of Hawaii, Honolulu.

———. 1973. Considerations regarding the phylogeny of cypriniform fishes, with special reference to structures associated with feeding. *Copeia* 1973:7761–876.

———. 1981. The evolution of the premaxillary protrusion system in some teleostean fish groups. *Journal of Zoology*, London 193:11–23.

———. 1987. Jaw structures and movements in higher teleostean fishes. *Japanese Journal of Ichthyology* 34:21–32.

———. 1997. Functional morphology of the caudal skeleton in teleostean fishes. *Ichthyological Research* 44:137–41.

Gotelli, N. J. 2001. *A primer of ecology*. 3rd ed. Sinauer Associates, Inc. Sunderland, Massachusetts.

Gotelli, N. J., and A. M. Ellison. 2004. *A primer of ecological statistics*. Sinauer Associates, Inc. Sunderland, Massachusetts.

Gotelli, N. J., and G. R. Graves. 1996. *Null models in ecology*. Smithsonian Institution Press, Washington, DC.

Gotelli, N. J., and D. J. McCabe. 2002. Species co-occurrence: A meta-analysis of J. M. Diamond's assembly rules model. *Ecology* 83:2091–96.

Gotelli, N. J., and M. Pyron. 1991. Life history variation in North American freshwater minnows: Effects of latitude and phylogeny. *Oikos* 62:30–40.

Gotelli, N. J., and C. M. Taylor. 1999a. Testing metapopulation models with stream-fish assemblages. *Evolutionary Ecology Research* 1:835–45.

———. 1999b. Testing macroecology models with stream-fish assemblages. *Evolutionary Ecology Research* 1:847–58.

Gould, S. J. 2002. *The structure of evolutionary theory*. Belknap Press of Harvard University Press, Cambridge, Massachusetts.

Gould, S. J., and E. S. Vrba. 1982. Exaptation—a missing term in the science of form. *Paleobiology* 8:4–15.

Gowan, C. M., and K. D. Fausch. 1996. Long-term demographic responses of trout populations to habitat manipulation in six Colorado streams. *Ecological Applications* 6:931–46.

Gowan, C. M., M. K. Young, K. D. Fausch, and S. C. Riley. 1994. Restricted movement in resident stream salmonids: A paradigm lost? *Canadian Journal of Fisheries and Aquatic Sciences.* 51:2626–37.

Gowaty, P. A. 1984. Cuckoldry: The limited scientific usefulness of a colloquial term. *Animal Behaviour* 32:924–25.

Gozlan, R. E. 2008. Introduction of non-native freshwater fish: Is it all bad? *Fish and Fisheries* 9:106–15.

———. 2009. Response by R Gozlan. Biodiversity crisis and the introduction of non-native fish: Solutions not scapegoats. *Fish and Fisheries* 10:109–10.

Gozlan, R. E., J. R. Britton, I. Cowx, and G. H. Copp. 2010. Current knowledge on non-native freshwater fish introductions. *Journal of Fish Biology* 76:751–86.

Grande, L. 1982. A revision of the fossil genus *Knightia*, with a description of a new genus from the Green River Formation (Teleostei, Clupeidae). *American Museum Novitates* 2731:1–22.

———. 1984. *Paleontology of the Green River Formation, with a review of the fish fauna*. 2nd ed. *Geological Survey of Wyoming*, Bulletin 63:1–333.

———. 1990. Vicariance biogeography, 448–51. In *Palaeobiology: A synthesis*. D. E. G. Briggs and P. Crowther (eds.). Blackwell Scientific Publications, Oxford, United Kingdom.

———. 1999. The first *Esox* (Esocidae: Teleostei) from the Eocene Green River Formation, and a brief review of esocid fishes. *Journal of Vertebrate Paleontology* 19:271–92.

———. 2001. An updated review of the fish faunas from the Green River Formation, the world's most productive freshwater Lagerstätten, 1–38. In *Eocene biodiversity: Unusual occurrences and rarely sampled habitats*. G. F. Gunnell (ed.). Kluwer Academic/Plenum Publishers, New York, New York.

Grande, L., and W. E. Bemis. 1991. Osteology and phylogenetic relationships of fossil and Recent paddlefishes (Polyodontidae) with comments on the interrelationships of Acipenseriformes. *Journal of Vertebrate Paleontology* 11, supplement 1:1–121.

Grande, L., J. T. Eastman, and T. M. Cavender. 1982. *Amyzon gosiutensis*, a new catostomid fish from the Green River Formation. *Copeia* 1982:523–32.

———. 1996. Interrelationships of Acipenseriformes, with comments on "Chondrostei," 85–115. In *Interrelationships of fishes*. M. Stiassny, L. Parenti, and G. D. Johnson (eds.). Academic Press, San Diego, California.

———. 1998. A comprehensive phylogenetic study of amiid fishes (Amiidae) based on comparative skeletal anatomy. An empirical search for interconnected patterns of natural history. *Journal of Vertebrate Paleontology* 19, supplement 1, memoir 4:1–690.

———. 1999. Historical biogeography and historical paleoecology of Amiidae and other halecomorph fishes, 413–24. In *Systematics and fossil record*. G. Arratia and H.-P. Schultze (eds.). Verlag Dr. Frierich Pfeil, Munchen Germany.

Grande, L., F. Jin, Y. Yabumoto, and W. E. Bemis. 2002. *Protopsephurus liui*, a well-preserved primitive paddlefish (Acipenseriformes: Polyodontidae) from the lower Cretaceous of China. *Journal of Vertebrate Paleontology* 22:209–37.

Grande, L., and J. G. Lundberg. 1988. Revision and redescription of the genus *Astephus* (Siluriformes; Ictaluridae) with a discussion of its phylogenetic relationships. *Journal of Vertebrate Paleontology* 8:139–71.

Grant, E. H. C., W. H. Lowe, and W. F. Fagan. 2007. Living in the branches: Population dynamics and ecological processes in dendritic networks. *Ecology Letters* 10:165–75.

Grant, J. W. A., D. L. G. Noakes, and K. M. Jonas. 1989. Spatial distribution of defence and foraging in young-off-the-year Brook Charr, *Salvelinus fontinalis*. *Journal of Animal Ecology* 58:773–84.

Graybill, J. P., R. L. Burgner, J. C. Gislason, P. E. Huffman, K. H. Wyman, Q. J. Stober, T. W. Fagnan, and A. P. Stayman. 1978. *Assessment of the reservoir-related effects of the Skagit Project on downstream fishery resources of the Skagit River, Washington*. University of Washington, Fisheries Research Institute, Interim Report for the City of Seattle, Department of Lighting, Seattle.

Green, R. H. 1971. A multivariate statistical approach to the Hutchinsonian niche: Bivalve molluscs of central Canada. *Ecology* 52:543–56.

Greenberg, L. A. 1991. Habitat use and feeding behavior of thirteen species of benthic stream fishes. *Environmental Biology of Fishes* 31:389–401.

Greenberg, L. A., B. Hernnäs, C. Brönmark, J. Dahl, A. Eklöv, and K. H. Olsén. 2002. Effects of kinship on growth and movements of Brown Trout in field enclosures. *Ecology of Freshwater Fish* 11:251–59.

Gregory, S., H. Li, and J. Li. 2002. The conceptual basis for ecological responses to dam removal. *BioScience* 52:713–23.

Grier, H. J. 1981. Cellular organization of the testis and spermatogenesis in fishes. *American Zoologist* 21:345–57.

Grier, H. J., D. P. Moody, and B. C. Cowell. 1990. Internal fertilization and sperm morphology in the Brook Silverside, *Labidesthes sicculus* (Cope). *Copeia* 1990:221–26.

Griffiths, D. 1980. Foraging costs and relative prey size. *The American Naturalist* 116:743–52.

———. 1994. The size structure of lacustrine Arctic Charr (Pisces: Salmonidae) populations. *Biological Journal of the Linnean Society* 51:337–57.

Griffiths, S. W. 2003. Learned recognition of conspecifics by fishes. *Fish and Fisheries* 4:256–68.

Griffiths, S. W., and J. D. Armstrong. 2001. The benefits of genetic diversity outweigh those of kin association in a territorial animal. *Proceedings of the Royal Society of London B* 268:1293–96.

Grinnell, J. 1904. The origin and distribution of the Chest-Nut-Backed Chickadee. *The Auk* 21:364–82.

———. 1917a. The niche-relationships of the California Thrasher. *The Auk* 34:427–33.

———. 1917b. Field tests of theories concerning distributional control. *The American Naturalist* 51:115–28.

Gross, E. L., P. J. Patchett, T. A. Dallegge, and J. E. Spencer. 2001. The Colorado River system and Neogene sedimentary formations along its course: Apparent Sr isotopic connections. *Journal of Geology* 109:449–61.

Gross, M. R. 1982. Sneakers, satellites and parentals: Polymorphic mating strategies in North American sunfishes. *Zeitschrift für Tierpsychologie* 60:1–26.

———. 1987. Evolution of diadromy in fishes. *American Fisheries Society Symposium* 1:14–25.

Gross, M. R., and E. L. Charnov. 1980. Alternative male life histories in Bluegill sunfish. *Proceedings of the National Academy of Sciences* 77:6937–40.

Gross, M. R., R. M. Coleman, R. M. McDowall. 1988. Aquatic productivity and the evolution of diadromous fish migration. *Science* 239:1291–93.

Gross, M. R., and A. M. Macmillan. 1981. Predation and the evolution of colonial nesting in Bluegill sunfish (*Lepomis macrochirus*). *Behavioral Ecology and Sociobiology* 8:163–74.

Gross, M. R., and R. C. Sargent. 1985. The evolution of male and female parental care in fishes. *American Zoologist* 25:807–22.

Grossman, G. D., and V. Boulé. 1991. Effects of Rosyside Dace (*Clinostomus funduloides*) on microhabitat use of Rainbow Trout (*Oncorhynchus mykiss*). *Canadian Journal of Fisheries and Aquatic Sciences* 48:1235–43.

Grossman, G. D., J. F. Dowd, and M. Crawford. 1990. Assemblage stability in stream fishes: A review. *Environmental Management* 14:661–71.

Grossman, G. D., and M. C. Freeman. 1987. Microhabitat use in a stream fish assemblage. *Journal of Zoology*, London 212:151–76.

Grossman, G. D., J. Hill, and J. T. Petty. 1995. Observations on habitat structure, population regulation, and habitat use with respect to evolutionarily significant units: A landscape perspective for lotic systems. *American Fisheries Society Symposium* 17:381–91.

Grossman, G. D., P. B. Moyle, and J. O. Whitaker, Jr. 1982. Stochasticity in structural and functional characteristics of an Indiana stream fish assemblage: A test of community theory. *The American Naturalist* 120:423–54.

Grossman, G. D., and R. E. Ratajczak, Jr. 1998. Long-term patterns of microhabitat use by fish in a southern Appalachian stream from 1983 to 1992: Effects of hydrologic period, season and fish length. *Ecology of Freshwater Fish* 7:108–31.

Grossman, G. D., R. E. Ratajczak, Jr., M. Crawford, and M. C. Freeman. 1998. Assemblage organization in stream fishes: Effects of environmental variation and interspecific interactions. *Ecological Monographs* 68:395–420.

Grossman, G. D., P. A. Rincon, M. D. Farr, and R. E. Ratajczak, Jr. 2002. A new optimal foraging model predicts habitat use by drift-feeding stream minnows. *Ecology of Freshwater Fish* 11:2–10.

Grutter, A. S. 2002. Cleaning symbioses from the parasites' perspective. *Parasitology* 124:S65–S81.

Gunderson, D. R. 1997. Trade-off between reproductive effort and adult survival in oviparous and viviparous fishes. *Canadian Journal of Fisheries and Aquatic Sciences* 54:990–98.

Gunning, G. E., and R. D. Suttkus. 1991. Species dominance in the fish populations of the Pearl River at two study areas in Mississippi and Louisiana:1966–1988. *Southeastern Fishes Council Proceedings* 23:7–15.

Guthrie, D. M., and W. R. A. Muntz. 1993. Role of vision in fish behaviour, 89–128. In *Behaviour of teleost fishes*. T. J. Pitcher (ed.). Chapman and Hall, New York, New York.

Gutowski, M. J., and J. R. Stauffer. 1993. Selective predation by *Noturus insignis* (Richardson) (Teleostei: Ictaluridae) in the Delaware River. *American Midland Naturalist* 129:309–18.

Gutreuter, S., A. D. Bartels, K. Irons, and M. B. Sandheinrich. 1999. Evaluation of the flood-pulse concept based on statistical models of growth of selected fishes of the Upper Mississippi River system. *Canadian Journal of Fisheries and Aquatic Sciences* 56:2282–91.

Hager, M. C., and G. S. Helfman. 1991. Safety in numbers: Shoal size choice by minnows under predatory threat. *Behavioral Ecology and Sociobiology* 29:271–76.

Hairston, N. G., Sr. 1989. Ecological experiments: Purpose, design, and execution. *Cambridge Studies in Ecology*. Cambridge University Press, New York, New York.

Halbgewachs, C. F., T. A. Marchant, R. C. Kusch, and D. P. Chivers. 2009. Epidermal club cells and the innate immune system of minnows. *Biological Journal of the Linnean Society* 98:891–97.

Halford, G. S., W. H. Wilson, and S. Phillips. 1998. Processing capacity defined by relational complexity: Implications for comparative, developmental, and cognitive psychology. *Behavioral and Brain Sciences* 21:803–65.

Hall, A. A., S. B. Rood, and P. S. Higgins. 2009. Resizing a river: A downscaled, seasonal flow regime promotes riparian restoration. *Restoration Ecology* 19:351–59.

Hall, D. J., W. E. Cooper, and E. E. Werner. 1970. An experimental approach to the production dynamics and structure of freshwater animal communities. *Limnology and Oceanography* 15:839–928.

Halyk, L. C., and E. K. Balon. 1983. Structure and ecological production of the fish taxocene of a small floodplain system. *Canadian Journal of Zoology* 61:2446–64.

Hambright, K. D. 1991. Experimental analysis of prey selection by Largemouth Bass: Role of predator mouth width and prey body depth. *Transactions of the American Fisheries Society* 120:500–508.

Hamilton, W. D. 1964a. The genetical evolution of social behaviour. I. *Journal of Theoretical Biology* 7:1–16.

———. 1964b. The genetical evolution of social behaviour. II. *Journal of Theoretical Biology* 7:17–52.

———. 1971. Geometry for the selfish herd. *Journal of Theoretical Biology* 31:295–311.

Hankinson, T. L. 1932. Observations on the breeding habitats of fishes in southern Michigan. *Papers of the Michigan Academy of Science, Arts, and Letters* 15:411–24.

Hansen, D. F. 1965. Further observations on nesting of the White Crappie, *Pomoxis annularis*. *Transactions of the American Fisheries Society* 94:182–84.

Hansen, M. J., and C. W. Ramm. 1994. Persistence and stability of fish community structure in a southwest New York stream. *American Midland Naturalist* 132:52–67.

Hanski, I., and M. E. Gilpin. 1991. Metapopulation dynamics: Brief history and conceptual domain. *Biological Journal of the Linnean Society* 42:3–16.

———. 1997. *Metapopulation biology: Ecology, genetics, and evolution*. Academic Press, New York, New York.

Hanski, I., and D. Simberloff. 1997. The metapopulation approach, its history, conceptual domain, and application to conservation, 5–26. In *Metapopulation biology: Ecology, genetics, and evolution*. I. Hanski and M. E. Gilpin (eds.). Academic Press, New York, New York.

Harden Jones, F. R. 1968. *Fish migration*. Edward Arnold Publisher, London, United Kingdom.

Harding, J. S., E. F. Benfield, P. V. Bolstad, G. S. Helfman, and E. B. D. Jones, III. 1998. Stream biodiversity: The ghost of land use past. *Proceedings of the National Academy of Sciences* 95:14843–47.

Hardisty, M. W., and I. C. Potter. 1971. The general biology of adult lampreys, 127–206. In *The biology of lampreys*. Vol. 1. M. W. Hardisty and I. C. Potter (eds.). Academic Press, New York, New York.

Hardy, J. D., Jr. 1978. *Development of fishes of the mid-Atlantic bight*. Vol. 2. *Anguillidae through Syngnathidae*. FWS/OBS-78/12, U.S. Fish and Wildlife Service, Biological Service Program.

Harig, A. L., and M. B. Bain. 1998. Defining and restoring biological integrity in wilderness lakes. *Ecology* 8:71–87.

Haro, A., W. Richkus, K. Whalen, A. Hoar, W.-D. Busch, S. Lary, T. Brish, and D. Dixon. 2000. Population decline of the American Eel: Implications for research and management. *Fisheries* 25:7–16.

Harper, D. G., and R. W. Blake. 1991. Prey capture and the fast-start performance of Northern Pike *Esox lucius*. *Journal of Experimental Biology* 155:175–92.

Harrington, R. W., Jr. 1955. The osteocranium of the American cyprinid fish, *Notropis bifrenatus*, with an annotated synonymy of teleost skull bones. *Copeia* 1955:267–90.

Harrington, R. W., Jr., and E. S. Harrington. 1961. Food selection among fishes invading a high subtropical salt marsh: From onset of flooding through the progress of a mosquito brood. *Ecology* 42:646–66.

Harrison, S., D. D. Murphy, and P. R. Ehrlich. 1988. Distribution of the Bay Checkerspot Butterfly, *Euphydryas editha bayensis*: Evidence for a metapopulation model. *The American Naturalist* 132:360–82.

Harrison, S., and A. D. Taylor. 1997. Empirical evidence for metapopulation dynamics, 27–42. In *Metapopulation biology, ecology, genetics, and evolution*. I. Hanski and M. E. Gilpin (eds.). Academic Press, San Diego, California.

Hart, D. D., T. E. Johnson, K. L. Bushaw-Newton, R. J. Horwitz, A. T. Bednarek, D. F. Charles, D. A. Kreeger, and D. J. Velinsky. 2002. Dam removal: Challenges and opportunities for Ecological Research and river restoration. *BioScience* 52:669–81.

Hart, W. S., J. Quade, D. B. Madsen, D. S. Kaufman, and C. G. Oviatt. 2004. The $^{87}Sr/^{86}Sr$ ratios of lacustrine carbonates and lake-level history of the Bonneville paleolake system. *Geological Society of America Bulletin* 116:1107–19.

Harvey, B. C. 1987. Susceptibility of young-of-the-year fishes to downstream displacement by flooding. *Transactions of the American Fisheries Society* 116:851–55.

———. 1991a. Interactions among stream fishes: Predator-induced habitat shifts and larval survival. *Oecologia* 87:29–36.

———. 1991b. Interaction of abiotic and biotic factors influences larval fish survival in an Oklahoma stream. *Canadian Journal of Fisheries and Aquatic Sciences* 48:1476–80.

Harvey, B. C., R. C. Cashner, and W. J. Matthews. 1988. Differential effects of Largemouth and Smallmouth bass on habitat use by Stoneroller minnows in stream pools. *Journal of Fish Biology* 33:481–87.

Harvey, B. C., R. J. Nakamoto, and J. L. White. 1999. Influence of large woody debris and a bankfull flood on movement of adult resident coastal Cutthroat Trout (*Oncorhynchus clarki*) during fall and winter. *Canadian Journal of Fisheries and Aquatic Sciences* 56:2161–66.

Harvey, B. C., and A. J. Stewart. 1991. Fish size and habitat depth relationships in headwater streams. *Oecologia* 87:336–42.

Harvey, B. C., J. L. White, and R J. Nakamoto. 2002. Habitat relationships and larval drift of native and nonindigenous fishes in neighboring tributaries of a coastal California river. *Transactions of the American Fisheries Society* 131:159–70.

Hasler, A. D. 1956. Perception of pathways by fishes in migration. *The Quarterly Review of Biology* 31:200–209.

Hasler, A. D., and A. T. Scholz. 1980. Artificial imprinting: A procedure for conserving salmon stocks, 179–99. In *Fish behavior and its use in the capture and culture of fishes.* J. E. Bardach, J. J. Magnuson, R. C. May, and J. M. Reinhart (eds.). *Physiological and behavioral manipulation of food fish as production and management tools*, Bellagio, Italy, 3–8 November. International Center for Living Aquatic Resources Management, Manila, Phillipines.

Hasler, A. D., and W. J. Wisby. 1951. Discrimination of stream odors by fishes and its relation to parent stream behavior. *The American Naturalist* 85:223–38.

Hastings, R. W., and R. W. Yerger. 1971. Ecology and life history of the Diamond Killifish, *Adinia xenica* (Jordan and Gilbert). *American Midland Naturalist* 86:276–91.

Havel, J. E. 2009. Stanley Ivan Dodson: A distinguished ecologist, naturalist, and teacher. *Hydrobiologia* 636:1–6.

Havel, J. E., C. E. Lee, and M. J. Vander Zanden. 2005. Do reservoirs facilitate invasions into landscapes? *BioScience* 55:518–25.

Hay, M. E., J. D. Parker, D. E. Burkepile, C. C. Caudill, A. E. Wilson, Z. P. Hallinan, and A. D. Chequer. 2004. Mutualisms and aquatic community structure: The enemy of my enemy is my friend. *Annual Review of Ecology and Systematics* 35:175–97.

Haynes, J. L., and R. C. Cashner. 1995. Life history and population dynamics of the Western Mosquitofish: A comparison of natural and introduced populations. *Journal of Fish Biology* 46:1026–41.

He, J. X., and D. J. Stewart. 2001. Age and size at first reproduction of fishes: Predictive models based only on growth trajectories. *Ecology* 82:784–91.

Healey, M. C., and C. Groot. 1987. Marine migration and orientation of ocean-type Chinook and Sockeye salmon, 298–312. In *Common strategies of anadromous and catadromous fishes.* M. J. Dadswell, R. J. Klauda, C. M. Moffitt, R. L. Saunders, R. A. Rulifson, and J. E. Cooper (eds.). *American Fisheries Society Symposium 1*, Bethesda, Maryland.

Hedges, S. B. 1996. Historical biogeography of West Indian vertebrates. *Annual Review of Ecology and Systematics* 27:163–96.

Heidinger, R. C. 1976. Synopsis of biological data on the Largemouth Bass, *Micropterus salmoides* (Lacepède) 1802. *FAO Fisheries Synopsis* 115:1–85.

Heins, D. C. 1990. Field evidence for multiple clutches in the Longnose Shiner. *Copeia* 1990:579–82.

———. 1991. Variation in reproductive investment among populations of the Longnose Shiner, *Notropis longirostris*, from contrasting environments. *Copeia* 1991:736–44.

Heins, D. C., and J. A. Baker. 1989. Growth, population structure, and reproduction of the percid fish *Percina vigil*. *Copeia* 1989:727–36.

———. 1993. Reproductive biology of the Brighteye Darter, *Etheostoma lynceum* (Teleostei: Percidae), from the Homochitto River, Mississippi. *Ichthyological Explorations of Freshwaters* 14:11–20.

Heins, D. C., J. A. Baker, and W. P. Dunlap. 1992. Yolk loading in oocytes of darters and its consequences for life-history study. *Copeia* 1992:404–12.

Heins, D. C., J. A. Baker, and J. M. Guill. 2004. Seasonal and interannual components of intrapopulation variation in clutch size and egg size of a darter. *Ecology of Freshwater Fish* 13:258–65.

Heins, D. C., J. A. Baker, and D. J. Tylicki. 1996. Reproductive season, clutch size, and egg size of the Rainbow Darter, *Etheostoma caeruleum*, from the Homochitto River, Mississippi, with an evaluation of data from the literature. *Copeia* 1996:1005–10.

Heins, D. C., and G. H. Clemmer. 1976. The reproductive biology, age and growth of the North American cyprinid, *Notropis longirostris* (Hay), in Mississippi. *American Midland Naturalist* 94:284–95.

Heins, D. C., and F. G. Rabito, Jr. 1986. Spawning performance in North American minnows: Direct evidence of the occurrence of multiple clutches in the genus *Notropis*. *Journal of Fish Biology* 28:343–57.

Heise, R. J., W. T. Slack, S. T. Ross, and M. A. Dugo. 2004. Spawning and associated movement patterns of Gulf Sturgeon in the Pascagoula River drainage, Mississippi. *Transactions of the American Fisheries Society* 133:221–30.

———. 2005. Gulf Sturgeon summer habitat use and fall migration in the Pascagoula River, Mississippi. *Journal of Applied Ichthyology* 21:461–68.

Helfield, J. M., and R. J. Naiman. 2001. Effects of salmon-derived nitrogen on riparian forest growth and implications for stream productivity. *Ecology* 82:2403–9.

Helfman, G. S. 1979. Twilight activities of Yellow Perch, *Perca flavescens*. *Journal of the Fisheries Research Board of Canada* 36:173–79.

———. 1981. Twilight activities and temporal structure in a freshwater fish community.

Canadian Journal of Fisheries and Aquatic Sciences 38:1405–20.

———. 1986. Behavioral responses of prey fishes during predator-prey interactions, 135–56. In *Predator-prey relationships*. M. E. Feder and G. V. Lauder (eds.). The University of Chicago Press, Chicago, Illinois.

———. 2007. *Fish Conservation: A guide to understanding and restoring global aquatic biodiversity and fishery resources.* Island Press, Washington, DC.

Helfman, G. S., and J. B. Clark. 1986. Rotational feeding: Overcoming gape-limited foraging in anguillid eels. *Copeia* 1986:679–85.

Helfman, G. S., B. B. Collette, D. E. Facey, and B. W. Bowen. 2009. *The diversity of fishes.* 2nd ed. Wiley-Blackwell, Chichester, West Sussex, United Kingdom.

Helfman, G. S., and D. L. Winkelman. 1991. Energy trade-offs and foraging mode choice in American Eels. *Ecology* 72:310–18.

Helms, B. S., D. C. Werneke, M. M. Gangloff, E. E. Hartfield, and J. W. Feminella. 2011. The influence of low-head dams on fish assemblages in streams across Alabama. *Journal of the North American Benthological Society* 30:1095–106.

Henderson, B. A., J. L. Wong, and S. J. Nepszy. 1996. Reproduction of Walleye in Lake Erie: Allocation of energy. *Canadian Journal of Fisheries and Aquatic Sciences* 53:127–33.

Hendry, A. P., and O. K. Berg. 1999. Secondary sexual characters, energy use, senescence, and the cost of reproduction in Sockeye Salmon. *Canadian Journal of Zoology* 77:1663–75.

Hennig, W. 1966. *Phylogenetic systematics.* University of Illinois Press, Urbana.

Henrich, S. 1988. Variation in offspring sizes of the poeciliid fish *Heterandria formosa* (Pisces: Poeciliidae) in relation to fitness. *Oikos* 51:13–18.

Henshall, J. A. 1881. *Book of the black bass, comprising its complete scientific and life history, together with a practical treatise on angling and fly fishing, and a full description of tools, tackle and implements.* Robert Clarke and Co., Cincinnati, Ohio.

Hernandez, L. P., L. Ferry-Graham, and A. C. Gibb. 2008. Morphology of a picky eater: A novel mechanism underlies premaxillary protrusion and retraction within cyprinodontiforms. *Zoology* 111:442–54.

Hernandez, L. P., A. C. Gibb, and L. Ferry-Graham. 2009. Trophic apparatus in cyprinodontiform fishes: Functional specializations for picking and scraping behaviors. *Journal of Morphology* 270:645–61.

Hesslein, R. H., K. A. Hallard, and P. Ramlal. 1993. Replacement of sulfur, carbon, and nitrogen in tissue of growing Broad Whitefish (*Coregonus nasus*) in response to a change in diet traced by δ^{34}S, δ^{13}C, and δ^{15}N. *Canadian Journal of Fisheries and Aquatic Sciences* 50:2071–76.

Higham, T. E., S. W. Day, and P. C. Wainwright. 2005. Sucking while swimming: Evaluating the effects of ram speed on suction generation in Bluegill sunfish *Lepomis macrochirus* using digital particle image velocimetry. *Journal of Experimental Biology* 208:2653–60.

———. 2006a. Multidimensional analysis of suction feeding performance in fishes: Fluid speed, acceleration, strike accuracy and the ingested volume of water. *Journal of Experimental Biology* 209:2713–25.

———. 2006b. The pressure of suction feeding: The relation between buccal pressure and induced fluid speed in centrarchid fishes. *Journal of Experimental Biology* 209:3281–87.

Hinch, S. G., N. C. Collins, and H. H. Harvey. 1991. Relative abundance of littoral zone fishes: Biotic interactions, abiotic factors, and postglacial colonization. *Ecology* 72:1314–24.

Hoare, D. J., G. D. Ruxton, J.-G. J. Godin, and J. Krause. 2000. The social organization of free-ranging fish shoals. *Oikos* 89:546–54.

Hobson, E. S. 1974. Feeding relationships of teleostean fishes on coral reefs in Kona, Hawaii. *Fishery Bulletin* 72:615–1031.

Hocking, M. D., and J. D. Reynolds. 2011. Impacts of salmon on riparian plant diversity. *Science* 331:1609–12.

Hocutt, C. H. 1987. Evolution of the Indian Ocean and the drift of India: A vicariant event. *Hydrobiologia* 150:203–23.

Hoetker, G. M., and K. W. Gobalet. 1999. Fossil Razorback Sucker (Pisces: Catostomidae, *Xyrauchen texanus*) from southeastern California. *Copeia* 1999:755–59.

Hoff, M. H., and C. R. Bronte. 1999. Structure and stability of the midsummer fish communities in Chequamegon Bay, Lake Superior, 1973–1996. *Transactions of the American Fisheries Society* 128:362–73.

Holland-Bartels, L. E., S. K. Littlejohn, and M. L. Huston. 1990. *A guide to larval fishes of the upper Mississippi River.* U.S. Fish and Wildlife Service, National Fisheries Research Center, LaCrosse, Wisconsin.

Holt, D. E., and C. E. Johnston. 2011. Hearing sensitivity in two black bass species using the auditory brainstem response approach. *Environmental Biology of Fishes* 91:121–26.

Holzman, R., D. C. Collar, S. W. Day, K. L. Bishop, and P. C. Wainwright. 2008c. Scaling of suction-induced flows in Bluegill: Morphological and kinematic predictors for the ontogeny of feeding performance. *Journal of Experimental Biology* 211:2658–68.

Holzman, R., S. W. Day, R. S. Mehta, and P. C. Wainwright. 2008a. Jaw protrusion enhances forces exerted on prey by suction feeding fishes. *Journal of the Royal Society*, Interface 5:1445–57.

———. 2008b. Integrating the determinants of suction feeding performance in centrarchids. *Journal of Experimental Biology* 211:3296–305.

Holzman, R., S. W. Day, and P. C. Wainwright. 2007. Timing is everything: Coordination of strike kinematics affects the force exerted by suction feeding fish on attached prey. *Journal of Experimental Biology* 210:3328–36.

Holzman, R., and P. C. Wainwright. 2009. How to surprise a copepod: Strike kinematics reduce hydrodynamic disturbance and increased stealth of suction-feeding fish. *Limnology and Oceanography* 54:2201–12.

Hoogenboezem, W., E. H. R. R. Lammens, P. J. MacGillavry, and F. A. Sibbing. 1993. Prey retention and sieve adjustment in filter-feeding bream (*Abramis brama*) (Cyprinidae). *Canadian Journal of Fisheries and Aquatic Sciences* 50:465–71.

Hoogenboezem, W., J. G. M. van den Boogaart, F. A. Sibbing, E. H. R. R Lammens, A. Terlouw, and J. W. M. Osse. 1991. A new model of particle retention and branchial sieve adjustment in filter-feeding Bream (*Abramis brama*, Cyprinidae). *Canadian Journal of Fisheries and Aquatic Sciences* 48:7–18.

Hoogland, R., D. Morris, and N. Tinbergen. 1956. The spines of sticklebacks (*Gasterosteus* and *Pygosteus*) as means of defence against predators (*Perca* and *Esox*). *Behaviour* 10:205–36.

Horn, M. H., and L. A. Ferry-Graham. 2006. Feeding mechanisms and trophic interactions, 387–410. In *Ecology of marine fishes: California and adjacent waters*. L. G. Allen, D. J. Pondella II, and M. H. Horn (eds.). University of California Press, Berkeley.

Horton, R. E. 1945. Erosional development of streams and their drainage basins. *Bulletin of the Geological Society of America* 56:275–370.

Horwitz, R. J. 1978. Temporal variability patterns and the distributional patterns of stream fishes. *Ecological Monographs* 48:307–21.

Howes, G. J. 1991. Systematics and biogeography: An overview, 1–33. In *Cyprinid fishes: Systematics, biology, and exploitation*. I. J. Winfield and J. S. Nelson (eds.). Chapman and Hall, London, United Kingdom.

Howick, G. L., and W. J. O'Brien. 1983. Piscivorous feeding behavior of Largemouth Bass: An experimental analysis. *Transactions of the American Fisheries Society* 112:508–16.

Hrbek, T., and A. Larson. 1999. The evolution of diapause in the killifish family Rivulidae (Atherinomorpha, Cyprinodontiformes): A molecular phylogenetic and biogeographic perspective. *Evolution* 53:1200–1216.

Hubbs, C. 1976. The diel reproductive pattern and fecundity of *Menidia audens Copeia* 1976:386–88.

———. 1985. Darter reproductive seasons. *Copeia* 1985:56–68.

Hubbs, C. L., and L. C. Hubbs. 1932. Apparent parthenogenesis in nature, in a form of fish of hybrid origin. *Science* 76:628–30.

———. 1946. Experimental breeding of the Amazon Molly. *The Aquarium Journal* 17:4–6.

Hubbs, C. L., R. R. Miller, and L. C. Hubbs. 1974. Hydrographic history and relict fishes of the North-Central Great Basin. *Memoirs of the California Academy of Sciences* 7:1–259.

Hubbs, C. L., and T. E. B. Pope. 1937. The spread of the Sea Lamprey through the Great Lakes. *Transactions of the American Fisheries Society* 66:172–76.

Hubbs, C. L., and E. C. Raney. 1946. *Endemic fish fauna of Lake Waccamaw, North Carolina*. Miscellaneous Publications Museum of Zoology, University of Michigan 65:1–30.

Hubbs, C., and K. Strawn. 1957. The effects of light and temperature on the fecundity of the Greenthroat Darter, *Etheostoma lepidum*. *Ecology* 38:596–602.

Huck, L. L., and G. E. Gunning. 1967. Behavior of the Longear Sunfish, *Lepomis megalotis* (Rafinesque). *Tulane Studies in Zoology* 14:121–31.

Huckins, C. J. F. 1997. Functional linkages among morphology, feeding performance, diet, and competitive ability in molluscivorous sunfish. *Ecology* 78:2401–14.

Hulsey, C. D., F. J. García de León, Y. S. Johnson, D. A. Hendrickson, and T. J. Near. 2004. Temporal diversification of Mesoamerican cichlid fishes across a major biogeographic boundary. *Molecular Phylogenetics and Evolution* 31:754–64.

Hulsey, C. D., D. A. Hendrickson, and F. J. García de León. 2005. Trophic morphology, feeding performance and prey use in the polymorphic fish *Herichthys minckleyi*. *Evolutionary Ecology Research* 7:303–24.

Hulsey, C. D., J. Marks, D. A. Hendrickson, C. A. Williamson, A. E. Cohen, and M. J. Stephens.

2006. Feeding specialization in *Herichthys minckleyi*: A trophically polymorphic fish. *Journal of Fish Biology* 68:1399–410.

Hurlbert, S. H. 1981. A gentle depilation of the niche: Dicean resource sets in resource hyperspace. *Evolutionary Theory* 5:177–84.

———. 1984. Pseudoreplication and the design of ecological field experiments. *Ecological Monographs* 54(2):187–211.

———. 2009. The ancient black art and transdisciplinary extent of pseudoreplication. *Journal of Comparative Psychology* 123:434–43.

Huston, M. 1979. A general hypothesis of species diversity. *The American Naturalist* 113:81–101.

Hutchings, J. A., and M. Festa-Bianchet. 2009. Canadian species at risk (2006–2008), with particular emphasis on fishes. *Environmental Reviews* 17:53–65.

Hutchinson, G. E. 1957a. Concluding remarks. *Cold Spring Harbor Symposium on Quantitative Biology* 22:415–27.

———. 1957b. *A treatise on limnology.* Vol. 1. John Wiley and Sons, Inc. New York New York.

———. 1978. *An introduction to population ecology.* Yale University Press, New Haven, Connecticut.

Hynes, H. B. N. 1950. The food of freshwater sticklebacks (*Gasterosteus aculeatus* and *Pygosteus pungitius*) with a review of methods used in studies of food of fishes. *Journal of Animal Ecology* 19:36–58.

———. 1970. *The ecology of running waters.* University of Toronto Press, Canada.

Inouye, B., and J. R. Stinchcombe. 2001. Relationships between ecological interaction modifications and diffuse coevolution: Similarities, differences, and causal links. *Oikos* 95:353–60.

International Nonindigenous Species Database Network. 2012. http://www.nisbase.org/nisbase/index.jsp.

Ioannou, C. C., F. Bartumeus, J. Krause, and G. D. Ruxton. 2011. Unified effects of aggregation reveal larger prey groups take longer to find. *Proceedings of the Royal Society B* 278:2985–90.

Ioannou, C. C., C. R. Tosh, L. Neville, and J. Krause. 2008. The confusion effect from neural networks to reduced predation risk. *Behavioral Ecology* 19:126–30.

Isermann, D. A., J. R. Meerbeek, G. D. Scholten, and D. W. Willis. 2003. Evaluation of three different structures used for Walleye age estimation with emphasis on removal and processing times. *North American Journal of Fisheries Management* 23:625–31.

Ivlev, V. S. 1961. *Experimental ecology of the feeding of fishes.* Yale University Press, New Haven, Connecticut.

Jablonski, D., and J. J. Sepkoski, Jr. 1996. Paleobiology, community ecology, and scales of ecological pattern. *Ecology* 77:1367–78.

Jackson, D. A., and H. H. Harvey. 1989. Biogeographic associations in fish assemblages: Local vs. regional processes. *Ecology* 70:1472–84.

Jackson, D. C. 2012. Headliners: Mississippi's Pascagoula River designated as a model river in America's Great Outdoor Rivers Program. *Fisheries* 37(7):294–95.

Jakes, A. F., J. W. Snodgrass, and J. Burger. 2007. *Castor canadensis* (Beaver) impoundment associated with geomorphology of southeastern streams. *Southeastern Naturalist* 6:271–82.

James, F. C., R. F. Johnston, N. O. Wamer, G. J. Niemi, and J. Boecklen. 1984. The Grinnellian niche of the Wood Thrush. *The American Naturalist* 124:17–47.

Janssen, J., J. E. Marsden, C. R. Bronte, D. J. Jude, S. P. Sitar, and F. W. Goetz. 2007. Challenges to deep-water reproduction by Lake Trout: Pertinence to restoration in Lake Michigan. *Journal of Great Lakes Research* 33:59–74.

Janvier, P. 1997a. Hyperoartia. Lampreys. Version 1, January 1997 (under construction). http://tolweb.org/Hyperoartia/14831/1997.01.01. *The Tree of Life Web Project.* http://tolweb.org/.

———. 1997b. Gnathostomata. Jawed Vertebrates. Version 1, January 1997 (under construction). http://tolweb.org/Gnathostomata/14843/1997.01.01. *The Tree of Life Web Project.* http://tolweb.org/.

Janvier, P., and R. Lund. 1983. *Hardistiella montanensis* N. Gen. et Sp. (Petromyzontidae) from the Lower Carboniferous of Montana, with remarks on the affinities of the lampreys. *Journal of Vertebrate Paleontology* 2:407–13.

Janzen, D. H. 1980. When is it coevolution? *Evolution* 34:611–12.

Jardine, T. D., S. A. McGeachy, C. M. Paton, M. Savoie, and R. A. Cunjak. 2003. Stable isotopes in aquatic systems: Sample preparation, analysis, and interpretation. *Canadian Manuscript Report of Fisheries and Aquatic Sciences* 2656:1–39.

Jeffres, C. A., J. J. Opperman, and P. B. Moyle. 2008. Ephemeral floodplain habitats provide best growth conditions for juvenile Chinook Salmon in a California river. *Environmental Biology of Fishes* 83:449–58.

Jelks, H. L., S. J. Walsh, N. M. Burkhead, S. Contreras-Balderas, E. Díaz-Pardo, D. A. Hendrickson, J. Lyons, et al. 2008. Conservation status of imperiled North American freshwater and diadromous fishes. *Fisheries* 33:372–407.

Jennings, C. A., and S. J. Zigler. 2009. Biology and life history of Paddlefish in North America: An update, 1–22. In *Paddlefish management, propagation, and conservation in the 21st century*. C. P. Paukert and G. D. Scholten (eds.). American Fisheries Society, Symposium 66, Bethesda, Maryland.

Jennings, M. J., and D. P. Philipp. 1992. Reproductive investment and somatic growth rates in Longear Sunfish. *Environmental Biology of Fishes* 35:257–71.

———. 2002. Alternative mating tactics in sunfishes (Centrarchidae): A mechanism for hybridization? *Copeia* 2002:1102–5.

Johansson, F., and J. Andersson. 2009. Scared fish get lazy, and lazy fish get fat. *Journal of Animal Ecology* 78:772–77.

John, K. R. 1964. Survival of fish in intermittent streams of the Chiricahua Mountains, Arizona. *Ecology* 45:112–19.

Johnson, D. L. 2000. Sound production in *Cyprinodon bifasciatus* (Cyprinodontiformes). *Environmental Biology of Fishes* 59:341–46.

Johnson, J. B. 2002. Evolution after the flood: Phylogeography of the desert fish Utah Chub. *Evolution* 56:948–60.

Johnson, J. B., T. E. Dowling, and M. C. Belk. 2004. Neglected taxonomy of rare desert fishes: Congruent evidence for two species of Leatherside Chub. *Systematic Biology* 53:841–55.

Johnson, J. H., and E. Z. Johnson. 1982. Observations on the eye-picking behavior of the Cutlips Minnow, *Exoglossum maxillingua*. *Copeia* 1982:711–12.

Johnson, J. H., and J. E. McKenna, Jr. 2007. Diel periodicity of drift of larval fishes in tributaries of Lake Ontario. *Journal of Freshwater Ecology* 22:347–50.

Johnson, L. 1994. Long-term experiments on the stability of two fish populations in previously unexploited arctic lakes. *Canadian Journal of Fisheries and Aquatic Science* 51:209–25.

Johnson, M. T. J., and J. R. Stinchcombe. 2007. An emerging synthesis between community ecology and evolutionary biology. *Trends in Ecology and Evolution* 22:250–57.

Johnson, M. T. J., M. Vellend, and J. R. Stinchcombe. 2009. Evolution in plant populations as a driver of ecological changes in arthropod communities. *Philosophical Transactions of the Royal Society B* 364:1593–605.

Johnson, P. C., and G. L. Vinyard. 1987. Filter-feeding behavior and particle retention efficiency of Sacramento Blackfish. *Transactions of the American Fisheries Society* 116:634–40.

Johnson, P. T. J., J. D. Olden, and M. J. Vander Zanden. 2008. Dam invaders: Impoundments facilitate biological invasions into freshwaters. *Frontiers in Ecology and the Environment* 6:357–63.

Johnson, T. B., M. H. Hoff, A. S. Trebitz, C. R. Bronte, T. D. Corry, J. F. Kitchell, S. J. Lozano, et al. 2004. Spatial patterns in assemblage structures of pelagic forage fish and zooplankton in western Lake Superior. *Journal of Great Lakes Research* 30:395–406.

Johnson, W. E. 1980. *Cyprinodon variegatus* Lacepède, Sheepshead Minnow, 504. In *Atlas of North American freshwater fishes*. D. S. Lee, C. R. Gilbert, C. H. Hocutt, R. E. Jenkins, D. E. McAllister, and J. R. Stauffer, Jr., et al. 1980. North Carolina State Museum of Natural History, Raleigh.

Johnston, C. E. 1994a. The benefit to some minnows of spawning in the nests of other species. *Environmental Biology of Fishes* 40:213–18.

———. 1994b. Nest association in fishes: Evidence for mutualism. *Behavioral Ecology* 35:379–83.

Johnston, C. E., M. K. Bolling, D. E. Holt, and C. T. Phillips. 2008. Production of acoustic signals during aggression in Coosa Bass, *Micropterus coosae*. *Environmental Biology of Fishes* 82:17–20.

Johnston, C. E., and D. L. Johnson. 2000. Sound production during the spawning season in cavity-nesting darters of the subgenus *Catonotus* (Percidae: *Etheostoma*). *Copeia* 2000:475–81.

Johnston, C. E., and K. J. Kleiner. 1994. Reproductive behavior of the Rainbow Shiner (*Notropis chrosomus*) and the Rough Shiner (*Notropis baileyi*), nest associates of the Bluehead Chub (*Nocomis leptocephalus*) (Pisces: Cyprinidae) in the Alabama River drainage. *Journal of the Alabama Academy of Science* 65:230–40.

Johnston, C. E., and M. J. Maceina. 2009. Fish assemblage shifts and species declines in Alabama, USA streams. *Ecology of Freshwater Fish* 18:33–40.

Johnston, C. E., and L. M. Page. 1992. The evolution of complex reproductive strategies in North American minnows (Cyprinidae), 600–621. In *Systematics, historical ecology, and North American freshwater fishes*. R. L. Mayden (ed.). Stanford University Press, Stanford, California.

Johnston, C. E., and C. T. Phillips. 2003. Sound production in sturgeon *Scaphirhynchus albus* and *S. platorynchus* (Acipenseridae). *Environmental Biology of Fishes* 68:59–64.

Johnston, C. E., and S. P. Vives. 2003. Sound production in *Codoma ornata* (Girard) (Cyprinidae). *Environmental Biology of Fishes* 68:81–85.

Johnston, T. A., M. N. Gaboury, R. A. Janusz, and L. R. Janusz. 1995. Larval fish drift in the Valley River, Manitoba: Influence of abiotic and biotic factors, and relationships with future year-class strengths. *Canadian Journal of Fisheries and Aquatic Sciences* 52:2423–31.

Jones, C. G., J. H. Lawton, and M. Shachak. 1994. Organisms as ecosystem engineers. *Oikos* 69:373–86.

———. 1997a. Positive and negative effects of organisms as physical ecosystem engineers. *Ecology* 78:1946–57.

———. 1997b. Ecosystem engineering by organisms: Why semantics matters. *Trends in Ecology and Evolution* 12:275.

Jones, C. G., J. L. Gutiérrez, J. E. Byers, J. A. Crooks, J. G. Lambrinos, and T. S. Talley. 2010. A framework for understanding physical ecosystem engineering by organisms. *Oikos* 119:1862–69.

Jones, E. A., K. S. Lucey, and D. J. Ellerby. 2007. Efficiency of labriform swimming in the Bluegill sunfish (*Lepomis macrochirus*). *Journal of Experimental Biology* 210:3422–29.

Jonsson, B., N. Jonsson, K. Hindar, T. G. Northcote, and S. Engen. 2008. Asymmetric competition drives lake use of coexisting salmonids. *Oecologia* 157:553–60.

Junk, W. J., P. B. Bayley, and R. E. Sparks. 1989. The flood pulse concept in river-floodplain systems. In *Proceedings of the international large river symposium.* D. P. Dodge (ed.). *Canadian Special Publication of Fish and Aquatic Sciences* 106:110–27.

Junk, W. J., and K. M. Wantzen. 2004. The flood pulse concept: New aspects, approaches and applications—an update, 117–40. In *Proceedings of the second international symposium on the management of large rivers for fisheries.* Vol. 2. R. L. Welcomme and T. Petr (eds.). Food and Agriculture Organization of the United Nations and the Mekong River Commission.

Jurasinski, G., V. Retzer, and C. Beierkuhnlein. 2009. Inventory, differentiation, and proportional diversity: A consistent terminology for quantifying species diversity. *Oecologia* 159:15–26.

Kallman, K. D., and R. W. Harrington, Jr. 1964. Evidence for the existence of homozygous clones in the self-fertilizing hermaphroditic teleost *Rivulus marmoratus* (Poey). *Biological Bulletin* 126:101–14.

Kanehl, P. D., J. Lyons, and J. E. Nelson. 1997. Changes in the habitat and fish community of the Milwaukee River, Wisconsin, following removal of the Woolen Mills Dam. *North American Journal of Fisheries Management* 17:387–400.

Karr, J. R. 1964. Age, growth, fecundity and food habits of Fantail Darters in Boone County, Iowa. *Iowa Academy of Science* 71:274–80.

———. 1981. Assessment of biotic integrity using fish communities. *Fisheries* 6(6):37–41.

Kasumyan, A. O. 2008. Sounds and sound production in fishes. *Journal of Ichthyology* 48:981–1030.

Katula, R. S., and L. M. Page. 1998. Nest association between a large predator, the Bowfin (*Amia calva*), and its prey, the Golden Shiner (*Notemigonus crysoleucas*). *Copeia* 1998:220–21.

Kay, L. K., R. Wallus, and B. L. Yeager. 1994. *Reproductive biology and early life history of fishes in the Ohio River drainage.* Vol. 2. Catostomidae. Tennessee Valley Authority, Chattanooga, Tennessee.

Keast, A. 1991. Panbiogeography: Then and now. *The Quarterly Review of Biology* 66:467–72.

Keast, A., and D. Webb. 1966. Mouth and body form relative to feeding ecology in the fish fauna of a small lake, Lake Opinicon, Ontario. *Journal of the Fisheries Research Board of Canada* 23:1845–74.

Keefer, M. L., D. C. Joosten, C. L. Williams, C. M. Nauman, M. A. Jepson, C. A. Peery, T. C. Bjornn, et al. 2008b. *Adult salmon and Steelhead passage through fishways and transition pools at Bonneville Dam, 1997–2002.* Report for project MPE-P-95-1 submitted to the U.S. Army Corps of Engineers, Portland and Walla Walla districts, and the Bonneville Power Administration, Portland, Oregon.

Keefer, M. L., C. A. Peery, and M. J. Heinrich. 2008a. Temperature-mediated en route migration mortality and travel rates of endangered Snake River Sockeye Salmon. *Ecology of Freshwater Fish* 17:136–45.

Keenleyside, M. H. A., and F. T. Yamamoto. 1962. Territorial behaviour of juvenile Atlantic Salmon (*Salmo salar* L.) *Behaviour* 19:139–69.

Kendall, J. L., K. S. Lucey, E. A. Jones, J. Wang, and D. J. Ellerby. 2007. Mechanical and energetic factors underlying gait transitions in Bluegill sunfish (*Lepomis macrochirus*). *Journal of Experimental Biology* 210:4265–71.

Kennedy, T. B., and G. L. Vinyard. 1997. Drift ecology of western catostomid larvae with emphasis on Warner Suckers, *Catostomus warnerensis* (Teleostei). *Environmental Biology of Fishes* 49:187–95.

Khagram, S. 2004. *Dams and development: Transnational struggles for water and power.* Cornell University Press, Ithaca, New York.

Kierl, N. C., and C. E. Johnston. 2010. Sound production in the Pygmy Sculpin *Cottus paulus*

(Cottidae) during courtship and agonistic behaviours. *Journal of Fish Biology* 77:1268–81.

Kiffney, P. M., G. R. Pess, J. H. Anderson, P. Faulds, K. Burton, and S. C. Riley. 2009. Changes in fish communities following recolonization of the Cedar River, WA, USA by Pacific salmon after 103 years of local extirpation. *River Research and Applications* 25:438–52.

Killingsworth, M. J., and J. S. Palmer. 1992. *Ecospeak: Rhetoric and environmental politics in America*. Southern Illinois University Press, Carbondale.

Kim, J.-O., and C. W. Mueller. 1978. *Introduction to factor analysis*. Sage University paper series on quantitative applications in the social sciences, series 07–013. Sage Publications, Beverly Hills, California.

King, A. J., P. Humphries, and P. S. Lake. 2003. Fish recruitment on floodplains: The roles of patterns of flooding and life history characteristics. *Canadian Journal of Fisheries and Aquatic Sciences* 60:773–86.

Kingsland, S. E. 1985. *Modeling nature, episodes in the history of population ecology*. The University of Chicago Press, Chicago, Illinois.

Kinsolving, A. D., and M. B. Bain. 1993. Fish assemblage recovery along a riverine disturbance gradient. *Ecological Applications* 3:531–44.

Kitchell, J. F., and S. R. Carpenter 1993a. Cascading trophic interactions, 1–14. In *The trophic cascade in lakes*. S. R. Carpenter and J. F. Kitchell (eds.). Cambridge University Press, United Kingdom.

———. 1993b. Synthesis and new directions, 332–50. In *The trophic cascade in lakes*. S. R. Carpenter and J. F. Kitchell (eds.). Cambridge University Press, United Kingdom.

Kitchell, J. F., M. G. Johnson, C. K. Minns, K. H. Loftus, L. Greig, and C. H. Olver. 1977. Percid habitat: The river analogy. *Journal of the Fisheries Research Board of Canada* 34:1936–40.

Kitching, R. L. 1986. Prey-predator interactions, 214–39. In *Community ecology: Pattern and process*. J. Kikkawa and D. J. Anderson (eds.). Blackwell Scientific Publications, Palo Alto, California.

Klug, H., and M. B. Bonsall. 2009. Life history and the evolution of parental care. *Evolution* 64:823–35.

Knapp, R. A., and R. C. Sargent. 1989. Egg mimicry as a mating strategy in the Fantail Darter, *Etheostoma flabellare*: Females prefer males with eggs. *Behavioral Ecology and Sociobiology* 25:321–26.

Kneib, R. T., and A. E. Stiven 1978. Growth, reproduction, and feeding of *Fundulus heteroclitus* (L.) on a North Carolina salt marsh. *Journal of Experimental Marine Biology and Ecology* 31:121–40.

Knight, J. G., and S. T. Ross 1992. Reproduction, age and growth of the Bayou Darter *Etheostoma rubrum* (Pisces, Percidae): An endemic of Bayou Pierre. *American Midland Naturalist* 127:91–105.

———. 1994. Feeding habits of the Bayou Darter. *Transactions of the American Fisheries Society* 123:794–802.

Knighton, D. 1984. *Fluvial forms and processes*. Edward Arnold, London.

Kodric-Brown, A. 1986. Satellites and sneakers: Opportunistic male breeding tactics in pupfish (*Cyprinodon pecosensis*). *Behavioral Ecology* and *Sociobiology* 19:425–32.

Kodric-Brown, A., and J. H. Brown. 2007. Native fishes, exotic mammals, and the conservation of desert springs. *Frontiers in Ecology and the Environment* 5:549–53.

Kodric-Brown, A., and U. Strecker. 2001. Responses of *Cyprinodon maya* and *C. labiosus* females to visual and olfactory cues of conspecific and heterospecific males. *Biological Journal of the Linnean Society* 74:541–48.

Koehl, M. A. R. 1996. When does morphology matter? *Annual Review of Ecology and Systematics* 27:501–42.

Koleff, P., K. J. Gaston, and J. J. Lennon. 2003. Measuring beta diversity for presence-absence data. *Journal of Animal Ecology* 72:367–82.

Kontula, T., and R. Väinölä. 2003. Relationships of Palearctic and Nearctic "glacial relict" *Myoxocephalus* sculpins from mitochondrial DNA data. *Molecular Ecology* 12:3179–84.

Kotliar, N. B., and J. A. Wiens. 1990. Multiple scales of patchiness and patch structure: A hierarchical framework for the study of heterogeneity. *Oikos* 59:253–60.

Kotrschal, K. 1991. Solitary chemosensory cells—taste, common chemical sense or what? *Reviews in Fish Biology and Fisheries* 1:3–22.

———. 2000. Taste(s) and olfaction(s) in fish: A review of specialized sub-systems and central integration. *European Journal of Physiology* 439, supplement: R178–R180.

Krabbenhoft, T. J., M. L. Collyer, and J. M. Quattro. 2009. Differing evolutionary patterns underlie convergence on elongate morphology in endemic fishes of Lake Waccamaw, North Carolina. *Biological Journal of the Linnean Society* 98:636–45.

Kramer, D. L. 1987. Dissolved oxygen and fish behavior. *Environmental Biology of Fishes* 18:81–92.

Kramer, R. H., and L. L. Smith, Jr. 1960a. First-year growth of the Largemouth Bass, *Micropterus salmoides* (Lacepède), and some related ecological factors. *Transactions of the American Fisheries Society* 89:222–33.

———. 1960b. Utilization of nests of Largemouth Bass (*Micropterus salmoides*) by Golden Shiners (*Notemigonus crysoleucas*). *Copeia* 1960:73–74.

Kraus, F. 1995. The conservation of unisexual vertebrate populations. *Conservation Biology* 9:956–59.

Krause, J., and J.-G. J. Godin. 1994. Shoal choice in the Banded Killifish (*Fundulus diaphanus*), Teleostei, Cyprinodontidae: Effects of predation risk, fish size, species composition and size of shoals. *Ethology* 98:128–36.

Krebs, C. J. 1985. *Ecology: The experimental analysis of distribution and abundance.* 3rd ed. Harper and Row, New York.

Kuehne, R. A. 1962. A classification of streams, illustrated by fish distribution in an eastern Kentucky creek. *Ecology* 43:608–14.

Kuehne, R. A., and R. W. Barbour. 1983. *The American darters.* University Press of Kentucky, Lexington.

Ladich, F. 1997. Agonistic behaviour and significance of sounds in vocalizing fish. *Marine and Freshwater Behaviour and Physiology* 29:87–108.

Lake, C. T. 1936. The life history of the Fan-tailed Darter, *Catonotus flabellaris flabellaris* (Rafinesque). *American Midland Naturalist* 17:816–30.

Lamberti, G. A., D. T. Chaloner, and A. E. Hershey. 2010. Linkages among aquatic ecosystems. *Journal of the North American Benthological Society* 29:245–63.

Lamberti, G. A., S. V. Gregory, L. R. Ashkenas, R. C. Wildman, and K. M. S. Moore. 1991. Stream ecosystem recovery following a catastrophic debris flow. *Canadian Journal of Fisheries and Aquatic Sciences* 48:196–208.

Lamouroux, N., N. L. Poff, and P. L. Angermeier. 2002. Intercontinental convergence of stream fish community traits along geomorphic and hydraulic gradients. *Ecology* 83:1792–807.

Landeau, L., and J. Terborgh. 1986. Oddity and the "confusion effect" in predation. *Animal Behaviour* 34:1372–80.

Lanigan, S. H., and H. M. Tyus. 1989. Population size and status of the Razorback Sucker in the Green River Basin, Utah and Colorado. *North American Journal of Fisheries Management* 9:68–73.

Larson, E. R., and J. D. Olden. 2011. The state of crayfish in the Pacific Northwest. *Fisheries* 36:60–73.

Lathrop, B. F. 1982. Ichthyoplankton density fluctuations in the lower Susquehanna River, Pennsylvania, from 1976 through 1980, 28–36. In *Fifth annual larval fish conference*, C. F. Bryan, J. V. Conner and F. M. Truesdale (eds.), Louisiana Cooperative Fisheries Research Unit, Baton Rouge, Louisiana.

Lauder, G. V. 1979. Feeding mechanics in primitive teleosts and in the halecomorph fish *Amia calva. Journal of Zoology,* London 187:543–78.

———. 1980. Evolution of the feeding mechanism in primitive actinopterygian fishes: A functional anatomical analysis of *Polypterus, Lepisosteus,* and *Amia. Journal of Morphology* 163:283–317.

———. 1982. Patterns of evolution in the feeding mechanism of actinopterygian fishes. *American Zoologist* 22:275–85.

———. 1983a. Neuromuscular patterns and the origin of trophic specialization in fishes. *Science* 219:1235–37.

———. 1983b. Functional design and evolution of the pharyngeal jaw apparatus in euteleostean fishes. *Zoological Journal of the Linnean Society* 77:1–38.

———. 1983c. Functional and morphological bases of trophic specialization in sunfishes (Teleostei, Centrarchidae). *Journal of Morphology* 178:1–21.

———. 1983d. Prey capture hydrodynamics in fishes: Experimental tests of two models. *Journal of Experimental Biology* 104:1–13.

———. 1986. Aquatic prey capture in fishes: Experimental and theoretical approaches. *Journal of Experimental Biology* 125:411–16.

———. 1989. Caudal fin locomotion in ray-finned fishes: Historical and functional analyses. *American Zoologist* 29:85–102.

Lauder, G. V., and E. G. Drucker. 2000. Morphology and experimental hydrodynamics of fish fin control surfaces. *IEEE Journal of Oceanic Engineering* 29:556–70.

Lauder, G. V., and K. F. Liem. 1981. Prey capture by *Luciocephalus pulcher*: Implications for models of jaw protrusion in teleost fishes. *Environmental Biology of Fishes* 6:257–68.

———. 1983. *The evolution and interrelationships of the Actinopterygian fishes.* Bulletin of the Museum of Comparative Zoology at Harvard College 150:95–197.

Lauder, G. V., and H. B. Shaffer. 1993. Design of feeding systems in aquatic vertebrates: Major patterns and their evolutionary interpretations, 113–49. In *The skull.* Vol. 3. *Functional and evolutionary mechanisms.* J. Hanken and B. K. Hall (eds.). The University of Chicago Press, Illinois.

Lauder, G. V., and E. D. Tytell. 2006. Hydrodynamics of undulatory propulsion, 425–68. In *Fish* physiology: Fish biomechanics. Vol. 23. R. E. Shadwick, and G. V. Lauder (eds.). Elsevier Academic Press, San Diego, California.

Lauder, G. V., and P. C. Wainwright. 1992. Function and history: The pharyngeal jaw apparatus in primitive ray-finned fishes, p.455–71. In *Systematics, historical ecology, and North American freshwater fishes*. R. L. Mayden (ed.). Stanford University Press, Stanford, California.

Lawrie, A. H. 1970. The Sea Lamprey in the Great Lakes. *Transactions of the American Fisheries Society* 99:766–75.

Leach, S. D., and E. D. Houde. 1999. Effects of environmental factors on survival, growth, and production of American Shad larvae. *Journal of Fish Biology* 54:767–86.

Leavy, T. R., and T. H. Bonner. 2009. Relationships among swimming ability, current velocity association, and morphology for freshwater lotic fishes. *North American Journal of Fisheries Management* 29:72–83.

Lee, D. S. 1980a. *Lepomis gibbosus* (Linnaeus) Pumpkinseed, p.593. In *Atlas of North American freshwater fishes*. D. S. Lee, C. R. Gilbert, C. H. Hocutt, R. E. Jenkins, D. E. McAllister, and J. R. Stauffer, Jr., et al. 1980. North Carolina State Museum of Natural History, Raleigh.

———. 1980b. *Lepomis microlophus* (Günther) Redear Sunfish, 601. In *Atlas of North American freshwater fishes*. D. S. Lee, C. R. Gilbert, C. H. Hocutt, R. E. Jenkins, D. E. McAllister, and J. R. Stauffer, Jr., et al. 1980. North Carolina State Museum of Natural History, Raleigh.

Legendre, P., and L. Legendre.1998. *Numerical ecology*. 2nd English ed. Elsevier, New York, New York.

Leggett, W. C. 1977. The ecology of fish migrations. *Annual Review of Ecology and Systematics* 8:285–308.

Leggett, W. C., and J. E. Carscadden. 1978. Latitudinal variation in reproductive characteristics of American Shad (*Alosa sapidissima*): Evidence for population specific life history strategies in fish. *Journal of the Fisheries Research Board of Canada* 35:1469–78.

Leopold, A. 1933. *Game management*. Charles Scribner's Sons, New York, New York.

———. 1949. *A sand county almanac*. Oxford University Press, New York, New York.

Leopold, L. B. 1994. *A view of the river*. Harvard University Press, Cambridge, Massachusetts.

Leopold, L. B., M. G. Wolman, and J. P. Miller. 1964. *Fluvial processes in geomorphology*. W. H. Freeman Co., San Francisco, California.

Lepori, F., and B. Malmqvist. 2009. Deterministic control on community assembly peaks at intermediate levels of disturbance. *Oikos* 118:471–79.

Leslie, J. 2005. *Deep water: The epic struggle over dams, displaced people, and the environment*. Farrar, Strauss, and Giroux, New York, New York.

Levin, S. A., and R. T. Paine. 1974. Disturbance, patch formation, and community structure. *Proceedings of the National Academy of Science, USA* 71:2744–47.

Levine, J. S., P. S. Lobel, and E. F. MacNichol, Jr. 1980. Visual communication in fishes, 447–76. In *Environmental physiology of fishes*. M. A. Ali (ed.). Plenum Press, New York, New York.

Levins, R. 1970. Extinction, 75–107. In *Some mathematical problems in biology*. M. Gerstenhaber (ed.). American Mathematical Society, Providence, Rhode Island.

Levitis, D. A. 2011. Before senescence: The evolutionary demography of ontogenesis. *Proceedings of the Royal Society B* 278:801–9.

Lewis, W. M., Jr. 1970. Morphological adaptations of cyprinodontids for inhabiting oxygen deficient waters. *Copeia* 1970:319–26.

Li, H. W., P. A. Rossignol, and G. Castillo. 1999. Risk analysis of species introductions: Insights from qualitative modeling, 431–47. In *Nonindigenous freshwater organisms, vectors, biology, and impacts*. R. Claudi and J. H. Leach (eds.). Lewis Publishers, Boca Raton, Florida.

Li, W., A. P. Scott, M. J. Siefkes, H. Yan, Q. Liu, S-S. Yun, and D. A. Gage. 2002. Bile acid secreted by male Sea Lamprey that acts as a sex pheromone. *Science* 296:138–41.

Lieberman, B. S. 2003. Paleobiogeography: The relevance of fossils to biogeography. *Annual Review of Ecology and Systematics* 34:51–69.

Liem, K. F. 1973. Evolutionary strategies and morphological innovations: Cichlid pharyngeal jaws. *Systematic Zoology* 22:425–41.

———. 1980a. Acquisition of energy by teleosts: Adaptive mechanisms and evolutionary patterns, 299–334. In *Environmental physiology of fishes*. M. A. Ali (ed.). Plenum Press, New York, New York.

———. 1980b. Adaptive significance of intra- and interspecific differences in the feeding repertoires of cichlid fishes. *American Zoologist* 20:295–314.

Liem, K. F., and L. S. Kaufman. 1984. Intraspecific macroevolution: Functional biology of the polymorphic cichlid species *Cichlasoma minckleyi*, 203–15. In *Evolution of fish species flocks*. A. A. Echelle and I. Kornfield (eds.). University of Maine at Orono Press.

Liem, K. F., and J. W. M. Osse. 1975. Biological
versatility, evolution, and food resource exploitation in African cichlid fishes. *American Zoologist*
15:427–54.
Liley, N. R. 1982. Chemical communication in fish.
*Canadian Journal of Fisheries and Aquatic Sci-
ences* 39:22–35.
Lima, S. L. 1995. Back to the basics of anti-
predatory vigilance: The group-size effect.
Animal Behaviour 49:11–20.
———. 1998. Nonlethal effects in the ecology of
predator-prey interactions. *BioScience* 48:25–34.
Lima, S. L., and L. M. Dill. 1990. Behavioral
decisions made under the risk of predation:
A review and prospectus. *Canadian Journal of
Zoology* 68:619–40.
Lindeman, R. L. 1942. The trophic-dynamic aspect
of ecology. *Ecology* 23:399–417.
Linder, 1970. Fossil sculpins (Cottidae) from Idaho.
Copeia 1970:755–56.
Lindsey, C. C. 1978. Form, function, and locomo-
tory habits in fish, 1–100. In *Fish physiology:
Locomotion.* Vol. 7. W. S. Hoar and D. J. Randall
(eds.). Academic Press, New York, New York.
Liow, L. H., L. Van Valen, and N. C. Stenseth.
2011. Red Queen: From populations to taxa and
communities. *Trends in Ecology and Evolution*
26:349–58.
Lively, C. M., C. Craddock, and R. C. Vrijenhoek.
1990. Red Queen hypothesis supported by
parasitism in sexual and clonal fish. *Nature*
344:864–66.
Loder, N. 2005. Point of no return. *Conservation
Practice* 6:29–34.
Lodge, D. M. 1993. Biological invasions: Lessons for
ecology. *Trends in Ecology and Evolution* 8:133–37.
Lodge, D. M., C. A. Taylor, D. M. Holdich, and
J. Skurdal. 2000. Nonindigenous crayfishes
threaten North American freshwater biodiversity: Lessons from Europe. *Fisheries* 25:7–20.
Loew, E. R., and H. Zhang. 2006. Propagation of
visual signals in the aquatic environment: An
interactive Windows-based model, 281–302. In
Communication in Fishes. Vol. 2. F. Ladich, S. P.
Collin, P. Moller, and B. G. Kapoor (eds.). Science Publishers, Enfield, New Hampshire.
Lohr, S. C., and K. D. Fausch. 1997. Multiscale
analysis of natural variability in stream fish
assemblages of a western Great Plains watershed. *Copeia* 1997:706–24.
Lomolino, M. V., B. R. Riddle, and J. H. Brown.
2006. *Biogeography.* 3rd ed. Sinauer Associates,
Inc. Sunderland, Massachusetts.
Losey, G. S., Jr. 1972. The ecological importance of
cleaning symbiosis. *Copeia* 1972:820–33.

Losos, C. J. C., J. D. Reynolds, and L. M. Dill. 2010.
Sex-selective predation by Threespine Stickle-
backs on Sea Lice: A novel cleaning behaviour.
Ethology 116:981–89.
Lotrich, V. A. 1973. Growth, production, and com-
munity composition of fishes inhabiting a first-,
second-, and third-order stream of eastern Ken-
tucky. *Ecological Monographs* 43:377–97.
Lowe, J. J., and M. J. C. Walker. 1997. *Reconstructing
Quaternary environments.* 2nd ed. Addison Wesley
Longman, Ltd., Essex, England, United Kingdom.
Lowe, S., M. Browne, S. Boudjelas, and M. De
Poorter. 2004. *100 of the world's worst invasive
alien species: A selection from the Global Invasive
Species Database.* The invasive species special-
ist group (ISSG), Species Survival Commission
(SSC) of the World Conservation Union (IUCN).
Hollands Printing Ltd., Aukland, New Zealand.
Lowry, W. R. 2003. *Dam politics: Restoring America's
rivers.* Georgetown University Press, Washington,
DC.
Lu, M., T. Koike, and N. Hayakawa. 1996. A distrib-
uted hydrological modeling system linking GIS
and hydrological models, 141–48. In *Application
of geographic information systems in hydrology and
water resources management.* K. Kovar and H. P.
Nachtnebel (eds.). International Association of
Hydrological Sciences, publication 235.
Lubinski, B. J., J. R. Jackson, and M. A. Eggleton.
2008. Relationships between floodplain lake fish
communities and environmental variables in a
large river-floodplain system. *Transactions of the
American Fisheries Society* 137:895–908.
Lucas, M. C., and E. Baras. 2001. *Migration of freshwater fishes.* Blackwell Science, Oxford, United
Kingdom.
Ludwig, J. A., and J. F. Reynolds. 1988. *Statistical ecology.* John Wiley and Sons, New York, New York.
Lugli, M., and M. L. Fine. 2003. Acoustic communication in two freshwater gobies: Ambient noise
and short-range propagation in shallow streams.
Journal of the Acoustical Society of America
114:512–21.
Lugli, M., H. Y. Yan, and M. L. Fine. 2003. Acoustic
communication in two freshwater gobies: The
relationship between ambient noise, hearing
thresholds and sound spectrum. *Journal of Comparative Physiology A* 189:309–20.
Lukas, J. A., and D. J. Orth. 1993. Reproductive
ecology of Redbreast Sunfish *Lepomis auritus* in
a Virginia stream. *Journal of Freshwater Ecology*
8:235–44.
Lundberg, J. G. 1992. The phylogeny of ictalurid
catfishes: A synthesis of recent work, 392–420. In
Systematics, historical ecology, and North American

freshwater fishes. R. L. Mayden (ed.). Stanford University Press, Stanford, California.

Lundberg, J. G., M. Kottelat, G. R. Smith, M. L. J. Stiassny, and A. C. Gill. 2000. So many fishes, so little time: An overview of recent ichthyological discovery in continental waters. *Annals of the Missouri Botanical Garden* 87:26–62.

Lydeard, C., M. C. Wooten, and A. Meyer. 1995. Molecules, morphology, and area cladograms: A cladistic and biogeographic analysis of *Gambusia* (Teleostei: Poeciliidae). *Systematic Biology* 44:21–236.

Lyons, J. 1989. Changes in the abundance of small littoral-zone fishes in Lake Mendota, Wisconsin. *Canadian Journal of Zoology* 67:2910–16.

Lyons, J., and J. J. Magnuson. 1987. Effects of Walleye predation on the population dynamics of small littoral-zone fishes in a northern Wisconsin lake. *Transactions of the American Fisheries Society* 116:29–39.

Lytle, D. A, and N. L. Poff. 2004. Adaptation to natural flow regimes. *Trends in Ecology and Evolution* 19:94–100.

MacArthur, R., and R. Levins. 1967. The limiting similarity, convergence, and divergence of coexisting species. *The American Naturalist* 101:377–85.

MacArthur, R. H., and E. O. Wilson. 1967. *The theory of island biogeography.* Princeton University Press, Princeton, New Jersey.

MacCrimmon, H. R. 1971. World distribution of Rainbow Trout (*Salmo gairdneri*). *Journal of the Fisheries Research Board of Canada* 28:663–704.

————. 1972. World distribution of Rainbow Trout (*Salmo gairdneri*): Further observations. *Journal of the Fisheries Research Board of Canada* 29:1788–91.

Mackiewicz, M., D. E. Fletcher, S. D. Wilkins, J. A. DeWoody, and J. C. Avise. 2002. A genetic assessment of parentage in a natural population of Dollar Sunfish (*Lepomis marginatus*) based on microsatellite markers. *Molecular Ecology* 11:1877–83.

MacRae, P. S. D., and D. A. Jackson. 2001. The influence of Smallmouth Bass (*Micropterus dolomieu*) predation and habitat complexity on the structure of littoral zone fish assemblages. *Canadian Journal of Fisheries and Aquatic Sciences* 58:342–51.

Maekawa, K., and S. Nakano. 2002. Latitudinal trends in adult body size of Dolly Varden, with special reference to the food availability hypothesis. *Population Ecology* 44:17–22.

Magnuson, J. J., B. J. Benson, and A. S. McLain. 1994. Insights on species richness and turnover from long-term *Ecological Research*: Fishes in north temperate lakes. *American Zoologist* 34:437–51.

Magnuson, J. J., and T. P. Quinn. 2005. Arthur D. Hasler: He showed us the way. *Environmental Biology of Fishes* 74:67–77.

Magurran, A. E., and A. Higham. 1988. Information transfer across fish shoals under predator threat. *Ethology* 78:153–58.

Magurran, A. E., P. W. Irving, and P. A. Henderson. 1996. Is there a fish alarm pheromone? A wild study and critique. *Proceedings of the Royal Society B* 263:1551–56.

Magurran, A. E., W. J. Oulton, and T. J. Pitcher. 1985. Vigilant behaviour and shoal size in minnows. *Zeitschrift für Tierpsychologie* 67:167–78.

Magurran, A. E., and T. J. Pitcher. 1987. Provenance, shoal size and the sociobiology of predator-evasion behaviour in minnow shoals. *Proceedings of the Royal Society B* 229:439–65.

Magurran, A. E., and H. L. Queiroz. 2003. Partner choice in Piranha shoals. *Behaviour* 140:289–99.

Maiorana, V. C. 1977. Density and competition among sunfish: Some alternatives. *Science* 195:94.

Major, P. F. 1978. Predator-prey interactions in two schooling fishes, *Caranx ignobilis* and *Stolephorus purpureus*. *Animal Behaviour* 26:760–77.

Mandrak, N. E. 1995. Biogeographic patterns of fish species richness in Ontario lakes in relation to historical and environmental factors. *Canadian Journal of Fisheries and Aquatic Sciences* 52:1462–74.

Mandrak, N. E., and E. J. Crossman. 1992. Postglacial dispersal of freshwater fishes into Ontario. *Canadian Journal of Zoology* 70:2247–59.

Mangel, M. 1996. Life history invariants, age at maturity and the ferox trout. *Evolutionary Ecology* 10:249–63.

Mank, J. E., and J. C. Avise. 2006. Comparative phylogenetic analysis of male alternative reproductive tactics in ray-finned fishes. *Evolution* 60:1311–16.

Mank, J. E., D. E. L. Promislow, and J. C. Avise. 2005. Phylogenetic perspectives in the evolution of parental care in ray-finned fishes. *Evolution* 59:1570–78.

Manly, B. F. J. 1986. *Multivariate statistical methods: A primer.* Chapman and Hall, New York, New York.

Marchetti, M. P., T. Light, J. Feliciano, T. Armstrong, Z. Hogan, J. Viers, and P. B. Moyle. 2001. Homogenization of California's fish fauna through abiotic change, 259–78. In *Biotic homogenization.* J. L. Lockwood and M. L.

McKinney (eds.). Kluwer/Academic Press, New York, New York.

Marchetti, M. P., J. L. Lockwood, and T. Light. 2006. Effects of urbanization on California's fish diversity: Differentiation, homogenization and the influence of spatial scale. *Biological Conservation* 127:310–18.

Marchetti, M. P., and P. B. Moyle. 2000. Spatial and temporal ecology of native and introduced fish larvae in lower Putah Creek, California. *Environmental Biology of Fishes* 58:75–87.

Marchetti, M. P., P. B. Moyle, and R. Levine. 2004a. Alien fishes in California watersheds: Characteristics of successful and failed invaders. *Ecological Applications* 14:587–96.

———. 2004b. Invasive species profiling? Exploring the characteristics of non-native fishes across invasion stages in California. *Freshwater Biology* 49:646–61.

Marks, J. C. 2007. Down go the dams. *Scientific American* 296:66–71.

Marsh, E. 1984. Egg size variation in central Texas populations of *Etheostoma spectabile* (Pisces: Percidae). *Copeia* 1984:291–301.

———. 1986. Effects of egg size on offspring fitness and maternal fecundity in the Orangethroat Darter, *Etheostoma spectabile* (Pisces: Percidae). *Copeia* 1986:18–30.

Marsh-Matthews, E., and W. J. Matthews. 2000. Geographic, terrestrial and aquatic factors: Which most influence the structure of stream fish assemblages in the midwestern United States? *Ecology of Freshwater Fish* 9:9–21.

———. 2002. Temporal stability of minnow species co-occurrence in streams of the central United States. *Transactions of the Kansas Academy of Science* 105:162–77.

Marsh-Matthews, E., W. J. Matthews, K. B. Gido, and R. L. Marsh. 2002. Reproduction by young-of-year Red Shiner (*Cyprinella lutrensis*) and its implications for invasion success. *The Southwestern Naturalist* 47:605–10.

Marshall, T. R., and P. A. Ryan. 1987. Abundance patterns and community attributes of fishes relative to environmental gradients. *Canadian Journal of Fisheries and Aquatic Sciences* 44:198–215.

Martin, C. H., and P. C. Wainwright. 2011. Trophic novelty is linked to exceptional rates of morphological diversification in two adaptive radiations of *Cyprinodon* pupfish. *Evolution* 65:2197–212.

Martin, R. 1999. *A story that stands like a dam: Glen Canyon and the struggle for the soul of the West.* University of Utah Press, Salt Lake City.

Matamoros, W. A., J. Schaefer, P. Mickle, W. Arthurs, J. Ikoma, and R. Ragsdale. 2009. First record of *Agonostomus monticola* (Family: Mugilidae) in Mississippi freshwaters with notes of its distribution in the southern USA. *Southeastern Naturalist* 8:1–4.

Mateos, M., O. I. Sanjur, and R. C. Vrijenhoek. 2002. Historical biogeography of the livebearing fish genus *Poeciliopsis* (Poeciliidae: Cyprinodontiformes). *Evolution* 56:972–84.

Mathis, A., and R. J. F. Smith. 1993a. Intraspecific and cross-superorder responses to chemical alarm signals by Brook Stickleback. *Ecology* 74:2395–404.

———. 1993b. Chemical alarm signals increase the survival time of Fathead Minnows (*Pimephales promelas*) during encounters with Northern Pike (*Esox lucius*). *Behavioral Ecology* 4:260–65.

Matthew, W. D. 1915. Climate and evolution. *Annals of the New York Academy of Science* 24:171–318.

Matthews, W. J. 1977. *Influence of physico chemical factors on habitat selection by Red Shiners*, Notropis lutrensis (*Pisces: Cyprinidae*). Unpublished PhD Dissertation, University of Oklahoma, Norman.

———. 1982. Small fish community structure in Ozark streams: Structured assembly patterns or random abundance of species? *American Midland Naturalist* 107:42–54.

———. 1985a. Critical current speeds and microhabitats of the benthic fishes *Percina roanoka* and *Etheostoma flabellare*. *Environmental Biology of Fishes* 12:303–8.

———. 1985b. Distribution of midwestern fishes on multivariate environmental gradients, with emphasis on *Notropis lutrensis*. *American Midland Naturalist* 113:225–37.

———. 1986a. Fish faunal breaks and stream order in the eastern and central United States. *Environmental Biology of Fishes* 17:81–92.

———. 1986b. Fish faunal structure in an Ozark stream: Stability, persistence and a catastrophic flood. *Copeia* 1986:388–97.

———. 1987. Physicochemical tolerance and selectivity of stream fishes as related to their geographic ranges and local distributions, 111–20. In *Community and evolutionary ecology of North American stream fishes.* W. J. Matthews and D. C. Heins (eds.). University of Oklahoma Press, Norman.

———. 1988. North American prairie streams as systems for ecology study. *Journal of the North American Benthological Society* 7:387–409.

———. 1998. *Patterns in freshwater fish ecology.* Chapman and Hall, New York, New York.

Matthews, W. J., R. C. Cashner, and F. P. Gelwick. 1988. Stability and persistence of fish faunas and assemblages in three midwestern streams. *Copeia* 1988:945–55.

Matthews, W. J., K. B. Gido, and E. Marsh-Matthews. 2001. Density-dependent overwinter survival and growth of Red Shiners from a southwestern river. *Transactions of the American Fisheries Society* 130:478–88.

Matthews, W. J., B. C. Harvey, and M. E. Power. 1994. Spatial and temporal patterns in the fish assemblages of individual pools in a midwestern stream (USA). *Environmental Biology of Fishes* 39:381–97.

Matthews, W. J., and E. Marsh-Matthews. 2006a. Persistence of fish species associations in pools of a small stream of the southern Great Plains. *Copeia* 2006:696–710.

———. 2006b. Temporal changes in replicated stream fish assemblages: Predictable or not? *Freshwater Biology* 51:1605–22.

Matthews, W. J., and H. W. Robison. 1998. Influence of drainage connectivity, drainage area, and regional species richness on fishes of the Interior Highlands in Arkansas. *American Midland Naturalist* 139:1–19.

Matthews, W. J., A. J. Stewart, and M. E. Power. 1987. Grazing fishes as components of North American stream ecosystems: Effects of *Campostoma anomalum*, 128–35. In *Community and evolutionary ecology of North American stream fishes*. W. J. Matthews and D. C. Heins (eds.). University of Oklahoma Press, Norman.

Matthews, W. J., E. Surat, and L. G. Hill. 1982. Heat death of the Orangethroat Darter *Etheostoma spectabile* (Percidae) in a natural environment. *Southwestern Naturalist* 27:216–17.

Matuszek, J. E., and G. L. Beggs. 1988. Fish species richness in relation to lake area, pH, and other abiotic factors. *Canadian Journal of Fisheries and Aquatic Sciences* 45:1931–41.

Maurakis, E. G., W. S. Woolcott, and J. T. Magee. 1990. Pebble-nests of four *Semotilus* species. *Proceedings of the Southeastern Fishes Council* 22:7–13.

Maurakis, E. G., W. S. Woolcott, and M. H. Sabaj. 1992. Water currents in spawning areas of pebble nests of *Nocomis leptocephalus* (Pisces: Cyprinidae). *Proceedings of the Southeastern Fishes Council* 25:1–3.

Mayden, R. L. 1985. Biogeography of Ouachita Highland fishes. *The Southwestern Naturalist* 30:195–211.

———. 1987a. Historical ecology and North American highland fishes: A research program in community ecology, 210–22. In *Community and evolutionary ecology of North American stream fishes*. W. J. Matthews and D. C. Heins (eds.). University of Oklahoma Press, Norman.

———. 1987b. Pleistocene glaciation and historical biogeography of North American central-highland fishes, 141–51. In *Quaternary environments of Kansas*. W. C. Johnson (ed.). Kansas Geological Survey, Guidebook Series 5.

———. 1988. Vicariance biogeography, parsimony, and evolution in North American freshwater fishes. *Systematic Zoology* 37:329–55.

———. 1989. Phylogenetic studies of North American minnows, with emphasis on the genus *Cyprinella* (Teleostei: Cypriniformes). *Miscellaneous Publications of the Museum of Natural History, University of Kansas* 80:1–189.

Mayden, R. L. (ed.). 1992. *Systematics, historical ecology, and North American freshwater fishes*. Stanford University Press, Stanford, California

Mayden, R. L., and A. M. Simons. 2002. Crevice spawning behavior in *Dionda dichroma*, with comments on the evolution of spawning modes in North American shiners (Teleostei: Cyprinidae). *Reviews in Fish Biology and Fisheries* 12:327–37.

Mayden, R. L., and E. O. Wiley. 1992. The fundamentals of phylogenetic systematics, 114–85. In *Systematics, historical ecology, and North American freshwater fishes*. R. L. Mayden (ed.). Stanford University Press, Stanford, California.

Mayden, R. L., and R. M. Wood. 1995. Systematics, species concepts, and the evolutionarily significant unit in biodiversity and conservation biology. *American Fisheries Society Symposium* 17:58–113.

Mayr, E., and P. D. Ashlock. 1991. *Principles of systematic biology*. 2nd ed. McGraw-Hill, Inc., New York, New York.

McAllister, D. E., S. P. Platania, F. W. Schueler, M. E. Baldwin, and D. S. Lee. 1986. Ichthyofaunal patterns on a geographic grid, 17–51. In *The zoogeography of North American freshwater fishes*. C. H. Hocutt and E. O. Wiley (eds.). John Wiley and Sons, New York, New York.

McCarthy, M. S., and W. L. Minckley. 1987. Age estimation for Razorback Sucker (Pisces: Catostomidae) from Lake Mohave, Arizona and Nevada. *Journal of the Arizona-Nevada Academy of Science* 21:87–97.

McComish, T. S. 1967. Food habits of Bigmouth and Smallmouth buffalo in Lewis and Clark Lake and the Missouri River. *Transactions of the American Fisheries Society* 96:70–74.

McCune, A. R., K. S. Thomson, and P. E. Olsen. 1984. Semionotid fishes from the Mesozoic great lakes of North America, 27–44. In *Evolution of species flocks*. A. A. Echelle and I. Kornfield (eds.). University of Maine at Orono Press, Orono.

McCusker, M. R., E. Parkinson, and E. B. Taylor. 2000. Mitochondrial DNA variation in Rainbow Trout (*Oncorhynchus mykiss*) across its native range: Testing biogeographical hypotheses and their relevance to conservation. *Molecular Ecology* 9:2089–108.

McDowall, R. M. 1987. Evolution and importance of diadromy. *American Fisheries Society Symposium* 1:1–13.

———. 1988. *Diadromy in fishes, migrations between freshwater and marine environments*. Croom Helm, London, United Kingdom.

———. 1992. Diadromy: Origins and definitions of terminology. *Copeia* 1992:248–51.

———. 1997. The evolution of diadromy in fishes (revisited) and its place in phylogenetic analysis. *Reviews in Fish Biology and Fisheries* 7:443–62.

———. 1999. Different kinds of diadromy: Different kinds of conservation problems. *ICES Journal of Marine Science* 56:410–13.

———. 2007. On amphidromy, a distinct form of diadromy in aquatic organisms. *Fish and Fisheries* 8:1–13.

McDowell, D. M., and R. J. Naiman,. 1986. Structure and function of a benthic invertebrate stream community as influenced by Beaver (*Castor canadensis*). *Oecologia* 68:481–89.

McFarland, W. N. 1986. Light in the sea—correlations with behaviors of fishes and invertebrates. *American Zoologist* 26:389–401.

McKaye, K. R., and N. M. McKaye. 1977. Communal care and kidnapping of young by parental cichlids. *Evolution* 31:674–81.

McKaye, K. R., D. J. Weil, and T. M. Lim. 1979. Comments on the breeding biology of *Gobiomorus dormitor* (Osteichthyes: Eleotridae) and the advantage of schooling behavior to its fry. *Copeia* 1979:542–44.

McKinney, M. L. 2005. Species introduced from nearby sources have a more homogenizing effect than species from distant sources: Evidence from plants and fishes in the USA. *Diversity and Distributions* 11:367–74.

McLaughlin, R. L., J. W. A. Grant, and D. L. G. Noakes. 2000. Living with failure: The prey capture success of young Brook Charr in streams. *Ecology of Freshwater Fish* 9:81–89.

McLennan, D. A. 2007. The umwelt of the Three-Spined Stickleback, 179–224. In *Biology of the Three-Spined Stickleback*. S. Östlund-Nilsson, I. Mayer, and F. A. Huntingford (eds.). CRC Press, Boca Raton, Florida.

McLennan, D. A., and J. D. McPhail. 1989. Experimental investigations of the evolutionary significance of sexually dimorphic nuptial coloration in *Gasterosteus aculeatus* (L.): Temporal changes in the structure of the male mosaic signal. *Canadian Journal of Zoology* 67:1767–77.

McNaughton, S. J. 1984. Grazing lawns: Animals in herds, plant form, and coevolution. *The American Naturalist* 124:863–86.

McNeely, D. L. 1987. Niche relations within an Ozark stream cyprinid assemblage. *Environmental Biology of Fishes* 18:195–208.

McNeely, D. L., W. Caire, A. L. N. Doss, V. M. Harris, and T. Rider. 2004. *Cyprinodon rubrofluviatilis*, Red River Pupfish (Teleostei: Cyprinodontidae), established in the Cimarron River in Oklahoma. *The Southwestern Naturalist* 49:85–87.

McNyset, K. M. 2005. Use of ecological niche modelling to predict distributions of freshwater fish species in Kansas. *Ecology of Freshwater Fish* 14:243–55.

McPhail, J. D. 1967. Distribution of freshwater fishes in western Washington. *Northwest Science* 41:1–11.

———. 1977. A possible function of the caudal spot in characid fishes. *Canadian Journal of Zoology* 55:1063–66.

———. 1994. Speciation and the evolution of reproductive isolation in the sticklebacks (*Gasterosteus*) of south-western British Columbia, 399–437. In *The evolutionary biology of the Threespine Stickleback*. M. A. Bell and S. A. Foster (eds.). Oxford University Press, New York, New York.

———. 1997. Status of the Nooksack Dace, *Rhinichthys* sp., in Canada. *Canadian Field Naturalist* 111:258–62.

McPhail, J. D., and C. C. Lindsey. 1970. *Freshwater fishes of northwestern Canada and Alaska*. Bulletin 173. Fisheries Research Board of Canada, Ottawa.

———. 1986. Zoogeography of the freshwater fishes of Cascadia (the Columbia System and rivers north to the Stikine), 615–37. In *The zoogeography of North American freshwater fishes*. C. H. Hocutt and E. O. Wiley (eds.). John Wiley and Sons, New York, New York.

McPhail, J. D., and E. B. Taylor. 1999. Morphological and genetic variation in northwestern Longnose Suckers, *Catostomus catostomus*: The Salish Sucker problem. *Copeia* 1999:884–93.

Meadows, G. A., S. D. Mackey, R. R. Goforth, D. M. Mickelson, T. B. Edil, J. Fuller, D. E. Guy, Jr., et al. 2005. Cumulative habitat impacts of nearshore engineering. *Journal of Great Lakes Research* 31:90–112.

Meffe, G. K. 1984. Effects of abiotic disturbance on coexistence of predator-prey fish species. *Ecology* 65:1525–34.

———. 1985. Predation and species replacement in American Southwestern fishes: A case study. *The Southwestern Naturalist* 30:173–87.

Meffe, G. K., and T. M. Berra. 1988. Temporal characteristics of fish assemblage structure in an Ohio stream. *Copeia* 1988:684–90.

Meffe, G. K., and C. R. Carroll. 1997. *Principles of conservation biology.* 2nd ed. Sinauer Associates, Inc. Publishers, Sunderland, Massachusetts.

Meffe, G. K., D. A. Hendrickson, W. L. Minckley, and J. N. Rinne. 1983. Factors resulting in decline of the endangered Sonoran Topminnow (Atheriniformes: Poeciliidae) in the United States. *Biological Conservation* 25:135–59.

Meffe, G. K., and W. L. Minckley. 1987. Persistence and stability of fish and invertebrate assemblages in a repeatedly disturbed Sonoran Desert stream. *American Midland Naturalist* 117:177–91.

Mendelson, J. 1975. Feeding relationships among species of *Notropis* (Pisces: Cyprinidae) in a Wisconsin stream. *Ecological Monographs* 45:199–230.

Menhinick, E. F. 1991. *The freshwater fishes of North Carolina.* North Carolina Wildlife Resources Commission, Raleigh, North Carolina.

Meretsky, V. J., D. L. Wegner, and L. E. Stevens. 2000. Balancing endangered species and ecosystems: A case study of adaptive management in Grand Canyon. *Environmental Management* 25:579–86.

Meronek, T. G., P. M. Bouchard, E. P. Buckner, T. M. Burri, K. K. Demmerly, D. C. Hatleli, R. A. Klumb, S. H. Schmidt, and D. W. Coble. 1996. A review of fish control projects. *North American Journal of Fisheries Management* 16:37–41.

Merz, J. E., and P. B. Moyle. 2006. Salmon, wildlife, and wine: Marine-derived nutrients in human-dominated ecosystems of central California. *Ecological Applications* 16:999–1009.

Meyer, A., and C. Lydeard. 1993. The evolution of copulatory organs, internal fertilization, placentae and viviparity in killifishes (Cyprinodontiformes) inferred from a DNA phylogeny of the tyrosine kinase gene *X-src. Proceedings of the Royal Society of London B* 254:153–62.

Milinski, M. 1985. Risk of predation of parasitized sticklebacks (*Gasterosteus aculeatus* L.) under competition for food. *Behaviour* 93:203–16.

Milinski, M., and T. C. M. Bakker. 1990. Female sticklebacks use male coloration in mate choice and hence avoid parasitized males. *Nature* 344:330–33.

Miller, G. A. 1956. The magical number seven, plus or minus two: Some limits on our capacity for processing information. *The Psychological Review* 63:81–97.

Miller, J. G. 1962. Occurrence of ripe Chain Pickerel in the fall. *Transactions of the American Fisheries Society* 91:323.

Miller, R. C. 1922. The significance of the gregarious habit. *Ecology* 3:122–26.

Miller, R. J. 1962. Reproductive behavior of the Stoneroller minnow, *Campostoma anomalum pullum. Copeia* 1962:407–17.

Miller, R. J., and H. W. Robison. 2004. *Fishes of Oklahoma.* University of Oklahoma Press, Norman.

Miller, R. R. 1981. Coevolution of deserts and pupfishes (genus *Cyprinodon*) in the American southwest, 39–94. In *Fishes in North American deserts.* R. J. Naiman and D. L. Soltz (eds.). John Wiley and Sons, New York, New York.

———. 1982. First fossil record (Plio-Pleistocene) of Threadfin Shad, *Dorosoma petenense*, from the Gatuña Formation of southeastern New Mexico. *Journal of Paleontology* 56:423–25.

———. 2005. *Freshwater fishes of Mexico.* With the collaboration of W. L. Minckley and S. M. Norris. University of Chicago Press, Chicago, Illinois.

Miller, R. R., and S. M. Norris. 2005. Family Cichlidae, 352–86. In *Freshwater fishes of Mexico.* R. R. Miller (with W. L. Minckley and S. M. Norris). University of Chicago Press, Chicago.

Miller, R. R., and G. R. Smith. 1981. Distribution and evolution of *Chasmistes* (Pisces: Catostomidae) in western North America. *Occasional Papers of the Museum of Zoology, University of Michigan* 696:1–46.

Miller, R. R., and M. L. Smith. 1986. Origin and geography of the fishes of central Mexico, 487–517. In *The zoogeography of North American freshwater fishes.* C. H. Hocutt and E. O. Wiley (eds.). John Wiley and Sons, New York, New York.

Miller, R. R., J. D. Williams, and J. E. Williams. 1989. Extinctions of North American fishes during the past century. *Fisheries* 14:22–38.

Mills, E. L., J. R. Chrisman, and K. T. Holeck. 1999. The role of canals in the spread of nonindigenous species in North America, 347–79. In *Nonindigenous freshwater organisms, vectors, biology, and impacts.* R. Claudi and J. H. Leach (eds.). Lewis Publishers, Boca Raton, Florida.

Mills, K. H., S. M. Chalanchuk, and D. J. Allan. 1987. Recovery of fish populations in Lake 223 from experimental acidification. *Canadian Journal of Fisheries and Aquatic Sciences* 57:192–204.

Mills, L. S., M. E. Soulé, and D. F. Doak. 1993. The keystone-species concept in ecology and conservation. *BioScience* 43:219–24.

Milner, A. M. 1987. Colonization and ecological development of new streams in Glacier Bay National Park, Alaska. *Freshwater Biology* 18:53–70.

Milner, A. M., E. E. Knudsen, C. Soiseth, A. L. Robertson, D. Schell, I. T. Phillips, and K. Magnusson. 2000. Colonization and development of stream communities across a 200-year gradient in Glacier Bay National Park, Alaska, U.S.A. *Canadian Journal of Fisheries and Aquatic Sciences* 57:2319–35.

Mims, M. C., J. D. Olden, Z. R. Shattuck, and N. L. Poff. 2010. Life history trait diversity of native freshwater fishes in North America. *Ecology of Freshwater Fish* 19:390–400.

Minckley, W. L., and M. E. Douglas. 1991. Discovery and extinction of western fishes: A blink of the eye in geologic time, 7–17. In *Battle against extinction, native fish management in the American west*. W. L. Minckley and J. E. Deacon (eds.). The University of Arizona Press, Tucson.

Minckley, W. L., D. A. Hendrickson, and C. E. Bond. 1986. Geography of western North American freshwater fishes: Description and relationships to intracontinental tectonism, 519–613. In *The zoogeography of North American freshwater fishes*. C. H. Hocutt and E. O. Wiley (eds.). John Wiley and Sons, New York.

Minckley, W. L., P. C. Marsh, J. E. Deacon, T. E. Dowling, P. W. Hedrick, W. J. Matthews, and G. Mueller. 2003. A conservation plan for native fishes of the Lower Colorado River. *BioScience* 53:219–34.

Minckley, W. L., R. R. Miller, C. D. Barbour, J. J. Schmitter Soto, and S. M. Norris. 2005. Historical ichthyogeography, 24–47. In *Freshwater fishes of Mexico*. R. R. Miller (with W. L. Minckley and S. M. Norris). University of Chicago Press, Illinois.

Miranda, L. E. 2005. Fish assemblages in oxbow lakes relative to connectivity with the Mississippi River. *Transactions of the American Fisheries Society* 134:1480–89.

Miranda, L. E., and P. W. Bettoli. 2007. Mortality, 229–77. In *Analysis and interpretation of freshwater fisheries data*. C. S. Guy and M. L. Brown (eds.). American Fisheries Society, Bethesda, Maryland.

Mittelbach, G. G. 1981. Foraging efficiency and body size: A study of optimal diet and habitat use by Bluegills. *Ecology* 62:1370–86.

———. 1988. Competition among refuging sunfishes and effects of fish density on littoral zone invertebrates. *Ecology* 69:614–23.

Mittelbach, G. G., A. M. Turner, D. J. Hall, and J. E. Rettig. 1995. Perturbation and resilience: A long-term, whole-lake study of predator extinction and reintroduction. *Ecology* 76:2347–60.

Miyazono, S., J. N. Aycock, L. E. Miranda, and T. E. Tietjen. 2010. Assemblage patterns of fish functional groups relative to habitat connectivity and conditions in floodplain lakes. *Ecology of Freshwater Fish* 19:578–85.

Mock, K. E., R. P. Evans, M. Crawford, B. L. Cardall, S. U. Janecke, and M. P. Miller. 2006. Rangewide molecular structuring in the Utah Sucker (*Catostomus ardens*). *Molecular Ecology* 15:2223–38.

Modarressie, R., I. P. Rick, and T. C. M. Bakker. 2006. UV matters in shoaling decisions. *Proceedings of the Royal Society B* 273:849–54.

Modde, T. C., K. P. Burnham, and E. J. Wick. 1996. Population status of the Razorback Sucker in the middle Green River (U.S.A.). *Conservation Biology* 10:110–19.

Modde, T. C., and D. B. Irving. 1998. Use of multiple spawning sites and seasonal movement by Razorback Suckers in the middle Green River, Utah. *North American Journal of Fisheries Management* 18:318–26.

Modde, T. C., R. T. Muth, and G. B. Haines. 2001. Floodplain wetland suitability, access, and potential use by juvenile Razorback Suckers in the middle Green River, Utah. *Transactions of the American Fisheries Society* 130:1095–105.

Moller, P. 2006. Electrocommunication: History, insights, and new questions, 579–98. In *Communication in fishes*. Vol. 2. F. Ladich, S. P. Collin, P. Moller, and B. G. Kapoor (eds.). Science Publishers, Enfield, New Hampshire.

Molles, M. C., Jr. 2010. *Ecology, concepts and applications*. 5th ed. McGraw-Hill, New York, New York.

Montgomery, D. R. 2003. *King of fish, the thousand-year run of salmon*. Westview Press, Boulder, Colorado.

Montgomery, D. R., J. M. Buffington, N. P. Peterson, D. Schuett-Hames, and T. P. Quinn. 1996. Stream-bed scour, egg burial depths, and the influence of salmonid spawning on bed surface mobility and embryo survival. *Canadian Journal of Fisheries and Aquatic Sciences* 53:1061–70.

Moodie, G. E. E., J. D. McPhail, and D. W. Hagen. 1973. Experimental demonstration of selective predation on *Gasterosteus aculeatus*. *Behaviour* 47:95–105.

Mooney, H. A., and E. E. Cleland. 2001. The evolutionary impact of invasive species. *Proceedings of the National Academy of Sciences* 98:5446–51.

Moore, J. W. 2006. Animal ecosystem engineers in streams. *BioScience* 56:237–46.

Moore, J. W., and D. E. Schindler. 2008. Biotic disturbance and benthic community dynamics in salmon-bearing streams. *Journal of Animal Ecology* 77:275–84.

Moore, W. S. 1984. Evolutionary ecology of unisexual fishes, 329–98. In *Evolutionary genetics of fishes*. B. J. Turner (ed.). Plenum Press, New York, New York.

Morel, J. 2011. *Public Service Company of New Mexico (PNM) fish passage facility: 2010 annual report*. Submitted to the San Juan Basin Recovery Implementation Program. U.S. Fish and Wildlife Service, Albuquerque, New Mexico.

Morgan, P., and C. A. Swanberg. 1985. On the Cenozoic uplift and tectonic stability of the Colorado Plateau. *Journal of Geodynamics* 3:39–63.

Moriarity, D. J. W. 1976. Quantitative studies on bacteria and algae in the food of the Mullet *Mugil cephalus* L. and the prawn *Metapenaeus bennettae* (Racek and Dall). *Journal of Experimental Marine Biology and Ecology* 22:131–43.

Moring, J. R. 2000. The creation of the first public salmon hatchery in the United States. *Fisheries* 25:6–12.

Morran, L. T., O. G. Schmidt, I. A. Gelarden, II, Parrish, R. C., and C. M. Lively. 2011. Running with the Red Queen: Host-parasite coevolution selects for biparental sex. *Science* 333:216–18.

Morrell, L. J., and R. James. 2007. Mechanisms for aggregation in animals: Rule success depends on ecological variables. *Behavioral Ecology* 19:193–201.

Moshenko, R. W., and J. H. Gee. 1973. Diet, time and place of spawning, and environments occupied by Creek Chub (*Semotilus atromaculatus*) in the Mink River, Manitoba. *Journal of the Fisheries Research Board of Canada* 30:357–62.

Motta, P. J. 1984. Mechanics and functions of jaw protrusion in teleost fishes: A review. *Copeia* 1984:1–18.

———. 1988. Functional morphology of the feeding apparatus of ten species of Pacific butterflyfishes (Perciformes, Chaetodontidae): An ecomorphological approach. *Environmental Biology of Fishes* 22:39–67.

Mount, J. F. 1995. *California rivers and streams*. University of California Press, Berkeley.

Moyer, G. R., M. Osborne, and T. F. Turner. 2005. Genetic and ecological dynamics of species replacement in an arid-land river system. *Molecular Ecology* 14:1263–73.

Moyle, P. B. 1973. Ecological segregation among three species of minnows (Cyprinidae) in a Minnesota lake. *Transactions of the American Fisheries Society* 102:794–805.

———. 2002. *Inland fishes of California*. University of California Press, Berkeley.

Moyle, P. B., and J. J. Cech, Jr. 2004. *Fishes, an introduction to ichthyology*. 5th ed. Prentice Hall, Upper Saddle River, New Jersey.

Moyle, P. B., P. K. Crain, and K. Whitener. 2007. Patterns in the use of a restored California floodplain by native and alien fishes. *San Francisco Estuary and Watershed Science* 5:1–27.

Moyle, P. B., and B. Herbold. 1987. Life-history patterns and community structure in stream fishes of western North America: Comparisons with eastern North America and Europe, 25–32. In *Community and evolutionary ecology of North American stream fishes*. W. J. Matthews and D. C. Heins (eds.). University of Oklahoma Press, Norman.

Moyle, P. B., and H. W. Li. 1979. Community ecology and predator prey relations in warmwater streams, 171–80. In *Predator prey systems in fisheries management*. H. Clepper (ed.). Sport Fishing Institute, Washington, DC.

Moyle, P. B., H. W. Li, and B. A. Barton. 1986. The Frankenstein effect: Impact of introduced fishes on native fishes in North America, 415–26. In *Fish culture in fisheries management*. R. H. Stroud (ed.). American Fisheries Society, Bethesda, Maryland.

Moyle, P. B., and T. Light. 1996a. Biological invasions of fresh water: Empirical rules and assembly theory. *Biological Conservation* 78:149–61.

Moyle, P. B., and B. Vondracek. 1985. Persistence and structure of the fish assemblage in a small California stream. *Ecology* 66:1–13.

———. 1996b. Fish invasions in California: Do abiotic factors determine success? *Ecology* 77:1666–70.

Moy-Thomas, J. A., and R. S. Miles. 1971. *Palaeozoic fishes*. W. B. Saunders Co., Philadelphia, Pennsylvania.

Mueller, G. A. 2005. Predatory fish removal and native fish recovery in the Colorado River mainstem: What have we learned? *Fisheries* 30:10–19.

Muller, E. H. 1965. Quaternary geology of New York, 99–112. In *The Quaternary of the United States*. H. E. Wright, Jr., and D. G. Frey (eds.). Princeton University Press, New Jersey.

Müller, U. K., J. Smit, E. J. Stamhuis, and J. J. Videler. 2001. How the body contributes to the wake in undulatory fish swimming: Flow fields of a swimming eel (*Anguilla anguilla*). *Journal of Experimental Biology* 204:2751–62.

Munz, F W., and W. N. McFarland. 1973. The significance of spectral position in the rhodopsins of tropical marine fishes. *Vision Research* 13:1829–74.

Murray, A. M. 2001a. The fossil record and biogeography of the Cichlidae. *Biological Journal of the Linnean Society* 74:517–32.

———. 2001b. The oldest fossil cichlids (Teleostei: Perciformes): Indication of a 45 million-year-old species flock. *Proceedings of the Royal Society of London B* 268:679–84.

Murray, A. M., and M. V. H. Wilson. 1996. A new Palaeocene genus and species of percopsiform (Teleostei: Paracanthopterygii) from the Paskapoo Formation, Smoky Tower, Alberta, Canada. *Journal of Earth Sciences* 33:429–38.

Musick, J. A., M. M. Harbin, S. A. Berkeley, G. H. Burgess, A. M. Eklund, L. Findley, R. G. Gilmore, et al. 2000. Marine, estuarine, and diadromous fish stocks at risk of extinction in North America (exclusive of Pacific salmonids). *Fisheries* 25(11):6–30.

Muth, R. T., and J. C. Schmulbach. 1984. Downstream transport of fish larvae in a shallow prairie river. *Transactions of the American Fisheries Society* 113:224–30.

Myers, G. S. 1949. Usage of anadromous, catadromous and allied terms for migratory fishes. *Copeia* 1949:89–97.

———. 1966. Derivation of the freshwater fish fauna of Central America. *Copeia* 1966:766–73.

Myrick, C. A., and J. J. Cech. 2004. Temperature effects on juvenile anadromous salmonids in California's Central Valley: What don't we know? *Reviews in Fish Biology and Fisheries* 14:113–23.

Naiman, R. J., C. A. Johnston, and J. C. Kelley. 1988. Alteration of North American streams by Beaver. *BioScience* 38:753–62.

Naiman, R. J., and D. L. Soltz. 1981. *Fishes in North American deserts*. John Wiley and Sons, New York, New York.

Natural Resources Canada. 2006. *Canada physiographic regions*. Ottawa, Canada. http://atlas.nrcan.gc.ca/sites/english/maps/reference/anniversary_maps/physiographicregions/map.pdf.

Near, T. J., and M. F. Benard. 2004. Rapid allopatric speciation in logperch darters (Percidae: *Percina*). *Evolution* 58:2798–808.

Near, T. J., D. I. Bolnick, and P. C. Wainwright. 2005. Fossil calibrations and molecular divergence time estimates in centrarchid fishes (Teleostei: Centrarchidae). *Evolution* 59:1768–82.

Near, T. J., T. W. Kassler, J. B. Koppelman, C. B. Dillman, and D. P. Philipp. 2003. Speciation in North American black basses, *Micropterus* (Actinopterygii: Centrarchidae). *Evolution* 57:1610–21.

Near, T. J., and B. P. Keck. 2005. Dispersal, vicariance, and timing of diversification in *Nothonotus* darters. *Molecular Ecology* 14:3485–96.

Near, T. J., L. M. Page, and R. L. Mayden. 2001. Intraspecific phylogeography of *Percina evides* (Percidae: Etheostomatinae): An additional test of the Central Highlands pre-Pleistocene vicariance hypothesis. *Molecular Ecology* 10:2235–40.

Neff, B. D., and E. L. Clare. 2008. Temporal variation in cuckoldry and paternity in two sunfish species (*Lepomis* spp.) with alternative reproductive tactics. *Canadian Journal of Zoology* 86:92–98.

Nelson, G., and N. Platnick. 1981. *Systematics and biogeography. Cladistics and vicariance*. Columbia University Press, New York, New York.

Nelson, J. S. 2006. *Fishes of the world*. 4th ed. John Wiley and Sons, Inc., Hoboken, New Jersey.

Nelson, J. S., E. J. Crossman, H. Espinosa-Pérez, L. T. Findley, C. R. Gilbert, R. N. Lea, and J. D. Williams. 2004. *Common and scientific names of fishes from the United States, Canada, and Mexico*. 6th ed. American Fisheries Society, Special Publication 29, Bethesda, Maryland.

Nilsson, C., C. A. Reidy, M. Dynesius, and C. Revenga. 2005. Fragmentation and flow regulation of the world's large river systems. *Science* 308:405–8.

Nilsson, P. A., and C. Brönmark. 2000. Prey vulnerability to a gape-size limited predator: Behavioural and morphological impacts on Northern Pike piscivory. *Oikos* 88:539–46.

Noble, R. A. A., I. G. Cowx, D. Goffaux, and P. Kestemont. 2007. Assessing the health of European rivers using functional ecological guilds of fish communities: Standardising species classification and approaches to metric selection. *Fisheries Management and Ecology* 14:381–92.

Nordeng, H. 1977. A pheromone hypothesis for homeward migration in anadromous salmonids. *Oikos* 28:155–59.

———. 2009. Char ecology. Natal homing in sympatric populations of anadromous Arctic Char *Salvelinus alpinus* (L.): Roles of pheromone recognition. *Ecology of Freshwater Fish* 18:41–51.

Nordlie, F. G. 2003. Fish communities of estuarine salt marshes of eastern North America, and comparisons with temperate estuaries of other continents. *Reviews in Fish Biology and Fisheries* 13:281–325.

Northcote, T. G. 1995. Confessions from a four decade affair with Dolly Varden: A synthesis and critique of experimental tests for interactive segregation between Dolly Varden Char (*Salvelinus malma*) and Cutthroat Trout (*Oncorhynchus clarki*) in British Columbia. *Nordic Journal of Freshwater Research* 71:49–67.

Nowell, R. M., and P. A. Jumars. 1984. Flow environments of aquatic benthos. *Annual Review of Ecology and Systematics* 15:303–28.

Nuismer, S. L., R. Gomulkiewicz, and M. T. Morgan. 2003. Coevolution in temporally variable environments. *The American Naturalist* 162:195–204.

Nummi, P., and A. Hahtola. 2008. The Beaver as an ecosystem engineer facilitates teal breeding. *Ecography* 31:519–24.

Nursall, J. R. 1973. Some behavioral interactions of Spottail Shiners (*Notropis hudsonius*), Yellow Perch (*Perca flavescens*), and Northern Pike (*Esox lucius*). *Journal of the Fisheries Research Board of Canada* 30:1161–78.

Nyberg, D. W. 1971. Prey capture in the Largemouth Bass. *American Midland Naturalist* 86:128–44.

Oakey, D. D., M. E. Douglas, and M. R. Douglas. 2004. Small fish in a large landscape: Diversification of *Rhinichthys osculus* (Cyprinidae) in western North America. *Copeia* 2004:207–21.

O'Brien, W. J. 1979. The predator-prey interaction of planktivorous fish and zooplankton. *American Scientist* 67:572–81.

O'Brien, W. J., B. I. Evans, and G. L. Howick. 1986. A new view of the predation cycle of a planktivorous fish, White Crappie (*Pomoxis annularis*). *Canadian Journal of Fisheries and Aquatic Sciences* 43:1894–99.

O'Connell, M. T. 2003. Direct exploitation of prey on an inundated floodplain by Cherryfin Shiners (*Lythrurus roseipinnis*) in a low order, blackwater stream. *Copeia* 2003:635–45.

Odum, W. E. 1970. Utilization of the direct grazing and plant detritus food chains by the Striped Mullet *Mugil cephalus*, 222–40. In *Marine food chains*. J. H. Steele (ed.). University of California Press, Berkeley.

Olden, J. D. 2006. Biotic homogenization: A new research agenda for conservation biogeography. *Journal of Biogeography* 33:2027–39.

Olden, J. D., and M. J. Kennard. 2010. Intercontinental comparison of fish life history strategies along a gradient of hydrologic variability, 83–107. In *Community ecology of stream fishes: Concepts, approaches, and techniques*. K. B. Gido and D. A. Jackson (eds.). American Fisheries Society, Symposium 73, Bethesda, Maryland.

Olden, J. D., and N. L. Poff. 2003. Toward a mechanistic understanding and prediction of biotic homogenization. *The American Naturalist* 162:442–60.

Olden, J. D., N. L. Poff, and K. R. Bestgen. 2006. Life-history strategies predict fish invasions and extirpations in the Colorado River Basin. *Ecological Monographs* 76:25–40.

Olden, J. D., N. L. Poff, M. R. Douglas, M. E. Douglas, and K. D. Fausch. 2004. Ecological and evolutionary consequences of biotic homogenization. *Trends in Ecology and Evolution* 19:18–24.

Olden, J. D., and T. P. Rooney. 2006. On defining and quantifying biotic homogenization. *Global Ecology and Biogeography* 15:113–20.

Olsén, K. H., M. Grahn, and J. Lohm. 2002. Influence of MHC on sibling discrimination in Arctic Charr, *Salvelinus alpinus* (L.). *Journal of Chemical Ecology* 28:783–95.

Olsén, K. H., M. Grahn, J. Lohm, and A. Langefors. 1998. MHC and kin discrimination in juvenile Arctic Charr, *Salvelinus alpinus* (L.). *Animal Behaviour* 56:319–27.

Olson, M. H., S. R. Carpenter, P. Cunningham, S. Gafny, B. R. Herwig, N. P. Nibbelink, T. Pellett, C. Storlie, A. S. Trebitz, and K. A. Wilson. 1998. Managing macrophytes to improve fish growth: A multi-lake experiment. *Fisheries* 23:6–12.

Ono, D. R., J. D. Williams, and A. Wagner. 1983. *Vanishing fishes of North America*. Stone Wall Press, Inc., Washington D. C.

Opperman, J. J., R. Luster, B. A. McKenney, M. Roberts, and A. W. Meadows. 2010. Ecologically functional floodplains: Connectivity, flow regime, and scale. *Journal of the American Water Resources Association* 46:211–26.

Orth, D. J., and O. E. Maughan. 1984. Community structure and seasonal changes in standing stocks of fish in a warm-water stream. *American Midland Naturalist* 112:369–78.

Orzack, S. H., and S. Tuljapurkar. 1989. Population dynamics in variable environments. VII. The demography and evolution of iteroparity. *The American Naturalist* 133:901–23.

Osmundson, D. B., R. J. Ryel, and T. E. Mourning. 1997. Growth and survival of Colorado squawfish in the upper Colorado River. *Transactions of the American Fisheries Society* 126:687–98.

Owen, L. A., R. C. Finkel, R. A. Minnich, and A. E. Perez. 2003. Extreme southwestern margin of late Quaternary glaciation in North America: Timing and controls. *Geology* 31:729–32.

Paavola, R., T. Muotka, R. Virtanen, J. Heino, D. Jackson, and A. Mäki-Petäys. 2006. Spatial scale affects community concordance among fishes, benthic macroinvertebrates, and bryophytes in streams. *Ecological Applications* 16:368–79.

Page, L. M. 1983. *Handbook of darters*. TFH Publications, Neptune City, New Jersey.

Page, L. M., and H. L. Bart. 1989. Egg mimics in darters (Pisces: Percidae). *Copeia* 1989:514–17.

Page, L. M., and B. M. Burr. 1991. *A field guide to freshwater fishes of North America north of Mexico*. Houghton Mifflin Co., Boston, Massachusetts.

Page, L. M., and C. E. Johnston. 1990. Spawning in the Creek Chubsucker, *Erimyzon oblongus*, with a review of spawning behavior in suckers (Catostomidae). *Environmental Biology of Fishes* 27:265–72.

Page, L. M., and J. H. Knouft. 2000. Variation in egg-mimic size in the Guardian Darter, *Etheostoma oophylax* (Percidae). *Copeia* 2000:782–85.

Page, L. M., and D. L. Swofford. 1984. Morphological correlates of ecological specialization in darters. *Environmental Biology of Fishes* 11:139–59.

Paine, R. T. 1969. A note on trophic complexity and community stability. *The American Naturalist* 103:91–93.

———. 1980. Food webs: Linkage, interaction strength and community infrastructure. *Journal of Animal Ecology* 49:666–85.

Paine, M. D., and E. K. Balon. 1986. Early development of Johnny Darter, *Etheostoma nigrum*, and Fantail Darter, *E. flabellare*, with a discussion of its ecological and evolutionary aspects. *Environmental Biology of Fishes* 15:191–220.

Palkovacs, E. P., M. C. Marshall, B. A. Lamphere, B. R. Lynch, D. J. Weese, D. F. Fraser, D. N. Reznick, C. M. Pringle, and M. T. Kinnison. 2009. Experimental evaluation of evolution and coevolution as agents of ecosystem change in Trinidadian streams. *Philosophical Transactions of the Royal Society B* 364:1617–28.

Paller, M. H. 1987. Distribution of larval fish between macrophyte beds and open channels in a southeastern floodplain swamp. *Journal of Freshwater Ecology* 4:191–200.

———. 1994. Relationships between fish assemblage structure and stream order in South Carolina coastal plain streams. *Transactions of the American Fisheries Society* 123:150–61.

———. 2002. Temporal variability in fish assemblages from disturbed and undisturbed streams.

Journal of Aquatic Ecosystem Stress and Recovery 9:149–58.

Panek, F. M. 1987. Biology and ecology of Carp. In *Carp in North America*. E. L. Cooper (ed.). American Fisheries Society, Bethesda, Maryland.

Paragamian, V. L., V. D. Wakkinen, and G. Kruse. 2002. Spawning locations and movement of Kootenai River White Sturgeon. *Journal of Applied Ichthyology* 18:608–16.

Parenti, L. R. 1981. A phylogenetic and biogeographic analysis of cyprinodontiform fishes (Teleostei, Atherinomorpha). *Bulletin of the American Museum of Natural History* 168:335–557.

Parker, A., and I. Kornfield. 1995. Molecular perspective on evolution and zoogeography of cyprinodontid killifishes (Teleostei; Atherinomorpha). *Copeia* 1995:8–21.

Parkos, J. J., III, and D. H. Wahl. 2010. Influence of body size and prey type on the willingness of age-0 fish to forage under predation risk. *Transactions of the American Fisheries Society* 139:969–75.

Partridge, B. L. 1982. The structure and function of fish schools. *Scientific American* 246:114–23.

Partridge, B. L., N. R. Liley, and N. E. Stacey. 1976. The role of pheromones in the sexual behaviour of the Goldfish. *Animal Behaviour* 24:291–99.

Partridge, L., and P. H. Harvey. 1988. The ecological context of life history evolution. *Science* 241:1449–55.

Patterson, C. 1981. The development of the North American fish fauna—a problem of historical biogeography, 265–81. In *The evolving biosphere*. P. L. Forey (ed.). Cambridge University Press.

Patzner, R. A. 2008. Reproductive strategies of fish, 311–50. In *Fish reproduction*. M. J. Rocha, A. Arukwe, and B. G. Kapoor (eds.). Science Publishers, Enfield, New Hampshire.

Pavlov, D. S. 1994. The downstream migration of young fishes in rivers: Mechanisms and distribution. *Folia Zoologica* 43:193–208.

Pearse, D. E., S. A. Hayes, M. H. Bond, C. V. Hanson, E. C. Anderson, R. B. Macfarlane, and J. C. Garza. 2009. Over the falls? Rapid evolution of ecotypic differentiation in Steelhead/Rainbow Trout (*Oncorhynchus mykiss*). *Journal of Heredity* 100:515–25.

Pearson, M. 2000. The biology and management of the Salish Sucker and Nooksack Dace, 619–24. In *Proceedings of a conference on the biology and management of species and habitats at risk,*

Kamloops, B. C. Vol. 2. L. M. Darling (ed.). British Columbia Ministry of Environment Lands and Parks, Victoria, and University College of the Cariboo, Kamloops, British Columbia.

Pearsons, T. N., H. W. Li, and G. A. Lamberti. 1992. Influence of habitat complexity on resistance to flooding and resilience of stream fish assemblages. *Transactions of the American Fisheries Society* 121:427–36.

Pease, A. A., J. J. Davis, M. S. Edwards, and T. F. Turner. 2006. Habitat and resource use by larval and juvenile fishes in an arid-land river (Rio Grande, New Mexico). *Freshwater Biology* 51:475–86.

Peckarsky, B. 2010. Resolution of respect. *Bulletin of the Ecological Society of America* 91:128–41.

Peckarsky, B. I. 1983. Biotic interactions or abiotic limitations? A model of lotic community structure, 303–23. In *Dynamics of lotic ecosystems.* T. D. Fontaine, III and S. M. Bartell (eds.). Ann Arbor Science, Michigan.

Peet, R. K. 1974. The measurement of species diversity. *Annual Review of Ecology and Systematics* 5:285–307.

Perga, M. E., and D. Gerdeaux. 2005. "Are fish what they eat" all year round? *Oecologia* 144:598–606.

Persson, A., and L.-A. Hansson. 1999. Diet shift in fish following competitive release. *Canadian Journal of Fisheries and Aquatic Sciences* 56:70–78.

Peterson, A. T. 2001. Predicting species' geographic distributions based on ecological niche modeling. *The Condor* 103:599 605.

Peterson, A. T., M. Papes, and M. Eaton. 2007. Transferability and model evaluation in ecological niche modeling: A comparison of GARP and Maxent. *Ecography* 30:440–560.

Peterson, B. J., and B. Fry. 1987. Stable isotopes in ecosystem studies. *Annual Review of Ecology and Systematics* 18:293–320.

Peterson, C. H., and M. L. Quammen. 1982. Siphon nipping: Its importance to small fishes and its impact on growth of the bivalve *Protothaca staminea* (Conrad). *Journal of Experimental Marine Biology and Ecology* 63:249–68.

Peterson, M. S. 2003. A conceptual view of environment-habitat-production linkages in tidal river estuaries. *Reviews in Fisheries Science* 11:291–313.

Peterson, M. S., and S. J. VanderKooy. 1995. Phenology and spatial and temporal distribution of larval fishes in a partially channelized warmwater stream. *Ecology of Freshwater Fish* 4:93–105.

Peterson, M. S., M. R. Weber, M. L. Partyka, and S. T. Ross. 2007. Integrating *in situ* quantitative geographic information tools and size-specific laboratory-based growth zones in a dynamic river-mouth estuary. *Aquatic Conservation: Marine and Freshwater Ecosystems* 17:602–18.

Peterson, T. L. 1996. Seasonal migration in the southern Hogchoker, *Trinectes maculatus fasciatus* (Achiridae). *Gulf Research Reports* 9:169–76.

Peuhkuri, N., and P. Seppä. 1998. Do Three-spined Sticklebacks group with kin? *Annales Zoologici Fennici* 35:21–27.

Pfeiffer, W. 1963. The fright reaction in North American fish. *Canadian Journal of Zoology* 41:69–77.

———. 1977. The distribution of fright reaction and alarm substance cells in fishes. *Copeia* 1977:653–65.

Pfennig, K. S, and D. W. Pfennig. 2009. Character displacement: Ecological and reproductive responses to a common evolutionary problem. *The Quarterly Review of Biology* 84:253–76.

Pflieger, W. L. 1971. A distributional study of Missouri fishes. *Museum of Natural History, University of Kansas* 20:225–570.

Philipp, D. P., and M. R. Gross. 1994. Genetic evidence for cuckoldry in Bluegill *Lepomis macrochirus. Molecular Ecology* 2:563–69.

———. 2008a. Sound production and associated behaviors in *Cyprinella galactura. Environmental Biology of Fishes* 82:265–75.

———. 2008b. Geographical divergence of acoustic signals in *Cyprinella galactura,* the Whitetail Shiner (Cyprinidae). *Animal Behaviour* 75:617–26.

———. 2009. Evolution of acoustic signals in *Cyprinella*: Degree of similarity in sister species. *Journal of Fish Biology* 74:120–32.

Phillips, C. T., C. E. Johnston, and A. R. Henderson. 2010. Sound production and spawning behavior in *Cyprinella lepida,* the Edwards Plateau Shiner. *The Southwestern Naturalist* 55:129–35.

Pianka, E. R. 1970. On r- and K-selection. *The American Naturalist* 204:592–97.

———. 1988. *Evolutionary ecology.* 4th ed. Harper and Rowe, New York, New York.

Piccolo, J. J., M. D. Adkison, and F. Rue. 2009. Linking Alaskan salmon fisheries management with ecosystem-based escapement goals: A review and prospectus. *Fisheries* 34:124–34.

Pickett, S. T. A., and P. S. White (eds.). 1985. *The ecology of natural disturbance and patch dynamics.* Academic Press, New York, New York.

Pielou, E. C. 1984. *The interpretation of ecological data, a primer on classification and ordination.* John Wiley and Sons, New York, New York.

———. 1991. *After the Ice Age, the return of life to glaciated North America.* University of Chicago Press, Illinois.

Pierce, C. L., J. B. Rasmussen, and W. C. Leggett. 1994. Littoral fish communities in southern Quebec lakes: Relationships with limnological and prey resource variables. *Canadian Journal of Fisheries and Aquatic Sciences* 51:1128–38.

Pierce, C. L., M. D. Sexton, M. E. Pelham, and J. G. Larscheid. 2001. Short-term variability and long-term change in the composition of the littoral zone fish community in Spirit Lake, Iowa. *American Midland Naturalist* 146:290–99.

Pilger, T. J., K. B. Gido, and D. L. Propst. 2010. Diet and trophic niche overlap of native and nonnative fishes in the Gila River, USA: Implications for native fish conservation. *Ecology of Freshwater Fish* 19:300–321.

Piller, K. R., H. L. Bart, Jr., and C. A. Walser. 2001. Morphological variation of the Redfin Darter, *Etheostoma whipplei,* with comments on the status of the subspecific populations. *Copeia* 2001:802–7.

Pitcher, T. J. 1986. Functions of shoaling behaviour, 294–337. In *Behaviour of teleost fishes,* T. J. Pitcher (ed.). Croom Helm, London, United Kingdom.

Pitcher, T. J., and J. K. Parrish. 1993. Functions of shoaling behaviour in teleosts, 363–439. In *Behaviour of teleost fishes.* 2nd ed. T. J. Pitcher (ed.). Chapman and Hall, New York, New York.

Pizzuto, J. 2002. Effects of dam removal on river form and process. *BioScience* 52:683–91.

Platania, S. P., and C. S. Altenbach. 1998. Reproductive strategies and egg types of seven Rio Grande basin cyprinids. *Copeia* 1998:559–69.

Plath, M., A. M. Makowicz, I. Schlupp, and M. Tobler. 2007. Sexual harassment in live-bearing fishes (Poeciliidae): Comparing courting and noncourting species. *Behavioral Ecology* 18:680–88.

Platts, W. S. 1979. Relationships among stream order, fish populations, and aquatic geomorphology in an Idaho River drainage. *Fisheries* 4:5–9.

Poff, N. L. 1997. Landscape filters and species traits: Towards a mechanistic understanding and prediction in stream ecology. *Journal of the North American Benthological Society* 16:391–409.

Poff, N. L., and J. D. Allan. 1995. Functional organization of stream fish assemblages in relation to hydrological variability. *Ecology* 76:606–27.

Poff, N. L., J. D. Allan, M. B. Bain, J. R. Karr, K. L. Prestegaard, B. D. Richter, R. E. Sparks, and J. C. Stromberg. 1997. The natural flow regime. *BioScience* 47:769–84.

Poff, N. L., J. D. Allan, M. A. Palmer, D. D. Hart, B. D. Richter, A. H. Arthington, K. H. Rogers, J. L. Meyer, and J. A. Stanford. 2003. River flows and water wars: Emerging science for environmental decision making. *Frontiers in Ecology and the Environment* 1:298–306.

Poff, N. L., B. P. Bledsoe, and C. O. Cuhaciyan. 2006. Hydrologic variation with land use across the contiguous United States: Geomorphic and ecological consequences for stream ecosystems. *Geomorphology* 79:264–85.

Poff, N. L., and D. D. Hart. 2002. How dams vary and why it matters for the emerging science of dam removal. *BioScience* 52:659–68.

Poff, N. L., B. D. Richter, A. H. Arthington, S. E. Bunn, R. J. Naiman, E. Kendy, M. Acreman, et al. 2010. The ecological limits of hydrologic alteration (ELOHA): A new framework for developing regional environmental flow standards. *Freshwater Biology* 55:147–70.

Poff, N. L., and J. K. H. Zimmerman. 2010. Ecological responses to altered flow regimes: A literature review to inform the science and management of environmental flows. *Freshwater Biology* 55:194–205.

Poizat, G., E. Rosecchi, and A. J. Crivelli. 1999. Empirical evidence of a trade-off between reproductive effort and expectation of future reproduction in female Three-Spined Sticklebacks. *Proceedings of the Royal Society of London B* 266:1543–48.

Policanksy, D., and J. J. Magnuson. 1998. Genetics, metapopulations, and ecosystem management of fisheries. *Ecological Applications,* supplement 8:S119–S123.

Polis, G. A., A. L. W. Sears, G. R. Huxel, D. R. Strong, and J. Maron. 2000. When is a trophic cascade a trophic cascade? *Trends in Ecology and Evolution* 15:473–75.

Pollock, M. M., M. Heim, and D. Werner. 2003. Hydrologic and geomorphic effects of Beaver dams and their influence on fishes, 213–33. In *The ecology and management of wood in world rivers.* S. Gregory, K. Boyer, and A. Gurnell (eds.). American Fisheries Society Symposium 37, Bethesda, Maryland.

Pool, T. K., and J. D. Olden. 2011. Taxonomic and functional homogenization of an endemic desert fish fauna. *Diversity and Distributions* 2011:1–11.

Poole, R. W. 1974. *An introduction to quantitative ecology.* McGraw-Hill, New York.

Porter, B. A., A. C. Fiumera, and J. C. Avise. 2002. Egg mimicry and alloparental care: Two mate-attracting tactics by which nesting Striped Darter (*Etheostoma virgatum*) males enhance reproductive success. *Behavioral Ecology and Sociobiology* 51:350–59.

Porter, H. T., and P. J. Motta. 2004. A comparison of strike and prey capture kinematics of three species of piscivorous fishes: Florida Gar (*Lepisosteus platyrhincus*), Redfin Needlefish (*Strongylura notata*), and Great Barracuda (*Sphyraena barracuda*). *Marine Biology* 145:989–1000.

Porter, J. H., and J. L. Dooley, Jr. 1993. Animal dispersal patterns: A reassessment of simple mathematical models. *Ecology* 74:2436–43.

Porter, S. C., and T. W. Swanson. 1998. Radiocarbon age constraints on rates of advance and retreat of the Puget Lobe of the Cordilleran Ice Sheet during the last glaciation. *Quaternary Research* 50:205–13.

Portz, D. E., and H. M. Tyus. 2004. Fish humps in two Colorado River fishes: A morphological response to cyprinid predation? *Environmental Biology of Fishes* 71:233–45.

Post, D. M. 2002. The long and short of food-chain length. *Trends in Ecology and Evolution* 17:269–77.

Post, D. M., M. L. Pace, and N. G. Hairston, Jr. 2000. Ecosystem size determines food-chain length in lakes. *Nature* 405:1047–49.

Post, J. R., and D. J. McQueen. 1988. Ontogenetic changes in the distribution of larval and juvenile Yellow Perch (*Perca flavescens*): A response to prey or predators? *Canadian Journal of Fisheries and Aquatic Sciences* 45:1820–26.

Potts, G. W. 1983. The predatory tactics of *Caranx melampygus* and the response of its prey, 181–91. In *Predators and prey in fishes*. D. L. G. Noakes, D. G. Lindquist, G. S. Helfman, and J. A. Ward (eds.). Dr. W. Junk Publishers, The Hague, Netherlands.

Powell, J. L. 2005. *Grand Canyon, solving the Earth's grandest puzzle*. Penguin Group, Inc., New York, New York.

Power, H. W., E. Litovich, and M. P. Lombardo. 1981. Male starlings delay incubation to avoid being cuckolded. *The Auk* 989:386–89.

Power, M., G. Power, J. D. Reist, and R. Bajno. 2009. Ecological and genetic differentiation among the Arctic Charr of Lake Aigneau, Northern Québec. *Ecology of Freshwater Fish* 18:445–60.

Power, M. E. 1987. Predator avoidance by grazing fishes in temperate and tropical streams: Importance of stream depth and prey size, 333–51. In *Predation: Direct and indirect impacts on aquatic communities*. W. C. Kerfoot and A. Sih (eds.). University Press of New England, Hanover, New Hampshire.

———. 1990. Effects of fish in river food webs. *Science* 250:811–14.

———. 1992a. Hydrologic and trophic controls of seasonal algal blooms in northern California. *Archiv für Hydrobiologie* 125:385–410.

———. 1992b. Top-down and bottom-up forces in food webs: Do plants have primacy? *Ecology* 73:733–46.

———. 1997. Ecosystem engineering by organisms: Why semantics matters—reply from M. Power. *Trends in Ecology and Evolution* 12:275–76.

Power, M. E., and W. E. Dietrich. 2002. Food webs in river networks. *Ecological Research* 17:451–71.

Power, M. E., and W. J. Matthews. 1983. Algae-grazing minnows (*Campostoma anomalum*), piscivorous bass (*Micropterus* spp.), and the distribution of attached algae in a small prairie-margin stream. *Oecologia* 60:328–32.

Power, M. E., W. J. Matthews, and A. J. Stewart. 1985. Grazing minnows, piscivorous bass, and stream algae: Dynamics of a strong interaction. *Ecology* 66:328–32.

Power, M. E., M. S. Parker, and W. E. Dietrich. 2008. Seasonal reassembly of a river food web: Floods, droughts, and impacts of fish. *Ecological Monographs* 78:263–82.

Power, M. E., A. Sun, G. Parker, W. E. Dietrich, and J. T. Wootton. 1995. Hydraulic food-chain models. *BioScience* 45:159–67.

Power, M. E., D. Tilman, J. A. Estes, B. A. Menge, W. J. Bond, L. S. Mills, G. Daily, J. C. Castilla, J. Lubchenco, and R. Paine. 1996. Challenges in the quest for keystones. *BioScience* 46:609–20.

Powles, P. M., and I. M. Sandeman. 2008. Growth, summer cohort output, and observations on the reproduction of Brook Silverside, *Labidesthes sicculus* (Cope) in the Kawartha Lakes, Ontario. *Environmental Biology of Fishes* 82:421–31.

Pratt, T. C., and M. G. Fox. 2002. Influence of predation risk on the overwinter mortality and energetic relationships of young-of-year Walleyes. *Transactions of the American Fisheries Society* 131:885–98.

Price, A. C., C. J. Weadick, J. Shim, and F. H. Rodd. 2008. Pigments, patterns, and fish behavior. *Zebrafish* 5:297–307.

Price, P. W. 1984. Communities of specialists: Vacant niches in ecological and evolutionary time, 510–23. In *Ecological communities-conceptual issues and the evidence*. D. R. Strong,

Jr., D. Simberloff, L. G. Abele, and A. B. Thistle (eds.). Princeton University Press, Princeton, New Jersey.

Propst D. L., and K. B. Gido. 2004. Responses of native and nonnative fishes to natural flow regime mimicry in the San Juan River. *Transactions of the American Fisheries Society* 133:922–31.

Puckett, K. J., and L. M. Dill. 1985. The energetics of feeding territoriality in juvenile Coho Salmon (*Oncorhynchus kisutch*). *Behaviour* 92:97–111.

Pulliam, H. R. 1988. Sources, sinks, and population regulation. *The American Naturalist* 132:651–52.

Purcell, E. M. 1977. Life at low Reynolds number. *American Journal of Physics* 45:3–11.

Purkett, C. A., Jr. 1963. The Paddlefish fishery of the Osage River and the Lake of the Ozarks, Missouri. *Transactions of the American Fisheries Society* 92:239–44.

Pyke, G. H., H. R. Pulliam, and E. L. Charnov. 1977. Optimal foraging: A selective review of theory and tests. *The Quarterly Review of Biology* 52:137–54.

Pyron, M. 1995. Mating patterns and a test for female mate choice in *Etheostoma spectabile* (Pisces, Percidae). *Behavioral Ecology* and Sociobiology 36:407–12.

Pyron, M., T. E. Lauer, and J. R. Gammon. 2006. Stability of the Wabash River fish assemblages from 1974 to 1998. *Freshwater Biology* 51:1789–97.

Quinn, T. P., and D. J. Adams. 1996. Environmental changes affecting the migratory timing of American Shad and Sockeye Salmon. *Ecology* 77:1151–62.

Quinn, T. P., E. L. Brannon, and R. P. Whitman. 1983. Pheromones and the water source preferences of adult Coho Salmon *Oncorhynchus kisutch* Walbaum. *Journal of Fish Biology* 22:677–84.

Quinn, T. P., and C. A. Busack. 1985. Chemosensory recognition of siblings in juvenile Coho Salmon (*Oncorhynchus kisutch*). *Animal Behaviour* 33:51–56.

Quinn, T. P., and G. M. Tolson. 1986. Evidence of chemically mediated population recognition in Coho Salmon (*Oncorhynchus kisutch*). *Canadian Journal of Zoology* 64:84–87.

Radomski, P. 2006. Historical changes in abundance of floating-leaf and emergent vegetation in Minnesota lakes. *North American Journal of Fisheries Management* 26:932–40.

Radomski, P. J., and T. J. Goeman. 1995. The homogenizing of Minnesota lake fish assemblages. *Fisheries* 20:20–23.

———. 2001. Consequences of human lakeshore development on emergent and floating-leaf vegetation abundance. *North American Journal of Fisheries Management* 21:46–61.

Rage, J-C., and Z. Rocek. 2003. Evolution of anuran assemblages in the Tertiary and Quaternary of Europe, in the context of palaeoclimate and palaeogeography. *Amphibia-Reptilia* 24:133–67.

Rahel, F. J. 1984. Factors structuring fish assemblages along a bog lake successional gradient. *Ecology* 65:1276–89.

———. 1986. Biogeographic influences on fish species composition of northern Wisconsin lakes with applications for lake acidification studies. *Canadian Journal of Zoology* 43:124–34.

———. 1990. The hierarchical nature of community persistence: A problem of scale. *The American Naturalist* 136:328–44.

———. 2000. Homogenization of fish faunas across the United States. *Science* 288:854–56.

———. 2002. Homogenization of freshwater faunas. *Annual Review of Ecology and Systematics* 33:291–315.

———. 2007. Biogeographic barriers, connectivity and homogenization of freshwater faunas: It's a small world after all. *Freshwater Biology* 52:696–710.

Rahel, F. J., J. D. Lyons, and P. A. Cochran. 1984. Stochastic or deterministic regulation of assemblage structure? It may depend on how the assemblage is defined. *The American Naturalist* 124: 583–89.

Rajakaruna, R. S., J. A. Brown, K. H. Kaukinen, and K. M. Miller. 2006. Major histocompatibility complex and kin discrimination in Atlantic Salmon and Brook Trout. *Molecular Ecology* 15:4569–75.

Rakocinski, C. 1991. Prey-size relationships and feeding tactics of primitive stream-dwelling darters. *Canadian Journal of Fisheries and Aquatic Sciences* 48:681–93.

Rand, D. M., and G. V. Lauder. 1981. Prey capture in the Chain Pickerel, *Esox niger*: Correlations between feeding and locomotor behavior. *Canadian Journal of Zoology* 59:1072–78.

Randall, R. G., M. C. Healey, and J. B. Dempson. 1987. Variability in length of freshwater residence of salmon, trout, and char, 27–41. In *Common strategies of anadromous and catadromous fishes*. M. J. Dadswell, R. J. Klauda, C. M. Moffitt, R. L. Saunders, R. A. Rulifson, and J. E. Cooper (eds.). American Fisheries Society Symposium 1, Bethesda, Maryland.

Rauchenberger, M. 1988. Historical biogeography of poeciliid fishes in the Caribbean. *Systematic Zoology* 37:356–65.

Redfern, R. 2001. *Origins, the evolution of continents, oceans and life*. University of Oklahoma Press, Norman.

Reed, J. R., and D. L. Pereira. 2009. Relationships between shoreline development and nest site selection by Black Crappie and Largemouth Bass. *North American Journal of Fisheries Management* 29:943–48.

Reheis, M. 1999. *Extent of Pleistocene Lakes in the western Great Basin: U.S. Geological Survey Miscellaneous Field Studies Map MF-2323*. U.S. Geological Survey, Denver, Colorado. http://geo-nsdi.er.usgs.gov/metadata/map-mf/2323/metadata.faq.html.

Reid, S. M., and N. E. Mandrak. 2009. Lake Erie beaches: Diel variation in fish assemblage structure and implications for monitoring. *Hydrobiologia* 618:139–48.

Reighard, J. E. 1903. *The natural history of* Amia, 57–109. Mark anniversary volume. G. H. Parker (ed.). Henry Holt and Co., New York, New York.

———. 1910. Methods of studying the habits of fishes with an account of the breeding habits of the Horned Dace. *Bulletin of the U.S. Bureau of Fisheries* 28:1111–36.

———. 1920. The breeding behavior of the suckers and minnows. *Biological Bulletin* 38:1–32.

Reimchen, T. E. 1989. Loss of nuptial color in Threespine Sticklebacks (*Gasterosteus aculeatus*). *Evolution* 43.450–60.

Rennie, M. D., C. F. Purchase, B. J. Shuter, N. C. Collins, P. A. Abrams, and G. E. Morgan. 2010. Prey life-history and bioenergetic responses across a predation gradient. *Journal of Fish Biology* 77:1230–51.

Resetarits, W. J., Jr. 1995. Limiting similarity and the intensity of competitive effects on the Mottled Sculpin, *Cottus bairdi*, in experimental stream communities. *Oecologia* 104:31–38.

———. 1997. Interspecific competition and qualitative competitive asymmetry between two benthic stream fish. *Oikos* 78:429–39.

Resh, V. H., A. G. Hildrew, B. Statzner, and C. R. Townsend. 1994. Theoretical habitat templets, species traits, and species richness: A synthesis of longterm ecological research on the Upper Rhône River in the context of concurrently developed ecological theory. *Freshwater Biology* 31:539–54.

Resh, V. J., A. V. Brown, A. P. Covich, M. E. Gurtz, H. W. Li, G. W. Minshall, S. R. Reice, A. L. Sheldon, J. B. Wallace, and R. C. Wissmar. 1988. The role of disturbance in stream ecology. *Journal of the North American Benthological Society* 7:433–55.

Rettig, J. E., and G. G. Mittelbach. 2002. Interactions between adult and larval Bluegill sunfish: Positive and negative effects. *Oecologia* 130:222–30.

Reznick, D., M. J. Bryant, and F. Bashey. 2002. r- and K-selection revisited: The role of population regulation in life-history evolution. *Ecology* 83:1509–20.

Ricciardi, A. 2007. Are modern biological invasions an unprecedented form of global change? *Conservation Biology* 21:329–36.

Ricciardi, A., and J. B. Rasmussen. 1999. Extinction rates of North American freshwater fauna. *Conservation Biology* 13:1220–22.

Rice, J. A., L. B. Crowder, and M. E. Holey. 1987. Exploration of mechanisms regulating larval survival in Lake Michigan Bloater: A recruitment analysis based on characteristics of individual larvae. *Transactions of the American Fisheries Society* 116:703–18.

Richard, B. A., and P. C. Wainwright. 1995. Scaling the feeding mechanism of Largemouth Bass (*Micropterus salmoides*): Kinematics of prey capture. *Journal of Experimental Biology* 198:419–33.

Richmond, G. M. 1965. Glaciation of the Rocky Mountains, 217–30. In *The Quaternary of the United States*. H. E. Wright, Jr., and D. G. Frey (eds.). Princeton University Press, New Jersey.

Richter, B. D. 2010. Re-thinking environmental flows: From allocations and reserves to sustainability boundaries. *River Research and Applications* 26:1052–63.

Richter, B. D., J. V. Baumgartner, J. Powell, and D. P. Braun. 1996. A method for assessing hydrologic alteration within ecosystems. *Conservation Biology* 10:1163–74.

Ricker, W. E. 1975. *Computation and interpretation of biological statistics of fish populations*. Bulletin 191, Fisheries Research Board of Canada.

Ricker, W. E. (ed.). 1968. *Methods for assessment of fish production in fresh waters*. IBP Handbook 3. Blackwell, London, England.

Ricklefs, R. E. 1987. Community diversity: Relative roles of local and regional processes. *Science* 235:167–71.

———. 1990. *Ecology*. 3rd ed. W. H. Freeman and Co., New York, New York.

Ricklefs, R. E., and J. Travis. 1980. A morphological approach to the study of avian community organization. *The Auk* 97:321–38.

Ridley, M. 1978. Paternal care. *Animal Behaviour* 26:904–32.

Rigley, L., and J. Muir. 1979. The role of sound production by the Brown Bullhead *Ictalurus nebulosus*. *Proceedings of the Pennsylvania Academy of Science* 53:132–34.

Roach, K. A., and K. O. Winemiller. 2011. Diel turnover of assemblages of fish and shrimp on sandbanks in a temperate floodplain river. *Transactions of the American Fisheries Society* 140:84–90.

Robinson, A. T., R. W. Clarkson, and R. E. Forrest. 1998. Dispersal of larval fishes in a regulated river tributary. *Transactions of the American Fisheries Society* 127:772–86.

Robinson, B. W. 2000. Trade offs in habitat-specific foraging efficiency and the nascent adaptive divergence of sticklebacks in lakes. *Behaviour* 137:865–88.

Robinson, B. W., A. J. Januszkiewicz, and J. C. Koblitz. 2007. Survival benefits and divergence of predator-induced behavior between Pumpkinseed sunfish ecomorphs. *Behavioral Ecology* 19:263–71.

Robinson, B. W., and D. Schluter. 2000. Natural selection and the evolution of adaptive genetic variation in northern freshwter fishes, 65–94. In *Adaptive genetic variation in the wild*. T. A. Mousseau, B. A. Sinervo, and J. A. Endler (eds.). Oxford University Press, New York, New York.

Robinson, B. W., and D. S. Wilson. 1994. Character release and displacement in fishes: A neglected literature. *The American Naturalist* 144:596–627.

———. 1996. Genetic variation and phenotypic plasticity in a trophically polymorphic population of Pumpkinseed sunfish (*Lepomis gibbosus*). *Evolutionary Ecology* 10:631–52.

Robinson, B. W., D. S. Wilson, and A. S. Margosian. 2000. A pluralistic analysis of character release in Pumpkinseed sunfish (*Lepomis gibbosus*). *Ecology* 81:2799–812.

Robinson, B. W., D. S. Wilson, A. S. Margosian, and P. T. Lotito. 1993. Ecological and morphological differentiation of Pumpkinseed sunfish in lakes without Bluegill sunfish. *Evolutionary Ecology* 7:451–64.

Robinson, B. W., D. S. Wilson, and G. O. Shea. 1996. Trade-offs of ecological specialization: An intraspecific comparison of Pumpkinseed sunfish phenotypes. *Ecology* 77:170–78.

Robinson, L. K., and W. M. Tonn. 1989. Influence of environmental factors and piscivory in structuring fish assemblages of small Alberta lakes. *Canadian Journal of Fisheries and Aquatic Sciences* 46:81–89.

Robison, H. W., and T. M. Buchanan. 1988. *Fishes of Arkansas*. University of Arkansas Press Fayetteville.

Rodríguez, M. A. 2002. Restricted movement in stream fish: The paradigm is incomplete, not lost. *Ecology* 83:1–13.

Root, R. B. 1967. The niche exploitation pattern of the Blue-Gray Gnatcatcher. *Ecological Monographs* 37:317–50.

Roper, B. B., and D. L. Scarnecchia. 1999. Emigration of age-0 Chinook Salmon (*Oncorhynchus tshawytscha*) smolts from the upper South Umpqua River basin, Oregon, U.S.A. *Canadian Journal of Fisheries and Aquatic Sciences* 56:939–46.

Roscoe, D. W., and S. G. Hinch. 2010. Effectiveness monitoring of fish passage facilities: Historical trends, geographic patterns and future directions. *Fish and Fisheries* 11:12–33.

Roscoe, D. W., S. G. Hinch, S. J. Cooke, and D. A. Patterson. 2010. Fishway passage and post-passage mortality of up-river migrating Sockeye Salmon in the Seton River, British Columbia. *River Research and Applications* 27:693–705.

Rosen, D. E. 1978. Vicariant patterns and historical explanation in biogeography. *Systematic Zoology* 27:159–88.

———. 1982. Teleostean interrelationships, morphological function and evolutionary inference. *American Zoologist* 22:261–73.

Rosen, R. A., and D. C. Hales. 1981. Feeding of Paddlefish, *Polyodon spathula*. *Copeia* 1981:441–45.

Rosenzweig, M. L. 2001. The four questions: What does the introduction of exotic species do to diversity? *Evolutionary Ecology* 3:361–67.

Roskowski, J. A., P. J. Patchett, P. A. Pearthree, J. E. Spencer, J. E. Faulds, and A. C. Reynolds. 2007. *A late Miocene-early Pliocene chain of lakes fed by the Colorado River: Evidence from Sr isotopes of the Bouse Formation between Grand Canyon and the Gulf of California*. Paper 160-14, Geological Society of America, Annual Meeting, Denver, Colorado.

Ross, M. R. 1977. Function of Creek Chub (*Semotilus atromaculatus*) nest-building. *Ohio Journal of Science* 77:36–37.

Ross, S. T. 1986. Resource partitioning in fish assemblages: A review of field studies. *Copeia* 1986:352–88.

———. 1991. Mechanisms structuring stream fish assemblages- are there lessons from introduced species? *Environmental Biology of Fishes* 30:359–68.

———. 2001. *Inland fishes of Mississippi*. University Press of Mississippi, Jackson.

Ross, S. T., and J. A. Baker. 1983. The response of fishes to periodic spring floods in a southeastern stream. *American Midland Naturalist* 109:1–14.

Ross, S. T., J. A. Baker, and K. E. Clark. 1987. Microhabitat partitioning of southeastern stream fishes: Temporal and spatial predictability, 42–51. In *Community and evolutionary ecology* of *North American stream fishes*. W. J. Matthews and D. C. Heins (eds.). University of Oklahoma Press, Norman.

Ross, S. T., D. C. Heins, and J. W. Burris. 1992a. Fishes of Okatoma Creek, a free-flowing stream in south-central Mississippi. *Southeastern Fishes Council Proceedings* 26:2–10.

Ross, S. T., J. Knight, and D. Wilkins. 1990. Longitudinal occurrence of the Bayou Darter (Percidae: *Etheostoma rubrum*) in Bayou Pierre—a response to stream order or habitat availability? *Polish Archives of Hydrobiology* 37:221–33.

Ross, S. T., J. G. Knight, and D. S. Wilkins. 1992b. Distribution and microhabitat dynamics of the threatened Bayou Darter, *Etheostoma rubrum*. *Copeia* 1992:658–71.

Ross, S. T., and W. J. Matthews. In press. Evolution and ecology of North American freshwater fish assemblages. In *North American freshwater fishes: Ecology, evolution, and behavior*. M. L. Warren, Jr., and B. M. Burr (eds.). Johns Hopkins University Press, Baltimore, Maryland.

Ross, S. T., W. J. Matthews, and A. A. Echelle. 1985. Persistence of stream fish assemblages: Effects of environmental change. *The American Naturalist* 126:24–40.

Ross, S. T., M. T. O'Connell, D. M. Patrick, C. A. Latorre, W. T. Slack, J. G. Knight, and S. D. Wilkins. 2001. Stream erosion and densities of *Etheostoma rubrum* (Percidae) and associated riffle-inhabiting fishes- biotic stability in a variable habitat. *Copeia* 2001:916–27.

Ross, S. T., W. T. Slack, R. J. Heise, M. A. Dugo, H. Rogillio, B. R. Bowen, P. Mickle, and R. W. Heard. 2009. Estuarine and coastal habitat use of Gulf Sturgeon (*Acipenser oxyrinchus desotoi*) in the north-central Gulf of Mexico. *Estuaries and Coasts* 32:360–74.

Ross, S. T., and S. D. Wilkins. 1993. Reproductive behavior and larval characteristics of the threatened Bayou Darter (*Etheostoma rubrum*) in Mississippi. *Copeia* 1993:1127–32.

Roughgarden, J. 1998. *Primer of ecological theory*. Prentice Hall, Upper Saddle River, New Jersey.

Rubenstein, D. I. 1981. Population density, resource patterning, and territoriality in the Everglades Pygmy Sunfish. *Animal Behaviour* 29:155–72.

Rubenstein, D. I., and M. A. R. Koehl. 1977. The mechanisms of filter feeding: Some theoretical considerations. *The American Naturalist* 111:981–94.

Ruehl, C. B., and T. J. DeWitt. 2005. Trophic plasticity and fine-grained resource variation in populations of Western Mosquitofish, *Gambusia affinis*. *Evolutionary Ecology Research* 7:801–19.

Ruesink, J. L. 2005. Global analysis of factors affecting the outcome of freshwater fish introductions. *Conservation Biology* 19:1883–93.

Rundle, H. D., L. Nagel, J. W. Boughman, and D. Schluter. 2000. Natural selection and parallel speciation in sympatric sticklebacks. *Science* 287:306–8.

Ruxton, G. D. 2009. Non-visual crypsis: A review of the empirical evidence for camouflage to senses other than vision. *Philosophical Transactions of the Royal Society B* 364:549–57.

Ruxton, G. D., M. P. Speed, and D. J. Kelly. 2004. What, if anything, is the adaptive function of countershading? *Animal Behaviour* 68:445–51.

Ryder, R. A., and S. R. Kerr. 1989. Environmental priorities: Placing habitat in hierarchic perspective, 1–12. In *Proceedings of the national workshop on effects of habitat alteration on salmonid stocks*. C. D. Levings, L. B. Holtby, and M. A. Hendersons (eds.). *Canadian Special Publications in Fisheries and Aquatic Science* 105.

Sada, D. W., and G. L. Vinyard. 2002. Anthropogenic changes in biogeography of Great Basin aquatic biota. *Smithsonian Contributions to the Earth Sciences* 33:277–93.

Salas, A., and E. B. Snyder. 2010. Diel fish habitat selection in a tributary stream. *American Midland Naturalist* 163:33–43.

Salathé, M., R. D. Kouyos, and S. Bonhoeffer. 2008. The state of affairs in the kingdom of the Red Queen. *Trends in Ecology and Evolution* 23:439–45.

Sanderson, S. L., and J. J. Cech, Jr. 1995. Particle retention during respiration and particulate feeding in the suspension-feeding Blackfish, *Orthodon microlepidotus*. *Canadian Journal of Fisheries and Aquatic Sciences* 52:2534–42.

Sanderson, S. L., J. J. Cech, Jr., and A. Y. Cheer. 1994. Paddlefish buccal flow velocity during ram suspension feeding and ram ventilation. *Journal of Experimental Biology* 186:145–56.

Sanderson, S. L., J. J. Cech, Jr., and M. R. Patterson. 1991. Fluid dynamics in suspension-feeding Blackfish. *Science* 251:1346–48.

Sanderson, S. L., A. Y. Cheer, J. S. Goodrich, J. D. Graziano, and W. T. Callan. 2001. Crossflow filtration in suspension-feeding fishes. *Nature* 412:439–41.

Sanderson, S. L., M. E. Mort, and J. J. Cech, Jr. 1998. Particle retention by non-suspension-feeding cyprinid fishes. *Canadian Journal of Fisheries and Aquatic Sciences* 55:861–68.

Sanderson, S. L., M. C. Stebar, K. L. Ackermann, S. H. Jones, J. E. Batjakas, and L. Kaufman. 1996. Mucus entrapment of particles by a suspension-feeding Tilapia (Pisces: Cichlidae). *Journal of Experimental Biology* 199:1743–56.

Sanderson, S. L., and R. Wassersug. 1993. Convergent and alternative designs for vertebrate suspension feeding, 37–112. In *The skull.* Vol. 3. *Functional and evolutionary mechanisms.* J. Hanken and B. K. Hall (eds.). The University of Chicago Press, Illinois.

Sanford, C. P. J., and P. C. Wainwright. 2002. Use of sonomicrometry demonstrates the link between prey capture kinematics and suction pressure in Largemouth Bass. *Journal of Experimental Biology* 205:3445–57.

Santucci, V. J., Jr., S. R. Gephard, and S. M. Pescitelli. 2005. Effects of multiple low-head dams on fish, macroinvertebrates, habitat, and water quality in the Fox River, Illinois. *North American Journal of Fisheries Management* 25:975–92.

Sargent, R. C., and M. R. Gross. 1993. William's principle: An explanation of parental care in teleost fishes, 333–61. In *Behaviour of teleost fishes.* 2nd ed. T. J. Pitcher (ed.). Chapman and Hall, New York, New York.

Sargent, R. C., P. D. Taylor, and M. R. Gross. 1987. Parental care and the evolution of egg size in fishes. *The American Naturalist* 129:32–46.

Sazima, I. 1983. Scale-eating in characoids and other fishes. *Environmental Biology of Fishes* 9:87–101.

Sazima, I., and F. A. Machado. 1990. Underwater observations of piranhas in western Brazil. *Environmental Biology of Fishes* 28:17–31.

Schaeffer, B., and D. E. Rosen. 1961. Major adaptive levels in the evolution of the actinopterygian feeding mechanism. *American Zoologist* 1:187–204.

Schafer, J. P., and J. H. Hartshorn. 1965. The Quaternary of New England, 113–27. In *The Quaternary of the United States.* H. E. Wright, Jr., and D. G. Frey (eds.). Princeton University Press, New Jersey.

Schemske, D. W., G. G. Mittelbach, H. V. Cornell, J. M. Sobel, and K. Roy. 2009. Is there a latitudinal gradient in the importance of biotic interactions? *Annual Review of Ecology, Evolution, and Systematics* 40:245–69.

Schermer, E. R., D. G. Howell, and D. L. Jones. 1984. The origin of allochthonous terranes: Perspectives on the growth and shaping of continents. *Annual Review of Earth and Planetary Sciences* 12:107–31.

Schindler, D. E., S. R. Carpenter, J. J. Cole, J. F. Kitchell, and M. L. Pace. 1997. Influence of food web structure on carbon exchange between lakes and the atmosphere. *Science* 277:248–51.

Schlosser, I. J. 1982. Fish community structure and function along two habitat gradients in a headwater stream. *Ecological Monographs* 52:395–414.

———. 1985. Flow regime, juvenile abundance, and the assemblage structure of stream fishes. *Ecology* 66:1484–90.

———. 1987. A conceptual framework for fish communities in small warmwater streams, 17–24. In *Community and evolutionary ecology of North American stream fishes.* W. J. Matthews and D. C. Heins (eds.). University of Oklahoma Press, Norman.

———. 1988. Predation risk and habitat selection by two size classes of a stream cyprinid: Experimental test of a hypothesis. *Oikos* 52:36–40.

———. 1991. Stream fish ecology: A landscape perspective. *BioScience* 41:704–12.

———. 1995a. Critical landscape attributes that influence fish population dynamics in headwater streams. *Hydrobiologia* 303:71–81.

———. 1995b. Dispersal, boundary processes, and trophic-level interactions in streams adjacent to Beaver ponds. *Ecology* 76:908–25.

Schluter, D. 1994. Experimental evidence that competition promotes divergence in adaptive radiation. *Science* 266:798–801.

———. 1996. Ecological speciation in postglacial fishes. *Philosophical Transactions of the Royal Society B* 351:807–14.

———. 2000. Ecological character displacement in adaptive radiation. *The American Naturalist* 156:S4–S16.

———. 2001. Ecology and the origin of species. *Trends in Ecology and Evolution* 16:372–80.

———. 2010. Resource competition and coevolution in sticklebacks. *Evolution: Education and Outreach* 3:54–61.

Schluter, D., and J. D. McPhail. 1992. Ecological character displacement and speciation in sticklebacks. *The American Naturalist* 140:85–108.

Schmidt, S. N., J. D. Olden, C. T. Solomon, and M. J. Vander Zanden. 2007. Quantitative approaches to the analysis of stable isotope food web data. *Ecology* 88:2793–802.

Schmidt-Nielsen, K. 1972. Locomotion: Energy cost of swimming, flying, and running. *Science* 177:222–28.

Schneider, D. W. 2000. Local knowledge, environmental politics, and the founding of ecology in the United States, Stephen Forbes and "The lake as a microcosm" (1887). *Isis* 91:681–705.

Schneider, H. 1962. The labyrinth of two species of drumfish (Sciaenidae). *Copeia* 1962:336–38.

Schneider, H., and A. D. Hasler. 1960. Laute und lauterzeugung beim süsswassertrommler *Aplodinotus grunniens* Rafinesque (Sciaenidae, Pisces). Zeitschrift für Vergleichende Physiologie 43:499–517.

Schoener, T. W. 1971. Theory of feeding strategies. *Annual Review of Ecology and Systematics* 2:369–404.

———. 1974. Resource partitioning in ecological communities. *Science* 185:27–39.

———. 1983. Field experiments on interspecific competition. *The American Naturalist* 122:240–85.

———. 1986. Resource partitioning, 91–126. In *Community ecology: Pattern and process.* J. Kikkawa and D. J. Anderson (eds.). Blackwell Scientific Publications. Oxford, United Kingdom.

———. 2009. Ecological niche, 3–13. In *The Princeton guide to ecology.* S. A. Levin (ed.). Princeton University Press, Princeton, New Jersey.

———. 2011. The newest synthesis: Understanding the interplay of evolutionary and ecological dynamics. *Science* 331:426–29.

Schultz, R. J. 1977. Evolution and ecology of unisexual fishes. *Evolutionary Biology* 10:277–331.

Scoppettone, G. G. 1988. Growth and longevity of the Cui-ui and longevity of other catostomids and cyprinids in western North America. *Transactions of the American Fisheries Society* 117:301–7.

Scotese, C. R. 2002. *Plate tectonic maps and continental drift animations.* PALEOMAP Project. http://www.scotese.com.

Scott, M. C., and G. S. Helfman. 2001. Native invasions, homogenization, and the mismeasure of integrity of fish assemblages. *Fisheries* 26:6–15.

Scott, R. J. 2001. Sensory drive and nuptial colour loss in the Three-Spined Stickleback. *Journal of Fish Biology* 59:1520–28.

Scott, R. J., and S. A. Foster. 2000. Field data do not support a textbook example of convergent character displacement. *Proceedings of the Royal Society of London B* 267:607–12.

Scott, W. B., and E. J. Crossman. 1973. *Freshwater fishes of Canada.* Fisheries Research Board of Canada, Bulletin 185:1–966.

Seghers, B. H. 1981. Facultative schooling behavior in the Spottail Shiner (*Notropis hudsonius*): Possible costs and benefits. *Environmental Biology of Fishes* 6:21–24.

Selbie, D. T., B. A. Lewis, J. P. Smol, and B. P. Finney. 2007. Long-term population dynamics of the endangered Snake River Sockeye Salmon: Evidence of past influences on stock decline and impediments to recovery. *Transactions of the American Fisheries Society* 136:800–821.

Selman, K., and R. A. Wallace. 1986. Gametogenesis in *Fundulus heteroclitus. American Zoologist* 26:173–92.

Shaw, E. 1978. Schooling fishes. *American Scientist* 66:166–75.

Sheer, M. B., and E. A. Steel. 2006. Lost watersheds: Barriers, aquatic habitat connectivity, and salmon persistence in the Willamette and lower Columbia River basins. *Transactions of the American Fisheries Society* 135:1654–69.

Sheldon, A. L. 1968. Species diversity and longitudinal succession in stream fishes. *Ecology* 49:193–98.

Shelford, V. C. 1911. Ecological succession. I. Stream fishes and the method of physiographic analysis. *Biological Bulletin* 21:9–35.

Shelton, W. L., and R. O. Smitherman. 1984. Exotic fishes in warmwater aquaculture, 262–301. In *Distribution, biology, and management of exotic fishes.* J. R. Stauffer and W. R. Courtenay, Jr. (eds.). Johns Hopkins University Press, Baltimore, Maryland.

Sherman, M. L., and P. A. Moore. 2001. Chemical orientation of Brown Bullheads, *Ameiurus nebulosus*, under different flow conditions. *Journal of Chemical Ecology* 27:2301–18.

Sherratt, T. N., A. Rashed, and C. D. Beatty. 2005. Hiding in plain sight. *Trends in Ecology and Evolution* 20:414–16.

Shiklomanov, I. A. 1993. World fresh water resources, 13–24. In *Water in crisis, a guide to the world's fresh water resources.* P. H. Gleick (ed.). Oxford University Press, New York, New York.

Shute, J. R., P. W. Schute, and D. G. Lindquist. 1981. *Fishes of the Waccamaw River drainage.* Brimleyana 6:1–24.

Sibbing, F. A. 1982. Pharyngeal mastication and food transport in the Carp (*Cyprinus carpio* L.): A cineradiographic and elctromyographic study. *Journal of Morphology* 172:223–58.

———. 1988. Specializations and limitations in the utilization of food resources by the Carp, *Cyprinus carpio*: A study of oral food processing. *Environmental Biology of Fishes.* 22:161–78.

———. 1991a. Food processing by mastication in cyprinid fish, 57–92. In *Feeding and the texture of food*. J. F. V. Vincent and P. J. Lillford (eds.). Cambridge University Press, New York, New York.

———. 1991b. Food capture and processing, 377–412. In *Cyprinid fishes, systematics, biology and exploitation*. I. J. Winfield and J. S. Nelson (eds.). Chapman and Hall, New York, New York.

Siebeck, U. E., G. S. Losey, and J. Marshall. 2006. UV communication in fish, 423–55. In *Communication in fishes*. Vol. 2. F. Ladich, S. P. Collin, P. Moller, and B. G. Kapoor (eds.). Science Publishers, Enfield, New Hampshire.

Siefert, R. E. 1968. Reproductive behavior, incubation and mortality of eggs, and postlarval food selection in the White Crappie. *Transactions of the American Fisheries Society* 97:252–59.

Siefkes, M. J., and W. Li. 2004. Electrophysiological evidence for detection and discrimination of pheromonal bile acids by the olfactory epithelium of female Sea Lampreys (*Petromyzon marinus*). *Journal of Comparative Physiology A* 190:193–99.

Sigler, W. F., and J. W. Sigler. 1987. *Fishes of the Great Basin*. University of Nevada Press, Reno.

Sih, A., and B. Christensen. 2001. Optimal diet theory: When does it work, and when and why does it fail? *Animal Behaviour* 61:379–90.

Simberloff, D. 2007. Given the stakes, our *modus operandi* in dealing with invasive species should be "guilty until proven innocent." *Conservation Magazine* 8:18–18.

———. 2011. How common are invasion-induced ecosystem impacts? *Biological Invasions* 13:1255–68.

Simon, T. P. 1997. Ontogeny of the darter subgenus *Doration* with comments on intrasubgeneric relationships. *Copeia* 1997:60–69.

Simons, A. M. 2004. Phylogenetic relationships in the genus *Erimystax* (Actinopterygii: Cyprinidae) based on the cytochrome *b* gene. *Copeia* 2004:351–56.

Skalski, G., and J. Gilliam. 2000. Modeling diffusive spread in a heterogeneous population: A movement study with stream fish. *Ecology* 81:1685–700.

Skelhorn, J., H. M. Rowland, and G. D. Ruxton. 2010. The evolution and ecology of masquerade. *Biological Journal of the Linnean Society* 99:1–8.

Slack, W. T. 1996. *Fringing floodplains and assemblage structure of fishes in the DeSoto National Forest, Mississippi*. PhD Dissertation submitted to the University of Southern Mississippi, Hattiesburg.

Slack, W. T., S. T. Ross, and J. A. Ewing, III. 1998. Relative abundance and the ecology of early life history stages of the Bayou Darter: Critical habitat and downstream transport. *Museum Technical Report* 63; Mississippi Department of Wildlife, Fisheries and Parks, *Wildlife Heritage Program* 48.

Slack, W. T., S. T. Ross, and J. A. Ewing, III. 2004. Ecology and population structure of the Bayou Darter, *Etheostoma rubrum*: Disjunct riffle habitats and downstream transport of larvae. *Environmental Biology of Fishes* 71:151–64.

Slobodkin, L. B., F. E. Smith, and N. G. Hairston. 1967. Regulation in terrestrial ecosystems, and the implied balance of nature. *The American Naturalist* 101:109–24.

Sluijs, I. van der, S. M. Gray, M. C. P. Amorim, I. Barber, U. Candolin, A. P. Hendry, R. Krahe, M E. Maan, A. C. Utne-Palm, H-J. Wagner, and B. B. M. Wong. 2011. Communication in troubled waters: Responses of fish communication systems to changing environments. *Evolutionary Ecology* 25:623–40.

Smith, A. G., D. G. Smith, and B. M. Funnell. 1994. *Atlas of Mesozoic and Cenozoic coastlines*. Cambridge University Press, Cambridge, United Kingdom.

Smith, C., and R. J. Wootton. 1995. The costs of parental care in teleost fishes. *Reviews in Fish Biology and Fisheries* 5:7–22.

Smith, C. L. 1967. Contribution to a theory of hermaphroditism. *Journal of Theoretical Biology* 17:76–90.

———. 1985. *The inland fishes of New York state*. New York State Department of Environmental Conservation, Albany, New York.

Smith, C. L., and C. R. Powell. 1971. The summer fish fauna of Brier Creek, Marshall County, Oklahoma. American Museum. *Novitates* 2458:1–30.

Smith, C. T., R. J. Nelson, C. C. Wood, and B. F. Koop. 2001. Glacial biogeography of North American Coho Salmon (*Oncorhynchus kisutch*). *Molecular Ecology* 10:2775–85.

Smith, G. R. 1978. Biogeography of intermountain fishes. *Great Basin Naturalist Memoirs* 2:17–42.

———. 1981. Late Cenozoic freshwater fishes of North America. *Annual Review of Ecology and Systematics* 12:163–93.

———. 1992. Phylogeny and biogeography of the Catostomidae, freshwater fishes of North America and Asia, 778–826. In *Systematics, historical ecology, and North American freshwater fishes.* R. L. Mayden (ed.). Stanford University Press, Stanford, California.

Smith, G. R., C. Badgley, T. P. Eiting, and P. S. Larson. 2010. Species diversity gradients in relation to geological history in North American freshwater fishes. *Evolutionary Ecology Research* 12:693–726.

Smith, G. R., T. E. Dowling, K. W. Gobalet, T. Lugaski, D. K. Shiozawa, and R. P. Evans. 2002. Biogeography and timing of evolutionary events among Great Basin fishes, p., 175–234. In *Great Basin aquatic systems history.* R. Hershler, D. B. Madsen, and D. R. Currey (eds.). *Smithsonian Contributions to Earth Science* 33.

Smith, G. R., and D. R. Fisher. 1970. Factor analysis of distribution patterns of Kansas fishes, 259–77. In *Pleistocene and Recent environments of the central Great Plains.* W. Dort, Jr., and J. K. Jones, Jr. (eds.). Department of Geology, University of Kansas, Lawrence, Special Publication Number 3.

Smith, G. R., and R. F. Stearley 1989. The classification and scientific names of Rainbow and Cutthroat trouts. *Fisheries* 14:4–10.

Smith, G. R., and T. N. Todd. 1984. Evolution of species flocks in northern temperate lakes, 45–68. In *Evolution of species flocks.* A. A. Echelle and I. Kornfield (eds.). University of Maine at Orono Press, Orono.

Smith, M. L. 1981. Late Cenozoic fishes in the warm deserts of North America: A reinterpretation of desert adaptations, 11–38. In *Fishes in North American deserts.* R. J. Naiman and D. L. Soltz (eds.). John Wiley and Sons, New York, New York.

Smith, R. J. F. 1979. Alarm reaction of Iowa and Johnny darters (*Etheostoma*, Percidae, Pisces) to chemicals from injured conspecifics. *Canadian Journal of Zoology* 57:1278–82.

———. 1982. Reaction of *Percina nigrofasciata*, *Ammocrypta beani*, and *Etheostoma swaini* (Percidae, Pisces) to conspecific and intergeneric skin extracts. *Canadian Journal of Zoology* 60:1067–72.

———. 1992. Alarm signals in fishes. *Reviews in Fish Biology and Fisheries* 2:33–63.

Smithson and Johnston. 1999. Movement patterns of stream fishes in a Ouachita Highlands stream: An examination of the restricted movement paradigm. *Transactions of the American Fisheries Society* 128:848–53.

Smokorowski, K. E., and T. C. Pratt. 2007. Effect of a change in physical structure and cover on fish and fish habitat in freshwater ecosystems—a review and meta-analysis. *Environmental Reviews* 15:15–41.

Sneath, P. H. A., and R. R. Sokal 1973. *Numerical taxonomy: The principles and practice of numerical classification.* W. H. Freeman and Company, San Francisco, California.

Snelson, F. F., Jr. 1985. Size and morphological variation in males of the Sailfin Molly, *Poecilia latipinna. Environmental Biology of Fishes* 13:35–47.

Snelson, F. F., Jr., J. D. Wetherington, and H. L. Large. 1986. The relationship between interbrood interval and yolk loading in a generalized poeciliid fish, *Poecilia latipinna. Copeia* 1986:295–304.

Snodgrass, J. W. 1997. Temporal and spatial dynamics of Beaver-created patches as influenced by management practices in a south-eastern American landscape. Journal of Applied *Ecology* 34:1043–56.

Snodgrass, J. W., and G. K. Meffe. 1998. Influence of Beavers on stream fish assemblages: Effects of pond age and watershed position. *Ecology* 79:928–42.

———. 1999. Habitat use and temporal dynamics of blackwater stream fishes in and adjacent to beaver ponds. *Copeia* 1999:628–39.

Soballe, D. M., and B. L. Kimmel. 1987. A large-scale comparison of factors influencing phytoplankton abundance in rivers, lakes, and impoundments. *Ecology* 68:1943–54.

Sokal, R. R., and F. J. Rohlf. 1995. *Biometry.* 3rd ed. Freeman, New York, New York.

Soltz, D. L., and R. J. Naiman. 1978. The natural history of native fishes in the Death Valley system. Natural History Museum of Los Angeles County, *Science Series* 30:1–76.

Sommer, T., B. Harrell, M. Nobriga, R. Brown, P. Moyle, W. Kimmerer, and L. Schemel. 2001. California's Yolo Bypass: Evidence that flood control can be compatible with fisheries, wetlands, wildlife, and agriculture. *Fisheries* 26:6–16.

Sommer, T., R. Baxter, and B. Herbold. 1997. Resilience of Splittail in the Sacramento-San Joaquin Estuary. *Transactions of the American Fisheries Society* 126:961–76.

Soranno, P. A., K. E. Webster, J. L. Riera, T. K. Kratz, J. S. Baron, P. A. Bukaveckas, G. W. Kling, D. S. White, N. Caine, R. C. Lathrop, and P. R. Leavitt. 1999. Spatial variation among lakes within landscapes: Ecological organization along lake chains. *Ecosystems* 2:395–410.

Sorensen, P. W., and J. Caprio. 1998. Chemoreception, 375–406. In *The physiology of fishes.* D. H. Evans (ed.). CRC Press, Boca Raton, Florida.

Sorensen, P. W., and F. W. Goetz. 1993. Pheromonal and reproductive function of F prostaglandins and their metabolites in teleost fish. *Journal of Lipid Mediators* 6:385–93.

Sorensen, P. W., and N. E. Stacey. 2004. Brief review of fish pheromones and discussion of their possible uses in the control of non-indigenous teleost fishes. *New Zealand Journal of Marine and Freshwater Research* 38:399–417.

Sorensen, P. W., and L. A. Vrieze. 2003. The chemical ecology and potential application of the Sea Lamprey migratory pheromone. *Journal of Great Lakes Research* 29, supplement 1:66–84.

Souchon, Y., C. Sabaton, R. Deibel, D. Reiser, J. Kershner, M. Gard, C. Katopodis, P. Leonard, N. L. Poff, W. J. Miller, and B. L. Lee. 2008. Detecting biological responses to flow management: Missed opportunities; future directions. *River Research and Applications* 24:506–18.

Soulé, M. E. 1985. What is conservation biology? *BioScience* 35:727–34.

Sousa, W. P. 1984. The role of disturbance in natural communities. *Annual Review of Ecology and Systematics* 15:353–91.

Southwood, T. R. E. 1977. Habitat, the templet for ecological strategies. *Journal of Animal Ecology* 46:337–65.

———. 1988. Tactics, strategies and templets. *Oikos* 52:3–18.

Speares, P., D. Holt, and C. Johnston. 2010. The relationship between ambient noise and dominant frequency of vocalizations in two species of darters (Percidae: *Etheostoma*). *Environmental Biology of Fishes* 90:103–10.

Spencer, J. E., and P. J. Patchett. 1997. Sr isotope evidence for a lacustrine origin for the upper Miocene to Pliocene Bouse Formation, lower Colorado River trough, and implications for timing of Colorado Plateau uplift. Geological Society of America, Bulletin 109:767–78.

Stacey N., and P. W. Sorensen. 2005. Reproductive pheromones, 359–412. In *Behaviour and physiology of fish*. Fish physiology. Vol. 24. K. A. Sloman, R. W. Wilson, and S. Balshine (eds.). Elsevier/Academic Press, New York, New York.

Stacey, N. E., and J. R. Cardwell. 1995. Hormones as sex pheromones in fish: Widespread distribution among freshwater species, 244–48. In *Proceedings of the fifth international symposium on the reproductive physiology of fish*. F. W. Goetz and P. Thomas (eds.). Symposium 95, Austin, Texas.

Stachowicz, J. J. 2001. Mutualism, facilitation, and the structure of ecological communities. *BioScience* 51:235–46.

Stamford, M. D., and E. B. Taylor. 2004. Phylogeographical lineages of Arctic Grayling (*Thymallus arcticus*) in North America: Divergence, origins and affinities with Eurasian *Thymallus*. *Molecular Ecology* 13:1533–49.

Stanley, E. H., and M. W. Doyle. 2003. Trading off: The ecological effects of dam removal. *Frontiers in Ecology and the Environment* 1:15–22.

Starret, W. C. 1950. Distribution of the fishes of Boone County, Iowa, with special reference to the minnows and darters. *American Midland Naturalist* 43:112–27.

Stearley, R. F. 1992. Historical ecology of Salmoninae, with special reference to *Oncorhynchus*, 622–58. In *Systematics, historical ecology, and North American freshwater fishes*. R. L. Mayden (ed.). Stanford University Press, Stanford, California.

Stearley, R. F., and G. R. Smith. 1993. Phylogeny of the Pacific trouts and salmons (*Oncorhynchus*) and genera of the family Salmonidae. *Transactions of the American Fisheries Society* 122:1–33.

Stearns, S. C. 1976. Life history tactics: A review of the ideas. *The Quarterly Review of Biology* 51:3–47.

———. 1977. The evolution of life history traits: A critique of the theory and a review of the data. *Annual Review of Ecology and Systematics* 8:145–71.

———. 1983. The evolution of life-history traits in mosquitofish since their introduction to Hawaii in 1905: Rates of evolution, heritabilities, and developmental plasticity. *American Zoologist* 23:65–75.

Stefan, H. G., X. Fang, and J. G. Eaton. 2001. Simulated fish habitat change in North American lakes in response to projected climate warming. *Transactions of the American Fisheries Society* 130:459–77.

Stein, R. A. 1977. Selective predation, optimal foraging, and the predator-prey interaction between fish and crayfish. *Ecology* 58:1237–53.

Steinmetz, J., D. A. Soluk, and S. L. Kohler. 2008. Facilitation between Herons and Smallmouth Bass foraging on common prey. *Environmental Biology of Fishes* 81:51–61.

Stell, W. K., S. E. Walker, K. S. Chohan, and A. K. Ball. 1984. The Goldfish terminalis: A luteinizing hormone-releasing hormone and molluscan cardioexcitatory peptide immunoreactive olfactoretinal pathway. *Proceedings of the National Academy of Sciences* 81:940–44.

Stevens, M. 2005. The role of eyespots as anti-predator mechanisms, principally demonstrated in the Lepidoptera. *Biological Reviews* 80:573–88.

Stevens, M., C. J. Hardman, and C. L. Stubbins. 2008. Conspicuousness, not eye mimicry, makes "eyespots" effective antipredator signals. *Behavioral Ecology* 19:525–31.

Stevens, M., and S. Merilaita. 2009. Animal camouflage: Current issues and new perspectives. *Philosophical Transactions of the Royal Society B* 364:423–27.

Stevenson, M. M., G. D. Schnell, and R. Black. 1974. Factor analysis of fish distribution patterns in western and central Oklahoma. *Systematic Zoology* 23:202–18.

Stewart, D. J., J. F. Kitchell, and L. B. Crowder. 1981. Forage fishes and their salmonid predators in Lake Michigan. *Transactions of the American Fisheries Society* 110:751–63.

Stewart, J. G., C. S. Schieble, R. C. Cashner, and V. A. Barko. 2005. Long-term trends in the Bogue Chitto River fish assemblage: A 27 year perspective. *Southeastern Naturalist* 4:261–72.

Stiassny, M. L. J. 1996. An overview of freshwater biodiversity: With some lessons from African Fishes. *Fisheries* 21(9):7–13.

Stockwell, D., and D. Peters. 1999. The GARP modeling system: Problems and solutions to automated spatial prediction. *International Journal of Geographic Information Science* 13:143–58.

Stokes, W. L. 1986. Geology of Utah. *Utah Museum of Natural History, Occasional Paper* 6:1–280.

Stone, L., T. Dayan, and D. Simberloff. 2000. On desert rodents, favored states, and unresolved issues: Scaling up and down regional assemblages and local communities. *The American Naturalist* 156:322–28.

Stout, J. F. 1975. Sound communication during the reproductive behavior of *Notropis analostanus* (Pisces: Cyprinidae). *American Midland Naturalist* 94:296–325.

Strahler, A. N. 1952. Hypsometric (area-altitude) analysis of erosional topography. *Bulletin of the Geological Society of America* 63:1117–42.

———. 1957. Quantitative analysis of watershed geomorphology. *Transactions of the American Geophysical Union* 38:913–20.

Strange, E. M., and T. C. Foin. 2001. Interaction of physical and biological processes in the assembly of stream fish communities, 311–37. In *Ecological assembly rules*. E. Weiher and P. Keddy (eds.). Cambridge University Press, Cambridge, United Kingdom.

Strange, E. M., P. B. Moyle, and T. C. Foin. 1992. Interactions between stochastic and deterministic processes in stream fish community assembly. *Environmental Biology of Fishes* 36:1–15.

Strange, R. M. 1993. Seasonal feeding ecology of the Fantail Darter, *Etheostoma flabellare*, from Stinking Fork, Indiana. *Journal of Freshwater Ecology* 8:13–18.

Strange, R. M., and B. M. Burr. 1997. Intraspecific phylogeography of North American Highland fishes: A test of the Pleistocene vicariance hypothesis. *Evolution* 51:885–97.

Strauss, R. E. 1987. The importance of phylogenetic constraints in comparisons of morphological structure among fish assemblages, 136–43. In *Community and evolutionary ecology of North American stream fishes*. W. J. Matthews and D. C. Heins (eds.). University of Oklahoma Press, Norman.

Strawn, K., and C. Hubbs. 1956. Observations on stripping small fishes for experimental purposes. *Copeia* 1956:114–16.

Strong, D. R., Jr. 1983. Natural variability and the manifold mechanisms of ecological communities. *The American Naturalist* 122:636–60.

Sulak, K. J., and M. Randall. 2002. Understanding sturgeon life history: Enigmas, myths, and insights from scientific studies. *Journal of Applied Ichthyology* 18:519–28.

Sullivan, J. A., and R. J. Schultz. 1986. Genetic and environmental basis of variable sex ratios in laboratory strains of *Poeciliopsis lucida*. *Evolution* 40:152–58.

Svanbäck, R., and D. I. Bolnick. 2007. Intraspecific competition drives increased resource use diversity within a natural population. *Proceedings of the Royal Society B* 274:839–44.

Svanbäck, R., P. C. Wainwright, and L. A. Ferry-Graham. 2002. Linking cranial kinematics, buccal pressure, and suction feeding performance in Largemouth Bass. *Physiological and Biochemical Zoology* 75:532–43.

Swain, D. P. 1992. Selective predation for vertebral phenotype in *Gasterosteus aculeatus*: Reversal in the direction of selection at different larval sizes. *Evolution* 46:998–1013.

Swanson, B. O., A. C. Gibb, J. C. Marks, and D. A. Hendrickson. 2003. Trophic polymorphism and behavioral differences decrease intraspecific competition in a cichlid, *Herichthys minckleyi*. *Ecology* 84:1441–46.

———. 2008. Variation in foraging behavior facilitates resource partitioning in a polymorphic cichlid, *Herichthys minckleyi*. *Environmental Biology of Fishes* 83:147–54.

Swenson, R. O., K. W. Whitener, and M. Eaton. 2003. Restoring floods to floodplains: Riparian and floodplain restoration at the Cosumnes River Preserve, 224–29. In *California riparian systems: Processes and floodplain management, ecology, and restoration.* 2001 Riparian Habitat and Floodplains Conference Proceedings. P. M. Faber (ed.). Riparian Habitat Joint Venture, Sacramento, California.

Syme, D. A. 2006. Functional properties of skeletal muscle, 179–240. In *Fish physiology: Fish biomechanics.* Vol. 23. R. E. Shadwick, and G. V. Lauder (eds.). Elsevier Academic Press, San Diego, California.

Taillon, D., and M. G. Fox. 2004. The influence of residential and cottage development on littoral zone fish communities in a mesotrophic north temperate lake. *Environmental Biology of Fishes* 71:275–85.

Taylor, C. A., M. L. Warren, Jr., J. F. Fitzpatrick, H. H. Hobbs, III, R. F. Jezerinac, W. L. Pflieger, and H. W. Robison. 1996. Conservation status of crayfishes of the United States and Canada. *Fisheries* 21:25–38.

Taylor, C. M. 1996. Abundance and distribution within a guild of benthic stream fishes: Local processes and regional patterns. *Freshwater Biology* 36:385–96.

Taylor, C. M., M. R. Winston, and W. J. Matthews. 1996. Temporal variation in tributary and mainstream fish assemblages in a Great Plains stream system. *Copeia* 1996:280–89.

Taylor, E. B. 2004. An analysis of homogenization and differentiation of Canadian freshwater fish faunas with an emphasis on British Columbia. *Canadian Journal of Fisheries and Aquatic Sciences* 61:68–79.

Taylor, E. B., and J. D. McPhail. 1999. Evolutionary history of adaptive radiation in species pairs of Threespine Sticklebacks (*Gasterosteus*): Insights from mitochondrial DNA. *Biological Journal of the Linnean Society* 66:271–91.

ter Braak, C. J. F. 1995. Ordination, 91–173. In *Data analysis in community and landscape ecology.* Jongman, R. H. G., C. J. F. ter Braak, and O. F. R. van Tongeren (eds.). Cambridge University Press, New York, New York.

Thiessen, D. D., and S. K. Sturdivant. 1977. Female pheromone in the Black Molly fish (*Mollienesia latipinna*): A possible metabolic correlate. *Journal of Chemical Ecology* 3:207–17.

Thomas, D. A., and D. B. Hayes. 2006. A comparison of fish community composition of headwater and adventitious streams in a coldwater river system. *Journal of Freshwater Ecology* 21:265–75.

Thomas, R. J., T. A. King, H. E. Forshaw, N. M. Marples, M. P. Speed, and J. Cable. 2010. The response of fish to novel prey: Evidence that dietary conservatism is not restricted to birds. *Behavioral Ecology* 21:669–75.

Thompson, D. B. A., and M. L. P. Thompson. 1985. Early warning and mixed species association: The Plover's-Page revisited. *Ibis* 127:559–62.

Thompson, J. N. 1982. *Interaction and coevolution.* John Wiley and Sons, New York.

———. 1994. *The coevolutionary process.* University of Chicago Press, Illinois.

———. 1999a. *The evolution of species interactions.* *Science* 284:2116–18.

———. 1999b. Specific hypotheses on the geographic mosaic of coevolution. *The American Naturalist* 153, supplement:S1–S14.

———. 2005. *The geographic mosaic of coevolution.* The University of Chicago Press, Illinois.

———. 2009. The coevolving web of life. *The American Naturalist* 173:125–40.

Thomsen, M. S., T. Wernberg, A. Altieri, F. Tuya, D. Gulbransen, K. J. McGlathery, M. Holmer, and B. R. Silliman. 2010. Habitat cascades: The conceptual context and global relevance of facilitation cascades via habitat formation and modification. *Integrative and Comparative Biology* 50:158–75.

Thorpe, J. E. 1988. *Salmon migration.* Scientific Progress, Oxford 72:345–70.

Tieszen, L. L., T. W. Boutton, K. G. Tesdahl, and N. A. Slade. 1983. Fractionation and turnover of stable carbon isotopes in animal tissues: Implications for $\delta^{13}C$ analysis of diet. *Oecologia* 57:32–37.

Timmerman, C. M., and L. J. Chapman. 2004. Behavioral and physiological compensation for chronic hypoxia in the Sailfin Molly (*Poecilia latipinna*). *Physiological and Biochemical Zoology* 77:601–10.

Tinbergen, N. 1952a. "Derived" activities: Their causation, biological significance, origin, and emancipation during evolution. *The Quarterly Review of Biology* 27:1–32.

———. 1952b. The curious behavior of the Stickleback. *Scientific American* 414:1–6.

Todd, J. H., J. Atema, and J. E. Bardach. 1967. Chemical communication in social behavior of a fish, the Yellow Bullhead (*Ictalurus natalis*). *Science* 158:672–73.

Todd, T. N. 1976. Pliocene occurrence of the Recent atherinid fish *Colpichthys regis* in Arizona. *Journal of Paleontology* 50:462–66.

Tollrian, R., and C. D. Harvell (eds.). 1998. *The ecology and evolution of inducible defenses.* Princeton University Press, Princeton, New Jersey.

Tonn, W. M., and J. J. Magnuson. 1982. Patterns in the species composition and richness of fish assemblages in northern Wisconsin lakes. *Ecology* 63:1149–66.

Tonn, W. M., J. J. Magnuson, M. Rask, and J. Toivonen. 1990. Intercontinental comparison of small-lake fish assemblages: The balance between local and regional processes. *The American Naturalist* 136:345–75.

Torsvik, T. H., and L. R. M. Cocks. 2004. Earth geography from 400 to 250 Ma: A palaeomagnetic, faunal and facies review. *Journal of the Geological Society* 161:555–72.

Torsvik, T. H., and R. Van der Voo. 2002. Refining Gondwana and Pangea palaeogeography: Estimates of Phanerozoic non-dipole (octupole) fields. *Geophysical Journal International* 151:771–94.

Torsvik, T. H., R. Van der Voo, J. G. Meert, J. Mosar, and H. J. Walderhaug. 2001. Reconstructions of the continents around the North Atlantic at about the 60th parallel. *Earth and Planetary Science Letters* 187:55–69.

Tosh, C. R., A. L. Jackson, and G. D. Ruxton. 2006. The confusion effect in predatory neural networks. *The American Naturalist* 167:E52–E65.

Townsend, C. R., and A. G. Hildrew. 1994. Species traits in relation to a habitat templet for river systems. *Freshwater Biology* 31:265–75.

Travnichek, V. H., M. B. Bain, and M. J. Maceina. 1995. Recovery of a warmwater fish assemblage after the initiation of a minimum-flow release downstream from a hydroelectric dam. *Transactions of the American Fisheries Society* 124:836–44.

Tronstad, L. M., R. O. Hall, Jr., T. M. Koel, and K. G. Gerow. 2010. Introduced Lake Trout produced a four-level trophic cascade in Yellowstone Lake. *Transactions of the American Fisheries Society* 139:1536–50.

Trush, W. J., S. M. McBain, and L. B. Leopold. 2000. Attributes of an alluvial river and their relation to water policy and management. *Proceedings of the National Academy of Sciences* 97:11858–63.

Tsolaki, E., and E. Diamadopoulos. 2010. Technologies for ballast water treatment: A review. *Journal of Chemical Technology and Biotechnology* 85:19–32.

Turgeon, J., A. Estoup, and L. Bernatchez. 1999. Species flock in the North American Great Lakes: Molecular ecology of Lake Nipigon ciscoes (Teleostei: Coregonidae: *Coregonus*). *Evolution* 53:1857–71.

Turner, M. G. 1989. Landscape ecology: The effect of pattern on process. *Annual Review of Ecology and Systematics* 20:171–97.

———. 2005. Landscape ecology: What is the state of the science. *Annual Review of Ecology and Systematics* 36:319–44.

Turner, R. E., and N. N. Rabalais. 2003. Linking landscape and water quality in the Mississippi River basin for 200 years. *BioScience* 53:563–72.

Turner, T. F., M. L. Collyer, and T. J. Krabbenhoft. 2010. A general hypothesis-testing framework for stable isotope ratios in ecological studies. *Ecology* 91:2227–33.

Turner, T. F., J. C. Trexler, G. L. Miller, and K. E. Toyer. 1994. Temporal and spatial dynamics of larval and juvenile fish abundance in a temperate floodplain river. *Copeia* 1994:174–83.

Tyler, A. V. 1963. A cleaning symbiosis between the Rainwater Killifish, *Lucania parva* and the Stickleback, *Apeltes quadracus. Chesapeake Science* 4:105–6.

Tytell, E. D., and G. V. Lauder. 2004. The hydrodynamics of eels swimming I. Wake structure. *Journal of Experimental Biology* 207:1825–41.

———. 2008. Hydrodynamics of the escape response in Bluegill Sunfish, *Lepomis macrochirus. Journal of Experimental Biology* 211:3359–69.

Tyus, H. M. 1990. Potamodromy and reproduction of Colorado Squawfish in the Green River Basin, Colorado and Utah. *Transactions of the American Fisheries Society* 119:1035–47.

Tyus, H. M., B. D. Burdick, R. A. Valdez, C. M. Haynes, T. A. Lytle, and C. R. Berry. 1982. Fishes of the upper Colorado River basin: Distribution, abundance, and status, 12–70. In *Fishes of the upper Colorado River system: Present and future.* W. H. Miller, H. M. Tyus, and C. A. Carlson (eds.). Western Division, American Fisheries Society, Bethesda, Maryland.

Tyus, H. M., and G. B. Haines. 1991. Distribution, habitat use, and growth of Age-0 Colorado Squawfish in the Green River basin, Colorado and Utah. *Transactions of the American Fisheries Society* 120:79–89.

Tyus, H. M., and C. A. Karp. 1990. Spawning and movements of Razorback Sucker, *Xyrauchen texanus*, in the Green River basin of Colorado and Utah. *The Southwestern Naturalist* 35:427–33.

Tyus, H. M., and C. W. McAda. 1984. Migration, movements and habitat preferences of Colorado Squawfish, *Ptychocheilus lucius*, in the Green, White and Yampa rivers, Colorado and Utah. *The Southwestern Naturalist* 29:289–99.

Tyus, H. M., and J. F. Saunders, III. 2000. Nonnative fish control and endangered fish recovery: Lessons from the Colorado River. *Fisheries* 25:17–24.

Underhill, J. C. 1986. The fish fauna of the Laurentian Great Lakes, the St. Lawrence lowlands, Newfoundland and Labrador, 105–59. In *The zoogeography of North American freshwater fishes.*

C. H. Hocutt and E. O. Wiley (eds.). John Wiley and Sons, New York, New York.

Underwood, A. J. 1978. An experimental evaluation of competition between three species of intertidal prosobranch gastropods. *Oecologia* 33:185–202.

———. 1986. The analysis of competition by field experiments, 240–68. In *Community ecology: Pattern and process*. J. Kikkawa and D. J. Anderson (eds.). Blackwell Scientific Publications, Palo Alto, California.

———. 1990. Experiments in ecology and management: Their logics, functions and interpretations. *Australian Journal of Ecology* 15:365–89.

———. 1997. *Experiments in ecology*. Cambridge University Press, Cambridge, United Kingdom.

———. 1998. Design, implementation, and analysis of ecological and environmental experiments, 325–49. In *Experimental ecology: Issues and perspectives*. W. J. Resetarits and J. Bernardo (eds.). New York, New York.

Unger, L. M. 1983. Nest defense by deceit in the Fathead Minnow, *Pimephales promelas*. *Behavioral Ecology and Sociobiology* 13:125–30.

USGS. 2007. http://waterdata.usgs.gov/nwis/rt.

Van den Berg, C., G. J. M. van Snik, J. G. M. van den Boogaart, F. A. Sibbing, and J. W. M. Osse. 1994. Comparative microanatomy of the branchial sieve in three sympatric cyprinid species, related to filter-feeding mechanisms. *Journal of Morphology* 219:73–87.

Vander Zanden, M. J., G. Cabana, and J. B. Rasmussen. 1997. Comparing trophic position of freshwater fish calculated using stable nitrogen isotope ratios (δ^{15}N) and literature dietary data. *Canadian Journal of Fisheries and Aquatic Sciences* 54:1142–58.

Vander Zanden, M. J., S. Chandra, B. C. Allen, J. E. Reuter, and C. R. Goldman. 2003. Historical food web structure and restoration of native aquatic communities in the Lake Tahoe (California-Nevada) basin. *Ecosystems* 6:274–88.

Vander Zanden, M. J., J. D. Olden, J. H. Thorne, and N. E. Mandrak. 2004. Predicting occurrences and impacts of Smallmouth Bass introductions in north temperate lakes. *Ecological Applications* 14:132–48.

Van Snik Gray, E., K. A. Kellogg, and J. R. Stauffer, Jr. 2005. Habitat shift of a native darter *Etheostoma olmstedi* (Teleostei: Percidae) in sympatry with a non-native darter *Etheostoma zonale*. *American Midland Naturalist* 154:166–77.

Van Snik Gray, E., and J. R. Stauffer, Jr. 2001. Substrate choice by three species of darters (Teleostei: Percidae) in an artificial stream: Effects of a nonnative species. *Copeia* 2001:254–61.

Van Valen, L. 1965. Morphological variation and width of ecological niche. *The American Naturalist* 99:377–90.

———. 1973. A new evolutionary law. *Evolutionary Theory* 1:1–30.

Vannote, R. L., G. W. Minshall, K. W. Cummins, J. R. Sedell, and C. E. Cushing. 1980. The river continuum concept. *Canadian Journal of Fisheries and Aquatic Sciences* 37:130–37.

Vences, M., J. Freyhof, R. Sonnenberg, J. Kosuch, and M. Veith. 2001. Reconciling fossils and molecules: Cenozoic divergence of cichlid fishes and the biogeography of Madagascar. *Journal of Biogeography* 28:1091–99.

Vermeij, G. J. 1982. Unsuccessful predation and evolution. *The American Naturalist* 120:701–20.

Vila-Gispert, A., R. Moreno-Amich, and E. García-Berthou. 2002. Gradients of life history variation: An intercontinental comparison of fishes. *Reviews in Fish Biology and Fisheries* 12:417–27.

Vitousek, P. M., H. A. Mooney, J. Lubchenco, and J. M. Melillo. 1997. Human domination of Earth's ecosystems. *Science* 277:494–99.

Vitule, J. R. S., C. A. Freire, and D. Simberloff. 2009. Introduction of non-native freshwater fish can certainly be bad. *Fish and Fisheries* 10:98–108.

Vives, S. P. 1990. Nesting ecology and behavior of Hornyhead Chub *Nocomis biguttatus*, a keystone species in Allequash Creek, Wisconsin. *American Midland Naturalist* 124:46–56.

Volpato, G. L., R. E. Barreto, A. L. Marcondes, P. S. A. Moreira, and M. F. de Barros Ferreira. 2009. Fish ladders select fish traits on migration—still a growing problem for natural fish populations. *Marine and Freshwater Behaviour and Physiology* 42:307–13.

von Frisch, K. 1938. Zur psychologie des fischschwarmes. *Die Naturwissenschaften* 26:601–6.

Vrieze, L. A., and P. W. Sorensen. 2001. Laboratory assessment of the role of a larval pheromone and natural stream odor in spawning stream localization by migratory Sea Lamprey (*Petromyzon marinus*). *Canadian Journal of Fisheries and Aquatic Sciences* 58:2374–85.

Vrijenhoek, R. C. 1978. Coexistence of clones in a heterogeneous environment. *Science* 199:549–52.

———. 1979. Factors affecting clonal diversity and coexistence. *American Zoologist* 19:787–97.

———. 1994. Unisexual fish: Model systems for studying ecology and evolution. *Annual Review of Ecology and Systematics* 25:71–96.

Vrijenhoek, R. C., R. M. Dawley, C. J. Cole, and J. P. Bogart. 1989. A list of the known unisexual verebrates, p.19–23. In *Evolution and ecology of unisexual vertebrates*. R. M. Dawley and J. P.

Bogart. Bulletin 466, New York State Museum, Albany, New York.

Vroman, M. E. 2011. Will Asian carp take over Great Lakes?, 375–86. In *Water and society*. D. W. Pepper and C. A. Brebbia (eds.). WIT Press, Southampton, United Kingdom.

Wahl, D. H., and R. A. Stein. 1988. Selective predation by three esocids: The role of prey behavior and morphology. *Transactions of the American Fisheries Society* 117:142–51.

Wainwright, P. C. 1996. Ecological explanation through functional morphology: The feeding biology of sunfishes. *Ecology* 77:1336–43.

———. 2006. *Functional morphology of the pharyngeal jaw apparatus*, 77–101. In *Fish biomechanics*. G. V. Lauder and R. E. Shadwick (eds.). Elsevier, New York, New York.

Wainwright, P. C., and D. R. Bellwood. 2002. Ecomorphology of feeding in coral reef fishes, 33–55. In *Coral reef fishes, dynamics and diversity in a complex ecosystem*. P. F. Sale (ed.). Academic Press, Orlando, Florida.

Wainwright, P. C., A. M. Carroll, D. C. Collar, S. W. Day, T. E. Higham, and R. A. Holzman. 2007. Suction feeding mechanics, performance, and diversity in fishes. *Integrative and Comparative Biology* 47:96–106.

Wainwright, P. C., and S. W. Day. 2007. The forces exerted by aquatic suction feeders on their prey. *Journal of the Royal Society Interface* 4:553–60.

Wainwright, P. C., and G. V. Lauder. 1992. The evolution of feeding biology in sunfishes (Centrarchidae), 472–91. In *Systematics, historical ecology, and North American freshwater fishes*. R. L. Mayden (ed.). Stanford University Press, Stanford, California.

Wainwright, P. C., C. W. Osenberg, and G. G. Mittelbach. 1991. Trophic polymorphism in the Pumpkinseed sunfish (*Lepomis gibbosus* Linnaeus): Effects of environment on ontogeny. *Functional Ecology* 5:40–55.

Wainwright, P. C., and B. A. Richard. 1995. Predicting patterns of prey use from morphology of fishes. *Environmental Biology of Fishes* 44:97–113.

Walburg, C. H., and W. R. Nelson. 1966. *Carp, River Carpsucker, Smallmouth Buffalo, and Bigmouth Buffalo in Lewis and Clark Lake, Missouri River*. Research Report No. 69, Bureau Sport Fisheries and Wildlife, U.S. Department of the Interior, Washington, DC.

Walker, J. A. 2004. Dynamics of pectoral fin rowing in a fish with an extreme rowing stroke: The Threespine Sticklebacks (*Gasterosteus aculeatus*). *Journal of Experimental Biology* 207:1925–39.

Walker, J. A., C. K. Ghalambor, O. L. Griset, D. McKenney, and D. N. Reznick. 2005. Do faster starts increase the probability of evading predators? *Functional Ecology* 19:808–15.

Walker, J. D., and J. W. Geissman, compilers. 2009. *Geologic Time Scale*. Geological Society of America. http://www.geosociety.org/science/timescale/timescl.pdf.

Wallace, A. R. 1876. *The geographical distribution of animals*. Macmillan, London.

Wallace, R. A., and K. Selman. 1979. Physiological aspects of oogenesis in two species of sticklebacks, *Gasterosteus aculeatus* L. and *Apeltes quadracus* (Mitchill). *Journal of Fish Biology* 14:551–64.

———. 1981. Cellular and dynamic aspects of oocyte growth in teleosts. *American Zoologist* 21:325–43.

Wallin, J. E. 1989. Bluehead Chub (*Nocomis leptocephalus*) nests used by Yellowfin Shiners (*Notropis lutipinnis*). *Copeia* 1989:1077–80.

———. 1992. The symbiotic nest association of Yellowfin Shiners, *Notropis lutipinnis*, and Bluehead Chubs, *Nocomis leptocephalus*. *Environmental Biology of Fishes* 33:287–92.

Walsh, S. J., and B. M. Burr. 1984. Life history of the Banded Pygmy Sunfish, *Elassoma zonatum* Jordan (Pisces: Centrarchidae), in western Kentucky. *Bulletin of the Alabama Museum of Natural History* 8:31–52.

Walter, E. E., J. P. Scandol, and M. C. Healey. 1997. A reappraisal of the ocean migration patterns of Fraser River Sockeye Salmon (*Oncorhynchus nerka*) by individual-based modeling. *Canadian Journal of Fisheries and Aquatic Sciences* 54:847–58.

Walter, R. C., and D. J. Merritts. 2008. Natural streams and the legacy of water-powered mills. *Science* 319:299–304.

Walters, D. M., D. S. Leigh, M. C. Freeman, B. J. Freeman, and C. M. Pringle. 2003. Geomorphology and fish assemblages in a Piedmont river basin, U.S.A. *Freshwater Biology* 48:1950–70.

Waples, R. S., D. W. Jensen, and M. McClure. 2010. Eco-evolutionary dynamics: Fluctuations in population growth rate reduce effective population size in Chinook Salmon. *Ecology* 91:902–14.

Waples, R. S., P. B. Aebersold, and G. A. Winans. 2011. Population genetic structure and life history variability in *Oncorhynchus nerka* from the Snake River basin. *Transactions of the American Fisheries Society* 140:716–33.

Waples, R. S., R. W. Zabel, M. D. Scheuerell, and B. L. Sanderson. 2007. Evolutionary responses by native species to major anthropogenic changes to their ecosystems: Pacific salmon in the Columbia River hydropower system. *Molecular Ecology* 17:84–96.

Ward, A. J. W., and P. J. B. Hart. 2003. The effects of kin and familiarity on interactions between fish. *Fish and Fisheries* 4:348–58.

Ward, A. J. W., J. E. Herbert-Read, D. J. T. Sumpter, and J. Krause. 2011. Fast and accurate decisions through collective vigilance in fish shoals. *Proceedings of the National Academy of Sciences* 108:2312–15.

Ward, A. J. W., M. M. Webster, and P. J. B. Hart. 2007. Social recognition in wild fish populations. *Proceedings of the Royal Society B* 274:1071–77.

Ward, D. L., A. A. Schultz, and P. G. Matson. 2003. Differences in swimming ability and behavior in response to high water velocities among native and nonnative fishes. *Environmental Biology of Fishes* 68:87–92.

Warren, M. L., Jr. 2009. Centrarchid identification and natural history, 375–533. In *Centrarchid fishes: Diversity, biology, and conservation*. S. J. Cooke and D. P. Philipp (eds.). Wiley-Blackwell, West Sussex, United Kingdom.

Warren, M. L., Jr., and B. M. Burr. 1994. Status of freshwater fishes of the United States: Overview of an imperiled fauna. *Fisheries* 19:1–18.

Warren, M. L., Jr., B. M. Burr, S. J. Walsh, H. L. Bart, Jr., R. C. Cashner, D. A. Etnier, B. J. Freeman, B. R. Kuhajda, R. L. Mayden, H. W. Robison, S. T. Ross, and W. C. Starnes. 2000. Diversity, distribution, and conservation status of the native freshwater fishes of the southern United States. *Fisheries* 25:7–31.

Warren, M. L., Jr., and M. G. Pardew. 1998. Road crossings as barriers to small-stream fish movement. *Transactions of the American Fisheries Society* 127:637–44.

Wartenberg, D., S. Ferson, and F. J. Rohlf. 1987. Putting things in order: A critique of detrended correspondence analysis. *The American Naturalist* 129:434–48.

Wayne, W. J., and J. H. Zumberge. 1965. Pleistocene geology of Indiana and Michigan, 63–84. In *The Quaternary of the United States*. H. E. Wright, Jr., and D. G. Frey (eds.). Princeton University Press, New Jersey.

Weatherly, A. H. 1972. *Growth and ecology of fish populations*. Academic Press, New York, New York.

Webb, C. O., D. D. Ackerly, M. A. McPeek, and M. J. Donoghue. 2002. Phylogenies and community ecology. *Annual Review of Ecology and Systematics* 33:475–505.

Webb, P. W. 1975. *Hydrodynamics and energetics of fish propulsion*. Bulletin of the Fisheries Research Board of Canada, Bulletin 190.

———. 1984. Body form, locomotion and foraging in aquatic vertebrates. *American Zoologist* 24:107–20.

———. 1986. Locomotion and predator-prey relationships, 24–41. In *Predator-prey relationships*. M. E. Feder and G. V. Lauder (eds.). The University of Chicago Press, Illinois.

———. 1989. Station-holding by three species of benthic fishes. *Journal of Experimental Biology* 145:303–20.

———. 1994. The biology of fish swimming, 45–62. In *Mechanics and physiology of animal swimming*. L. Maddock, Q. Bone, and J. M. V. Rayner (eds.). Cambridge University Press, New York, New York.

———. 2002. Control of posture, depth, and swimming trajectories of fishes. *Integrative and Comparative Biology* 42:94–101.

———. 2006. Stability and maneuverability, 281–332. In *Fish physiology: Fish biomechanics*. Vol. 23. R. E. Shadwick, and G. V. Lauder (eds.). Elsevier Academic Press, San Diego, California.

Webb, P. W., C. L. Gerstner, and S. T. Minton. 1996. Station-holding by the Mottled Sculpin, *Cottus bairdi* (Teleostei: Cottidae), and other fishes. *Copeia* 1996:488–93.

Webb, P. W., and J. M. Skadsen. 1980. Strike tactics of *Esox*. *Canadian Journal of Zoology* 58:1462–69.

Webb, P. W., and D. Weihs. 1986. Functional locomotor morphology of early life history stages of fishes. *Transactions of the American Fisheries Society* 115:115–27.

Webb, S. A., J. A. Graves, C. Macias-Garcia, A. E. Magurran, D. Ó. Foighil, and M. G. Ritchie. 2004. Molecular phylogeny of the livebearing Goodeidae (Cyprinodontiformes). *Molecular Phylogenetics and Evolution* 30:527–44.

Weber, M. J., and M. L. Brown. 2009. Effects of common carp on aquatic ecosystems 80 years after "Carp as a dominant": Ecological insights for fisheries management. *Reviews in Fisheries Science* 17:524–37.

Weddle, G. K., and B. M. Burr. 1991. Fecundity and the dynamics of multiple spawning in darters: An in-stream study of *Etheostoma rafinesquei*. *Copeia* 1991:419–33.

Weeks, S. C., C. Benvenuto, and S. K. Reed. 2006. When males and hermaphrodites coexist: A review of androdioecy in animals. *Integrative and Comparative Biology* 46:449–64.

Weibel, A. C., T. E. Dowling, and B. J. Turner. 1999. Evidence that an outcrossing population is a derived lineage in a hermaphroditic fish (*Rivulus marmoratus*). *Evolution* 53:1217–25.

Weidel, B. C., D. C. Josephson, and C. E. Kraft. 2007. Littoral fish community response to Smallmouth Bass removal from an Adirondack lake. *Transactions of the American Fisheries Society* 136:778–89.

Weiher, E., and P. A. Keddy. 1995. Assembly rules, null models, and trait dispersion: New questions from old patterns. *Oikos* 74:159–64.

Welcomme, R. L. 1979. *Fisheries ecology of floodplain rivers*. Longman, New York, New York.

Welty, J. C. 1934. Experiments in group behavior of fishes. *Physiological Zoology* 7:85–128.

Werner, E. E. 1977. Species packing and niche complementarity in three sunfishes. *The American Naturalist* 111:553–78.

Werner, E. E., and J. F. Gilliam. 1984. The ontogenetic niche and species interactions in size-structured populations. *Annual Review of Ecology and Systematics* 15:393–425.

Werner, E. E., J. F. Gilliam, D. J. Hall, and G. G. Mittelbach. 1983a. An experimental test of the effects of predation risk on habitat use in fish. *Ecology* 64:1540–48.

Werner, E. E., and D. J. Hall. 1974. Optimal foraging and the size selection of prey by the Bluegill sunfish (*Lepomis macrochirus*). *Ecology* 55:1042–52.

———. 1976. Niche shifts in sunfishes: Experimental evidence and significance. *Science* 191:404–6.

———. 1977a. Density and competition among sunfish: Some alternatives. *Science* 195:94–95.

———. 1977b. Competition and habitat shift in two sunfishes (Centrarchidae). *Ecology* 58:869–76.

———. 1988. Ontogenetic habitat shifts in Bluegill: The foraging rate-predation risk trade-off. *Ecology* 69:1352–66.

Werner, E. E., D. J. Hall, D. R. Laughlin, D. J. Wagner, L. A. Wilsmann, and F. C. Funk. 1977. Habitat partitioning in a freshwater fish community. *Journal of the Fisheries Research Board of Canada* 34:360–70.

Werner, E. E., G. G. Mittelbach, D. J. Hall, and J. F. Gilliam. 1983b. Experimental tests of optimal habitat use in fish: The role of relative habitat profitability. *Ecology* 64:1525–39.

West, G. B., J. H. Brown, and B. J. Enquist. 2001. A general model for ontogenetic growth. *Nature* 413:628–31.

Westneat, M. W. 2001. Ingestion in fish, 1–6. In *Encyclopedia of life sciences*. Macmillan Publishing, Ltd. Basingstoke, Hants, United Kingdom.

———. 2004. Evolution of levers and linkages in the feeding mechanisms of fishes. *Integrative and Comparative Biology* 44:378–89.

Wetzel, R. G. 2001. *Limnology: Lake and river ecosystems*. 3rd ed. Academic Press, San Diego, California.

Whang, A., and J. Janssen. 1994. Sound production through the substrate during reproduction in the Mottled Sculpin, *Cottus bairdi* (Cottidae). *Environmental Biology of Fishes* 40:141–48.

Whitaker, J. O., Jr. 1976. Fish community change at one Vigo County, Indiana locality over a twelve year period. *Proceedings of the Indiana Academy of Science* 85:191–207.

White, J. L., and B. C. Harvey. 2003. Basin-scale patterns in the drift of embryonic and larval fishes and lamprey ammocoetes in two coastal rivers. *Environmental Biology of Fishes* 67:369–78.

White, P. S., and S. T. A. Pickett. 1985. Natural disturbance and patch dynamics: An introduction, 3–13. In *The ecology of natural disturbance and patch dynamics*. S. T. A. Pickett and P. S. White (eds.). Academic Press, New York, New York.

Whitehead, D. R. 1973. Late-Wisconsin vegetational changes in unglaciated eastern North America. *Quaternary Research* 3:621–31.

Whiteley, A. R., C. A. Bergstrom, T. Linderoth, and D. A. Tallmon. 2011. The spectre of past spectral conditions: Colour plasticity, crypsis and predation risk in freshwater sculpin from newly deglaciated streams. *Ecology of Freshwater Fish* 20:80–91.

Whiteley, A. R., S. M. Gende, A. J. Gharrett, and D. A. Tallmon. 2009. Background matching and color-change plasticity in colonizing freshwater sculpin populations following rapid deglaciation. *Evolution* 63:1519–29.

Whiteside, B. G., and R. M. McNatt. 1972. Fish species diversity in relation to stream order and physicochemical conditions in the Plum Creek drainage basin. *American Midland Naturalist* 88:90–101.

Whittaker, R. H. 1972. Evolution and measurement of species diversity. *Taxon* 21:213–51.

Whittier, T. R., and T. M. Kincaid. 1999. Introduced fish in northeastern USA lakes: Regional extent, dominance, and effect on native species richness. *Transactions of the American Fisheries Society* 128:769–83.

Wiens, J. A. 1977. On competition and variable environments. *American Scientist* 65:590–97.

———. 1997. Metapopulation dynamics and landscape ecology, 43–62. In *Metapopulation biology, ecology, genetics, and evolution*. I. Hanski and M. E. Gilpin (eds.). Academic Press, San Diego, California.

———. 2002. Riverine landscapes: Taking landscape ecology into the water. *Freshwater Biology* 47:501–15.

Wiley, E. O. 1976. The phylogeny and biogeography of fossil and recent gars (Actinopterygii: Lepisosteidae). *Miscellaneous Publication—Museum of Natural History, University of Kansas* 64:1–111.

———. 1981. *Phylogenetics: The theory and practice of phylogenetic systematics*. John Wiley and Sons, New York, New York.

———. 1986. A study of the evolutionary relationships of *Fundulus* topminnows (Teleostei: Fundulidae). *American Zoologist* 26:121–30.

———. 1992. Phylogenetic relationships of the Percidae (Teleostei: Perciformes): A preliminary hypothesis, 247–67. In *Systematics, historical ecology, and North American freshwater fishes*. R. L. Mayden (ed.). Stanford University Press, Stanford, California.

Wiley, E. O., and R. L. Mayden. 1985. Species and speciation in phylogenetic systematics, with examples from the North American fish fauna. *Annals of the Missouri Botanical Garden* 72:596–635.

Wiley, E. O., K. M. McNyset, A. T. Peterson, C. R. Robins, and A. M. Stewart. 2003. Niche modeling and geographic range predictions in the marine environment using a machine-learning algorithm. *Oceanography* 16:120–27.

Williams, G. C. 1992. *Natural selection: Domains, levels, and challenges*. Oxford University Press, New York, New York.

Williams, J. E., J. E. Johnson, D. A. Hendrickson, S. Contreras-Balderas, J. D. Williams, M. Navarro-Mendoza, D. E. McAllister, and J. E. Deacon. 1989. Fishes of North America endangered, threatened or of special concern. *Fisheries* 14:2–20.

Williams, J. G., R. W. Zabel, R. S. Waples, J. A. Hutchings, and W. P. Connor. 2008. Potential for anthropogenic disturbances to influence evolutionary change in the life history of a threatened salmonid. *Evolutionary Applications* 1:271–85.

Williams, T. H., and T. C. Mendelson. 2010. Behavioral isolation based on visual signals in a sympatric pair of darter species. *Ethology* 116:1038–49.

Williamson, K. S., and B. May. 2005. Homogenization of fall-run Chinook Salmon gene pools in the Central Valley of California, USA. *North American Journal of Fisheries Management* 25:993–1009.

Williamson, M., and A. Fitter. 1996. The varying success of invaders. *Ecology* 77:1661–66.

Willig, M. R., D. M. Kaufman, and R. D. Stevens. 2003. Latitudinal gradients of biodiversity: Pattern, process, scale, and synthesis. *Annual Review of Ecology, Evolution, and Systematics* 34:273–309.

Wilson, E. O., and W. H. Bossert. 1971. *A primer of population biology*. Sinauer Associates, Inc. Stamford, Connecticut.

Wilson, H. F., and M. A. Xenopoulos. 2008. Landscape influences on stream fish assemblages across spatial scales in a northern Great Plains ecoregion. *Canadian Journal of Fisheries and Aquatic Sciences* 65:245–57.

Wilson, J. B. 1999. Guilds, functional types and ecological groups. *Oikos* 86:507–22.

Wilson, M. V., and A. Shmida. 1984. Measuring beta diversity with presence-absence data. *Journal of Ecology* 72:1055–64.

Wilson, M. V. H. 1992. Importance for phylogeny of single and multiple stem-group fossil species with examples from freshwater fishes. *Systematic Biology* 41:462–70.

Wilson, M. V. H., and R. R. G. Williams. 1992. Phylogenetic, biogeographic, and ecological significance of early fossil records of North American freshwater teleostean fishes, 224–44. In *Systematics, historical ecology, and North American freshwater fishes*. R. L. Mayden (ed.). Stanford University Press, Stanford, California.

Windell, J. T. 1971. Food analysis and rate of digestion, 215–26. In *Methods for assessment of fish production in freshwaters*. W. E. Ricker (ed.). Blackwell Scientific Publications, Oxford, United Kingdom.

Winemiller, K. O. 1989. Patterns of variation in life history among South American fishes in seasonal environments. *Oecologia* 81:225–41.

———. 1991. Ecomorphological diversification in lowland freshwater fish assemblages from five biotic regions. *Ecological Monographs* 61:343–65.

———. 1992. Life-history strategies and the effectiveness of sexual selection. *Oikos* 63:318–27.

———. 1995. Fish ecology, 49–65. In *Encyclopedia of environmental biology*. Vol. 2. W. M. Nierenberg (ed.). Academic Press, San Diego, California.

———. 1996. Factors driving temporal and spatial variation in aquatic floodplain food webs, 298–312. In *Food webs: Integration of patterns and dynamics*. G. A. Polis and K. O. Winemiller (eds.). Chapman and Hall, New York, New York.

———. 2005. Life history strategies, population regulation, and implications for fisheries management. *Canadian Journal of Fisheries and Aquatic Sciences* 62:872–85.

Winemiller, K. O., and K. A. Rose. 1992. Patterns of life-history diversification in North American fishes: Implications for population regulation. *Canadian Journal of Fisheries and Aquatic Sciences* 49:2196–218.

———. 1993. Why do most fish produce so many tiny offspring? *The American Naturalist* 142:585–603.

Winemiller, K. O., D. C. Taphorn, and A. Barbarino-Duque. 1997. Ecology of *Cichla* (Cichlidae) in two blackwater rivers of southern Venezuela. *Copeia* 1997:690–96.

Winn, H. E., and J. F. Stout. 1960. Sound production by the Satinfin Shiner, *Notropis analostanus*. *Science* 132:222–23.

Winnell, M. H., and D. J. Jude. 1991. Northern large-river benthic and larval fish drift: St. Marys River, USA/Canada. *Journal of Great Lakes Research* 17:168–82.

Winston, M. R. 1995. Co-occurrence of morphologically similar species of stream fishes. *The American Naturalist* 145:527–45.

Winston, M. R., C. M. Taylor, and J. Pigg. 1991. Upstream extirpation of four minnow species due to damming of a prairie stream. *Transactions of the American Fisheries Society* 120:98–105.

Winterbottom, R. 1974. A descriptive synonymy of the striated muscles of the Teleostei. *Proceedings of the Academy of Natural Sciences of Philadelphia* 125:225–317.

Wipfli, M. S., and C. V. Baxter. 2010. Linking ecosystems, food webs, and fish production: Subsidies in salmonid watersheds. *Fisheries* 35:373–87.

Wisby, W. J., and A. D. Hasler. 1954. Effect of olfactory occlusion on migrating Silver Salmon (*Oncorhynchus kisutch*). *Journal of the Fisheries Research Board of Canada* 11:472–78.

Wisenden, B. D. 1999. Alloparental care in fishes. *Reviews in Fish Biology and Fisheries* 9:45–70.

———. 2008. Active space of chemical alarm cue in natural fish populations. *Behaviour* 145:391–407.

Wisenden, B. D., C. L. Binstock, K. E. Knoll, A. J. Linke, and B. S. Demuth. 2010. Risk-sensitive information gathering by cyprinids following release of chemical alarm cues. *Animal Behaviour* 79:1101–7.

Wisenden, B. D., J. Karst, J Miller, S Miller, and L. Fuselier. 2007. Anti-predator behaviour in response to conspecific chemical alarm cues in an esociform fish, *Umbra limi* (Kirtland 1840). *Environmental Biology of Fishes* 82:85–92.

Witte, C. C., M. L. Wildhaber, A. Arab, and D. B. Noltie. 2009. Substrate choice of territorial male Topeka Shiners (*Notropis topeka*) in the absence of sunfish (*Lepomis* sp.). *Ecology of Freshwater Fish* 18:350–59.

Wood, B. M., and M. B. Bain. 1995. Morphology and microhabitat use in stream fish. *Canadian Journal of Fisheries and Aquatic Sciences* 52:1487–98.

Woodward, G., and A. G. Hildrew. 2002. Food web structure in riverine landscapes. *Freshwater Biology* 47:777–98.

Wootton, J. T., M. S. Parker, and M. E. Power. 1996. Effects of disturbance on river food webs. *Science* 273:1558–61.

Wootton, R. J. 1984. Introduction: Strategies and tactics in fish reproduction, 1–12. In *Fish reproduction: Strategies and tactics*. G. W. Potts and R. J. Wootton (eds.). Academic Press, Orlando, Florida.

———. 1994. Energy allocation in the Threespine Stickleback, 114–43. In *The evolutionary biology of the Threespine Stickleback*. M. A. Bell and S. A. Foster (eds.). Oxford University Press, New York, New York.

———. 1998. *Ecology of teleost fishes*. 2nd ed. Kluwer Academic Publishers, Boston.

Wren, D. G., G. R. Davidson, W. G. Walker, and S. J. Galicki. 2008. The evolution of an oxbow lake in the Mississippi alluvial floodplain. *Journal of Soil and Water Conservation* 63:129–35.

Wright, J. P. 2009. Linking populations to landscapes: Richness scenarios resulting from changes in the dynamics of an ecosystem engineer. *Ecology* 90:3418–29.

Wright, J. P., and C. G. Jones. 2006. The concept of organisms as ecosystem engineers ten years on: Progress, limitations, and challenges. *BioScience* 56:203–9.

Wu, J., and O. L. Loucks. 1995. From balance of nature to hierarchical patch dynamics: A paradigm shift in ecology. *The Quarterly Review of Biology* 70:439–66.

Wund, M. A., J. A. Baker, B. Clancy, J. L. Golub, and S. A. Foster. 2008. A test of the "flexible stem" model of evolution: Ancestral plasticity, genetic accommodation, and morphological divergence in the Threespine Stickleback radiation. *The American Naturalist* 172:449–62.

Yamamoto, Y., and H. Ueda. 2009. Behavioral responses by migratory Chum Salmon to amino acids in natal stream water. *Zoological Science* 26:778–82.

Yant, P. R., J. R. Karr, and P. L. Angermeier. 1984. Stochasticity in stream fish communities: An alternative interpretation. *The American Naturalist* 124:573–82.

Young, P. S., and J. J. Cech, Jr. Environmental tolerances and requirements of Splittail. *Transactions of the American Fisheries Society* 125:664–78.

Yount, J. D., and G. J. Niemi. 1990. Recovery of lotic communities and ecosystems from disturbance—a narrative review of case studies. *Environmental Management* 14:547–69.

Zaret, T. M. 1980. *Predation and freshwater communities*. Yale University Press, New Haven, Connecticut.

Zeug, S. C., and K. O. Winemiller. 2008a. Evidence supporting the importance of terrestrial carbon in a large-river food web. *Ecology* 89:1733–43.

———. 2008b. Relationships between hydrology, spatial heterogeneity, and fish recruitment dynamics in a temperate floodplain river. *River Research and Applications* 24:90–102.

Zhang, C., S. B. Brown, and T. J. Hara. 2001. Biochemical and physiological evidence that bile acids produced and released by Lake Char (*Salvelinus namaycush*) function as chemical signals. *Journal of Comparative Physiology B* 171:161–71.

Zhang, C., and T. J. Hara. 2009. Lake Char (*Salvelinus namaycush*) olfactory neurons are highly sensitive and specific to bile acids. *Journal of Comparative Physiology A* 195:203–15.

Zigler, S. J., M. R. Dewey, B. C. Knights, A. L. Runstrom, and M. T. Steingraeber. 2004. Hydrologic and hydraulic factors affecting passage of Paddlefish through dams in the Upper Mississippi River. *Transactions of the American Fisheries Society* 133:160–72.

Zimmerman, E. G., and M. C. Richmond. 1981. Increased heterozygosity at the MDH-B locus in fish inhabiting a rapidly fluctuating thermal environment. *Transactions of the American Fisheries Society* 110:410–16.

Zimmerman, M. S. 2007. A field study of Brook Stickleback morphology: Multiple predators and multiple traits. *Canadian Journal of Zoology* 85:250–60.

Zimmerman, M. S., and C. C. Krueger. 2009. An ecosystem perspective on re-establishing native deepwater fishes in the Laurentian Great Lakes. *North American Journal of Fisheries Management* 29:1352–71.

Zwick, P. 2000. Phylogenetic system and zoogeography of the Plecoptera. *Annual Review of Entomology* 45:709–46.

INDEX

Note: The notations (b), (f), and (t) indicate boxes, figures, and tables, respectively.

American Shad (*Alosa sapidissima*), 187, 263(t), 315

Amia calva (Bowfin), 128, 133–34, 146, 145(f), 149–50, 152, 196, 262(t), 300, 344–45, 354

Amiidae (family), 14–18, 22, 129, 198(t), 262(t)

Amiiformes (order), 6(f), 128, 133–34, 146(t)

Amiinae (subfamily), 17–18

ammocoetes, 21, 130, 215, 363–64, 369

Ammocrypta beani (Naked Sand Darter), 50

amphidromy, 188–90, 190(f), 201, 369, 370

Amyzon (extinct genus), 23

anadromy, 22–23, 89, 101, 135, 187–91, 189(f), 190(f), 201, 238, 292, 294–95, 297, 304, 316, 324–25, 329, 332–37, 349, 363, 369, 370, 372

Anguilla rostrata (American Eel), 130, 132, 147
    catadromy of, 188
    feeding and rotational shaking of, 159–60, 263(t)
    semelparity of, 187

Apeltes quadracus (Fourspine Stickleback), 184(t)

Aphredoderidae (family), 14, 147, 198(t), 261(t)

Aphredoderus sayanus (Pirate Perch), 147, 199(t), 261(t), 320

Aplodinotus grunniens (Freshwater Drum), 61, 224(t), 263(t), 344

Appalachian Mountains, 12–13, 38, 234, 318

aquatic insects, 58, 101, 234, 250–51, 257–59, 261(t), 262(t)

aquatic surface respiration (ASR), 99

Archoplites (genus), 24

Archoplites interruptus (Sacramento Perch), 197–98

Arctic Char (*Salvelinus alpinus*), 111, 210, 214, 238–40, 240(f)

Ardea alba (Heron), 290–91

area cladograms, 235–36, 235(f), 369

Ariidae (family), 198(t)

Army Corps of Engineers, 366

artificial stream systems, 97–8, 115, 302

Ash Meadows spring complex (Nevada), 351, 353(f)

ASR (aquatic surface respiration), 99

assemblages
    a posteriori models, 60
    a priori models, 54–60
    change, 95
    colonization of, 72–90
    conceptual models, 2(f), 54
    environment and, 54–66
    formation, maintenance, and persistence of, 13, 45–116
    habitat template, 54–55
    historical effects, 75
    hydrologic cycles and, 312–28
    landscape filters, 55–57
    lentic vs. lotic, 10
    local vs. regional effects on, 66–68
    movement and, 78–89
    multivariate statistics, 60–66
    null models of species assembly, 72
    persistence of, 91–116

river continuum concept (RCC), 57–60
    rules, 72–76
    species characteristics and, 75
    species tolerances and physical habitat, 75
    statistical models, 54
    structured vs. random, 72
    traits, 54, 55(f)
    turnover, 94

Astephus (extinct genus), 23

Atheriniformes (order), 6(f), 135, 150(t), 162(f)

Atherinopsidae (family), 7(t), 14–15, 27, 87(t), 130, 150(t), 162(f), 184(t), 195(t), 198(t), 201, 262(t)

Atlantic Refugium, 42, 42(f)

Atlantic Salmon (*Salmo salar*), 23, 188, 209–13, 215, 244

Atlantic Silverside (*Menidia menidia*), 184(t), 186, 194

atmospheric carbon, 284–85

Atractosteus (genus), 158

Atractosteus spatula (Alligator Gar), 4, 262(t)

Atractosteus tropicus (Tropical Gar), 4

Australian realm, 4

autecology, 231, 232(b), 369

autogenic ecosystem engineers, 292, 369

Awaous banana (River Goby), 190

background noise, 206, 221, 225

backwater areas, 89, 236

Bald Eagle (*Haliaeetus leucocephalus*), 298

ballast water, 360, 366

Banded Darter (*Etheostoma zonale*), 40(t), 219, 237

Banded Killifish (*Fundulus diaphanus*), 135, 272, 276

Banded Sculpin (*Cottus carolinae*), 39, 74–75, 262(t), 330

Bannerfin Shiner (*Cyprinella leedsi*), 182, 184(t), 184–85, 185(f)

basic metabolic model, 180, 201

basioccipital bones, 163, 369

Basking Shark (*Cetorhinus maximus*), 158

Bayou Darter (*Nothonotus rubrum*), 50, 50(f), 86, 97, 97(f), 184(t)

Bayou Pierre (Mississippi), 50, 50(f), 53, 97

BCF. *See* body and caudal fin (BCF) locomotion

Bear Lake, 35

Bear River, 32(f), 34–35

Beaver (*Castor canadensis*)
    as an allogenic engineer, 292–93
    impoundments by, 53, 292–94

Beaverdam Creek (Mississippi), 266, 314(b), 314(f), 319

bed-scour, 295

behavioral regulation hypothesis, 195

Beluga Sturgeon (*Huso huso*), 22

Beringia, 18(f), 20–22, 24, 41–42, 42(f)

Beringian Refugium, 9–10

bet hedging, 182, 186–87, 324, 369

Bidahochi Basin, 31, 33

Bigeye Shiner (*Notropis boops*), 40, 40(t), 236, 261(t), 302

Bighead Carp (*Hypophthalmichthys nobilis*), 66, 366

Bigmouth Buffalo (*Ictiobus cyprinellus*), 60, 260(t), 261(t)

Bigmouth Sleeper (*Gobiomorus dormitor*), 190

bile acids, 209, 214

biogeography, 12–13
  biogeographic filters, 2(f), 355(f)
  island biogeography theory, 39
  plate tectonics and, 14(b)
  theories of, 16(b)
  vicariance, 13, 16(b)

biotic homogenization, 352–65, 367–68
  basic measures of, 356(b)
  nonindigenous species and, 360
  patterns of, 357

Blackbanded Darter (*Percina nigrofasciata*), 57, 261(t), 266, 270, 271(f), 321

Black Bear (*Ursus americanus*), 298

Black Buffalo (*Ictiobus niger*), 60, 260(t), 263(t), 354

Black Crappie (*Pomoxis nigromaculatus*), 234, 263(t), 321, 344, 350–51

Black Creek (Mississippi), 94, 104(t), 319–21

Blackfin Cisco (*Coregonus nigripinnis*), 348

blackflies, 292–93

Blacknose Dace (*Rhinichthys atratulus*), 100, 260(t)

Blackspotted Topminnow (*Fundulus olivaceus*), 82, 262(t), 320

Blacktail Shiner (*Cyprinella venusta*), 40(t), 50, 104, 102, 123(f), 131(t), 234, 261(t), 278, 290, 302, 330

blackwater streams, 234, 290, 293

Bloater (*Coregonus hoyi*), 110, 261(t), 317(b), 348

Blotchside Logperch (*Percina burtoni*), 302

Blue Catfish (*Ictalurus furcatus*), 10, 263(t)

Bluefin Killifish (*Lucania goodei*), 217–19

Bluegill (*Lepomis macrochirus*), 61, 129, 134, 136, 152–54, 158–59, 192–94, 197

Bluehead Chub (*Nocomis leptocephalus*), 82, 260(t), 293, 295, 296(f), 300

Bluehead Sucker (*Catostomus discobolus*), 31, 88, 94, 130, 131(t), 260(t), 315, 316(f), 332

Bluntface Shiner (*Cyprinella camura*), 65–66, 65(f), 131(t)

Bluntnose Minnow (*Pimephales notatus*), 131(t), 184(t), 185, 212, 223(t), 260(t), 290

body and caudal fin (BCF) locomotion, 125, 141
  anguilliform, 125–27
  carangiform, 125–28, 126(t), 132, 141
  common modes of, 126(t)
  evolution of, 132–33
  gaits, 129–32
  nonanguilliform, 127–28, 128(f), 133
  ostraciiform, 125
  subcarangiform, 125–28, 126(t), 132
  thunniform, 125

body shape, 122–25
  deep-bodied, 130, 154, 267, 277
  limnetic (fusiform), 137–38

streamlined, 119–21, 125, 129, 135, 154, 277, 372
terete, 239, 267, 372

bog lakes, 61

Bonneville Basin (Utah), 32(f), 33–35, 41

Bonneville Dam, 333

Bonneville Salt Flats, 34

Bonytail Chub (*Gila elegans*), 31, 33, 48, 127, 260(t), 261(t), 326

Bony Tongues (Hiodontidae), 147, 148(f)

Bosmina (genus), 251

Bouse Embayment, 31

Bowfin (*Amia calva*), 128, 133–34, 145(f), 146, 149–50, 152, 196, 262(t), 300, 344–45, 354

Brazos River (Texas), 321–22

Bridgelip Sucker (*Catostomus columbianus*), 110, 260(t)

Bridge River (British Columbia), 101, 328–29

Brier Creek (Oklahoma), 79–80, 81(f), 108(t), 99, 103–10, 114, 115(f), 282, 285, 287

Brighteye Darter (*Etheostoma lynceum*), 184(t), 186, 266

brood pouch, 198(t), 199(t), 200

brood survivorship, 300–301

Brook Silverside (*Labidesthes sicculus*), 262(t), 317(b)

Brook Stickleback (*Culaea inconstans*), 261(t), 277, 279, 283, 366

Brook Trout (*Salvelinus fontinalis*), 80, 110, 210–11, 244, 261(t), 263(t), 267

Brown Bear (*Ursus arctos*), 298

Brown Bullhead (*Ameiurus nebulosus*), 56–57, 210, 223(t), 260(t), 272, 354

Brown Trout (*Salmo trutta*), 76, 80, 110, 244, 261(t), 263(t), 353, 358, 365, 367

buccal cavities, 145, 149, 151, 369

buoyant lift, 124, 372

burrowers, 130

burst-and-coast swimming, 127, 130, 132

CA (correspondence analysis), 62, 63(b). *See also* detrended correspondence analysis (DCA)

caddisflies, 251, 292–93

Caddisfly (*Dicosmoecus*), 285, 286(f)

California Roach (*Lavinia symmetricus*), 285, 286(f)

Cambrian Period, 12, 30

Campostoma anomalum (Central Stoneroller), 235–36, 260(t), 281, 285–87, 290, 300, 302

Campostoma spadiceum (Highland Stoneroller), 40, 40(t)

cannibalism, 210, 218, 249, 255, 281
  Branded Darter (*Etheostoma zonale*), 40(t), 219, 237

Carassius auratus (Goldfish), 154–55
  diet of, 260(t), 277
  as a nonindigenous species, 352–53, 358
  sex and mating of, 208–9
  suction feeding by, 157

Carassius carassius (Crucian Carp), 269(f), 278

Diamond Killifish (*Adinia xenica*), 160
diatoms, 157, 260(t), 263, 285
Dicks Creek (Virginia), 99–100, 100(f)
Dicosmoecus (Caddisfly), 285, 286(f)
Didelphis virginiana (Opossum), 298–99
diet
    coloration and, 215
    isotopic ratios in, 228–30, 229(b)
    onotogenetic changes, 258
    optimal vs. actual, 270, 271(f)
    specialized, 164
    variation, 248
digital particle image velocimetry (DPIV), 144(b)
Dionda episcopa (Roundnose Minnow), 131(t)
Dipnoi (subclass), 3
discriminant functions, 62(b)–63(b)
dispersal model, 16(b)
distance weighting, 82–84, 83(b)
Dodson, S. I., 191, 264
Dollar Sunfish (*Lepomis marginatus*), 194, 293
Dolly Varden (*Salvelinus malma*), 191, 244–46, 245(f), 252
dominant species, 291, 292(f), 305
Dorosoma cepedianum (Gizzard Shad), 157, 260(t), 277, 283, 321–22, 326, 344
Dorosoma petenense (Threadfin Shad), 22, 111, 147, 148(f), 261(t), 283, 345(t)
dorsal fins, 122–24, 128, 135, 139, 215–16, 219–20, 220(f)
dorsal saddles, 274
DPIV (digital particle image velocimetry), 144(b)
drainage
    areas, 59(b), 293, 311, 312(f)
    changes due to volcanism, 34
    patterns, 29, 35, 37–40, 38(f), 41
Dreissena polymorpha (Zebra Mussel), 352, 360
drought, 31, 92, 95, 104(t)–97(t), 108(t)
    in colonization models, 77, 79–80, 81(f)
    diet changes and, 285
    movement and, 316
    prehistorical, 33
    resilience and, 103–10, 114
Dusky Darter (*Percina sciera*), 270, 271(f)
Dusky Shiner (*Notropis cummingsae*), 301
Dwarf Molly (*Poecillia chica*), 209
dynamic equilibrium model, 93
dynamic lift, 121, 370

Eastern Highlands, 37, 39, 225, 225(f)
Eastern Mosquitofish (*Gambusia holbrooki*), 274–76, 293
eco-evolution, 303–4
"ecological explosions," 352
ecological niches, 231–33, 232(b)
Ecological Society of America, 319
ecological time, 11, 51, 71, 93, 238, 252, 303–6
ecomorphological hypothesis, 138–41

"ecospecies," 114
ecosystem engineers, 291, 292–97, 292(f), 305, 363, 367, 369
ectoparasites, 159
egg mimicking, 219–20
Eichornia crassipes (Water Hyacinth), 351
Elassoma evergladei (Everglades Pygmy Sunfish), 216
Elassomatidae (family), 7(t), 184(t), 198(t)
Elassoma zonatum (Pygmy Sunfish), 184(t)
electrofishing, 3, 343–44
electronic tags, 3
Eleotridae (family), 7(t), 190, 198(t)
Elopomorpha (superorder), 146(t), 147
Elwha River (Washington), 334, 336
Embiotocidae (family), 198(t)
Emerald Shiner (*Notropis atherinoides*), 95–96, 104, 131(t), 261(t), 330
endemism, 6–7, 33–34
energy maximizers, 268, 270
energy webs, 249–50, 250(f)
Enos Lake (British Columbia), 137
envelope of detection, 272, 370
environmental change
    dealing with, 95–111
    predator impacts, 282–87
    resilience against, 99–102
    resistance to, 95–99
    responses to, 92–116
environmental flows, 328–38
environmental sex determination (ESD), 194
Eocene epoch, 15, 21–25, 31
Eosalmo (extinct genus), 23
epaxial muscle, 134, 145–46, 145(f), 153–54, 154(f), 163, 370
epaxial-neurocranial pathway, 145–46, 145(f)
epilimnion, 317(b), 323, 348, 370
epineural tendons, 126–27, 126(t), 370
epiphytic algae, 161, 258
epipleural tendons, 126–27, 126(t), 370
equilibrium species, 177–78, 180, 201, 344
Eretmodus (genus), 150(t)
Erimystax (genus), 39
Erimyzon oblongus (Creek Chubsucker), 262(t), 293
Erimyzon sucetta (Lake Chubsucker), 262(t), 293
ESD (environmental sex determination), 194
Esocidae (family), 14, 17, 22, 25, 360
    morphology of, 129–30, 147, 163
    predatory behavior of, 157–58, 263(t), 276–80
Esox americanus (Redfin Pickerel), 293
Esox lucius (Northern Pike), 22, 349
    assemblages, 61, 283–84, 346
    diet of, 230, 263
    predatory behavior of, 158, 265, 267, 269(f), 277, 279–80, 301–2
Esox masquinongy (Muskellunge), 22, 158, 263(t), 277
Esox niger (Chain Pickerel), 158, 182, 263(t), 293

Esox tiemani (extinct), 22

estuarine systems, 26, 31, 47, 75, 86, 256, 258, 297, 331

Etheostoma artesiae (Redspot Darter), 40(t), 219(f)

Etheostoma barrenense (Splendid Darter), 219

Etheostoma bellum (Orangefin Darter), 194

Etheostoma caeruleum (Rainbow Darter), 41, 186, 219, 219(f), 262(t), 279

Etheostoma collettei (Creole Darter), 40, 40(t)

Etheostoma crossopterum (Fringe Darter), 224(t), 225

Etheostoma flabellare (Fantail Darter), 169, 170(b), 171(t), 173–75, 174(t), 220, 220(f), 225, 243–44, 262(t)

Etheostoma fricksium (Savannah Darter), 293

Etheostoma grahami (Rio Grande Darter), 182, 184(t)

Etheostoma lepidum (Greenthroat Darter), 182, 184(t)

Etheostoma lynceum (Brighteye Darter), 184(t), 186, 266

Etheostoma olmstedi (Tessellated Darter), 135, 136(f), 184(t), 194, 237–38

Etheostoma perlongum (Waccamaw Darter), 135, 136(f)

Etheostoma radiosum (Orangebelly Darter), 40, 40(t), 262(t)

Etheostoma rafinesquei (Kentucky Snubnose Darter), 184(t), 185–86

Etheostoma serrifer (Sawcheek Darter), 293

Etheostoma spectabile (Orangethroat Darter, 74–75, 79–80, 184(t), 219, 262(t)

Etheostoma stigmaeum (Speckled Darter), 40(t), 262(t), 266, 320, 330

Etheostoma virgatum (Striped Darter), 220

Etheostoma zonale (Banded Darter), 40(t), 219, 237

Ethiopian realm, 4

ethmoid bone, 147, 370

Eunice Lake (British Columbia), 245, 245(f)

Eurasian Milfoil (*Myriophyllum spicatum*), 351

European Minnow (*Phoxinus phoxinus*), 216, 269(f), 276, 301–2

Euteleostei (subdivision), 146(t), 277

Everglades Pygmy Sunfish (*Elassoma evergladei*), 216

evolution
    allopatric speciation, 74, 369
    caudal fins, 134(b)
    coevolutionary proceses, 303–5
    directional selection, 304
    form and function, 132–38
    homoplasies, 17(b)
    landscape filters, 55–57
    natural selection, 135–38
    Paleozoic and Mesozoic landmarks, 12(f)
    pheromones, 207–8
    rapid genetic change, 304, 337
    of sticklebacks, 135–36
    of sunfishes, 136
    trends in trophic morphology, 144–51

exaptations, 280, 370

exogenous feeding, 174, 258, 317(b)

exoparasitism, 256–57, 287

experiments
    design of, 73(b), 137, 231, 240
    field and mecocosm, 240–49
    natural, 237–40
    observational studies, 233–37
    replication of, 240, 242, 244, 247

extinction, 9
    environmental change and, 30, 92
    predation-induced, 365
    probability of, 52
    rates, 31, 34, 307

extirpation, 9, 30–31, 251, 283
    man-made impoundments and, 85–86, 328
    migrations and, 85–86, 191
    nonnative species and, 352–53, 367
    patterns of biotic homogenziation, 357, 364–65
    repopulation after, 53, 101–2
    resistance to, 95, 113(t)
    types of perturbations, 92

facilitation, 203, 232(b), 244, 290–303
    direct mutualisms, 297
    ecosystem engineers, 291, 292–97, 292(f), 305, 363, 367, 369
    indirect mutualisms, 297–301, 299(f)
    piscine engineers, 294–97
    species associations and, 301–2

factor analysis, 60, 62(b), 63(b), 139

FAH (food availability hypothesis), 190–91

Fantail Darter (*Etheostoma flabellare*), 169, 170(b), 171(t), 173–75, 174(t), 220, 220(f), 225, 243–44, 262(t)

Fathead Minnow (*Pimephales promelas*), 102, 184(t), 185, 208–9, 216, 260(t), 267, 272, 275–77, 279–80, 280(f), 366–67

feeding
    ambush predators, 157
    biting, 159–60
    crepuscular, 234, 258
    crossflow filtration, 157
    dead-end filters, 156
    efficiency, 137, 281
    foraging habitats, 242, 243(f), 246, 257
    form and function in, 143–66
    guilds, 257–58, 259(f), 260(t), 264(t)
    hydrosol (sticky) filters, 156–15
    jaw protrusion and, 148–51
    manipulation, 159–61
    nocturnal, 234, 258
    optimal foraging, 268–70
    picking and scraping, 160–61
    processing, 162–65
    ram feeding, 157–59
    rate, 137, 234–35, 282
    siphon cropping, 256
    suction feeding, 151–55
    suspension feeding, 155–57, 157–59

Ictiobus cyprinellus (Bigmouth Buffalo), 60, 260(t), 261(t)

Ictiobus niger (Black Buffalo), 60, 260(t), 263(t), 354

impoundments, 110–11, 290, 291, 322, 327, 371
    Beaver, 53, 292–94
    man-made, 85, 86, 308, 341, 366. *See also* dams

imprinting hypothesis, 212–13

inclusive fitness, 211, 211(b), 371

Index of Biotic Integrity, 257

inducible defenses, 277–78, 371

inferior mouth, 371

Inland Silverside (*Menidia beryllina*), 135

Innuitian ice sheets, 36, 37(f)

insect drift, 234

insectivores, 58(f), 163, 251, 257

insolation, 92, 371

interactions, 203–305
    interspecific, 234–36, 291(b), 305
    intraspecific, 246–49
    between larval and adult fishes, 248

interaction web, 249–50, 250(f), 253, 285

interference competition, 230, 247
    asymmetrical, 74–75, 244
    examples of, 237–38, 242

Interior Highlands, 38–39

introduced species, 68, 244, 354, 358–62

invasion success, 76–77, 318

invertivores, 58(f), 59, 371

Ironcolor Shiner (*Notropis chalybaeus*), 293

island biogeography theory, 39

island-mainland metapopulation model, 52, 69

isotopic ratios
    carbon (^{13}C), 229, 236, 240(f)
    nitrogen (^{15}N), 76, 77(f), 228-9, 251

iteroparity, 176(b), 182, 186–88, 189(f), 201, 371

Jaccard coefficient (C_j), 356(b)

James River (Virginia), 99, 100(f), 111, 315, 318(f)

Japan, 318

jaws
    jawless fishes, 12, 12(f), 21
    modes of prey capture and, 151–65
    pharyngeal, 163–65
    protrusion, 148–51
    toothed pharyngeal, 162
    toothless maxilla, 147

John Day drainage (Oregon), 110

Johnson, Roswell H., 232(b)

June Sucker (*Chasmistes liorus*), 35

Jurassic Period, 12–13, 20, 30, 30(f)

Katherine Lake (British Columbia), 245, 245(f)

Kentucky Snubnose Darter (*Etheostoma rafinesquei*), 184(t), 185–86

Kerocottus (genus), 10

kettle lakes, 37

keystone modifiers, 291, 292–97, 292(f), 305, 363, 367, 369

kin recognition, 210–11

kleptogamy, 192, 192(b)

Kokanee Salmon, 325

Kryptolebias (genus), 149, 162(f)

Kryptolebias (genus syn. *Rivulus*), 149, 150(t), 162(f), 196

Kryptolebias marmoratus (Mangrove Rivulus), 196, 201

Labidesthes sicculus (Brook Silverside), 262(t), 317(b)

Labrador Sea, 21

Lahontan Basin, 33–24, 34(f)

Lake Aigneau (Quebec), 239–40

Lake Baikal (Russia), 7

Lake Bonneville (Utah), 34–35. *See also* Bonneville Basin

Lake Chapala (Mexico), 8–9

Lake Chubsucker (*Erimyzon sucetta*), 262(t), 293

Lake Herring. See *Coregonus artedi* (Cisco)

Lake Idaho, 32(f), 33

Lake Mendota (Wisconsin), 110, 112(t)

Lake Okeechobee (Florida), 341

Lake Opinicon (Ontario), 193

lakes, 371
    bog, 61
    connectivity and, 342
    domestication of, 349
    floodplain, 159, 308, 342–45, 345(t)
    inland saline, 4
    kettle, 37
    lacustrine species flocks, 10
    meltwater, 79
    polymictic, 246
    proglacial, 43, 43(f)

Lake Sturgeon (*Acipenser fulvescens*), 89, 262(t)

Lake Tahoe (California/Nevada), 77, 251

Lake Texoma (Texas/Oklahoma), 10, 79, 81(f), 103, 111, 112(t), 290

Lake Trout (*Salvelinus namaycush*), 102, 110, 209, 251, 263(t), 287, 346, 348, 363–64, 367

lampreys. *See also* Petromyzontidae (family)
    ammocoetes, 21, 130, 215, 363–64, 369
    Ichthyomyzon gagei (Southern Brook Lamprey), 320
    Mayomyzon pieckoensis (extinct lamprey), 21
    Petromyzon marinus (Sea Lamprey), 110, 209, 363

land bridges, 14(b)

landforms, 6, 30

Landsburg Dam, 332

landscape ecology, 48–53
    metapopulations, 51–53
    patches, 49–51

Largemouth Bass. See *Micropterus salmoides* (Largemouth Bass)

larvae
    anguilliform locomotion and, 127
    chironomid, 239, 246, 258, 273, 281

Milwaukee River (Wisconsin), 334

Mimic Shiner (*Notropis volucellus*), 260(t), 272, 290

minnows, 4, 15, 18. *See* Cyprinidae (family)

Miocene epoch, 14–15, 21–28, 30–31, 33, 35–36, 39, 92

Mississippi Alluvial Plain, 343(f), 344–45

Mississippi Embayment, 13

Mississippi Refugium, 41–42, 42(f)

Mississippi River, 9–10, 38–41, 38(f), 41, 50, 207, 225, 321, 326, 330, 342–43, 345(t)

Mississippi Silverside (*Menidia audens*), 184(t), 262(t), 331

Mississippi Silvery Minnow (*Hybognathus nuchalis*), 131(t), 260(t)

Missouri Refugium, 42, 42(f)

mitochondrial DNA, 17(b), 27, 35, 238

Mobulidae (family), 158

molecular clock analysis, 13–14, 26

molluscivores, 162–63

mollusks, 154, 164–5, 234

monophyletic taxa, 16(b), 21

mooneyes (*Hiodon* spp.), 15, 30, 44, 147, 261(t)

Morone chrysops (White Bass), 263(t), 321, 344, 345(t), 361

Morone mississippiensis (Yellow Bass), 263(t), 344, 354

Morone saxatilis (Striped Bass), 111, 263(t), 327, 376

Moronidae (family), 124, 188, 263(t)

morpholine, 213

morphology
body shape. *See* body shape
ecology and, 138–41
evolutionary trends in trophic, 144–51
fin form/placement. *See* fins
gill-rakers. *See* gills
pigmentation. *See* coloration
swimbladders, 124–25, 130, 139, 220–21, 223(t), 224(t), 226, 371, 372

morphotypes
of Arctic Char, 238
deepwater, 248
limnetic, 137
of Threespine Stickleback, 181

mosquitofishes. See *Gambusia* spp.

Mountain Madtom (*Noturus eleutherus*), 40, 40(t)

Mountain Mullet (*Agonostomus monticola*), 189–90

Mountain Redbelly Dace (*Chrosomus oreas*), 99–100

movement
adult stage, 88–89
fertilized egg stage, 85
larval stage, 85–88
leptokurtic, 82, 371

Moxostoma carinatum (River Redhorse), 262(t), 326

Moyle and Light hypothesis, 75–76

MPF (median or paired fins), 125, 128–30, 132–34, 141, 158

Mud Sunfish (*Acantharchus pomotis*), 293

Mugil cephalus (Striped Mullet), 188, 258–59, 263

Mugilidae (family), 149, 150(t)

Mugiliformes (order), 5(f), 150(t)

multivariate statistics, 60–66

muscle
epaxial, 134, 145–46, 145(f), 153–54, 154(f), 163, 370
hypochordal longitudinalis (HL), 133, 133(f)
levator operculi, 145(f), 146
pink vs. white, 129
retractor dorsalis, 163–64

museum specimens, 64–65, 69, 138, 252, 289–90, 349

Muskellunge (*Esox masquinongy*), 22, 158, 263(t), 277

mussels, 334, 352, 360

mutualisms, 297–301
direct, 297
indirect, 297–301, 299(f)
nonsymbiotic, 291(b), 297
symbiotic, 291(b), 297

myosepta, 125, 126(t), 371

Myoxocephalus (genus), 10, 27

Myoxocephalus thompsonii (Deepwater Sculpin), 348

Myriophyllum spicatum (Eurasian Milfoil), 351

Mysis relicta (Opossum Shrimp), 251, 348

Myxini (class), 3, 6(f), 21

Naked Sand Darter (*Ammocrypta beani*), 50

"Narcissus Effect," 72

nares, 212, 213(f)

National Wildlife Refuge, 344, 351

natural selection, 135–38, 167–68, 176(b), 201, 206, 211(b), 237, 256–57, 271, 277, 280, 303, 370

Nature Conservancy, The (TNC), 311

Navajo Dam (New Mexico), 322, 323(f). *See also* San Juan River

Nearctic realm, 4

Neogene Period, 14

Neo-Pangea, 354(b). *See also* Pangea

Neopterygii (subclass), 22, 146, 146(t)

Neoteleostei, 146(t), 163–64

Neotropical realm, 4, 25, 88(t), 256, 274

nesting
construction, 167–68, 179, 187–88, 191–93, 196–97, 199(t), 200–201, 217, 295–96
guarding, 167–68, 173, 179, 187–88, 217, 297, 305
holes, 230
nest associates, 294, 299–301, 301(f), 305
oral brooding, 167–68, 199(t)

network analysis, 59(b)

neurocranium, 152, 369, 371–72

Newton's first law, 127

New World silversides. *See* Atherinopsidae (family)

niche breadth, 138, 248

niche modeling, 64–66, 233

Nile Tilapia (*Oreochromis niloticus*), 157

Ninespine Stickleback (*Pungitius pungitius*), 267–68, 269(f)

Redear Sunfish (*Lepomis microlophus*), 154, 164, 192, 262(t), 330

Redface Topminnow (*Fundulus rubrifrons*), 160, 161(f)

Redfin Pickerel (*Esox americanus*), 293

Redfin Shiner (*Lythrurus umbratilis*), 40(t), 131(t), 260(t), 261(t), 267, 300–301, 301(f)

Redfish Lake (Idaho), 324–25

Red Queen hypothesis, 196, 256

Red River, 38(f), 39–40, 61, 74, 79, 81(f), 103, 334

Red River Pupfish (*Cyprinodon rubrofluviatilis*), 61

Red River Shiner (*Notropis bairdi*), 61

Red Shiner (*Cyprinella lutrensis*), 102, 131(t), 184(t), 236–37, 246–47, 247(f), 261(t), 283, 304, 322

Redspot Darter (*Etheostoma artesiae*), 40(t), 219(f)

Redspotted Sunfish (common), 155(f)

reforestation, 349

refuge-seeking behavior, 95, 96–7

regression line, 115(f), 170(b), 180, 311

reintroduced species, 110, 112(t), 113(t), 336(b), 356(b)

relict dams, 327

reproduction
    age-specific, 174–75
    avoidance of inbreeding, 210
    behavioral regulation hypothesis, 195

reproduction (*continued*)
    clutch production, 181–82
    communication in, 208–9, 216–20
    congenerics, 194–95, 370
    fecundity, 181
    lifetime, 186–88
    melanistic males, 218
    offspring number and size, 181–88
    parental care, 196–201
    pelagic eggs, 197
    reproductive effort, 181, 191–201
    reproductive life span, 101, 168, 180
    semelparity vs. iteroparity, 186–88
    sex and mating, 191–201
    sexual selection, 219, 337
    single season, 182–86
    spawning bars, 315
    spawning frequency, 181–88
    spawning migrations, 188–91
    spawning seasons, 102

resistance minimizers, 129

resources
    competitive exclusion and, 252
    limitation and competition, 230–49
    overlaps in, 138, 231
    resource partitioning, 233–37
    spatial, 227, 230, 233
    use, 54, 76, 138, 140, 228, 232(b)

resource-utilization niche, 231–33

restricted movement paradigm, 81–82

Reynolds number (R$_e$), 120, 127

Rhincodon typus (Whale Shark), 158

Rhinichthys atratulus (Blacknose Dace), 100, 260(t)

Rhinichthys cataractae (Longnose Dace), 43, 130, 352

Rhinichthys osculus (Speckled Dace), 31, 33, 76, 86, 94, 110, 262(t), 315, 316(f)

Rio Grande Bluntnose Shiner (*Notropis simus simus*), 85

Rio Grande Cichlid (*Cichlasoma cyanoguttatum*), 25, 261(t)

Rio Grande Darter (*Etheostoma grahami*), 182, 184(t)

Rio Grande River, 85, 236, 252, 311, 312(f)

Rio Grande Shiner (*Notropis jemezanus*), 85

Rio Grande Silvery Minnow (*Hybognathus amarus*), 85, 252, 365, 365(f)

Rio San Juan (Costa Rica), 4

River Carpsucker (*Carpiodes carpio*), 60, 134(b), 343

River Darter (*Percina shumardi*), 130, 262(t)

River Goby (*Awaous banana*), 190

River Redhorse (*Moxostoma carinatum*), 262(t), 326

rivers
    "River Analogy," 10
    river continuum concept (RCC), 48, 54, 57–60, 58(f), 69
    river-floodplain systems, 313–15, 321
    riverine fishes, 308, 321, 334, 342

River Shiner (*Notropis blennius*), 131(t)

Rivulidae (now family Aplocheilidae). See *Kryptolebias* (genus)

Rivulus hartii (Giant Rivulus), 304

r-K selection model, 175, 176(b)–177(b), 177

Roanoke River (Virginia), 140

Rock Bass (*Ambloplites rupestris*), 216, 263(t), 272, 283, 350

rockfish (common), 327

rock-ramp fish passages, 334

Rogue River (Oregon), 323

rostrum, 148

Rosyside Dace (*Clinostomus funduloides*), 82–84, 100, 222(t), 234–35, 290, 300

rotifers, 236, 258

Roundnose Minnow (*Dionda episcopa*), 131(t)

Roundtail Chub (*Gila robusta*), 31, 94, 260(t), 261(t)

Round Whitefish (*Prosopium cylindraceum*), 348

Rudd (*Scardinius erythrophthalmus*), 269(f)

runoff, 44, 61, 92, 106(t), 108(t), 249, 315, 322

run-of-river dams, 322

Rutilus rutilus (Roach), 228, 269(f)

Sacramento Blackfish (*Orthodon microlepidotus*), 158, 166, 331

Sacramento Perch (*Archoplites interruptus*), 197–98

Sacramento River (California), 157, 331

Sacramento-San Joaquin River system (California), 32–33, 101, 315, 330

Saddleback Darter (*Percina vigil*), 184(t), 186

Sagehen Creek (California), 77–78, 79(f)

Sailfin Molly (*Poecilia latipinna*), 99, 184(t), 209, 261(t), 274

Salangen River (Norway), 214